Microstructure of Cement-Based Systems/
Bonding and Interfaces in Cementitious Materials

MATERIALS RESEARCH SOCIETY SYMPOSIUM PROCEEDINGS VOLUME 370

Microstructure of Cement-Based Systems/ Bonding and Interfaces in Cementitious Materials

Symposia held November 28–December 1, 1994, Boston, Massachusetts, U.S.A.

EDITORS:

Sidney Diamond

Purdue University
West Lafayette, Indiana, U.S.A.

Sidney Mindess

University of British Columbia
Vancouver, British Columbia, Canada

F.P. Glasser

University of Aberdeen
Aberdeen, Scotland

Lawrence W. Roberts

W.R. Grace and Co.
Cambridge, Massachusetts, U.S.A.

Jan P. Skalny

Timonium, Maryland, U.S.A.

Lillian D. Wakeley

U.S. Army Waterways Experiment Station
Vicksburg, Mississippi, U.S.A.

MATERIALS RESEARCH SOCIETY
Pittsburgh, Pennsylvania

Single article reprints from this publication are available through
University Microfilms Inc., 300 North Zeeb Road, Ann Arbor, Michigan 48106

CODEN: MRSPDH

Published by:

Materials Research Society
9800 McKnight Road
Pittsburgh, Pennsylvania 15237
Telephone (412) 367-3003
Fax (412) 367-4373

Library of Congress Cataloging in Publication Data

Microstructure of cement-based systems/Bonding and interfaces in cementitious
    materials : symposia held November 28–December 1, 1994, Boston, Massachusetts,
    U.S.A. / editors, Sidney Diamond, Sidney Mindess, F.P. Glasser, Lawrence W.
    Roberts, Jan P. Skalny, Lillian D. Wakeley
        p. cm.—(Materials Research Society symposium proceedings ; v. 370)
    Includes bibliographical references and index.
    ISBN 1-55899-272-3 (acid-free paper)
        1. Cement—Surfaces—Congresses. 2. Cement—Mechanical properties—
Congresses. 3. Microstructure—Congresses. 4. Concrete—Congresses. I. Diamond,
Sidney II. Mindess, Sidney III. Glasser, F.P. IV. Roberts, Lawrence W.
V. Skalny, Jan P. VI. Wakeley, Lillian D. VII. Title: Bonding and interfaces in
cementitious materials. VIII. Series: Materials Research Society symposium
proceedings ; v. 370
TA435.M47 1995   95-6615
620.1'3592—dc20   CIP

Manufactured in the United States of America

Contents

PART I: MICROSTRUCTURE OF CEMENT-BASED SYSTEMS

*Invited Paper

Preface

This volume contains the published proceedings of two related symposia held at the 1994 MRS Fall Meeting in Boston. Symposium Va was concerned with "Microstructure of Cement-Based Systems" and constitutes the first part of this volume. Symposium Vb was devoted to "Bonding and Interfaces in Cementitious Materials" and constitutes the second section of this publication. A joint session on "Interfacial Microstructures" constituted the temporal interface between the two symposia, papers of which are also included in the second part of the volume. These symposia continue the MRS series on characteristics and properties of cementitious materials, the most recent of which was published in 1992 as MRS Symposium Proceedings Vol. 245, "Advanced Cementitious Systems: Mechanisms and Properties."

Recent and current advances in microstructure and related characterization of cement-based systems are well documented in Part I of the present volume. Sessions were held on microstructures of "normal" cement systems, on microstructure of "unusual" cement systems, on image analysis, modelling, and fractal analysis applications, and on assessment of pore structures. Nearly all of the papers presented at the symposium are included in this volume, which also includes several papers that, for one reason or another, could not be physically presented.

Part II contains the papers presented at the aforementioned Joint Session, as well as papers presented at individual sessions on elastic and fracture properties, on transport properties, and in particular, on the effects of interfaces in fiber-reinforced systems.

The two symposia were organized as part of a cooperative effort from the beginning, and it was felt that joint publication of the results would be both beneficial and cost-effective. We hope that prospective readers will find useful material in both symposia.

Sidney Diamond
Sidney Mindess
F.P. Glasser
Lawrence W. Roberts
Jan P. Skalny
Lillian D. Wakeley

January 1995

Acknowledgments

We are happy to acknowledge the financial support of these symposia by the Portland Cement Association.

We thank the Session chairmen, who in addition to the editors of this volume included G.T. McCarthy, Arnon Bentur, and Mark Alexander. Especially we thank our many colleagues who participated effectively in the paper review process, and the authors who cheerfully and expeditiously complied with the revisions suggested or required by the reviewers, and who executed the modifications in time to meet our short publication deadlines.

The international character of the symposia is evident from the list of affiliations of the various authors. We are grateful to the many who attended these sessions from around the world. We are particularly grateful to all of those, from whatever locale, who participated so vigorously in the often trenchant, but always interesting, discussions that marked both symposia.

Finally, we are grateful to the highly efficient and effective MRS staff for their efforts in conjunction both with planning and holding symposia, and with publication of this volume.

MATERIALS RESEARCH SOCIETY SYMPOSIUM PROCEEDINGS

MATERIALS RESEARCH SOCIETY SYMPOSIUM PROCEEDINGS

Prior Materials Research Society Symposium Proceedings available by contacting Materials Research Society

PART I

Microstructure of Cement-Based Systems

APPLICATION OF AUTOMATED IMAGE ANALYSIS TO THE STUDY OF CEMENT PASTE MICROSTRUCTURE

DAVID DARWIN AND MOHAMED NAGIB ABOU-ZEID
Department of Civil Engineering, University of Kansas, Lawrence, KS 66045

ABSTRACT

Digital acquisition and analysis of backscattered electron images provide powerful tools for the study of cement-based materials. The techniques can provide useful information on hydration phases, size distributions of unhydrated particles and voids, effects of changes in the water-cementitious material ratio and the use of mineral admixtures, and the distribution of microcracks. The results of automated analyses of cement pastes with different water-cement ratios and pastes containing silica fume are presented. The analyses demonstrate that microstructural data vary significantly from image to image, requiring multiple images to limit the effects of scatter. The analyses also indicate that, although the pastes exhibit different degrees of hydration, the size distributions of the unhydrated cement particles are nearly identical. In contrast, the size distribution of larger voids differs significantly as a function of water-cementitious material ratio and with the use of silica fume as a partial replacement for cement. The calcium hydroxide content obtained based on image analysis exceeds but generally parallels that obtained with thermogravimetric analysis. The majority of microcracks in both nonloaded and loaded specimens occur through or adjacent to the lowest density hydration phase.

INTRODUCTION

Backscattered electron imaging of polished surfaces is an important tool in studying the microstructure of cementitious materials[1-5]. It offers an advantage over secondary electron imaging of fracture surfaces, since it allows for the evaluation of a representative cross-section, rather than a "specimen selected" surface that represents the weakest portion of the microstructure. In spite of this advantage, backscattered electron (BSE) imaging is still subject to the same weaknesses as other imaging techniques using the scanning electron microscope (SEM), including poor selection of SEM settings, leading to an inadvertent loss of information, and the human tendency to emphasize "interesting" features for study.

Automated image analysis offers the potential of fully utilizing the capabilities of backscattered electron imaging, providing the power to process the large amounts of data needed to characterize this highly heterogeneous material. Taking full advantage of this technique requires the development of strategies for image acquisition and analysis to insure that a minimum of information is lost and that the data obtained from the analysis is representative of the material being studied.

The procedures discussed and this paper can provide information on hydration phases, size distributions of unhydrated particles and voids, effects of changes in the water-cementitious material ratio and the use of admixtures, and the size and orientation distribution of microcracks. This paper describes the approach taken in a large-scale study of cement paste microstructure and presents results for three cement pastes.

TEST SPECIMENS

The cement pastes evaluated in this study were made using a portland cement with a Bogue-calculated composition of C_3S 60 percent, C_2S 17 percent, C_3A 5.5 percent, and C_4AF 9.3 percent. Two of the pastes consisted of cement and water, with water-cement ratios of 0.5 and 0.3. The third paste had 15 percent mass replacement of cement by silica fume and a water-cementitious material ratio of 0.3. [Note: w/c will be used to represent both water-cement ratio and

3

water-cementitious material ratio.] The silica fume (SF) was a condensed powder containing 95.3 percent SiO_2 with a surface area in the range of 21-23 m^2/g. A high molecular weight sodium napthelene sulfonate superplasticizer, supplied in a powder form, was used to make a 20 percent aqueous solution. The water in the superplasticizer was taken into account when calculating the w/c ratio. Superplasticizer was added at the rate of 0.71, 1.74, and 2.19 g/kg of cementitious material for the 0.5, 0.3 and 0.3 w/SF pastes, respectively.

Prismatic specimens (25 x 25 x 127 mm) were cast vertically, stored horizontally for the first 24 hours, and then cured in lime-saturated water for an additional 27 days. Some specimens were loaded in compression to selected strains as high as 6,000 $\mu\varepsilon$.

Preparation for SEM viewing started with the removal of transverse and longitudinal 1 mm wafers. The wafers were cleaned for 2 minutes in a sonicator bath containing ethanol and then dried in an oven for 24 hours at 105° C. Specimens were cooled in a desiccator cabinet and impregnated with an ultralow viscosity epoxy within 48 hours of sectioning[6]. The wafers were then cut into four equal SEM specimens, polished manually with progressively finer grades of silicon carbide paper and diamond paste (final polishing with 0.5 μm paste), and coated with a 20 nm layer of gold palladium. Forty digital BSE images, evenly spaced, were obtained from each of two diagonally opposed SEM specimens cut from each wafer. The other two specimens were stored to provide backup.

IMAGE ACQUISITION STRATEGY

In the current study, specimens were studied using a Philips 515 scanning electron microscope with a backscattered electron detector consisting of four tilted, solid-state detectors on a swing-away arm. Images were acquired using an ELMDAS microcomputer-SEM interface installed in a 486/66 microcomputer. Effective digital image acquisition with this or any other instrumentation requires the development of a strategy to take full advantage of the capabilities of the instrumentation and to insure that the images accurately portray the material under study. The main considerations involve spatial resolution, feature contrast, and the degree of statistical certainty.

Spatial Resolution

The level of spatial resolution that can be obtained with a scanning electron microscope is a function of the energy of the beam electrons and the spot or probe diameter. In digital image analysis, it also a function of the spacing of the picture elements, or pixels, which directly correspond to specific sampling points on the surface under study.

For bulk specimens, such as cement paste, the volume from which backscattered electrons are generated is principally a function of the accelerating voltage, V_0. The higher the voltage, the greater the interaction volume and, thus, the greater the information volume. As V_0 increases, the energy of the primary electrons increases, resulting in greater penetration and increased lateral scattering of the electrons as they interact with the specimen. Backscattered electrons are high-energy electrons and, thus, provide information from a greater depth than do secondary electrons. The higher the atomic number, the shallower the depth of penetration of the primary electrons and the smaller the information volume. For a particular probe diameter, there is a direct relationship between the accelerating voltage and the beam current: An increase in the accelerating voltage will provide a higher signal. Selection of the accelerating voltage for a particular application requires both consideration of the size of the information volume and a comparison of images produced at different values of V_0, with the understanding that the sharpness of spatial resolution may decrease as V_0 increases.

Probe diameter and pixel spacing also affect spatial resolution. Studies have demonstrated that for probe diameters up to about twice the pixel spacing, pixel spacing will govern[7]. Probe diameters larger than twice the pixel spacing result in artificial smoothing of the specimen surface and a condition known as hollow magnification[7]. Thus, spatial resolution is usually limited by the larger of the diameter of the information volume, one-half of the probe diameter, or the pixel spacing. For backscattered electron imaging of edge regions between materials of greatly different backscattered electron coefficients, the latter two will govern.

In this study, $V_0 = 25$ kV, resulting in a depth of information of 2-3 μm and a lateral spatial resolution for hydration products of about 0.5 μm, the effective diameter of the information volume. Boundaries of cracks and voids that are filled with epoxy, however, can be distinguished much more sharply because of the low backscatter coefficient of the epoxy compared to the hydration products. In the current study, a probe diameter of approximately 100 nm and a pixel spacing of 77 nm convert to a resolution of crack and void boundaries of about 80 nm (the larger of the pixel spacing or one-half of the probe diameter).

Feature Contrast

The ability to distinguish features is dependent upon the differences in signal level between features compared to the noise inherent in the signal. An increase in the signal-to-noise ratio (which is proportional to the square root of the signal) will improve the ability to distinguish differences in signals. For backscattered electron imaging, the inherent contrast that is available, C, depends on the difference in backscattered electron coefficients, η_1 and η_2, compared to the higher coefficient η_1.

$$C = \frac{\eta_1 - \eta_2}{\eta_1} \tag{1}$$

The total signal available is a function of the beam current, i_B, (the current provided by the primary electrons), the efficiency of signal collection, ε, and the pixel dwell time, τ (the time during which the electron beam strikes the pixel of interest). The relationship between these parameters and the contrast (based on the ability of humans to distinguish differences in gray level) is expressed by the threshold equation[7]:

$$i_B = \frac{4 \times 10^{-18}}{\varepsilon C^2 \tau} \qquad A \tag{2}$$

For a given specimen and detector, the efficiency, ε, and the level of contrast, C, are fixed, so the imaging strategy must include the combination of pixel dwell time and beam current that will provide the desired results. An increase in the beam current, i_B, will result in an increase in the probe diameter for a fixed value of V_0 or an increase in V_0 for a fixed value of probe diameter, both of which may affect the spatial resolution. For lower values of i_B, greater image pixel swell times, τ, are required.

For best results, an image acquisition strategy should include a calculation of the desired minimum contrast, C, and an estimation of the signal collection efficiency, ε, to determine the required combination of i_B and τ.

In the current study, a beam current of 0.5 nA and a pixel dwell time of 400 μs were selected to provide the desired level of feature contrast.

Overall, images consisted of 1024 pixels per line and 960 lines, producing image dimensions of 79 x 74 μm. The total image acquisition time was 8 minutes and 19 seconds, consisting of 6 minutes and 35 seconds of dwell time and 1 minute and 44 seconds of computer overhead.

Limiting the potential for scatter and setting limits on the scatter in the data obtained from SEM analysis of bulk material requires both 1) a strategy for setting up the instrumentation and 2) the selection of the total number of images required to produce the desired level of statistical certainty.

In the development of an imaging strategy, a main element of uncertainty involves operator set-up of the instrumentation. To limit variations between SEM sessions and to account for minor variations in the performance of the instrumentation, it is highly desirable to use a standard to establish SEM and IAS settings. In this study, a silicon/magnesium standard[4,6] was used to insure that the contrast and brightness settings were consistent from session to session on both the SEM and IAS.

Another, perhaps more insidious, problem with SEM analysis involves the tendency of operators to isolate interesting features or areas on the surface and to use these features to characterize the material. Since hydrated cement paste is a highly variable material, even a simple statistical analysis will demonstrate the need for multiple images to properly characterize a specimen[4,6]. The current study demonstrates that information, such as the surface area occupied by specific phases or the density of cracks, varies significantly from image to image. However, if enough images are obtained, the average value for these parameters can be accepted with a desirable level of certainty. The number of images, n, required to provide 95 percent confidence that the average obtained in the analysis is within δ of the true average is

$$n = \left(\frac{1.96\sigma}{\delta} \right)^2 \tag{3}$$

in which σ is the estimated population deviation of the variable from image to image[8].

The number of images (80) taken of each SEM specimen in this study was selected based on total crack density, D_c, with $\delta = 0.10\,D_c$. [Note: For most specimens, n, based on Eq. 3, was in the range of 40 to 70.]

IMAGE ANALYSIS TECHNIQUES

Phases can be distinguished in a BSE image of epoxy-impregnated cement paste based on differences in signal intensity. High density phases, unhydrated particles (UH) and calcium hydroxide (CH), are the brightest, while low density phases, epoxy-filled cracks and voids, are the darkest. Crack identification, however, requires special techniques because, since cracks are narrow, the intensity of the signal varies with the intensity of adjacent or underlying phases.

Phase Identification

In the current study, phases within cement paste are identified based on gray level. The image acquisition system is adjusted to provide maximum contrast between the high and low density phases. The eight-byte graphics card typically used for image acquisition provides 256 discrete gray level intensities. If the full range is not used, for example if low density phases are adjusted to too high a gray level and high density phases are adjusted to too low a gray level, then the ability to use the full contrast, C, provided by the signal is lost. The selection of the actual settings depends on the information that is desired. In the current study, the key information involves the hydration products. Therefore, the contrast and brightness settings on the image acquisition system were set to maximize, as much as practical, the difference in gray level between UH and voids. As a result, it was not possible to distinguish between the different phases within UH.

In addition to UH and CH and voids, calcium silicate hydrate within the original boundary of cement grains (CSH-IP, often referred to as inner product) and undesignated product (UDP, the hydration products outside the well-defined phases of UH, CH, and CSH-IP) are identified based on gray level. Slightly different gray level ranges were required to identify the phases for water-cementitious material ratios of 0.3 and 0.5. For w/c = 0.3, the gray level ranges are: UH 255-210, CH 209-160, CSH-IP 159-135, UDP 134-27, and voids 26-0. The gray level of silica fume particles, 142-116, overlaps those of CSH-IP and UDP. This is not unexpected since, based on the work of Bonen and Diamond[5], the material in these particles consists of calcium silicate hydrate or other hydration products that have formed due to diffusion of calcium and other elements into the original silica fume grains. This overlap prevents silica fume particles from being distinguished in the analysis.

For w/c = 0.5, the gray level ranges are: UH 255-215, CH 214-174, CSH-IP 173-144, UDP 143-27, and voids 26-0.

Crack Identification

Crack identification procedures cannot be based on gray level alone 1) because of the dependence of the backscattered electron signal from epoxy-filled cracks upon the signal from adjacent and underlying phases and 2) because of the similarity in gray level between cracks and voids, which are also epoxy filled. Cracks, however, can be identified based on local differences in gray level between cracks and adjacent features, the gradient in gray level at the edge of a crack and selected geometric criteria (the latter to distinguish cracks from voids)[6]. The procedures used to identify cracks are presented in detail in References 6 and 9.

In addition to identifying cracks, the techniques used in this study also identify the phases adjacent to the cracks. This procedure includes consideration of the fact that cracks at the boundary between phases tend to form totally within the softer phase, preventing the direct assignment of a phase to a crack pixel based on the phases immediately in contact with that pixel. As a general rule, cracks occurring totally within a soft phase, but within 0.5 μm of a harder phase, would be characterized as boundary cracks[6,9].

RESULTS

Solid Phases and Voids

The results of the areal analyses of 640 images each for the pastes without silica fume and 320 images for the paste with silica fume are presented in Table I. The table shows the area percent and image-to-image standard deviation for each phase, along with the calculated 95 percent confidence interval, δ (see Eq. 3). Table I demonstrates that the area percents obtained from these images are, with a high degree of confidence, close to the true values.

A useful application of the analysis technique is to determine the degree of hydration of the cement. For the 0.5, 0.3, and 0.3 w/SF pastes, cement represented 38.9, 51.4, and 44.7 percent of the initial volume. At 28 days, unhydrated particles represented 10.0, 16.6, and 15.4 percent of the image area (= volume), corresponding to 74.4, 67.6, and 65.5 percent hydration, respectively. As expected, the material with the higher water-cementitious material ratio exhibited the greater degree of hydration. The lower degree of hydration exhibited by the paste with silica fume compared to the other paste with w/c = 0.3 is likely due to a greater tendency towards self-desiccation[10]. Fig. 1 shows the areal size distributions for UH particles in the three pastes, which, despite the differences in degree of hydration, are nearly identical.

The CH content in Table I can be compared with that obtained from thermogravimetric analysis (TGA). Assuming that TGA identifies 98 percent of the CH, calculated CH contents of 19.7, 18.8, and 11.2 percent, by mass, were obtained for the 0.5, 0.3, and 0.3 w/SF pastes, respectively. This compares to imaged volumes of 20.1, 20.7, and 12.4 percent and calculated mass contents based on the imaged volumes of 24.6, 22.0, and 13.5 percent, respectively.

Table I. Area percent, standard deviation σ, and 95 percent confidence interval ± δ, for phases in three cement pastes

	w/c = 0.5*			w/c = 0.3*			w/c = 0.3 w/SF**		
Phase	Area %	σ***	± δ	Area %	σ***	± δ	Area %	σ***	± δ
UH	10.0	4.4	0.3	16.6	5.0	0.4	15.4	4.9	0.5
CH	20.1	5.1	0.4	20.7	8.1	0.6	12.4	8.3	0.9
CSH-IP	17.5	2.9	0.2	20.8	4.0	0.3	30.9	5.7	0.6
UDP	45.8	7.9	0.6	40.3	10.8	0.8	39.8	12.3	1.4
Voids	5.8	2.3	0.2	1.4	4.2	0.3	1.2	1.0	0.1

*640 images
**320 images
***Between images

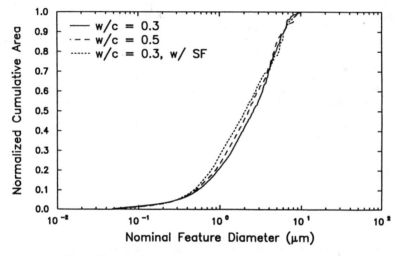

Fig. 1 Size distributions of UH particles in three cement pastes

The higher CH contents obtained from the images may be due in part to quantities of amorphous and microcrystalline CH that are not measured by TGA[11,12]. The ratio of the CH content determined by TGA to that calculated based on image analysis is 0.80, 0.85, and 0.83 for the three pastes, respectively.

Fig. 2 illustrates the cumulative area (volume) distribution of voids as a function of nominal feature diameter for the three pastes. These voids represent large pores and not total porosity. The results illustrated in Fig. 2, along with the information in Table I, show that the volume of these voids is significantly higher in the pastes with w/c = 0.5 than in the two lower w/c ratio pastes. The largest voids in the cement paste with silica fume (5.5 μm) are less than one-third the size of the largest voids in the cement pastes without silica fume (17.5 μm). Strength is affected by both total porosity and the size of individual voids, with larger voids expected to result in a greater reduction in strength than smaller voids. In the current study, the w/c = 0.5 paste had, as expected, the lowest strength followed by the w/c = 0.3 and w/c = 0.3 w/SF pastes.

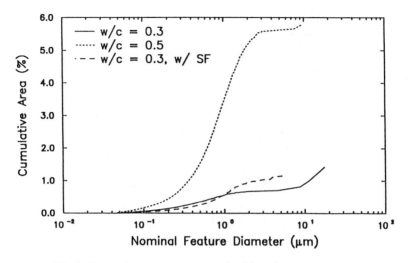

Fig. 2 Cumulative area percentages of voids in three cement pastes

Cracks

A micrograph of cement paste (with w/c = 0.5 loaded to 4000 µε) is shown in Fig. 3, along with the cracks identified using the automated image analysis techniques[6,9]. The number of cracks and crack density can be highly variable from image to image, as illustrated in Fig. 4, which shows the crack densities for individual images taken from transverse surfaces of cement paste specimens with w/c = 0.3 loaded to 0 and 6000 µε. The high variability from image to image necessitates the requirement of 80 images per specimen to provide the desired confidence in the results [as per Eq. 3]. Fig. 4 also illustrates another trend in the data, that is a measurable difference in crack density between the two wafers imaged per specimen. Images 1 through 40 were obtained from the wafer that was closer to the side of the specimen that was on the bottom during the first 24 hours of curing, while images 41 through 80 were obtained from the top wafer. The top wafer consistently exhibits a greater crack density. Finally, Fig. 4 shows that the application of compressive stress results in a significant increase in total crack density for cement paste.

Fig. 5 illustrates the cumulative crack densities for transverse surfaces as a function of crack length for the three pastes studied. Initially, cement paste with w/c = 0.3 exhibits the lowest crack density, followed by w/c = 0.3 w/SF and w/c = 0.5. With the application of 6000 µε in compressive, w/c = 0.5 exhibits both the greatest increase and the greatest total microcrack density. The increase in total crack density for w/c = 0.5 is nearly matched by w/c = 0.3, which results in total crack density that is greater than that of w/c = 0.3 w/SF. At 6000 µε, the cement paste with silica fume exhibits the lowest increase in crack density and the lowest total crack density of the three pastes.

A final example of the data that can be obtained from automated image analysis of cracks is shown in Fig. 6 [Fig. 3.33] which illustrates the increase in crack density per phase and phase boundary versus strain for cement pastes with w/c = 0.3. The trends shown in Fig. 6 match those observed for the other pastes in which the greatest amount of cracking occurs through UDP, the lowest density hydration product. A major portion of the cracking occurs either totally within UDP or within UDP at the boundary with the harder phases. The next greatest portion of the cracking occurs either totally within CSH-IP or within CSH-IP at the boundary with the harder phases. The lowest amounts of cracking occur within CH and UH, the hardest phases within the paste.

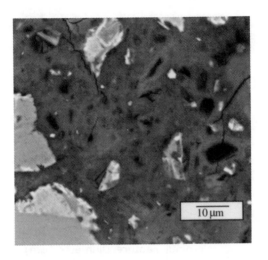

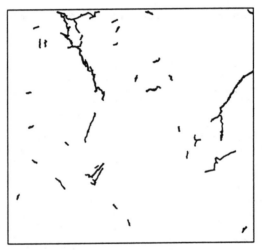

Fig. 3 Micrograph of cement paste with w/c = 0.5 loaded to 4000 με and cracks identified using the automated technique.

CONCLUSIONS

The following conclusions are based on the work described in this paper.

1. Automated image analysis of backscattered electron images of cement paste can provide useful information on hydration phases, size distributions of unhydrated particles and voids, effects of changes in water-cementitious ratio and the use of mineral admixtures, and the distribution of microcracks.

2. The analyses demonstrate that microstructural data can vary significantly from image to image, requiring multiple images to limit the effects of scatter.

3. Although the pastes in this study exhibit different degrees of hydration, the size distribu-

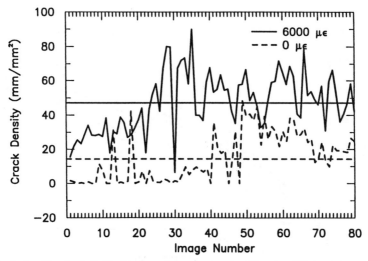

Fig. 4 Crack densities for individual images for cement paste with w/c = 0.3; transverse surface; 0 and 6000 με

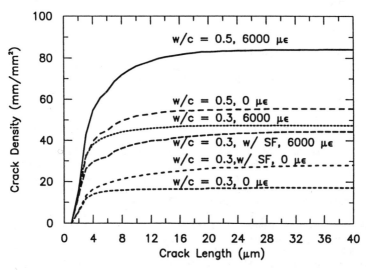

Fig. 5 Cumulative crack density versus crack length for three cement pastes; transverse surface; 0 and 6000 με

tions of the unhydrated cement particles are nearly identical.

4. The size distribution of large voids is strongly affected by the water-cementitious material ratio and by the use of silica fume as a partial replacement for cement.

5. The calcium hydroxide content obtained based on image analysis exceeds, but generally parallels, that obtained using thermogravimetric analysis.

6. The majority of microcracks in both nonloaded and loaded specimens occur through or adjacent to the lowest density hydration phase.

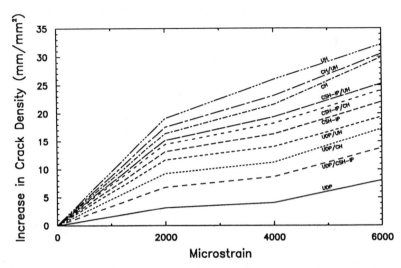

Fig. 6 Increase in crack density per phase and phase boundary versus strain for cement paste with w/c = 0.3

ACKNOWLEDGEMENTS

Image acquisition software was customized for use in this study by the ELMDAS Company. Cement was donated by the Ash Grove Cement Company, and the thermogravimetric analysis was performed at the Ash Grove Research Laboratory in Kansas City, Kansas.

REFERENCES

1. K. L. Scrivener and P. C. Pratt in *Proc.*, 6th Intl. Conf. on Cement Microscopy (Albuquerque, NM, 1984).
2. K. L. Scrivener and E. M. Gartner in *Bonding in Cementitious Composites,* edited by S. Mindess and S. P. Shah (Mater. Res. Soc. Proc. **114**, Pittsburgh, PA, 1988) pp. 77-85.
3. K. L. Scrivener in *Pore Structure and Permeability of Cementitious Materials*, edited by L. R. Roberts and J. P. Skalny (Mater. Res. Soc. Proc. **137**, Pittsburgh, PA, 1989) pp. 129-140.
4. H. Zhao and D. Darwin, *Cem. Concr. Res.* **22**, (4), 695 (1992).
5. D. Bonen and S. Diamond, *Cem. Concr. Res.* **22**, (6), 1059 (1992).
6. K. W. Ketcham, F. A. Romero, D. Darwin, S. Gong, M. N. Abou-Zeid, and J. L. Martin, *SM Report* No. 34, (University of Kansas Center for Res. 1993).
7. J. I. Goldstein, D. E. Newbury, P. Echlin, D. C. Joy, A. D. Romig, C. Fiori, and E. Lifshin, *Scanning Electron Microscopy and X-Ray Microanalysis*, 2nd ed. (Plenum Press, New York and London, 1992).
8. B. Ostle and L. C. Malone, *Statistics in Research*, 4th ed., (Iowa State University Press, 1988).
9. D. Darwin, M. N. Abou-Zeid, K. W. Ketcham in *Proc., 16th Intl Conf. on Cement Microscopy* (Richmond, VA, 1994).
10. L. Hjorth in *Microsilica in Concrete* **1**, paper 9 (Aalborg Portland, Denmark, 1982).
11. G. W. Groves, *Cem. Concr. Res.* **11** (5/6), 713 (1981).
12. K. H. Khayat and P. C. Aïtcin in *Fly Ash, Silica Fume, Slag, and Natural Pozzolans in Concrete* (Amer. Concr. Inst., **SP-132**, **II**, Detroit, MI, 1993) pp. 835-872.

A RE-EVALUATION OF HARDENED CEMENT PASTE MICROSTRUCTURE BASED ON BACKSCATTER SEM INVESTIGATIONS

SIDNEY DIAMOND* AND DAVID BONEN**
*School of Civil Engineering, Purdue University, West Lafayette, IN 47907
**Department of Civil Engineering, Northwestern University, Evanston, IL 60208

ABSTRACT

Backscattered electron imaging of polished cement paste specimens permits a re-evaluation of structural details of hydrated cement paste at the μm level. The primarily microstructural units comprise a highly porous groundmass and large distinct grains ("phenograins") set in it. The groundmass is composed of several kinds of fine particles, with a significant content of easily detected gross pores. Phenograins are primarily large clinker grains hydrating in-situ, but may be distinct deposits of CH, or may be mineral admixture grains. Detailed EDS analyses indicated that hydrating cement in phenograins has a highly consistent composition, interpreted as C-S-H with a small but regular incorporation of sub-μm CH and calcium monosulfoaluminate. Groundmass particles are highly variable in composition, but appear to consist of C-S-H with variable and occasionally major contents of other hydration products on a sub-μm scale. Incorporation of fly ash does not appear to change the basic microstructure, but silica fume incorporated with superplasticizer drastically modifies the character of the groundmass. Some attempts at quantification of these features by application of image analysis are briefly described.

INTRODUCTION

The development of backscatter electron image (BEI) scanning electron microscopy using split quadrant detectors on plane polished surface specimens has opened up new potentials for investigations of hardened cement paste (hcp). When coupled with appropriate image analysis systems and programs, facets of hcp microstructure not previously accessible may now be examined with ease. As a result of such examinations carried out over several years, we have developed an interpretation of hcp microstructure different in emphasis and more definite than the relatively vague picture based in individual particle morphology that is commonly used. This interpretation is based on recognition of a primary dichotomy that can be recognized between a groundmass of fine particles and recognizable pores, and the larger, individually distinct grains irrespective of their origin [1].

In particular we are concerned with microstructure on a relatively gross scale, i.e.1 to 10s of μm, as is easily observed with backscatter detectors. In BEI images at this scale residual unhydrated cement grains, C-S-H hydration product, and calcium hydroxide (CH) are all readily distinguishable from each other by differences in gray level. Pores (usually filled with epoxy resin) are also readily apparent. On plane surfaces the geometrical characteristics of the various features are evident, and their spatial relationships easy to visualize. Thus the texture of the hcp in any arbitrary plane can be clearly established.

Once these geometrical features are recognized on a consistent basis, a possibility exists for quantification by image analysis. Quantitative comparisons among different pastes are then possible, which may lead to the possibility of relating quantitative microstructural features to properties of the material.

It should be again pointed out that the scale of the microstructural features amenable to study using these techniques is limited. Finer-scale microstructural features that can readily be imaged on fracture surfaces using secondary electron mode SEM may not be visible at the magnification possible in BEI, to say nothing of even finer-scale features that can be imaged only in TEM. Indeed, the preparation of plane polished surfaces for study may obscure certain details even at the magnifications used. For example, individual ettringite needles in the groundmass are not usually distinguished. The proposed picture is thus not meant to be taken as a complete description of hcp microstructure.

INSTRUMENTATION AND SPECIMEN PREPARATION

The examinations described in this work were carried out using an Akashi Beam Technology 55A SEM operated at 15 keV, and equipped with a GW Electronics 30-A backscatter detector and a Tracor Northern 5505 EDXA system. In subsequent image analysis work referred to later, the SEM was coupled to a Princeton Gamma Tech Imagist image analysis system.

The pastes studied were prepared from a normal ASTM Type I portland cement, mixed with de-ionized water in an evacuated chamber mounted on a paint shaker to prevent entrapment of air voids. Specimens were prepared at w:c ratios normal in conventional concrete (0.4 to 0.6), and were hydrated in sealed containers for 1 day and subsequently in $Ca(OH)_2$-saturated water for periods ranging up to 1 year.

After the hydration period, individual slices were immersed in acetone, impregnated in ultra low viscosity epoxy mixture, evacuated, and oven cured at 95°C for about 3 hours. The hardened impregnated specimens were again sliced, and the surface to be examined was polished using successively finer grades of diamond grit down to 15 μm. Final polishing was carried out with great care using 3 μm and 1 μm impregnated diamond cloth. Final polishing is a sensitive operation; it is necessary to avoid over-polishing at each stage if the delicate groundmass structures are to be properly revealed.

After polishing the specimens were sputter coated with a 5 nm-thick layer of Pd.

THE PHENOGRAIN-GROUNDMASS CONCEPT

The primary feature of the concept of hcp microstructure developed here is quite simple. Examination of hcp, especially at young ages, leads to the conclusion that the field observed is readily divisible into two types of structural units - a "groundmass" of intermingle fine particles and visible pores, and placed within this are a number of larger, visibly solid grains quite distinct from the groundmass. We have used the term *phenograins* for these larger distinct grains (*pheno*, from the Greek for "distinct"). The derivation is from the petrographic term "phenocrysts" - for large crystals embedded in a finer matrix in rocks having a porphyritic texture.

Fig. 1. provides an illustration of this concept. The micrograph is that of a 3-day old w:c 0.4 paste. A section of the porous groundmass is included within the circle in the lower left portion of the figure. The distinct grains obviously comprise several different types. The brightest features are residual ground cement grains, generally surrounded by thin shells of less bright (but not visibly porous) C-S-H hydration product. The entire composite grain is considered a single phenograin. At the upper right corner is a slightly smaller grain entirely of the same gray level and smooth appearance as the hydration product shells, obviously representing a cement grain that has fully hydrated in situ.

Phenograins

It can be inferred from Fig. 1. that most phenograins are derived from the larger individual ground clinker grains. The extent of in-situ hydration varies from grain to grain, some individual grains remaining almost entirely unhydrated. Others, especially the smaller (but still distinct) phenograins may be fully hydrated, with no central core of bright unhydrated clinker mineral. As hydration proceeds, one expects more fully hydrated phenograins and thicker shells of hydration product around residual cores.

Fig. 2 shows a 3-day old w:c 0.40 paste at lower magnification, displaying essentially similar features. However, here at least two of the smaller phenograins (circled) show gaps between the central unhydrated core and the outer shell, i.e. fall into the category of hollow-shell hydration grains. This variant pattern is represented in many fields, but is never even locally dominant.

Many partly hydrated grains show a shell of hydration product around the unhydrated core that is reasonably complete and of more or less uniform thickness around the periphery of the grain. However, others do not, and examples of hydration shells that only encompass part of the periphery of the unhydrated clinker grain are common, as are grains where the thickness of the hydration shell is not even approximately uniform around the periphery of the grain.

Fig. 1. Backscatter electron image (BEI) micrograph of a 3-day old w:c 0.4 paste, illustrating the concept of various types of phenograins set in a obviously porous groundmass.

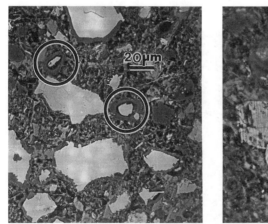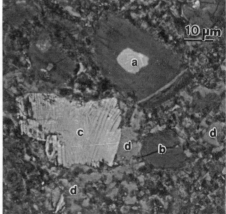

Fig. 2. (Left) BEI of 3-day old w:c 0.40 paste, showing gaps between hydration shells and residual clinker cores
Fig. 3. (Right) BEI of 28-day old w:c 0.40 paste, showing various features.

Fig. 3 shows something of other variations that are found. The field is a from a 28-day old w:c 0.4 paste. The phenograin designated 'a' shows the pattern of a symmetrical hydration shell, except for the brighter unhydrated ferrite "stripe". The phenograin marked 'b' (with a crack running through it) is a smaller, obviously fully hydrated grain. The grain marked 'c' shows a variant pattern of hydration. This is a belite grain (as is obvious from the morphological pattern and was confirmed by EDS). Here hydration has not produced a hydration shell at all; rather, it has dissolved some of the clinker material and exposed the underlying twin lamellae.

The phenograins marked 'd' are seen to be of a gray level intermediate between that of the hydration shells and the unhydrated cores. These phenograins are unlike those discussed previously in that they do not represent relicts of specific clinker grains, but rather are deposits of calcium hydroxide (CH) obviously precipitated from solution. Their boundaries are usually more difficult to delineate with precision than those of the phenograins derived from clinker particles.

It should be apparent to the reader that we have defined the term "phenograin" to include any grain large enough to be morphologically **distinct** from the porous groundmass, whatever its origin. Thus, particles of ground limestone (found in some cements) or fly ash (added to some blended cements) would also be distinct from the groundmass and so would be considered phenograins by this definition.

How to set a lower limit of size for phenograins is something of a quandary. To be consistent with the definition, the lower limit would depend on the texture of the groundmass - i.e. the size of the smallest grain that can be isolated as an individual visually distinct from the particular groundmass. Our original estimate was that perhaps 10 μm might serve as a reasonable lower limit, but in exploring the application of image analysis to pastes, it appears that the boundary should be set lower - to ca. 4 or 5 μm.

In exploring pastes of increasing age and degree of hydration, we find generally speaking that as expected, the number of fully hydrated phenograins progressively increases, and that the hydration shells in other phenograins grow progressively thicker. However, even at 1 year of hydration of these thin prism-shaped specimens in saturated $Ca(OH)_2$ solution, many grains were found with shells of only minimal thickness leaving extensive cores, frequently of C_3S, that remain largely untouched by the hydration process. An example is seen in Fig. 5.

The Groundmass

Fig. 4 provides an indication of the structure of the porous groundmass, as seen in a 1-day old

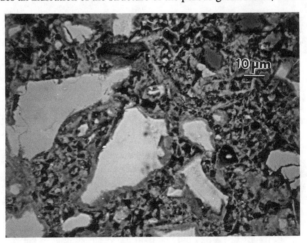

Fig. 4. BEI of 3-day old w:c 0.40 paste, showing groundmass features.

w:c 0.40 paste. The sponge-like character of this portion of the paste is evident, as are the larger than expected sizes of the pores constituting a significant proportion of the groundmass area.

In morphological terms we have divided the small solid particles in the groundmass into thin-walled *skeletal* grains and relatively nondescript and often fuzzy *amorphic* grains.

It is not very distinct in Fig. 4, but some of the larger voids entirely surrounded by thin-walled structures can be seen to be relicts of small, fully hydrated hollow shell hydration grains. This appears to be a general pattern.

As hydration proceeds, it might be expected that the texture of the groundmass should thickens and the sizes of the pores diminish. Again, these expectations are partly met, but even after 1 year for w:c 0.40 paste, the distinction between groundmass and phenograins remains evident, and the porous character of the former is preserved.

Fig. 5, taken from a 1-year old w:c paste, illustrates this distinction, along with other pertinent features. The very large central grain is a nearly entirely unhydrated clinker grain,, preserving not only a large mass of unhydrated C3S but much of the interstitial C3A-C4AF complex. Only a thin and incomplete shell of hydration product exists. As indicated previously, the porous groundmass can still be easily separated from the thin hydration shell of the phenograin.

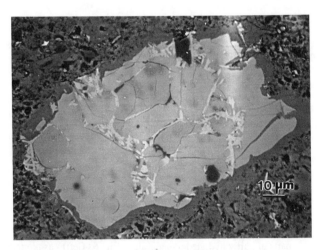

Fig. 5. BEI of 1 year old w:c o.40 paste, illustrating the persistence of distinguishable groundmass and the limited hydration of some large phenograins.

THE CHEMICAL COMPOSITIONS OF HYDRATED PHENOGRAINS AND GROUNDMASS PARTICLES

Examination of the preceding micrographs suggests that, except where CH is present, the gray level of the particles in the groundmass and that of the in-situ hydration shells or the fully hydrated phenograins is approximately the same. The implication is that they have a similar chemical composition.

It seemed important to establish the degree to which the groundmass particles are chemically identical to the hydration rims or the fully hydrated phenograins.

Accordingly, after careful calibration a series of more than 200 fully ZAF-corrected analyses of distinguishable particles was performed on a single 3-day old w:c 0.40 paste. Each reported analysis is the average of at least two determinations on the same particle, usually taken as far

17

apart as feasible considering the possibility of influence by neighboring particles. Details have been described previously [2].

The results of 48 individual analyses on fully hydrated phenograins and on the hydrated part of incompletely hydrated phenograins are shown in two ternary fields constituting Fig. 6. In Fig. 6 we plot atom compositions based on anhydrous element compositions obtained by the ZAF analyses. These typically totaled to about 80% or less in the analyses (bound water not being detectable), but they were recalculated to 100%. Atom percentages of individual metal atoms were calculated from these oxide analyses for plotting purposes.

Fig. 6a is a plot of atom percentage compositions on the Ca - (Al+Fe) - Si triangular field, and Fig. 6b a similar plot on the Ca - S - (Al + Fe) triangular field. On this plot, we indicate the points corresponding to the compositions of stoichiometric ettringite, calcium monosulfoaluminate, and C4AF, and the dashed lines are tie-lines connecting representative C-S-H composition to these points. .

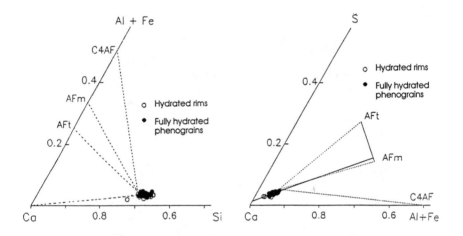

Fig. 6. Plots of compositional analyses for fully hydrated and for the hydrated parts of partly hydrated phenograins. Fig. 6a (left) - as plotted on a Ca - (Al + Fe) - Si field. Fig. 6b (right) - as plotted on a Ca - S - (Al +Fe) field.

It is evident that the range of compositional variation is quite limited, and that for all practical purposes the composition of the material present in hydrating phenograins is similar from grain to grain.

The average Ca:Si ratio for these analyses is close to 2.10, rather higher than expected for "pure" C-S-H. It is evident that the C-S-H is not free of other constituents, small contents of (Al + Fe) and of S being also indicated by the locations of the analyses above the base line in Figs. 6a and 6 b, respectively. As expected, the combined (Al + Fe) plotted is actually mostly Al, with only minor amounts of Fe present. The mean S/Si ratio for these analyses is about 0.08; the mean (Al + Fe)/Si ratio about 0.09.

Interpretation of these analyses is not straightforward. We observe that what little variation there is in Fig. 6a would be more or less on a tie line connecting the "general" C-S-H composition with the Ca corner of the triangle, i.e. the compositional location for CH. In Fig. 6b the small range in composition seems to be on a line directed almost precisely toward "AFm", i.e. stoichiometric calcium monosulfolaluminate.

We infer from these findings that a consistent composition of the hydrated phenograin material exists; that this material incorporates a small and only slightly variable proportion of both CH and calcium sulfoaluminate along with the C-S-H. We also infer that the mixtures are on a size scale much finer than that being examined, since the gray level is sensibly homogeneous in the hydration shells and the fully hydrated phenograins.

It is noteworthy that Richardson and Groves [3], using very high resolution TEM, have recently reported and depicted the existence of very small inclusions of both phases in so-called "inner product" (i.e. phenograin) C-S-H.

It was found that analyses for groundmass particles were very much more variable than those for the hydrated phenograin material. Many of the analyses for both "skeletal" and "amorphic" groundmass particles plot in the same general region as the phenograin analyses of Fig. 6, albeit over a somewhat wider range. A few were reasonably close to pure CH. Several others plotted close to theoretical analysis for calcium monosulfoaluminate on the Ca - (Al + Fe) - Si field (but were notably deficient in S). These analyses could be taken as representing individual particles of CH and calcium monosulfoaluminate components. But many of the analyses plotted along (or closely around) tie lines connecting the general C-S-H composition to points representing the locations of other hydration product phases.

For example a number of analysis plotted quite closely along the tie line connecting average composition of amorphous groundmass grains (representing C-S-H with characteristic minor incorporation of other hydration products) to the CH corner in the Ca - (Al + Fe) - Si field, as shown in Fig. 7a. Others plot along the tie line to the theoretical ettringite composition, as seen in Fig. 7b. Still others (not shown) plot around a corresponding tie line to the calcium monosulfoaluminate composition. It appears that these analyses represent individual groundmass particles that are composites of C-S-H with the other component over a wide range in proportions, from nearly all C-S-H to nearly all CH or ettringite or calcium monosulfoaluminate.

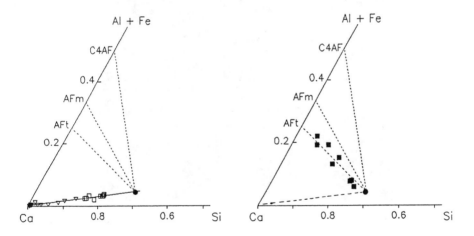

Fig. 7. Plots of compositional analyses for selected groundmass particles.
Fig. 7a (left), particles falling along C-S-H - Ca(OH)2 tie line.
Fig. 7b (right), particles falling along C-S-H - ettringite tie line.

Three further complications need to be mentioned. One is that many (but not all) of the particles deemed to be partly ettringite or partly monosulfate are deficient in S contents, sometimes substantially so. Secondly, some analyses seem to point to the simultaneous occurrence of more than two phases in the same particle - e. g. particles with substantial contents of both CH and ettringite along with the C-S-H .Finally, a few analyses show substantial contents of Al and Fe but very little S, and seem to have compositions indicative of C-S-H mixed with C4AF.

The general result of these analyses appears to be that the groundmass particles are much more highly variable in particle-by-particle chemical composition than is the hydration product in fully or partly hydrated phenograins.

EFFECTS OF INCORPORATION OF FLY ASH OR SILICA FUME

Since concretes containing either fly ash or silica fume are of widespread current concern, it is of interest to establish the degree to which the microstructural concepts discussed in this work need to be modified for cement pastes containing these components.

Fly Ash

Fig. 8. is a representative BEI micrograph of a w: cm 0.40 fly ash-bearing paste hydrated for 14 days. The fly ash content is 10% by weight of solids. The specific fly ash is a Class C material of high CaO content. Its particle size distribution has a weight mean diameter of 11 μm, slightly smaller that of the cement. It appears from Fig. 8 that the only microstructural change induced by the incorporation of the fly ash is that the fly ash particles become small phenograins, easily recognized because of their spherical shape.

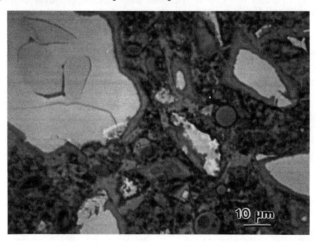

Fig. 8.   BEI of 14 day old w:cm 0.40 fly ash-bearing cement paste.

Silica Fume

Since silica fume is mostly used in combination with superplasticizer, the influence of the combined treatment on microstructural characteristics of paste was investigated. Fig. 9 is a representative micrograph of a 7-day old paste containing 10% silica fume by weight of solids. The paste was mixed at a w:cm ratio of 0.30, with the mix water incorporating a 1% dose of naphthalene sulfonate superplasticizer by weight of solid.

It is clear that the microstructure depicted in Fig. 9 is radically different from those depicted previously. Individual partly- and fully hydrated phenograins are still detectable, but the nature of the groundmass has changed considerably. Only a few isolated pores are detected within the 1 μm to 10 μm scale; these are relics of fully-hydrated hollow shell hydration grains. Skeletal groundmass particles are no longer observed, and the groundmass is appears to be a relatively dense conglomeration containing pores finer than can be detected at the magnification used.

This change in the character of the paste groundmass appears to be fundamental. It appears

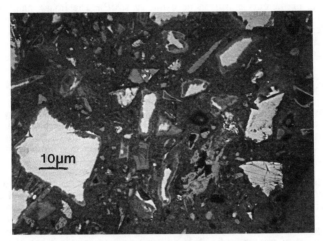

Fig. 9 BEI of a 7-day old w:cm 0.30 silica fume-bearing cement paste incorporating 1% naphthalene sulfonate superplasticizer.

that interpretations of the effects of silica fume in concrete that stress the well-known effects of silica fume on the interfacial transition zone are not entirely correct. The character of the "bulk" cement paste is obviously drastically modified as well.

QUANTITATIVE INVESTIGATIONS OF HCP MICROSTRUCTURE BY IMAGE ANALYSIS

The microstructure of hcp depicted in these micrographs is sufficiently complex that attempts at quantitative treatment are intrinsically difficult. Nevertheless, such investigations have been pursued and considerable progress has been made in quantitative assessment both of pore features and certain features of the microstructure of the solids. Papers reporting these assessments were presented at this Symposium and are published in this volume [4,5].

In an extension of the concepts of hcp microstructure presented here, image analysis procedures have been used to generate quantitative assessments of various phenograin and groundmass features [4]. It has been possible to binary segment and directly assess the fractional area (hence the fractional volume) of CH phenograins, of pore space, and of unhydrated or residual clinker particle cores.

Using a somewhat different approach, specific multi-pixel particle "features" can be individually recognized, and geometric parameters (size, shape, perimeter characteristics) automatically assessed. This kind of information has been derived for hydration shells and fully hydrated phenograins.

A method has been found to combine the bright cores and gray shells of individual partly hydrated grains to reconstitute each combined phenograin as an individual entity, and size and shape parameters of these have been calculated.

The fraction of the total field area in phenograins can be been tallied, not directly, but by summing the areas of the individual phenograin features. If this area fraction is added to the area fractions of CH phenograins and of pores, the area fraction of groundmass particles can be obtained by difference.

Image analysis also can provide a way of measuring pore size distribution, at least for the groundmass pores of sufficient size to be recognized as individual features. A procedure has been developed to determine cumulative quantitative pore size distributions from image analysis treatment of BEI micrographs [5]. The results can be expressed in a form directly comparable to the usual results of mercury intrusion porosimetry (MIP). The image analysis pore size distributions are incomplete, being limited to those pores larger than about 0.8 μm. However, they show extensive pore volume in the roughly 1μm to 10 μm size range, as is visible in the various hcp micrographs in the present paper. MIP analysis, carried out on the same pastes, mis-assigns these pores to sizes mostly less than 0.1 μm.

CONCLUSIONS

1. A treatment of the microstructure of hcp based on BEI is proposed in which individual distinct grains or "phenograins" of multi-μm sizes are distinguished from the highly porous groundmass surrounding them. The phenograins may represent fully or partly hydrated clinker grains, dense CH deposits, or solid particles of materials (fly ash, ground limestone, etc.) incorporated with the cement.

2 The hydrated portions of clinker-derived phenograins are seen to have a quite uniform composition indicative of the presence of small contents of incorporated CH and calcium monosulfoaluminate or ettringite within C-S-H as the host phase. The composition of individual small particles in the groundmass is much more variable. Some are similar to that of the hydrated phenograins, and a few are indicative of nearly pure CH or calcium sulfoaluminates, but most are indicative of mixtures of C-S-H and one or more of the other phases, with the proportions of the phases within individual particles varying widely.

3. Incorporation of fly ash does not modify the basic structure appreciably. Incorporation of silica fume (with superplasticizer) results in a drastic change to the microstructure of the groundmass. The pore structure is changed significantly, and skeletal groundmass particles are not found.

4. Image analysis can be applied to quantify the proportions of the phenograins of the several types that are present, and the content of detectable pores of the groundmass. The proportion of groundmass solid particles can only be estimated by difference. Sizes and shape parameters of the various phenograin types can be established, and the size distribution of the groundmass pores established.

ACKNOWLEDGMENTS

This paper is a contribution from the Purdue University component of the NSF Science and Technology Center for Advanced Cement Based Materials. The continued support of this Center by the National Science Foundation is gratefully acknowledged.
We thank Mr. Yuting Wang, who provided Figures 8 and 9.

REFERENCES

1. S. Diamond and D. Bonen, J. Amer. Ceram. Soc. 76, 2993 (1993).
2. D. Bonen and S. Diamond, J. Amer. Ceram. Soc. 77, 1875 (1994).
3. I. G. Richardson and G. W. Groves, J. Mater. Sci. 28, 265 (1993).
4. Y. Wang and S. Diamond, This Symposium (1995).
5. S. Diamond and M. E. Leeman, This Symposium (1995)

AN APPROACH TO QUANTITATIVE IMAGE ANALYSIS FOR CEMENT PASTES

YUTING WANG AND SIDNEY DIAMOND
School of Civil Engineering, Purdue University, West Lafayette, IN47907

ABSTRACT

Cement paste microstructure as revealed in backscatter SEM presents a number of inherent difficulties that interfere with implementing quantitative image analysis. An approach to overcoming these difficulties is presented, involving gray scale segmentation coupled with application of a hole filling algorithm. Using this approach it is possible to isolate the unhydrated and hydrated portions of phenograins separately, and to combine them for analysis of combined phenograins. Pores and coarse calcium hydroxide masses may also be isolated for feature analysis. Results are reported on mature cement pastes prepared at two water:cement ratios (w:c 0.45 and w:c 0.30) and with and without superplasticizer. It was found that superplasticizer greatly reduced the content and the average size of "visible pores" and increased the content and the average size of coarse CH particles compared to corresponding plain pastes. The area per hydrated phenograin was much smaller in the lower w:c ratio pastes and higher in superplasticized pastes. Among the solid features measured, unhydrated cement particles had the smallest circularity values (at about 2.7) and were the most circular features; Hydrated phenograins had the largest circularity values (at 3.5) and were the most elongated features.

INTRODUCTION

Image analysis of backscattered electron images of polished surfaces has been applied to cements, and have shown some promise in characterizing the microstruictures developed in hardened cement pastes. Scrivener et al [1] described the general method and quantified the volume fraction of unreacted cement components, porosity, and calcium hydroxide in hcp. Their measurements of the contents of unreacted cement and of porosity were found to correlate well with results obtained by other methods. More recent studies include those of Zhao and Darwin [2]. Kjellsen [3], Saada et al. [4], and Scrivener [5]. While interesting results have been obtained, image analysis of hcp presents a severe challenge.

Image analysis was developed primarily for the study of metal structures, within which the individual features (crystals) are geometrically easy to define, and porosity within and between features is not a problem. Such is not the case with hcp. Indeed, perhaps the most difficult step in image analysis of hcp is recognizing and bounding the individual features or microstructural units on a consistent basis.

Diamond and Bonen [6] have recently classified hcp microstructural units into "phenograins" - distinct individual features large enough to be individually recognized, and "groundmass" - a highly porous conglomerate of finer hydrated cement particles and large capillary pores. Most phenograins are composite, in that they consist of both residual unhydrated cement and relatively dense C-S-H developed in-situ. Calcium hydroxide occurs both as fine individual or composite particles within the groundmass and as large but sometimes ill-defined phenograins. Small hollow-shell hydration grains form a recognizable part of the groundmass structure. Fig. 1 provides indications of some of these characteristic features of hcp.

In the present work we attempted to quantify and apply geometric image analysis to some of these features as they occur in mature cement pastes of two w:c ratios (0.45 and 0.30). The specific features evaluated were (a) pores (in sizes above about 0.8 μm), (b) separately distinguishable calcium hydroxide particles, (c) residual unhydrated cement (normally within composite phenograins), (d) hydrated phenograins, consisting of both hydration shells and fully hydrated cement particles, (e) the combined composite phenograins, i.e. individual features of (c) plus (d), and (f) the C-S-H in the groundmass. In the remainder of this paper we refer to these features as, respectively, "pores", "CH", "UH" (for unhydrated cement), "HP" (for hydrated phenograins), "CP" for the combined composite phenograins and, and "GM" (for the solid constituents in the groundmass). The solid portion of the groundmass is quantitatively evaluated by difference, and not susceptible to geometric analysis.

The analyses assayed included area fraction measurement (usual taken as equivalent to volume fraction measurement), and geometrical measurement of the individual features (particles or voids).

Finally, we attempted to evaluate the effects of incorporation of superplasticzer on all of the parameters measured.

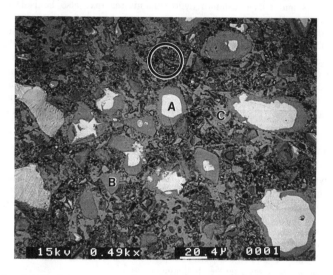

Fig.1 BEI of polished plain cement paste (w:c 0.45, 100days) showing the groundmass and various recognizable phenograins. The encircled area is visibly porous groundmass. **A** is a partly-hydrated composite phenograin (the bright area is the unhydrated portion, the gray area is the hydrated portion). **B** is a smaller, fully hydrated phenograin. **C** is an irregular CH phenograin.

SPECIMEN PREPARATION

Cement pastes of two w:c ratios (0.45 and 0.30) each with and without superplasticizer, were prepared following ASTM C 305. The pastes were all made from a single Type I portland cement, and the superplasticizer was naphthalene sulfonate used at a dosage level of 1% (solid superplasticizer) by weight of cement.

All of the pastes were cured in saturated lime water at room temperature for 100 days, and then individual slices were cut from the middle of the paste cylinders. These slices were dried using acetone replacement, followed by vacuum drying, and finally by overnight oven drying.

The dried slices were impregnated with ultralow viscosity epoxy resin under vacuum, and polished with diamond paste of progressively finer particle size, spread over Texmet polishing cloth. The finest diamond particle size used was 1 μm.

IMAGE ANALYSIS PROCEDURES

The SEM - image analysis system consisted of a Akashi Beam Technology 55A scanning electron microscope and a Princeton Gamma Tech IMIX Version 7 image analysis system. Backscattered electron images were acquired at 15 KeV, using a 256 point per pixel averaging system. The images were acquired at 512 x 512 pixels, with the usual division into 256 gray levels. A standard magnification of 500x was used for quantifying all of the features except the capillary pores. For these, a higher magnification of 1000x was used, permitting tallying of pores down to approximately 0.8 μm diameters in the feature analysis.

For area fraction measurement, a sufficient number of fields must be measured so that the measured average area fraction of one phase approximates the true mean area fraction of that phase within acceptable error and within acceptable confidence level. In these calculations we assumed an underlying normal distribution of repeat measurements. A pilot study of 60 repeat measurements of each phase were carried out for a single specimen. It was determined from this study that an acceptable estimate of standard deviation of repeat measurements could be obtained using 10 to 15 fields. Standard deviations were then measured for each phase within each specimen. Using normal statistical theory, the number of fields required for a 95% confidence that the average area fraction of each phase obtained by measurement is within 5% of the true mean area fraction of that phase was then calculated. Table I shows the results of such calculation. Where a range is indicated, the underlying standard deviations were different for different specimens.

Table I No of Fields Required for 5% Error in Area Fraction Measurement at 95% Confidence

Phase	Magnification	No of Fields
Pores	1000	15
UH	500	30-50
CH	500	10-20
HP	500	15
CP	500	15

An effective binary segmentation of each phase to be recorded is the most important step in image analysis. This fundamental requirement is extremely difficult to fulfill for cement pastes. In these analyses, the UH was always easy to segment from the other features because of its extreme brightness. The hydrated products vary in brightness, with CH possessing the highest gray levels of these phases. Usually CH forms a small peak or shoulder on the large broad gray level histogram associated with the hydrated products. The lower gray levels of the CH features may overlap to some extent with the higher gray levels of the C-S-H, making accurate segmentation difficult. At the low end of the histogram, The pores were not usually sharply distinguished from the lowest gray levels of the C-S-H. Thus to accomplish the required binary segmentation in these instances, a manual procedure, guided by the instrument operator's judgment , was required to position the bounding threshold levels. While this process is necessarily subjective, experience and caution produce good reproducibility and satisfactory results. Also in order to minimize quantitative variations. all fields within a given specimen were acquired using the same contrast and brightness settings, so that once a threshold for a phase was chosen, it remained the same for all fields within that specimen.

The separation of the hydrated part of the phenograins (HP) required several steps. The successive procedures used are illustrated in Fig. 2 and are described as follows:

a) The first step in segmenting the HP features was a gray scale thresholding. The resulting binary image contained both the hydrated phenograins and the many small particles of groundmass C-S-H, as shown in Fig. 2B.

b) The smallest particles were removed by applying only a single cycle of erosion and dilation of moderate severity on the binary image obtained in step (a) above, as shown in Fig. 2C.

c) The remaining small particles were removed by a hole filling program. This program was originally designed to fill holes within individual particles. However, by choosing background gray level as the particle gray level, the entire field was considered by the program to be a single "particle", and the original particles were considered to be holes within it. A threshold size of 4μm was selected and the smaller C-S-H particles were "filled", that is, removed from the image, as shown in Fig. 2D. The residual binary image was then taken as the binary image of hydrated phenograins (HP).

d) Binary images of the unhydrated residual cement particles (UH) were obtained by simple gray scale thresholding. Again particles smaller than 4 μm in average diameter were removed by the hole filling operation, The resulting image can be seen in Fig. 2E.

e) Combined binary images containing both hydrated and unhydrated portions of phenograins were constructed by combining the binary images of HP and UH using image math. The binary image of combined phenograins is shown in Fig. 2F.

Both area fraction measurements and feature measurements were then performed on binary images of the various phases.

Certain specific points need to be discussed with respect to the analyses of the features, as distinguished from area fraction measurements. In feature measurement, the geometry of each individual particle or pore "feature" is determined, necessarily in two dimensions. Measurements included area, average diameter (of a set of 12 "diameters" taken after successive 30 degree rotations) , "form factor" (a measurement of how convoluted the perimeter of a feature is) and "circularity" (a measurement of how elongated a feature is). The form factor is defined as $4\pi(area)/(perimeter)^2$. The value for a circular feature is 1.00; the more convoluted the perimeter, the lower the form factor. The circularity is defined as $\pi(max.diameter)^2/4(area)$. The value for a circular feature is also 1, but in this case the less circular (more elongate) the feature is, the higher the circularity value.

Phenograins are particularly complex features, and the geometric analysis provided for them may not be completely appropriate. For example, as indicated in Fig. 2C, some (but far from the majority) of hydrated phenograins appear geometrically as complete rings. In measuring the diamaters of such features, the values obtained are the same as would be obtained if the rings were completely filled. However, when measuring the areas, only the actual area of the ring is tallied. Thus for such particles, the diameters of the hydrated phenograins and the combined phenograins are the same, but the areas are very different.

A further complication can be seen in the grain midway up the picture near the left hand edge in Fig. 2A. It is seen that a single unhydrated core is almost, but not completely surrounded by hydration product. In Fig. 2E, after completion of the processing to isolate the HP features, it is seen that the products of the single clinker grain are subdivided into four distinct HP features, rather than being a single feature. This is not a defect of image processing, but is a true assessment of the separation of the hydration products visible in Fig. 2A. Not many such particles are observed, but their occurrence is a complicating feature that should not be overlooked.

Average values were compiled for each of the above feature analysis parameters for each phase analyzed and for all of the fields of a given specimen. Features touching the field boundary at any point were removed.

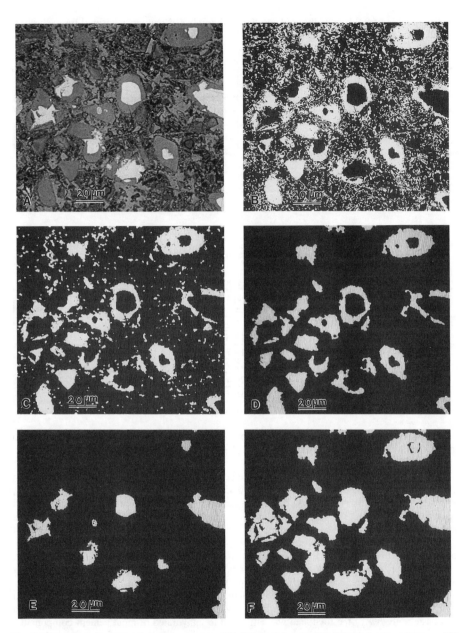

Fig. 2   Procedure in separating phenograins (A) Original image, (B) Binary image of image (A) after gray scale segmentation for hydrated phenograins, (C) Binary image after applying erosion and dilation on image (B), (D) Binary image of hydrated phenograins after applying hole filling operation on image (C), (E) Binary image of unhydrated cement particles, (F) Binary image of combined phenograins after combining the binary images of HP and UH features.

The number of fields actually used and the total number of features measured for each phase in each paste are shown in Table II.

Table II   No. of Fields and No. of Features Recorded in Feature Analyses

Phase	Magnification	No. of Fields	No. of Features
Pores	1000	5	3000-4000
UH	500	30-50	1000-2000
CH	500	10-20	1000-2000
HP	500	15	1000-1500
CP	500	15	1000-1500

RESULTS AND DISCUSSION

Results from area fraction measurement

The results of the various area fraction measurements are given in Table III.

Table III. Average area fraction of various phases in cement paste samples

	plain paste w:c 0.45	superplasticized w:c 0.45 paste	plain paste w:c 0.30	superplasticized w:c 0.30 paste
Pore	15.6	7.2	11.2	7.5
CH	12.1	18.6	13.2	18.1
UH	6.0	14.7	17.4	16.8
HP	13.1	13.8	10.1	11.1
CP	19.2	29.4	28.3	27.6
GM*	53.2	45.7	48.9	46.5

* The area fraction of GM was calculated by difference.

In area measurement, any pixel of the appropriately dark gray scale value is attributed to pores. The individual pixel area at 1,000 x is 0.25 μm x 0.25 μm; thus only pores of this size and larger are tallied.

The area percentages observed for detectable pores in Table III ranges from 7% to more than 15%. The reduction in pore area between w:c 0.45 and w:c 0.30 pastes was expected. However, an even greater difference was observed between pastes without superplasticizer and superplasticized pastes. The latter showed pore area percentages between 7% and 8% at both w:c ratios, very much smaller values than those for either of the plain pastes .

Area percentages for CH provided another surprise. CH area was determined by direct binary segmentation. For this phase, the effect of the superplasticizer was again found to be dominant. Pastes with superplasticizer showed area fractions of about 18%; compared to 12% or 13% for the plain pastes. CH area fraction difference in pastes of different w:c ratios were unimportant.

The tallies for the residual unhydrated cement (UH) provided a rather different effect from those seen previously. UH was also determined by direct binary segmentation. For the plain w:c 0.45 paste, only 6% of the total area was attributed to unhydrated residual cement components; for the companion superplasticized paste the figure was almost 15%, more than twice as large. It would appear that hydration was severely depressed by the superplasticizer, even at 100 days. However, superplasticizer seemed not to affect the UH contents for the w:c 0.30 pastes, both superplasticized and plain pastes showed about 17% UH.

Area percentages for the hydrated portions of phenograins (HP) were compiled by invoking the more elaborate processing described previously, which removed fine particles from this category. It was found that for this tally there was no significant effect of the superplasticizer. However, the w:c 0.45 pastes showed slightly larger area percentages than the w:c 0.30 pastes, about 13% compared to about 10%.

To characterize the area percentages of combined hydrated portions of phenograins with the residual unhydrated cement components, image math was invoked to combine the two binary images (HP and UH) previously obtained. The area obtained is slightly higher than the individual areas, due to a necessary filling in of some gaps between the two sets of individual features. The combined phenograin area is only 19% for the plain w:c 0.45 paste, as compared to around 28% for the other specimens, primarily due to the lower content of UH in the plain w:c 0.45 paste

Areas not assigned to any of the previously defined categories were tallied by difference as solid groundmass particles. In most cases this amounted to roughly half of the total area. The percentages observed varied only slightly between pastes, the plain pastes showing in each case slightly higher areas than the corresponding superplasticized pastes.

Results from feature analysis

The results of the feature analysis procedures carried out on the various samples are provided in Table IV.

The average diameter of the visible pores (i.e. those distinguished by feature analysis as > 0.8μm) was on the order of 2μm, but varied somewhat among the pastes. The effect of the superplasticizer was to reduce average visible pore size, from 2.3μm to 1.9μm for w:c 0.45 pastes, and from 1.9μm to 1.7μm for w:c 0.30 pastes. Similar reductions in average area of the individual pores were observed, from 3.3μm^2 to 1.6 μm^2 for w:c 0.45, and 1.9μm^2 to 1.5 μm^2 for w:c 0.30. The form factor and circularity values for pores were very similar for all pastes except the superplasticized w:c 0.45 paste, which had a low form factor (0.54 as compared to about 0.61), and a high circularity (3.3 as compared to about 2.7). Thus the pores in this paste were thus measurably more convoluted and more elongated than those of the other pastes. In fact as seen in Fig. 3 there were many microcracks in the superplasticized w:c 0.45 paste.

Somewhat more pronounced effects were observed for CH. As indicated in Table IV, the incorporation of superplastizeer increased both the average diameter and the average area of the individual CH features significantly. Average diameters for CH features detected were between about 6μm and about 8 μm. The average area per CH feature varied from as low as about 15μm^2 for plain w:c 0.30 paste to as much as 40μm^2 for superplasticized w:c 0.45 paste. Form factors for CH features were similar in all pastes at about 0.45, and circularities were also similar, at about 3.2.

Residual unhydrated cement (UH) features were significantly larger than CH features. The range in average diameter was from 10μm to 12μm, roughly twice those of the CH features. Incorporation of superplasticizer resulted in an increased average diameter of the UH features in the w:c 0.45 pastes, from 6.7μm to 8.1μm, but did not influence that of the w:c 0.30 pastes. Similar effects were recorded for the average area per feature. The unhydrated cement features were about as convoluted as the CH features (form factor about 0.42) but had somewhat smaller circularity values, about 2.7; and were thus less elongated.

The hydrated phenograin (HP) features were similar in average diameter to the unhydrated cement (UH) features (about 11μm in the overall average), but had very much smaller areas, due to the occurrence of ring-shaped HP features. There was a great difference in the average area per HP feature between w:c 0.45 and w:c 0.30 pastes, the former averaging about 70 μm^2, the latter only about 40 μm². It was observed visually that for the lower w:c paste the hydrated rims

Table IV. Results of feature measurement for various phases in cement pastes

	plain paste w:c 0.45	superplasticized w:c 0.45 paste	plain paste w:c 0.30	superplasticized w:c 0.30 paste
Pores				
Avg D,um	2.32	1.91	1.89	1.69
Avg area, um²	3.30	1.64	1.90	1.48
Form factor	0.61	0.54	0.60	0.61
Circularity	2.72	3.34	2.79	2.71
CH				
Avg D,um	6.69	8.11	5.76	6.53
Avg area, um²	21.21	40.08	15.73	19.55
Form factor	0.44	0.42	0.46	0.45
Circularity	3.18	3.23	3.24	3.37
UH				
Avg D,um	10.66	12.47	11.04	10.84
Avg area, um²	94.52	123.71	107.70	95.74
Form factor	0.38	0.41	0.43	0.44
Circularity	2.94	2.71	2.64	2.75
HP				
Avg D,um	11.41	12.98	9.11	10.01
Avg area, um²	64.89	77.59	36.55	43.09
Form factor	0.34	0.34	0.37	0.34
Circularity	3.39	3.52	3.51	3.74
CP				
Avg D,um	11.73	13.60	11.15	11.19
Avg area, um²	97.57	147.68	96.23	93.96
Form factor	0.36	0.41	0.43	0.41
Circularity	3.19	2.79	2.87	3.00

were definitely thinner, and there were fewer fully-hydrated grains. There is a modest effect of superplasticizer on area per HP feature, superplasticizer incorporation increasing it somewhat. The form factors recorded for HP features were not influenced by either w:c ratio or superplasticizer; neither were the circularity values. The form factors averaged 0.35, a significantly lower value than the other classes of feature, and the circularity value averaged 3.5, a higher value than the other features. Thus hydrated phenograins are more convoluted and more elongated than other features present in these pastes.

The combined phenograins (CP) were found to have slightly larger average diameters than either the hydrated phenograins or the unhydrated cement grains individually, as might be expected. The average diameters were slightly greater for w:c 0.45 than for w:c 0.30 pastes, and slightly greater for superplasticized as compared to plain pastes.

With respect to the average area per CP feature, the superplasticized w:c 0.45 paste had by far the highest value, 148μm², as compared to only about 95μm² for the other pastes.

It is of interest to compare the average area per feature of the combined phenograins with those of the component parts, that is, the hydrated phenograins and the unhydrated phenograins.

One would expect the combined phenograins to have larger area per feature than either of the individual components. This is true for a comparison with hydrated phenograins, but not true for unhydrated phenograins. Table IV indicates that the average area per CP feature is similar or lower than the average area per UH feature. This is apparently due to the inclusion of many small fully hydrated individuals in the CP tally, bring the average near or even below that of the residual UH grains.

The form factors for CP features are similar to those for UH features and slightly larger than those for HP features, all indicate geometrically complex perimeters. CP features appear to have circularity values between those of the HP and UH features.

Discussions

The results of the area fraction analyses on plain pastes suggested area fractions (equivalent to volume fractions) of around 12% for CH, around 12% for hydrated phenograins, and about 50% for groundmass solids. The last value is obtained by difference and may not be very accurate, in particular it must include a significant proportion of pores too fine to be tallied as pore space at the magnification used. Nevertheless, the implication seems to be that much more of the total hydration product is contained in the groundmass than in the dense phenograin structures.

As expected, the visible pore area fraction is substantially higher for w:c 0.45 plain paste than for w:c 0.30 plain paste. Less expected is the major effect of w:c on the area fraction of residual unhydrated cement. The area fraction was only 6% at w:c 0.45 compared to 17% at w:c 0.30. It appears that hydration of the larger cement grains is much more effective at the higher w:c ratio.

Some of the feature analysis results for plain pastes deserve mention. It is interesting that the average diameter for combined phenograins, about 12 μm, is not very different from the mean particle size expected in a modern Type I portland cement.

It is also of interest that the average pore diameter tallied is of the order of 2 μm, very much larger than would be indicated by mercury intrusion porosimetry for either paste. Diamond and Leeman [7] present detailed analyses contrasting pore size distributions obtained using image analysis with those obtained using MIP.

One of the unexpected results of this investigation is the major effect of the incorporation of superplasticizer on the features of the mature paste, especially at w:c 0.45. A representative micrograph of this paste is shown as Fig. 3.

The image analysis results previously tabulated indicated that the superplasticizer reduced both the area percentage of pores and the average size of the pores detected as individual features, that it increased both area percentage and average size of the observed CH features, and that it significantly increased the average area per hydrated phenograin feature. The circularity value of the observed pores also increased significantly, suggesting more elongated pore shape.

Some of these effects can be seen qualitatively in Fig. 3. The regions of porous groundmass are smaller and less common than in the representation of the plain w:c 0.45 paste shown in Fig. 1. Larger CH grains, for example that indicated at A, are common. Image analysis pore size distributions, not reported here for lack of space, indicate fewer large pores. Also, as seen in Fig. 3, there seems to be a significant amount of microcracking. The microcracks would be tallied as highly elongated pores, which could account for the high pore circularity value found for the superplasticized paste.

Finally it should be remarked that the analyses of features reported here is restricted to number average values of the various parameters. For the size-related parameters, additional insight can be obtained from analysis of the distributions of the size measurements, rather than the simple comparisons of average values discussed in this report.

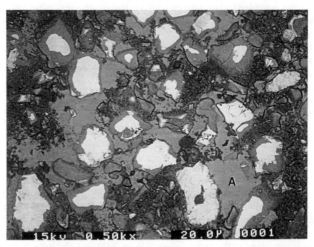

Fig. 3 BEI of superplasticized w:c 0.45 paste showing massive CH,
smaller area of porous regions and smaller visible pores, and microcracks

CONCLUSIONS

1. Quantitative image analysis of hydrated cement pastes can yield interesting, and sometimes unexpected insights into comparative microstructures.
2. Area fraction analyses of recognized microstructural subdivisions can readily be accomplished for pores, unhydrated cement, and CH. Area fractions of hydrated phenograin features can be measured, but only after fairly complex processing to achieve an adequate binary segmentation for these features.
3. Geometric feature analysis can be accomplished for coarse capillary pores, for residual unhydrated cement grains, for calcium hydroxide phenograins, for C-S-H in phenograins, and for "composite" phenograins, i.e. combining both unhydrated and hydrated portions of individual grains. Groundmass particles are too fine for such analysis with present equipment.

REFERENCES

1. K. L. Scrivener, H. H. Patel, P. L. Pratt, and L. J. Parrot, in <u>Microstructural Development during Hydration of Cement</u>, edited by L. J. Struble and P. W. Brown (Mater. Res. Soc. Proc. **85**, Boston, MA, 1986) pp.67-76.
2. H. Zhao and D. Darwin, Cem. and Concr. Res. **22**, 695 (1992).
3. K. O. Kjellsen, R. J. Detwiler, and O. E. Gjorv, Cem Concr. Res. **20**. 308 (1990); 20, 929 (1990).
4. R. Saada, E. Ringot, and M. Barrioulet, Cem. Concr. Res. **21**, 1058 (1991).
5. K. L. Scrivener, Cem. Concr. Res. **22**, 1224 (1992).
6. S. Diamond and D. Bonen, J. Amer. Ceram. Soc. **76**, 2993 (1993).
7. S. Diamond and M. E. Leeman, This Symposium.

MULTI-SCALE DIGITAL-IMAGE-BASED MODELLING OF CEMENT-BASED MATERIALS

D.P. BENTZ*, E.J. GARBOCZI*, H.M. JENNINGS+, AND D.A. QUENARD**
* Building and Fire Research Laboratory, Building 226, Room B-350, National Institute of Standards and Technology, Gaithersburg, MD 20899 USA
+ Department of Materials Science and Engineering, Northwestern University, Evanston, IL 60208 USA
** Centre Scientifique et Technique du Batiment, Saint-Martin d'Heres, FRANCE

ABSTRACT

Computer modelling of the properties and performance of cement-based materials is complicated by the large range of relevant size scales. Processes occurring in the nanometer-sized pores ultimately affect the performance of these materials at the structural level of meters and larger. One approach to alleviating this complication is the development of a suite of models, consisting of individual digital-image-based structural models for the calcium silicate hydrate gel at the nanometer level, the hydrated cement paste at the micrometer level, and a mortar or concrete at the millimeter level. Computations performed at one level provide input properties to be used in simulations of performance at the next higher level. This methodology is demonstrated for the property of ionic diffusivity in saturated concrete. The more complicated problem of drying shrinkage is also addressed.

INTRODUCTION

Predicting the performance of concrete structures is made difficult by the complexity of the material, which has a complex microstructure [1] and exhibits composite behavior at a series of length scales. At the scale of millimeters, one can view the concrete as a composite of aggregates and air voids embedded in a continuous cement paste matrix. Even here, one may need to account for the difference in cement paste microstructure and properties in the interfacial zones surrounding each inclusion [2, 3, 4, 5, 6]. At the scale of micrometers, the cement paste is a composite of unhydrated cement particles, hydration products (crystalline and amorphous), and capillary porosity. Finally, the major hydration product of calcium silicate cements, calcium silicate hydrate (C-S-H) gel, is itself a composite of nanometer-sized "particles" and pores.

A better understanding of the physical processes occurring at each of these scales and their interactions is necessary to increase the predictability of the performance of concrete. For example, it is the capillary forces which develop in the capillary and gel pores that are mainly responsible for the drying shrinkage of concrete. Thus, to reliably predict the drying shrinkage of a field concrete, one must understand the material at the scale of micrometers and even nanometers [7]. While this large scale range cannot be easily incorporated into a single model for concrete microstructure, it is possible to develop an integrated model in

33

which information gained at one scale level is used in computing characteristics at the next higher level [2, 8, 9], as is demonstrated in this paper.

COMPUTER MODELLING TECHNIQUES

At each scale of interest, a method for generating appropriate representative three-dimensional microstructures must be chosen. Computational techniques are then employed to compute the physical properties (diffusivity, elastic moduli, etc.) of these composite microstructures. It should be noted that application of these property computation techniques is not limited to model microstructures, as they can be equally applied to 3-D microstructural representations of real materials obtained by tomographic [10] or microscopic [11] techniques.

At the scale of nanometers, in the study described here, the C-S-H gel is modelled as a two-level structure of partially overlapping spherical particles [7]. At the macro level, the larger 40 nm spherical agglomerates shown in the top right image of Fig. 1 are each composed of smaller 5 nm diameter micro level particles, as shown in the top left image of Fig. 1. The particle sizes and the total porosity at each level have been chosen to be consistent with experimental data from small angle neutron scattering [12] and sorption measurements [13]. The models are generated in continuum space (in a three-dimensional cube with periodic boundaries) and subsequently digitized into a 3-D digital image for the evaluation of properties.

At the scale of micrometers, a cellular-automata based computer model is used to model the microstructural development of cement paste during hydration [14]. Three-dimensional versions of this computer model have been developed both for tricalcium silicate (the major component of portland cement) [15] and for portland cement, consisting of tricalcium silicate, dicalcium silicate, tricalcium aluminate, tetracalcium aluminoferrite, and gypsum [14]. The cement particles are modelled as digitized spheres following a user input size distribution, with the allowable particle diameter range being from about 1 to about $40 \mu m$. The hydration process is modelled in discrete cycles consisting of steps for random dissolution, diffusion, and precipitation/reaction. Volume stoichiometry is explicitly maintained by creating the appropriate number of hydration product volume units (pixels) for each cement pixel that dissolves and reacts. Three-dimensional digital-image-based microstructures generated using the models can be subsequently analyzed to determine phase percolation [16] or diffusion coefficients [17] as a function of water-to-cement (w/c) ratio and degree of hydration (α). The middle portion of Fig. 1 shows a typical two-dimensional microstructure (w/c=0.5, α=87%) generated using the tricalcium silicate hydration model. The unhydrated cement particles are white, capillary porosity is black, calcium hydroxide is light grey, and the C-S-H gel is dark grey.

At the scale of millimeters, mortar or concrete is modelled as a continuum of cement paste containing rigid (nonoverlapping) spherical [18] or ellipsoidal [19] aggregate particles following a size distribution corresponding to that of a real mortar or concrete. Each aggregate can be surrounded by an interfacial zone of some specified thickness, as shown in the bottom image of Fig. 1. Here, the aggregates are shown in dark grey, bulk cement paste is black, the interfacial zones are white, and the single rebar is light grey. One can clearly see the modification of the aggregate distribution in the vicinity of the labelled rebar as well

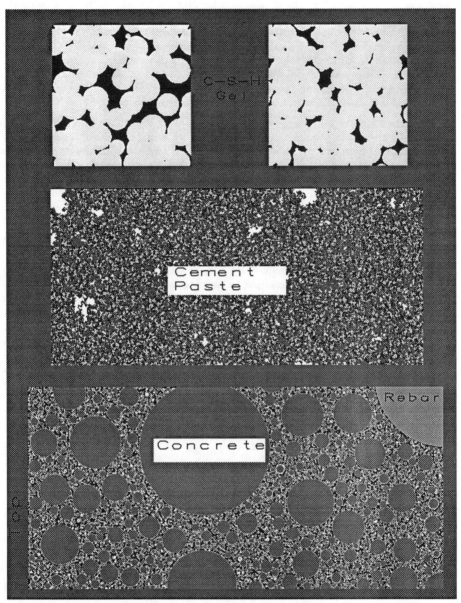

Figure 1: Two-dimensional slices from three-dimensional model microstructures of the C-S-H gel at the scale of nanometers (top image: left is micro or 25 nm by 25 nm and right is macro model or 250 nm by 250 nm), the hydrated cement paste (middle image: 250 μm by 540 μm), and a concrete (bottom image: 3 cm by 6 cm) as described in the text.

as at the top surface of the concrete and near the larger aggregate particles. Additionally, the interfacial zone cement paste is seen to occupy a large fraction of the total cement paste volume. Typically, we model mortar using a sample volume 1x1x1 cm^3, while for concrete, a sample volume of 3x3x3 cm^3 is employed. Aggregate particle diameters typically range from 75 μm to 19.05 mm (3/4"). With this range of particle sizes, a concrete simulation can require as many as 800,000 separate particles. For this reason, supercomputers are often employed for generating these structures. Once again, this continuum structure can be digitized into a 3-D digital image for the computation of properties.

Properties are generally computed using finite difference or finite element techniques [17, 20]. That is, each pixel in the 3-D microstructure is mapped into a node in either a finite element or finite difference analysis. The resultant physical equations are then solved utilizing conjugate gradient or other fast solution techniques to determine the property of interest such as an elastic modulus or diffusivity. For determining diffusivity, the Nernst-Einstein equation is used to determine the diffusivity by solving the equivalent electrical problem for the electrical conductivity of the composite material [17]. Alternatively, random walk simulations may be employed to estimate the diffusivity of a species in a composite medium, as has been utilized to estimate the diffusivity of a mortar relative to its component cement paste [3, 4, 5].

To utilize these solution techniques, the properties of each phase in a microstructure must be known or assumed. For example, to compute an elastic modulus of a cement paste microstructure, one must know the elastic properties of the unhydrated cement, the calcium hydroxide, and the C-S-H gel. Often, our approach is to utilize properties computed at one scale level of modelling as input into the computation procedures employed at a higher scale. For example, as is demonstrated in the results, the conductivity/diffusivity computed for the C-S-H gel nanostructure model can be used as an input property into the cement paste microstructure model, so that the conductivity/diffusivity of cement paste can be computed as a function of w/c and α. Likewise, the diffusivity computed for cement paste can be utilized in the model of concrete at the level of millimeters, to compute the diffusivity of a concrete, the value of actual interest to a designer or structural engineer.

RESULTS AND DISCUSSION

Transport Property Example: Diffusivity

The overall approach is demonstrated first for the case of computing the relative diffusivity of concrete. The relative diffusivity is defined as the ratio of the diffusivity of ions in a composite medium to their diffusivity in bulk water. Based on the Nernst-Einstein relation, the relative diffusivity (D/D_0) is equivalent to the relative conductivity (σ/σ_0). The results of this computation are summarized in Fig. 2, which is referred to in the discussion which follows.

The first step is the computation of the relative diffusivity of the C-S-H gel. Using the electrical analogy and the nanostructural model shown in the upper right of Fig. 1, a value of 1/300 is computed for the relative diffusivity of the gel. Here, we assume that diffusive transport occurs only in the cluster-level pores and not in the much smaller pores shown in

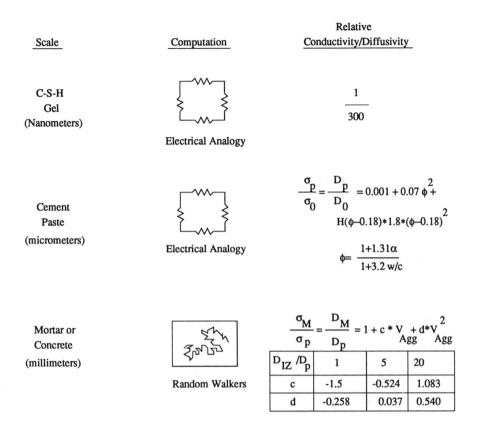

Scale	Computation	Relative Conductivity/Diffusivity
C-S-H Gel (Nanometers)	Electrical Analogy	$\dfrac{1}{300}$
Cement Paste (micrometers)	Electrical Analogy	$\dfrac{\sigma_p}{\sigma_0} = \dfrac{D_p}{D_0} = 0.001 + 0.07\,\phi^2 + H(\phi-0.18)*1.8*(\phi-0.18)^2$ $\phi = \dfrac{1+1.31\alpha}{1+3.2\,w/c}$
Mortar or Concrete (millimeters)	Random Walkers	$\dfrac{\sigma_M}{\sigma_p} = \dfrac{D_M}{D_p} = 1 + c*V_{Agg} + d*V_{Agg}^2$

D_{IZ}/D_p	1	5	20
c	-1.5	-0.524	1.083
d	-0.258	0.037	0.540

Figure 2: Relative diffusivity computed using structural models at each of the three scale levels shown in Fig. 1.

the particle-level model in the upper left of Fig. 1. This is a reasonable assumption, since these smaller, nanometer size pores are on the order of the size of a water molecule and would be virtually inaccessible to many diffusing ions. A further assumption is that this relative diffusivity value is characteristic of all C-S-H gel regardless of cement composition or when during the hydration process the C-S-H is produced.

The second step is to use the value computed for the relative diffusivity of the C-S-H gel in a computation of the relative diffusivity of cement pastes of various w/c ratios and degrees of hydration. A diffusivity value of 1.0 is assigned to the capillary porosity, while the unhydrated cement and calcium hydroxide are assigned diffusivities of 0.0 since they contain no porosity. Again using the electrical analogy, computations performed for a variety of microstructures have resulted in the development of an equation which relates relative diffusivity to the capillary porosity of the cement paste as shown in Fig. 2. Good agreement has been observed in comparing these computed model values to ones measured experimentally both for chloride ion diffusivity [17, 21] and cement paste conductivity [22].

Finally, the relative diffusivity values computed for cement paste can be used as input into a structural model for mortar to determine the effect of aggregates and their surrounding interfacial zones on the diffusivity of the mortar (D_M). At this level, one must select the thickness of the interfacial zone paste and the value of its diffusivity (D_{IZ}) compared to that of the bulk paste (D_P). Fig. 2 provides equations which approximate the relative diffusivity of a specific **model** mortar as a function of aggregate content for three ratios of (D_{IZ}) to (D_P) and an interfacial zone thickness of $20\mu m$ [3, 4, 5]. The mortar is considered to be a model mortar in that only four different size aggregates were utilized in generating its structure. However, the computational techniques described in [3] would be equally applicable to a real mortar specimen, although the computational times will increase in proportion to the number of particles in the simulation.

Mechanical Property Example: Drying Shrinkage

The problem of drying shrinkage in concrete is much more complex from a modelling standpoint than that of ion diffusion under saturated conditions. Because much of the drying shrinkage is induced by capillary forces in the pore water [23, 24], one must be able to calculate the locations of water in a microstructure at a given relative humidity, as well as the response of this microstructure to the stresses induced by the capillary tension in the water and changes in surface energy of the adsorbed water layer [25]. A computer algorithm has been developed to simulate the adsorption/desorption of water in a two-dimensional [26] or a three-dimensional [7] porous structure based on a geometrical analysis according to the Kelvin-Laplace equation, which relates a critical pore size for capillary condensation to a partial pressure (relative humidity): $ln(P/P_0) = (2\gamma V_m)/(rRT)$. Using this equation, one can study the effects of a change in relative humidity, temperature, or liquid surface tension on the critical pore radius, r, below which capillary condensation will occur. For a fixed set of these parameters, a geometrical algorithm can be utilized to determine the locations of all liquid water in a given microstructure [7] if the drying/wetting history of the sample is known.

Once the locations of all "liquid" water in a microstructure have been determined, capillary-tension-induced stresses ($\psi = 2\gamma/r$) can be located at all solid-liquid interfaces. Then, finite element techniques can be used to compute the drying shrinkage which would be caused by these capillary forces. As shown in Fig. 3, once again, a hierarchical computation can be employed to relate the processes occurring in the C-S-H gel pores to the actual shrinkage measured on a concrete structure. At the concrete level, preliminary two-dimensional investigations have been conducted on the role of the interfacial zone in influencing elastic and shrinkage properties [6]. Depending on its local environment, at this level, one may also need to consider the distributions of temperature and moisture within a concrete structure.

CONCLUSIONS

The use of a set of multi-scale microstructural models to investigate transport and mechanical properties of concrete has been demonstrated. By relating microstructure to prop-

Scale	Shrinkage components	Needs
Micro C-S-H Gel	Water-to-solid interface stresses 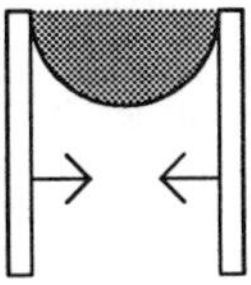	Elastic moduli of C-S-H particles
Macro C-S-H Gel	Interface stresses Bulk shrinkage of C-S-H agglomerates 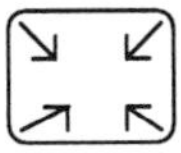	Elastic moduli of C-S-H agglomerates
Cement Paste	Interface stresses Bulk shrinkage of C-S-H	Elastic moduli of C-S-H, calcium hydroxide, and unhydrated cement
Mortar	Bulk shrinkage of bulk paste Bulk shrinkage of IZ paste	Elastic moduli of bulk cement paste, IZ cement paste, and aggregate Temperature, moisture distributions

Figure 3: Approach to computing drying shrinkage using structural models at each of the three scale levels shown in Fig. 1.

erties at all relevant size scales, a more complete understanding of the influence of microstructure and the underlying physical processes on the performance of these composite materials can be obtained. For the case of chloride diffusivity, a complete set of results has been presented. For the more complicated phenomenon of drying shrinkage, much work is still needed, but the basic conceptual groundwork for modelling has been developed. As the structural modelling and computational capabilities described herein continue to evolve, their use as a tool in the design of a concrete with desired properties and service life should become a reality.

NOTATION

Symbols
α - degree of hydration
γ - liquid surface tension

ϕ - porosity
ψ - stress
σ - conductivity
c, d - constants in equation to estimate mortar relative diffusivity
D - diffusivity
$H(x)$ - heaviside function = (1: x $\geq$ 0; 0 otherwise)
P - pressure
r - pore radius
R - universal gas constant
T - absolute temperature
V - volume fraction
w/c - water to cement ratio

Subscripts
Agg - aggregate
0 - bulk solution
IZ - interfacial zone cement paste
m - molar
M- mortar
P - bulk cement paste

ACKNOWLEDGMENTS

The authors would like to thank the National Science Foundation Science and Technology Center for Advanced Cement-Based Materials for partial financial support of much of the work described in this paper.

REFERENCES

[1] Scrivener, K.L., in Materials Science of Concrete I, edited by J.P. Skalny (American Ceramic Society, Westerville, OH, 1989), p. 127.

[2] Bentz, D.P., Schlangen, E., and Garboczi, E.J., in Materials Science of Concrete IV, edited by J.P. Skalny and S. Mindess (American Ceramic Society, Westerville, OH, 1994).

[3] Garboczi, E.J., Schwartz, L.M., and Bentz, D.P., "Modelling the Influence of the Interfacial Zone on the Conductivity and Diffusivity of Mortar," submitted to J. of Advanced Cement-Based Mat.

[4] Garboczi, E.J., Schwartz, L.M., and Bentz, D.P., "Modelling the D.C. Electrical Conductivity of Mortar," these proceedings.

[5] Schwartz, L.M., Garboczi, E.J., and Bentz, D.P., "Interfacial Transport in Porous Media: Application to D.C. Electrical Conductivity of Mortars," submitted to Phys. Rev. B.

[6] Neubauer, C.M., Jennings, H.M., and Garboczi, E.J., "Modelling the Effect of Interfacial Zone Microstructure and Properties on the Elastic Drying Shrinkage of Mortar," submitted to J. of Advanced Cement-Based Mat.

[7] Bentz, D.P., Quenard, D.A., Baroghel-Bouny, V., Garboczi, E.J., and Jennings, H.M., "Modelling Drying Shrinkage of Cement Paste and Mortar: Part 1. Structural Models from Nanometers to Millimeters," to appear in *Mat. and Struc.*

[8] Jennings, H.M., and Xi, Y., in Creep and Shrinkage of Concrete, edited by Z.P. Bazant and I. Carol (E & F Spon, London, 1993), p. 85.

[9] Huet, C., in Micromechanics of Concrete and Cementitious Composites, edited by C. Huet (Presses Polytechniques et Universitaires Romandes, Lausanne, 1993), p. 117.

[10] Bentz, D.P., Martys, N.S., Stutzman, P.E., Levenson, M.S., Garboczi, E.J., Dunsmuir, J., and Schwartz, L.M., "X-Ray Microtomography of an ASTM C109 Mortar Exposed to Sulfate Attack," these proceedings.

[11] Stutzman, P.E., *Ceramic Trans.* 16, 237 (1991).

[12] Allen, A.J., Oberthur, R.C., Pearson, D., Schofield, P., and Wilding, C.R., *Phil. Mag. B* 56 (3), 263 (1987).

[13] Baroghel-Bouny, V., PhD thesis, L'ecole Nationale des Ponts et Chaussees, Paris, France 1994.

[14] Bentz, D.P., Coveney, P.V., Garboczi, E.J., Kleyn, M.F., Stutzman, P.E., *Modelling and Sim. in Mat. Sci. and Eng.* 2 (4), 783 (1994).

[15] Bentz, D.P., and Garboczi, E.J., "Guide to Using HYDRA3D: A Three-Dimensional Digital-Image-Based Cement Microstructure Model," NISTIR 4746, U.S. Department of Commerce (1992).

[16] Bentz, D.P., and Garboczi, E.J., *Cem. and Conc. Res.* 21, 325 (1991).

[17] Garboczi, E.J., and Bentz, D.P., *J. of Mat. Sci.* 27 2083 (1992).

[18] Winslow, D.N., Cohen, M.D., Bentz, D.P., Snyder, K.A., and Garboczi, E.J., *Cem. and Conc. Res.* 24 (1), 25 (1994).

[19] Bentz, D.P., Hwang, J.T.G., Hagwood, C., Garboczi, E.J., Snyder, K.A., Buenfeld, N., Scrivener, K.L., "Interfacial Zone Percolation in Concrete: Effects of Interfacial Zone Thickness and Aggregate Shape," these proceedings.

[20] Garboczi, E.J., and Day, A. R., "An Algorithm for Computing the Effective Linear Elastic Properties of Heterogeneous Materials: 3-D Results for Composites with Equal Phase Poisson Ratios," submitted to *J. of Appl. Phys.*

[21] Halamickova, P., Detwiler, R.J., Bentz, D.P., and Garboczi, E.J., "Water Permeability and Chloride Ion Diffusion in Portland Cement Mortars: Relationship to Sand Content and Critical Pore Diameter," accepted by *Cem. and Conc. Res.*

[22] Olson, R.A., Christensen, B.J., Coverdale, R.T., Ford, S.J., Moss, G.M., Jennings, H.M., Mason, T.O., and Garboczi, E.J., "Microstructural Analysis of Freezing Cement Paste Using Impedance Spectroscopy," submitted to *J. of Mat. Sci.*

[23] Bentur, A., Berger, R.L., Lawrence, Jr., F.V., Milestone, N.B., Mindess, S., and Young, J.F., *Cem. and Conc. Res.* 9, 83 (1979).

[24] Fu, Y., Gu, P., Xie, P., and Beaudoin, J.J., *Cem. and Conc. Res.* 24 (6), 1085 (1994).

[25] Wittmann, F.H., in Creep and Shrinkage in Concrete Structures, edited by Z.P. Bazant and F.H. Wittmann (John H. Wiley & Sons, Ltd., New York, 1982) p. 129.

[26] Quenard, D.A., Bentz, D.P., and Garboczi, E.J., in Drying '92, edited by A.S. Mujumdar (Elsevier Science, 1992) p. 253.

MICROCRACK STUDY OF CEMENT-BASED MATERIALS
BY MEANS OF IMAGE ANALYSIS

YAHIA ALHASSANI, ALAIN BASCOUL, ERICK RINGOT
L.M.D.C. INSA-UPS Genie Civil, complexe scientifique de rangueil, 31077 Toulouse CEDEX,
France

ABSTRACT

The relationship between microscopical damaging of concrete and its mechanical parameters such as Young's modulus and degree of reversibility is a basic issue. In the same time, transport properties and of course the durability of civil engineering and hydraulic structures are linked to the state of microcracking.

The replica technique has been developed in our laboratory in order to study microcracks with a scanning electron microscope. This non-destructive method allows observation of the imprint of the concrete surface and to follow the evolution of the alterations with time. An objective quantitative analysis requires numerous replicas and many microscopic fields within each replica.

After a brief state of the art of the automatic method of extraction of lines by image processing, we propose a specific algorithm of image analysis which allows the cleaning of the pictures and the extraction of the microcrack skeleton. Today, this procedure has been validated on cement paste samples. It has to be improved to be applied on concrete.

The measured parameters are the microcrack specific lengths and their orientation. The objective values measured by means of this procedure compare favorably to results obtained by hand drawing.

INTRODUCTION

The aim of our work is to achieve an objective and systematic way to characterize microcracking and cracking in mortar and concrete. We have two goals. First, we try to link the microscopic damage of concrete to its mechanical parameters such as Young's modulus and the degree of reversibility [1]. Second, we wish to quantify the transport properties which are factors of major concern with regard to structural reliability. These properties are connected to the microcracking due to external as well as internal phenomenon. Another important issue is to improve our knowledge of the microcracking morphology in order to set up predictive models with a good degree of accuracy.

We are working on these topics in our laboratory by means of the replica technique in conjunction with scanning electron microscopy (SEM) and image analysis. Until now, the replica technique was semi-automatic because hand drawing of the microcrack skeleton preceded computer analysis. In this paper we present a new stage of our investigation by replacing the manual operation with automatic extraction by image processing techniques.

REPLICA TECHNIQUE

The replica technique gives the imprint of the polished surface of concrete[2, 3, 4, 5]. This imprint can be observed under SEM, because it is made of a material insensitive to the

43

atmosphere of the SEM. Moreover, the replica method is not destructive and it allows to follow the evolution of the microstructure under loading by taking successive imprints of the surface. At usual magnification (x100- x400), the minimum microcrack opening that is detected is around 0.1 - 0.25 µm. Higher magnifications are used when there are some doubts about the interpretation of pictures.

The existing program MINDA allows the recording of the hand extracted microcrack lines. The cumulative network of cracking can be reconstructed from the set of views observed on a replica. These data are quantified using the total projection stereological technique [6]. The results are reported in a polar curve called rose of projection which is significant of the orientation and the density of microcracking.

AUTOMATIC EXTRACTION OF MICROCRACKING

A magnification of x200 constitutes a low boundary for a sufficient visibility of the details and good sharpness of the digitized pictures. Under these conditions the quantification of the whole 2x2 cm² replica requires the study of 320 views. The need for an automatic procedure becomes obvious through this example.

Some references

The method proposed for the recognition of the cracks is inspired by early works which were achieved for biomedical purposes, as well as metallurgy and petrography.Meyer and Serra[7] developed a filter based on the morphologic center to preserve the small vessels in angiography pictures. They get a strong filtering of the noise without fuzzying the eye veins.Kurdy and Jeulin [8] improved the detection of grain boundaries in metallographic images by means of digital and binary directional filters. Lafon et al [9] have used morphological erosion and dilation with an isotropic volumic structuring element. Their procedure allows a good filtering of the noise in their pictures of volcanic rocks and the emphasis of the details before the analysis.Civil engineering is also interested by image analysis. For instance Salomon and Panetier [10] developed automatic extraction of crack patterns from impregnated sections of concrete in order to study damage created by the alkali reaction.

We have tried to use these tools for our purpose, but we did not get quite satisfying results. So we have adapted these procedures for our own need of crack recognition.

Algorithm of our procedure

The images from SEM are digitized in an hexagonal lattice which introduces less anisotropy than a square one. The lattice has 512 lines of 625 pixels. The magnification is x200 and the distance between two consecutive lines is 0.937 µm. Thus the distance between two neighbouring pixels is 1.082 µm. Each view is 675 µm wide and 480 µm height.

On our pictures, microcracks and cracks appear as more or less narrow bright lines which spread over a non-uniform background. The goal of the procedure was to extract a binary trace of the cracks.

The processing image procedure is made of five basic stages:
- Filtering

An opening followed by a closing modifies the image grey level. So the opening cuts off the sharp peaks, and the closing fills in the narrow valleys.

The difference between the initial image and the filtered one eliminates the wide areas with uniform brightness. In our pictures these areas often result from artifacts.

- Top hat operation

A small dilation increases the line widths and somewhat restores the continuity which may be altered by the previous filtering.

High top hat transformation keeps out the tops which belong to narrow strips whose width is smaller than the size of the structuring element. Moreover this operator makes the dark background more uniform.

- Binarization

A linear anamorphose is applied in order to expand the grey histogram. It makes it possible to obtain similar average grey levels for a lot of views taken up from the same sample. Next all of the pixels whose brightness is lower than a given threshold are removed in order to achieve the binarization of the picture.

- Cleaning

Artifacts which still remain are definitively removed by an erosion by a small isotropic structuring element followed by a geodesic reconstruction.

- End of the procedure

A final dilation is necessary to ensure the continuity of the cracks. Then the skeletonization completed by the removing of the artificial short branches enables us to find the final simple trace of the crack.

ANALYSIS

We measure three parameters for a sample constituted by a set of views.
        - the total crack length by pixels counting L,
        - the specific crack length by dividing L per the area A of the surface $L_A=L/A$,
        - the degree of orientation by applying the method of the secants:
          six neighbourhood configurations determining six directions are detected in our
          digitized images according to table A-b, [11].

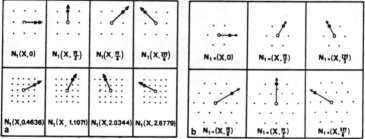

Table A : Neighbourhood configurations which allow to determine the different directions for -a) square lattice, -b) hexagonal lattice.

For each direction the number of configurations $N_1(\theta)$ is counted and weighted by the distance dD between two lines. So we get the projected length in the direction $(\theta+\Pi/2)$ according to the equation (1).

$$D_2(\theta) = N_1(\theta).dD(\theta) \qquad (1)$$

After dividing by the sample area, the results are presented in a polar curve called (Rose of intercepts).

SOME RESULTS

We present in this paper results which illustrate the different stages of the procedure and its efficiency. The tested specimens are made of cement paste and were microcracked by heating.

Fig.1 illustrates the study of a typical view(675x480)μm². This view was chosen because of the existence of a microcrack running across. Fig.1-a gives the initial picture such as it can be observed through SEM. Fig.1-b gives the filtered image. The result of the top hat operation is represented in fig.1-c. Binarization leads to the black and white pattern of fig.1d. At this stage cleaning is still necessary(fig.1-e). Finally fig.1-f gives the skeleton of the microcrack. This picture may be regarded as the one which could be subjectively hand drawn.

The procedure of analysis and quantification is illustrated in fig.2. It concerns the global study of an area(2025x1920)μm² by means of 12 views at x200 magnification. Fig.2-a gives the study by means of MINDA program, whereas the automatic recognition is presented in fig.2-b. The microcrack skeleton extracted by image analysis compares well with the hand drawn one. The rose of projection by MINDA program appears as a smooth polar curve, because the projected specific lengths can be calculated by steps of five degrees. On the other hand the rose of the intercepts has a rough shape which is due to the fact that only six neighbourhood configurations have been exploited. This leads to some imprecision in the determination of the directions of smaller and greater orientation. Moreover the comparison shows that the automatic procedure gives specific length values of the same order as the ones evaluated by MINDA program. However they are higher. Two facts may explain these differences. First, the hand drawing simplifies the trace of the microcrack. Second, short branches remain in the automatic procedure. They may be either legitimate parts of the crack or artifacts. All these considerations lead to conclude that hand drawn curves are propably too short.

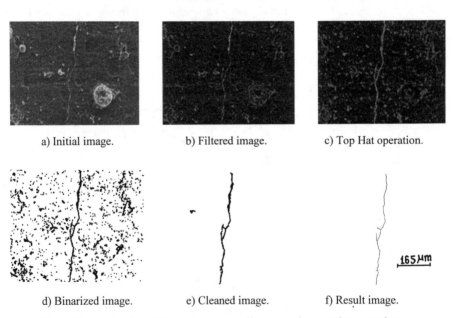

a) Initial image. b) Filtered image. c) Top Hat operation.

d) Binarized image. e) Cleaned image. f) Result image.

Fig.1 : Study of a view(675x480)μm², according to crack extraction procedure.

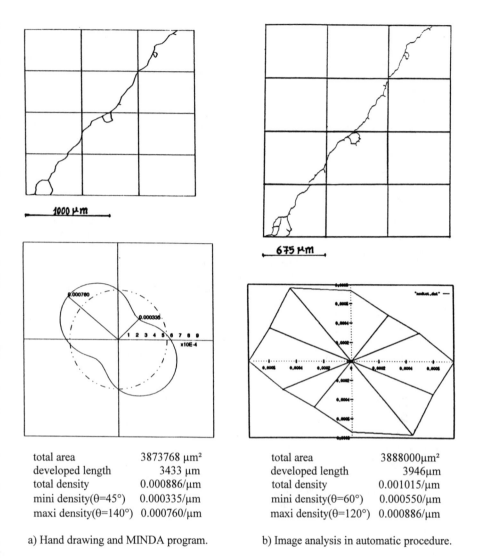

total area	3873768 μm²	total area	3888000μm²
developed length	3433 μm	developed length	3946μm
total density	0.000886/μm	total density	0.001015/μm
mini density(θ=45°)	0.000335/μm	mini density(θ=60°)	0.000550/μm
maxi density(θ=140°)	0.000760/μm	maxi density(θ=120°)	0.000886/μm

a) Hand drawing and MINDA program.   b) Image analysis in automatic procedure.

Fig.2. Microcracking quantification.

CONCLUSION

We have been working for many years about the observation and the characterization of concrete microcracking. The replica technique we use with SEM allows very fine observations. However it is limited to qualitative quantifications without an automatic procedure of microcrack extraction. A new step in our research is to develop such a procedure by means of image analysis.

The method we present in this paper gives good results on specimens made of cement paste. For such specimens microcracks appear with a sufficient contrast as bright lines which spread over a non-uniform background. The examples show that our automatic procedure is able to give microcrack skeletons similar to subjective hand drawings. It allows the evaluation of the total length of microcracks by cumulative measures on a set of views.The rose of the intercepts gives specific lengths projected in different directions which compare with the early MINDA program.

Now, the automatic procedure has to be improved to be applied on mortar and concrete surfaces. For these materials the pictures are more difficult to analyse. First, the interfaces between paste and aggregate may be mistaken for microcracks. Second the topography of the replicas is more complex that results from the polishing of various constituent parts with different toughnesses.

REFERENCES

1. A. BASCOUL, J.C. MASO. Influence of the intermediate stress on the mechanical behaviour of concrete under biaxial compression. Materials and building Research, Annales de l'Tnstitut Technique du Bâtiment et des Travaux Publics, supp.n° 351, june (1977).
2. J.P. OLLIVIER, A non destructive procedure to observe the microcracks of concrete by scanning electron microscopy, Cement and Concrete Research, Vol.15, 1055-1060, (1985).
3. E. RINGOT, J.P. OLLIVIER, J.C. MASO, Caracterization of the initial state of Concrete with regard to microcracking, Cement and Concrete Research, Vol.17, 411-419, (1987).
4. A. BASCOUL, J.P. OLLIVIER, M. POUSHANCHI, Stable microcracking of Concrete subjected to the tensile Gradient, Cement and Concrete Research, Vol.9, 81-88, (1989).
5. M. MASSAT, J.P. OLLIVIER, E. RINGOT, Microscopic analysis of microcracking damage in Concrete and durability, 3rd Euroseminar on microscopy applied to bulding materials, (1997).
6. E. RINGOT, Automatic quantification of microcracks network by stereological method of total projections in mortars and Concrete, Cement and Concrete Research, Vol.18, 35-43, (1988).
7. F. MEYER, J.SERRA, Filters; From theory to practice, Acta Stereologica, (1989), 8/2, 503-508.
8. M.B. KURDY, D. JEULIN, Directional mathematical morphology operations, Acta Stereologica (1989), 8/2, 473-480.
9. D. LAFON, P. BOIVIN, P. BONTON, A. PROVOST, Image analysis applied to the study of volcanic formations, Acta Stereologica, (1989), 8/2, 615-620.
10. M. SALOMON, PANETIER, Quantification du degré d'avancement de l'alcali-réaction dans les bétons et la néofissuration associée, 3rd CANNET/ACI, international conference on durability of Concrete, Nice, (1994), 383-401.
11. M. COSTER, J.L. CHERMANT, Precis d'Analyse d'Images, presses du CNRS, (1989), 271.

MONITORING OF CONCRETE QUALITY IN HIGH PERFORMANCE CIVIL ENGINEERING CONSTRUCTIONS

ANDERS HENRICHSEN AND PETER LAUGESEN
Dansk Beton Teknik A/S, Helleruplund Alle 21, DK-2900, Denmark.

ABSTRACT

Modern high performance concretes (HPCs) are well characterized by their materials, compositions, and other characteristics, and usually documented by 'test samples' from laboratory specimens and trial castings. The characteristics are applied for construction design and estimations of service life time. However, in these young constructions, the internal structure of the HPC often reveals serious differences from that of the 'test sample concrete'. Accordingly, the characteristics of the constructional concrete may be dramatically different from the expectations, often leading to increased rate of deterioration. A series of examples of such failures in the internal structures from major European HPC civil engineering constructions are presented. A test method is presented for the evaluation of the internal structure of the concrete as it exists in the actual structure. This test method is based on reground vacuum impregnated full sections of concrete core samples, cored from the construction at the moment of stripping the form work. Thus it is made possible to assess the structure during production. The cost of applying the test method approximates 0.1 % of the concrete value.

INTRODUCTION

During QA work on high performance concrete from major European civil engineering constructions, it has become evident that the internal structure of the concrete material possesses a substantial number of defects. The main defects are listed below in order of decreasing extent:

- Bonding defects at coarse aggregate and reinforcement, mainly evolved as highly porous zones of varying thickness.
- Uneven homogeneity of the cement paste, i.e. varied capillary porosity.
- Cracks, both macroscopic (e.g. from exposed surfaces) and microscopic (e.g. bundles of plastic cracks internally in the concrete volume).
- Uneven distribution of coarse aggregate.
- Poor quality of the air void system, mainly high spacing factor.
- Lack of hydration of cement locally in the cement paste.

The defects are caused by a variety of conditions in both the fresh and the hardening concrete. However the two main factors are concrete composition, and inappropriate handling, primarily excessive vibration. The bulk volume of concrete in the constructions is placed and compacted by use of high frequency poker vibrators, which may rapidly induce segregation phenomena of all orders of magnitude in the fresh concrete. Thus the initial vibration of the HPC is a delicate matter, which is often performed with too long duration.

Due to demands of high strength and low permeability it has become general practice to specify a very low capillary porosity of the cement paste. This is achieved by lowering the W-C ratio and by adding microsilica (silica fume). The latter induces a pronounced shrinkage of the concrete (i.e. crack formation) if not incorporated at modest levels. Both approaches demand application of superplasticizer (water reducing agent).

At present the large amounts of superplasticizer applied for high performance concrete with high slump (> 150 mm), are directly involved in the formation of several of the defects registered, e.g. lack of cement hydration[1] and low quality of the air void systems. Figure 1 contains an illustration of local failure of cement to hydrate in a heavily superlasticized silica-fume bearing HPC.

A crucial parameter for highly plasticized concretes is the pronounced loss of slump

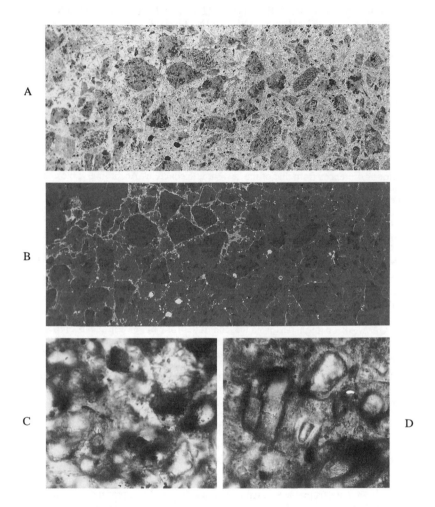

Figure 1. Fluorescent epoxy impregnated plane section of core taken from a pillar. (A) as seen in normal light, (B) as seen in UV light. The length of the core is 200 mm. The core contains highly porous zones, which appear bright in B due to the penetration of the fluorescent epoxy in these regions. C and D are micrographs from petrographic thin sections covering porous (C) and dense (D) areas. The fields of view in C and D are 0.100 mm x 0.120mm.

By thin section analysis, the W/C ratio of the porous zones was determined to be identical to that of the neighboring dense areas. However, the cement grains were virtually not hydrated in the porous zones (C), but showed normal hydration in the dense zones (D).

The W/C ratio of the concrete according to the mix design was 0.37; the concrete incorporated microsilica (silica fume) in slurry form, and a high content of superplasticizer.

experienced in the fresh concrete, eventually causing a variety of plastic defects. The maturity functions of these concretes may also differ markedly from that expected.

TEST METHODS

For almost a decade, it has been a common practice in many European countries to document the internal structure of concrete by petrographic thin section analysis[2,3] and air void analysis[4] on samples cored from the constructions. This practice has proven somewhat insufficient when applied for documentation of HPC constructions. By supplementing the traditional air void analysis and petrographic analysis of hardened concrete with a test method comprising large scale microstructure analysis, it has become possible to document the internal structures of HPC construction concrete. The quality monitoring, which is initiated at the moment of form work stripping (when coring is possible) comprises the following steps: a) preparation of fluorescent-epoxy impregnated reground plane sections (IRPS), for UV microstructure analysis of full size samples[5], b) preparation of contrast-impregnated polished sections for air void analysis, and c) preparation of fluorescent-epoxy impregnated thin sections for petrographic analysis and micro UV-analysis. The different steps can be independently and progressively implemented, depending on aim and results and thereby securing a cost/benefit optimized testing program.

Impregnated reground plane sections

The IRPS test method is suitable for detecting the extent and typology of defects in the internal structure of concrete materials. The possible defects detected by the method include permeable zones, bands of cracks and paste separations[1], porous zones or bonding defects along coarse aggregate interfaces, interconnecting crack systems, and others. The distinct features of these defect types can (and should) be more thoroughly studied in thin sections covering specific areas of interest, whereas the extent and connectivity of the defects are properly described only by the large scale microstructure analysis, facilitated with the IRPS method.

Test specimens can be cores, cylinders or tiles of hardened concrete with a maturity of at least 7 days. No dimensions of the sample should be less than 100 mm. Samples longer than 250 mm are divided in subsamples. Characteristics regarding type of construction, production details, orientation, and placing of sample must be taken into account when defining the plane to be analyzed. The orientation of cores or other samples is of great importance.

The equipment comprises a diamond disc grinder (100 micron diamonds) securing an accuracy in the grinding plane of +/- 0.05 mm, a heating closet accurate to +/- 2 °, a two component epoxy system and dissolvable fluorescent dye, a vacuum set up - with the required safety facilities - for epoxy impregnation of large samples, UV lamps, a UV microscope with barrier filters, and a Photographic set up with normal light and UV light.

Procedures

Whole core impregnation is major option. If visual inspection of the sample indicate weaknesses or evident defects in the material, or if joints are represented in the sample, the whole sample is dried at 35 °C and subsequently vacuum impregnated in fluorescent epoxy.

A plane parallel slab of hardened concrete is cut on the diamond saw, exposing the plane to be further investigated. If the sample was an impregnated whole core, a UV photo of the sectioned plane is taken. This plane is ground by diamond disc grinder until at least 1 mm is removed to ensure a perfect plane and a smooth surface without sawing marks and tearings. The tolerance of the surface is +/- 0.05 mm.

The slab is dried at 35 °C and vacuum impregnated with fluorescent epoxy. On the hardening of the epoxy, the ground surface is placed upside down on a sheet of thick glass covered with plastic, to minimize and even the epoxy thickness. The hardening is carried out at no more than 35 °C.

After hardening of the epoxy, the plane is reground to a depth of exactly 1 mm (+/- 0.05 mm) below the impregnated plane. Great care is taken to ensure plane parallel grinding. The sample is then dried at 35 °C. The '1 mm plane' is analyzed in both UV light and normal light, and the observed structures are documented by photos taken likewise in both normal light and UV light.

The surface is subsequently again reground to a depth of exactly 2 mm (+/- 0.05 mm) below the plane of impregnation, also ensuring parallel grinding. The '2 mm plane' is analyzed, and the observed structures of the '2 mm plane' are documented by photos.

Analysis of both the 1 mm plane and the 2 mm plane is carried out in normal light and UV light (+ barrier filters) with the aid of a microscope or hand lens. The observed structures or defects may be characterized as cracks, bundles of interconnecting (micro) cracks and/or paste separations, porous zones and debonding areas along aggregates, long interconnecting microcracks, and larger interconnecting permeable zones and bands.

Other features such as for example air void agglomerates or air inclusions are also revealed, and must be distinguished from the above-mentioned defects. Note that coarser air voids will be impregnated to at least the 1 mm plane, and may thus cause some disturbance of the view, especially as seen on the photo documentation.

Full sample photo documentation serving as an overview, is supported by relevant close ups.

INTERNAL DEFECTS IN HPC STRUCTURES

When describing and interpreting internal defects in the concrete in HPC constructions, it is important to distinguish between low slump concretes, e.g. used for concrete road pavements, and medium to high slump concretes used for constructions as bridges, high rise structures, etc. The need for distinction is caused by the marked differences in performance and composition of the fresh concrete used. Accordingly, the defects introduced in the two types of HPC constructions are of different natures.

Defects related to compaction

On casting the body of concrete it is essential to introduce the compaction energy needed for making the concrete and its cement paste flow. This assures both a dense packing and an adequate internal bonding between constituents of the concrete. However, it is of utmost importance that the amount of energy be restricted, in order to retain a stable internal structure of the concrete, and thus prevent the constituents from segregating. A given concrete composition may not achieve the needed flow without at the same time causing segregation, thus calling for adjustments of the mix design or changes in production.

Defects related to low slump concretes

In concrete pavements, insufficient compaction most often results in an open structure of the concrete, i.e. air inclusions of irregular shape, 2-30 mm in diameter, constituting a porosity of 5-25%. Also the cement paste may contain smaller openings between the cement grains, 0.1-0.5 mm wide, due to lack of flow and hence, lack of paste compaction. The bonding between cement paste and aggregate may contain defects of both orders of size, i.e. anywhere from 0.1mm to 30 mm. These compaction defects result in reduced density and lower strength of the concrete.

Insufficient compaction may be seen in a less grave mode, resulting in bodies of entrapped air of the size of 3mm to 20 mm, constituting a porosity of 2-5%. In this context, the concrete does flow, but the content of entrapped air involves a lowering of the compressive strength of the concrete.

Pavements vibrated with a vibrator beam may show adequate compaction of the upper part of the slab, e.g. at upper half of the layer thickness, whereas the lower part may contain the above mentioned compaction defects. The reduced internal cohesion in an insufficient compacted fresh concrete slab may cause enlarged setting, mainly observed in the margins of the pavement slab, causing longitudinal, vertical, plastic cracks of widths of 0.2 mm to 1.5 mm. In slip-formed concrete slabs, similar defects will arise if there is any slump in the concrete.

If the effective compaction of the pavement varies across the slab, the resulting concrete height may vary. Especially in air entrained concrete, the surface may be seen to raise to a varying extent after the passage of the vibrator beam or the surface leveller. Similar phenomena may arise if the air content of the concrete varies from batch to batch.

Excessive compaction may induce a low content of coarse aggregate in the top 10-20 mm in concrete pavements. If the concrete is air entrained, the air void structure may be of reduced quality and content, due to excessive compaction, e.g. from the poker vibrators dragged through the concrete. The air content can thus be drastically changed - e.g. from 8% to 1% -across the full concrete slab thickness, at the placing of the poker vibrators.

In roller compacted pavements, the compaction may induce bonding failures with widths of 0.02-0.2 mm along the coarse aggregate, caused by the rotation action of the aggregate during heavy compaction in the very 'dry' concrete used.

Defects related to moderate or high slump concretes

For moderate and high slump concretes, insufficient compaction mainly involves the formation of bodies of entrapped air (blow holes) at vertical surfaces being cast against form work. These blow holes are of often 5mm to 50 mm in diameter, and may have a depth of 5 mm to 15 mm. The defects are most often a cosmetic problem, but may be of importance with respect to protection of reinforcement if the effective thickness of the cover layer is diminished. The compaction around the rebars and the general bonding between concrete and rebars may be of low quality resulting in irregular cavities along the rebars, of a length of 3mm to 50 mm along the rebar and extending 2mm to 20 mm from the rebar.

More serious, but also quite rare, results of insufficient compaction, are the formation of porous and permeable zones or bands at casting joints. The thickness of such bands may vary between 2 mm and 20 mm.

Excessive compaction of medium and high slump concretes induces segregation phenomena of varying orders of magnitude in the fresh concrete, comprising a) settling of coarse aggregate, b) micro bleeding in the paste, c) porosities at grains and densified paste between grains, d) larger bleeding structures, i.e. porous areas of local very high W/C ratio, and e) porous bands of high W/C ratio along rebars.

The segregations involve settling of coarse aggregate lead to increased paste contents in the stone-depleted region. The content of coarse aggregate may vary from 20 vol. % in the depleted upper zone to 60 vol. % in the enriched lower zone. The upper depleted zone has an increased content of paste and will thus have increased shrinkage on hydration and increased tendency of crack formation. Such cracks may form map patterns, and will typically be 0.01 mm to 0.5 mm wide, and be perpendicular to the surface.

Micro bleeding in the paste due to excessive vibration will cause the water to be pressed out of the paste and placed along sand grains and stones. Consequently, the paste between the sand grains will be densified due to closer packing of cement grains. The hardened paste may develop local differences in capillary porosity corresponding to differences in equivalent W/C ratio of 0.20-0.30.

With intensified vibration, the internal structure of the concrete will form larger highly porous local regions below aggregate particles due to bleeding. These zones may have volumes equivalent to the area of the coarse aggregate particle, in a layer 0.5 mm to 2.0 mm in thickness. The porous volumes may well form partly interconnecting bands in the concrete.

Large interconnecting porous volumes may be found around the reinforcement. They are mainly formed below horizontal rebars and surrounding the vertical rebars. The cement paste locally may be highly porous - equivalent to a W/C ratio of 0.60 or more - in a 5 mm to 20 mm wide zone following the rebars, sometimes apparently extending to almost the full rebar length. An illustration of this effect is provided in Figure 2. Defects like these may form especially when the pokers are placed on the reinforcement during vibration.

In air entrained concretes the air void system may be spoiled by excessive poker vibration. An Inhomogeneous air void system may be produced where the single air voids clump together into

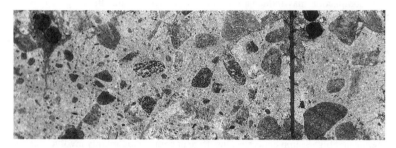

A

B

Figure 2. Fluorescent epoxy impregnated plane section of a core taken horizontally through a slip-formed wall, as seen in normal light (A) and in UV light (B). The core contains both horizontal rebars (hr) and a vertical rebar (rb). It should be noted that the black and white photos here and in Figure 1 do not fully display the detail observed in color in the fluorescent light.

The core contains highly porous zones (appearing bright in B) placed along and below the rebars. By thin section analysis the local W/C ratio of the paste in the porous zones was determined to be 0.70, whereas the general W/C of the paste was determined to be approximately 0.40. The mix design called for a W/C ratio of 0.42.

The porous zones were seen to be interconnecting, and were at several places exposed to the surface at both ends of the core.

The microstructure of this concrete segregated due to excessive compaction with poker vibrators placed on the reinforcement and exhibited internal bleeding. No visible (surface) bleeding was reported from the site, and no defects were seen on the finished wall.

The inclined cracks seen in the middle of the core are caused by the formwork lifting.

agglomerates that may in turn collapse into macro air inclusions. These may in turn be vibrated out of the concrete. Eventually the air void content of the concrete will be lowered, sometimes from 6% to as little as 1%. Furthermore, the overburden pressure influences the air void system. The result is generally an increase of the spacing factor.

A crucial parameter for highly plasticized concretes is the pronounced rate of loss of slump experienced in the fresh concrete. Manipulations of these concretes after true plasticity is lost or set is started will introduce plastic defects that are unlikely to heal. Such defects may be introduced in several ways:

• On re-vibrating previously placed, already set layers of concrete in order to secure well compacted joints. Defects produced in this manner comprise 0.01 mm to 0.1 mm wide plastic cracks in the paste and along the aggregate, and paste separations involving locally reduced hydration of cement grains.

• In slip-formed walls, if the speed of form work lifting exceeds the capability of the placed concrete to sustain itself. The defects introduced may be paste separations and 0.01mm to 0.2 mm wide plastic cracks, mainly parallel to the surface. The cracks may be concentrated in the outer portions of the wall.

• In slip-formed walls if the form work poses too high friction at the lifts due to skew form work lifts, setting of the concrete on the form work (mainly at too low lift speed), or to little spreading action at the bottom of the form work. The defects introduced will be 0.05 mm to 0.5 mm wide plastic cracks, mainly of horizontal or skew orientation and of listric shape.

• If concrete is cast on a sloping bottom, or in bottom plates without top form work when cast in-one with walls.

Vibration of HPC that has suffered from loss of slump, thus being no longer truly plastic, will inevitably result in Inhomogeneous cement paste (paste separations); plastic cracks; and bonding defects. Similar plastic defects may arise, as well in concretes of conventional design, if early form work collapse occurs, or if form work is stripped too early. The plastic cracks will in such cases be wider and more widespread.

DURABILITY ASPECTS OF THE HPC DEFECTS

The performance characteristics of a given HPC composition are usually based on laboratory testing of test block samples. This testing supplies values of strength, shrinkage, creep, heat development, frost resistance, resistance to chloride ingress, carbonation rate etc., all of which are used in both the static calculations and in durability considerations (i.e. service life time calculations). However, since the internal structure of the test block samples is usually without defects, the performance characteristics of the actual structures may well be radically different from those the test blocks.

The influence of internal defects in HPC on a series of parameters comprising compressive strength, rate of carbonation, and frost resistance have been revealed during the QA-monitoring. The compressive strength appears to be influenced only by the large scale defects, whereas parameters such as the rate of carbonation and the frost resistance are markedly related to type and extent of all defects[6]. Thus, the actual durability of the constructions must be expected to reduced if significant defects occur.

CONCLUDING REMARKS

Assessing the quality of HPC construction is possible by monitoring the internal structure. However, this is feasible only if the relevance of carrying out corrections on concreting practices, workmanship, or composition is recognized and approved by the parties involved. At present, the need of such evaluation appears to be urgent for HPC structures, so as to inhibit large scale future durability problems in current and planned constructions. The test methods presented here, revealing the internal structure of newly cast concrete, may form valuable and indeed highly cost effective tools in these HPC evaluations.

REFERENCES

1. P. Laugesen, "Micro defects in slipformed concrete", Unpublished report, Dansk Beton Teknik A/S, July 1990.
2. A. Damgaard Jensen, et al. "Petrographic analysis of concrete", Danish Building Export Council Ltd.
3. TI-BS: "Analysis of the micro structure of hardened concrete", DTI-Byggeteknik, 1987.
4. ASTM C 457: "Microscopical determination of air void content and parameters of the air void system in the hardened concrete".
5. DBT-LAB-45: "Impregnated reground plane sections", Unpublished laboratory test method, Dansk Beton Teknik A/S, 1990.
6. P. Laugesen, Unpublished reports, Dansk Beton Teknik A/S, 1992-1994.

MICROSTRUCTURE OF STEAM CURED CONCRETES DETERIORATED BY ALKALI-SILICA REACTION

JEAN PIERRE BOURNAZEL AND MICHELINE MORANVILLE - REGOURD
Laboratoire de Mécanique et Technologie, ENS Cachan / CNRS / Université Paris VI, 61 avenue du Président Wilson, 94 235 Cachan Cedex, France

ABSTRACT

In order to know the main cause of the cracking of concrete ties we tried to reproduce the deterioration process in the laboratory, using the same thermal cycle and the same materials. Model concretes were first steam cured then examined for ASR using the CSA-A-23-2-14A accelerated test. Linear expansion of concrete prisms were measured and fracture surfaces of concrete after treatment were observed under scanning electron microscope. They showed ASR gels and secondary ettringite. A simultaneous thermal mechanical computation gave the global microcracking of concrete induced by steam curing. A second computation showed the accelerating role of temperature on the local development of ASR products.

INTRODUCTION

Steam cured concretes ties have exhibited superficial cracks after several years in service on differents sites. They were produced with the same aggregates containing alkali reactive minerals but with various cements which differed from each other mainly in SO_3 contents.

Under the microscope, deteriorated concretes exhibited cracks within aggregates and cement paste. Microcracks in aggregates were like those already described in granite and more particularly in zones of distorted quartz and along planes of foliated gneiss [1]. These cracks radiated from aggregates to the cement paste in tortuous paths. Alkali silica gels filled the cracks in aggregates and partly filled those in the cement paste. Ettringite was found at the cement paste - aggregate interface and in cracks or pores in the cement paste.

The cause of expansion of steam cured concretes with alkali reactive aggregates has long been discussed. Some authors are attribute the expansion to delayed ettringite formation [2]. Others attribute expansion to both alkali-silica gel and ettringite but in many cases ASR appears as the first cause of expansion [3, 4, 5, 6]. Expansion of the cement paste resulting in the formation of gaps around the aggregates may be induced by heat treatment at temperatures higher than 60°C or to the formation of secondary ettringite [7].

In order to clarify the behavior of these site concretes, model concretes were prepared with the same materials. The influence of temperature during the steam curing on the cement paste and on the ASR was calculated using a computer model. First results are presented here.

EXPERIMENTAL STUDY

Materials

Model concretes were made with five different cements (table 1). Their behavior regarding the alkali-aggregate reaction was tested after thermal treatment and compared to those of standard samples kept at 20°C before the AAR tests. The mix design of the concretes is in table 2 :

Table 1 : Chemical composition of cements

Cement	A	B	C	D	E
C3A	8.1	8.1	10.6	8.2	8.2
SO₃	4.23	3.21	4.21	3.05	4.25
Na₂O	0.860	0.830	0.810	0.369	0.369

Table 2 : Mix design of concretes

Cement	:	446	kg/m3
Water	:	179.4	kg/m3
Coarse aggregates	:	1028	kg/m3
Sand	:	706	kg/m3
Retarder	:	928	ml/m3
Air entraining agent	:	77	ml/m3

Water/Cement ratio : 0.4

<u>Thermal cycles of the steam cured concretes</u>

Two different thermal cycles of the steam curing were applied to concrete prisms of $7 \times 7 \times 28$ cm. The maximun temperature was 60°C for the first cycle and 80°C for the second one (fig 1)

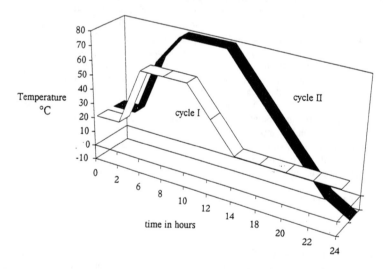

Figure 1 : Thermal cycles used

Alkali - Aggregate Reaction

The alkali aggregate reactivity was evaluated with a special equipment standardized in France ; it is a modified version of the Canadian A23-2-14A accelerated test. This equipment maintains the relative humidity at 100% and the temperature at 38°C with a double regulation system [8].

Linear expansion

Linear expansions have been measured up to 270 days (table 3).The results show that all the cements behaved in the same way. However, the expansions of cements A, B and D exceed the limit of 0.4%.

Table 3 : Linear expansion (%), 270 days

	A	B	C	D	E
first cycle	0.65 %	0.95 %	0.74 %	0.75 %	0.85 %
second cycle	0.78 %	0.55 %	0.28 %	0.74 %	0.36 %
standard	0.48 %	0.41 %	0.27 %	0.41 %	0.22 %

NUMERICAL STUDIES

In order to simulate the effect of the thermal treatment on concrete ties we have used two models developed at our laboratory. The first evaluates the predamaging zones induced by the thermomechanical effect due to the cement hydration. This model is built in the framework of thermodynamics of irreversible processes [9]. The second one evaluates the expansion of concrete bars affected by alkali silica reaction. In this model, the temperature effect on kinetics of reaction is represented by Arrhenius function. This model is built using a probabilistic approach [10]

Thermomechanical computations

During the hydration of the cement paste, some phenomena occur (heat generation and volume changes). They can result in the cracking of a concrete structure. From the mechanical point of view, one one hand, the Poisson's ratio, which is 0.5 during the dormant period, decreases during the setting and hardening periods. On the other hand, the Young's modulus increases. Simultaneously, strains appear due to self-dessiccation, thermal gradient, environmental effects (humidity, temperature) and concrete composition.

Using the thermodynamics of irreversible processes we have developed a global model capable of describing this main phenomena. The starting point is the free energy potential which can be expressed as [9] :

$$\rho\psi = \frac{1}{2}\Lambda(M)(1-D):(\varepsilon - (\varepsilon^{th} + \varepsilon^{sh} + \varepsilon^{c})):(\varepsilon - (\varepsilon^{th} + \varepsilon^{sh} + \varepsilon^{c})) + \rho\psi_M + \rho\psi_T \qquad (1)$$

where
$\rho\psi$ is the free energy potential, $\rho\psi_M$ is the free energy due to maturation, $\rho\psi_T$ is the free energy due to thermal effects, $\Lambda(M)$ is the tensor of elastics characteristics affected by maturity, M is maturity and D is damage considered, as isotropic.

Equation (2) expresses the hypothesis of strains partition

$$\varepsilon = \varepsilon^e + \varepsilon^{th} + \varepsilon^{sh} + \varepsilon^c \qquad (2)$$

where

$\varepsilon^{th} + \varepsilon^{sh}$ are the volumic strains due to thermal variations and autogeneous shrinkage, ε^c is the maturation creep strain and ε^e is the elastic strain

It is possible to deduce the expression of stress from the free energy according to the first state law defined by equation (3)

$$\sigma = \frac{\partial \rho \psi}{\partial \varepsilon} = \Lambda(M)(1-D):(\varepsilon - (\varepsilon^{th} + \varepsilon^{sh} + \varepsilon^c)) \qquad (3)$$

This model was applied to the case of steam cured concrete ties ; the mesh used is described in figure 2. The numerical computations, realised with the finite elements CESAR LCPC code [11], gave a damage map of the structure due to thermal effects. The results are presented in figure 3 and show that after three days the tie presents, large damaged zones in its thicker part. There are, however, no external indications of damage. the microporosity and microcracking functins change, thereby modifying the permeability of the material which, in turn, affects fluid transport.

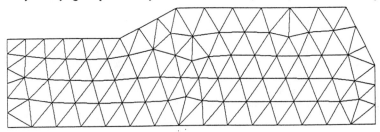

Figure 2 : Mesh used for numerical computation of concrete ties

Damage

de 0.000E+00 a 0.167E+00
de 0.167E+00 a 0.333E+00
de 0.333E+00 a 0.500E+00
de 0.500E+00 a 0.667E+00
de 0.667E+00 a 0.833E+00
de 0.833E+00 a 0.100E+01

Maximun value D = 0.85

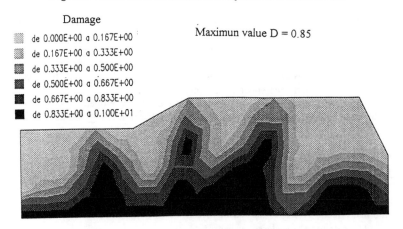

Figure 3 : Damage in a concrete tie after three days

Probabilistic computations

The mechanical modelling of alkali aggregate reaction is made with a deterministic approach. Usually concrete is considered as homogeneous and chemical reactions as uniform in the volume. However, chemical processes are mainly induced by local supersaturation and not by average values of concentration [13]. Moreover, considering the only average values would result in omitting the local gradients of concentration which are the driving force for reaction. In this study, the calculation takes into account on a random reactive site distribution. The model considers the average values and standard deviation of all parameters considered i.e. particle size, alkali concentrations.... (fig 4).

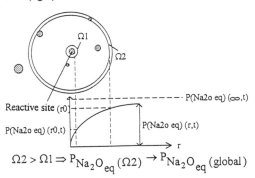

$$\Omega2 > \Omega1 \Rightarrow P_{Na_2O_{eq}}(\Omega2) \rightarrow P_{Na_2O_{eq}}(global)$$

Figure 4 : Evolution of the probability of Na$_2$O eq versus r, size of reactive silica particle

Some studies have shown that the kinetics of reaction are influenced by temperature and follow the Ahrrenius law. This was taken into account in numerical modelling. The computations (figure 5) showed that an increase of temperature accelerated the expansion of concrete. In this example a temperature of 70 °C induced the onset of expansion ten times faster than a concrete kept at room temperature. This numerical simulations are in agreement with experimental results by Ong and Diamond [14]

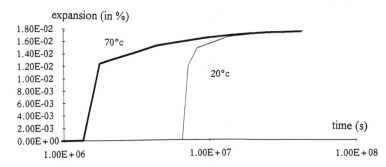

Figure 5 : Simulated expansions of two concretes cured at different temperatures (20°C and 70°C)

MICROSTRUCTURE OF MODEL CONCRETES

After 250 days of the AAR accelerated test, all concretes with the reactive aggregate presented the same external signs of alteration (pop - outs, gel exudations, cracks). In all concretes which expanded, ASR gels and ettringite were observed on surface fractures under SEM, whatever the type of cement or the thermal cycle, as follows :

- coexistence of ASR gel and ettringite (fig 6)
- ettringite as a thin layer on aggregate surface (fig 7), in fibers at the cement paste - aggregate interface (fig 8), and in veins in the cement paste, filling previously formed cracks (fig 9).

Same observations were already published [15, 16, 17]. The redistribution of SO_4^- ions in the pore solution forms ettringite through solution, possibly as microcrystals in the CSH [18]. Where the cement paste expands, ettringite recrystallises in the cracks [19, 20].

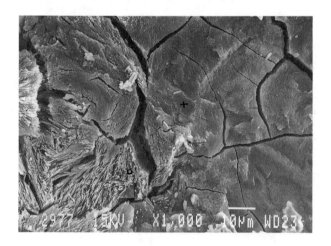

Figure 6 : Cement B, cycle I, 100 d at 38°C. Coexistence of AAR gel (+) and secondary ettringite (o)

Figure 7 : Cement B, 2d at 20°C. secondary ettringite (o) at the interface of a granite aggregate

Figure 8 : Cement A, 2d at 20°C 100d at 38°C, secondary ettringite (o) at the interface matrix - aggregate

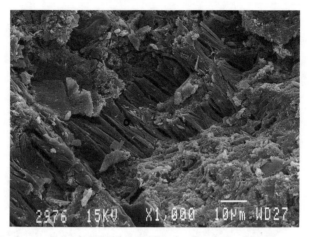

Figure 9 : Cement E, cycle I, 270d at 38°C, vein of ettringite (o) in the cement paste

DISCUSSION

The results of computer calculation show that :

1. the steam curing treatment (temperature > 60°C) damages the cement paste of concrete.
2. the steam curing treatment (temperature > 60°C) accelerates the ASR and the resulting expansion

The SEM observations of site and model concrete are similar : both clearly show ASR product and secondary ettringite. In cement A the ASR gel existing even at 20°C developed a grainy texture related to an increase in Ca^{++} ions as observed at three months (fig 10). The cements D and E contained only 1.65 Na_2Oeq /m3. With this low amount of Na_2Oeq no ASR usually occurs. The aggregate was treated chemically for evaluating an eventual solubilisation of alkalies. It was found that 0.63% K_2O and 0.12 Na_2O (0.53 Na_2Oeq) were released besides SiO_2 (2.20 %) Al_2O_3 (1.95 %) and Fe_2O_3 (2.10 %), probably from unstable minerals like micas and feldspars. So the aggregate was a source of alkalies in the concrete.

The same concretes made with a non reactive aggregate (limestone) did not expand either after the first cycle or after the second cycle of steam curing followed by the AAR accelerated tests. So the internal damaging induced by the thermal treatment, which was the first step of deterioration as stated by the holistic model presented by Mehta [21] did not result in visible cracks.

With an alkali - reactive aggregate, the concrete followed the second step of the Mehta's model corresponding to expansion and cracking.

The ettringite occurred as massive areas coexisting with ASR gels at the cement - aggregate interfaces or in veins in the matrix. This ettringite appears as a secondary ettringite was observed in all the concretes whatever the amount of SO_3 in the cement used in this study.

Figure 10 : Cement A, cycle I, 100d at 38°C, Coexistence of AAR gel grainy textured (+) and secondary ettringite (o)

CONCLUSION

The alkali-silica reaction was as the main cause of cracking of concrete prims made with alkali reactive aggregates. The deterioration process is reproducible.

- Steam curing damaged the cement paste and accelerated the ASR.
- The ettringite coexisting with ASR gels was a secondary ettringite able to enhance expansion.
- The same deterioration occurred for cement low and high SO_3 (3.05 - 4.25 %)
- There was no expansion and no visible cracks when alkali reactive aggregates were replaced by inert aggregates, whatever the type of cement.

REFERENCES

[1] G.M. Idorn, V. Johassen and N. Thaulow, in Material Science of Concrete III, edited by J. Skalny (Am. Ceram. Soc), pp. 71-104, (1992)
[2] D. Heinz and U. Ludwig, in K. and B. Matter Conference on Durability, ACI, SP 100, Vol 2, pp 2059 - 2071 (1987)
[3] T.W. Jones and A.B. Poole, in 7th Int Alkali Conf., publ. Noyes, pp 446-450 (1986)
[4] R.E. Oberholster, H. Maree and J.H.B Brandt, in 9th Int. Conf AAR, London, pp 739 - 749, (1992)
[5] A. Shayan and G.W. Quick, Adv. Cen. Res 4, 16, pp 149 - 157 (1992)
[6] S. Diamond and S Ong, Ceramic transactions Vol 40, Cement Technology, pp 79 - 90 (1993)
[7] V. Johansen, N. Thaulow and J. Skalny, Adv. Cem. Res. 5, 17 pp 23 - 29 (1993)
[8] R. Ranc, B. Cariou and D. Sorrentino, Ministry of Transportation, Canada. Engineering Material Report 92, March 1990.
[9] J.P. Bournazel and M. Moranville Regourd, ACI, SP 144, pp 233 - 249, (1994)

[10] A. Sellier, J.P. Bournazel and A. Mebarki, to appear in Mat. and Struct, RILEM
[11] CESAR - LCPC, Manuel d'utilisation, LCPC, (1987)
[13] J.P. Bournazel, B. Capra, A. Mebarki and A. Sellier, Proc. of EURO-C, ed Bicanic, De Borst and Mang, pp 491 - 499 (1994)
[14] S. Ong and S. Diamond, Cem Con Res, Vol 24, 7, pp 1305 - 1310 (1994)
[15] A. Shayan and G.W. Quick, ACI Materials Journal, V 89, 4, pp 348- 361 (1992)
[16] M. Regourd, H. Hornain and P. Poitevin, in Vth int. Conf. AAR, Cape Town, Paper 252/35 (1981)
[17] E. Soers and M. Meyskens, Annales des Travaux Publics de Belgique, 4 pp 339 - 343 (1989)
[18] I. Scrivener and H.F.W Taylor, adv. Cem. Res. 5, n°20, pp 139 - 146, (1993)
[19] H.F Taylor, Ceramic transactions, vol 40, Cement Technology, pp 61 - 78 (1993)
[20] H.F. Taylor, in Adv in Cem and Concr, ed M.W. Grutzeck and S.L Sarkar, ASCE, pp 122 - 131 (1994)
[21] P.K. Mehta, ACI, SP 144, pp 1 - 30 (1994)

HEAT CURING AND DELAYED ETTRINGITE FORMATION

M.C. LEWIS*, KAREN L. SCRIVENER* AND S. KELHAM**
*Imperial College of Science, Technology and Medicine, London, U.K
**Blue Circle Industries PLC, Kent, U.K

ABSTRACT

This paper reports some preliminary results from a study of the effect of elevated temperature curing on mortars and the phenomenon of delayed ettringite formation (DEF). Mortars made from cements with sulphate levels of 3%, 4%, and 5% and with 5% sulphate and added alkali were cured at 20 and 90° C and subsequently stored in water. Expansion measurements showed a pessimum effect with increasing SO_3 content. Mortars which expanded showed a corresponding decrease in strength. X-ray diffraction (XRD) studies indicated that no ettringite is present after heat treatment but re-forms over time within the material. However, the ultimate levels of ettringite reached do not correspond to the magnitude of expansion observed. X-ray microanalysis shows that immediately after the heat treatment the aluminate species and most of the sulphate species are incorporated within the C-S-H gel. The concentrations of these species decrease during expansion, such that at the end of expansion the amounts remaining correspond to the presence of AFm phase mixed with C-S-H.

INTRODUCTION

The heat curing of concrete is a widely used industrial techniques for the production of concrete structural members. This accelerates the rate of strength development, allowing the components to be demoulded and put into service more quickly.

In recent years the deterioration of a number of structures which have undergone some form of elevated temperature curing and subsequent exposure to moisture has been observed. The damage is characterised by substantial surface cracking and erosion of the corners and edges. Internal examination of the material shows a heavily cracked paste with ettringite deposited in cracks and pores and in uniform rims around the aggregate particles. Damaged structures all seem to have undergone long term exposure to moisture and the macroscopic signs of deterioration become apparent after a number of years in service.

Consequently a number of investigations have been carried out on the effect of heat treatment on cement pastes, mortars and concretes and on the subsequent formation of ettringite (e.g. refs 1-6). It is now accepted that if hydrating Portland cement is heated above a certain temperature, ettringite no longer forms and some or all of that which is formed prior to heat treatment is destroyed. If the material is subsequently kept in water or saturated air at ordinary temperature, ettringite is again formed. This ettringite subsequently recrystallizes around the aggregate particles as uniform rims as well as infilling voids and pores. In parallel with the reformation of ettringite, a general expansion of the materials is observed. This phenomena is generally referred to as delayed ettringite formation (DEF).

Despite the considerable amount of research in this area there is, as yet no generally agreed mechanism linking the observed expansion with processes occurring in the cement paste. Originally, it was proposed that expansion was caused by the crystallisation pressure of ettringite at the cement paste/aggregate interfaces. However, very high degrees of supersaturation would be necessary to produce the necessary crystallisation pressures, for which pore solution analyses provide no evidence. It has also been observed that the width of the rims around the aggregate particles is roughly proportional to the size of the aggregate[7] which would only be consistent with a general expansion of the cement paste.

The results reported here are preliminary findings from a study aimed at elucidating the mechanisms of expansion in heat cured pastes.

EXPERIMENTAL

The cements used in this work were laboratory prepared grinds using a normal production clinker whose chemical composition is given in Table I. The grinding was carried out cold and the cements then heated at 105°C for 4 hours to give partial gypsum dehydration.

Table I. Chemical composition of clinker used.

SiO$_2$	Al$_2$O$_3$	Fe$_2$O$_3$	Mn$_2$O$_3$	P$_2$O$_5$	TiO$_2$	CaO	MgO	SO$_3$	LOI	K$_2$O	Na$_2$O
20.7	6.0	2.4	0.07	0.08	0.24	66.8	0.7	1.5	0.4	0.59	0.22

The sulphate levels of the cements were adjusted by varying the addition of natural anhydrite during grinding. All cements were ground to a specific surface area of 450 m^2/kg. In the case of the mortar containing added alkali, KOH was added dissolved in the mixing water to increase the alkali level by the equivalent of 0.8 wt% Na$_2$O.

Specimen were cast as 1″ (25.4 mm) cubes and 16 × 16 × 160 mm expansion bars using a water/cement ratio of 0.5 and a sand/cement ratio of 3. After casting the specimens were given a 4 hour precure at room temperature and then either placed in a saturated atmosphere at 20°C or put in polythene bags and placed just above the water level in a heating tank. The temperature was increased to 90°C over 1 hour, maintained for a further 12 hours and then allowed to fall to room temperature. The samples were then demoulded and immediately placed into water.

For strength measurements, X-ray diffraction, examination in the SEM and microanalysis, characterisation times were chosen to correspond to similar points on the expansion curves of the different cements as illustrated in Figure 1: one day after heat treatment (1); just before expansion (2); midway through expansion (3); and at the end of expansion (4).

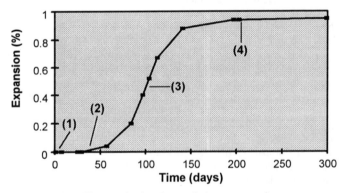

Figure 1. Characterisation times relative to expansion curves

Compressive strength tests were carried out using an Instron machine with a 100 kN load cell at a rate of 1 mm/s. Four cubes were tested for each cement at each period.

One cube from each set was retained, crushed and sifted to a fine powder for X-ray diffraction (this removes most of the sand grains). 10% rutile was added as a standard and the mixture reground before running the analysis. A Philips 1700 series diffractometer was used over an angular range of 5° to 50° 2θ, step size 0.04° at 2s per step.

For microscopy, the samples were cut in half and freeze dried to stop hydration by removing any free water. One half was resin impregnated, lapped, polished to ¼ μm and carbon coated for examination in the SEM. They were examined in a JEOL 35-CF SEM equipped with a solid state backscattered detector and a LINK AN10000 energy dispersive X-ray analysis system.

RESULTS

Expansive Behaviour

None of the mortars cured at 20°C throughout showed any significant expansion. Figure 2 shows the expansion behaviour of the various mortars initially cured at 90°C. With 3% SO₃ no significant expansion is observed. An SO₃ content of 4% leads to an ultimate expansion of just under 1%. However, increasing the SO₃ content further to 5% results in a reduced expansion of about 0.8%. On the other hand the addition of KOH to the cement with 5% SO₃ almost triples the ultimate expansion.

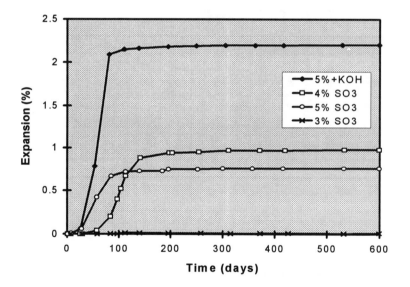

Figure 2. Expansions of the different mortars.

Strength Development

The variation in strengths of the mortars are shown in Figure 3. In all cases the heat treated mortars reached lower ultimate compressive strengths than their counterparts cured at 20°C. The cement with 3% SO₃, which did not expand following the 90°C cure, shows no decrease in strength. All the other mortars show a similar trend: an initial increase in strength followed by a decrease. The time at which the strength starts to decrease corresponds well with the onset of expansion, but there is no simple relationship between the degree of the strength decrease and the amount of expansion.

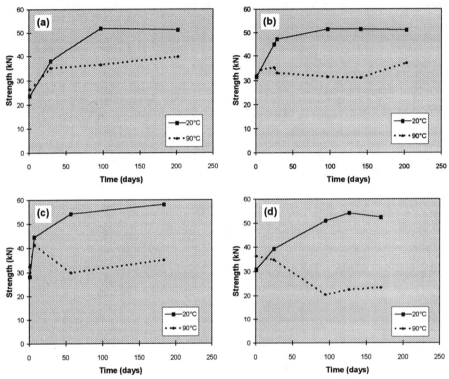

Figure 3. Strength development of mortars: (a) 3% SO₃; (b) 4% SO₃; (c) 5% SO₃; (d) 5% SO₃ + KOH

X-ray Diffraction

Tables II and III show the relative amounts of ettringite present in the mortars for the 20°C and 90°C cures respectively, calculated semi-quantitatively from the ettringite peak at 5.61Å.

X-ray patterns for the cements cured at 20°C show the peaks of the expected cementitious phases i.e. anhydrous phases, ettringite (AFt), calcium-aluminate monosulphate (AFm), CH, and some SiO₂ from the sand. The ettringite peak heights remain fairly constant over the testing period (Table II), although most show a slight decrease at the second characterisation time (2) followed by an increase at times (3) & (4). Typical diffraction traces for the mortar with 4% SO_3 are shown in Figure 4. The semi-quantitative data also indicates that, with the higher sulphate contents, the mortar can contain a considerable amount of ettringite without undergoing expansion.

Table II. Relative amounts of ettringite in the mortars cured at 20°C

Characterisation Time	3% SO_3	4% SO_3	5% SO_3	5% SO_3 + KOH
(1)	0.55	1.97	1.59	2.52
(2)	0.35	1.46	1.03	1.66
(3)	0.88	1.40	1.66	1.93
(4)	0.97	1.38	2.06	2.53

Table III. Relative amounts of ettringite in the mortars cured at 90°C

Characterisation Time	3% SO_3	4% SO_3	5% SO_3	5% SO_3 + KOH
(1)	0	0	0	0
(2)	0	0.78	0.32	0.71
(3)	0.5	1.05	1.73	2.17
(4)	0.9	2.35	3.04	3.57

The cements cured at 90°C show a very different pattern. (Figure 5 shows the traces for the mortar with 4% SO_3, which is typical). At the first characterisation time, one day after the heat treatment, no ettringite peaks were observed, regardless of whether the cements subsequently expanded or not. However, some small peaks corresponding to AFm could be observed at this time. Subsequently, the three mortars which expand, show large increases in the amounts of ettringite present throughout the expansion period. After expansion (4) the amounts of ettringite are all greater than the amounts in the equivalent mortars cured at 20°C throughout. Ettringite also forms in the mortar containing only 3% SO_3, but the amount is much smaller.

However, the semi-quantitative data shown in the tables indicates that the degree of expansion is not directly related to the amount of ettringite formed. The mortar made from the cement with 5% SO_3 and added KOH contains only slightly more ettringite at the end of expansion than that containing 5% SO_3 alone; yet the former expands almost three times as much. Furthermore, the amount of ettringite in the mortar produced with the cement containing 4% SO_3 is significantly lower than the amount in the mortar containing 5% SO_3 yet the former expands significantly more.

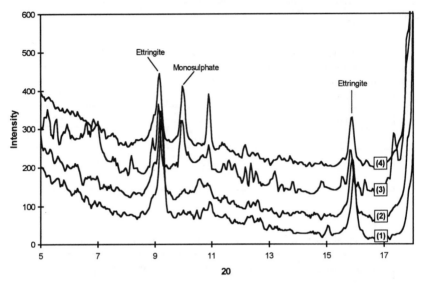

Figure 4. XRD patterns of 20°C, 4% SO_3 for periods (1-4)

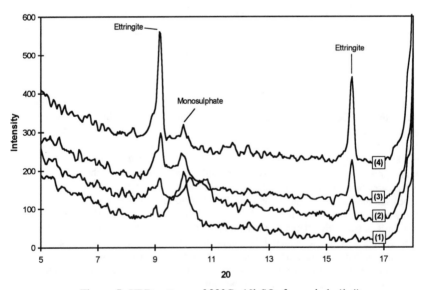

Figure 5. XRD patterns of 90°C, 4% SO_3 for periods (1-4)

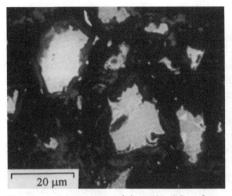

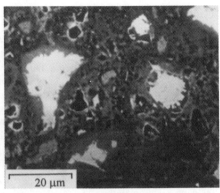

Figure 6. Mortar containing 4% SO₃ after one day at 20°C.

Figure 7. Mortar containing 4% SO₃ cured at 90°C, after one day.

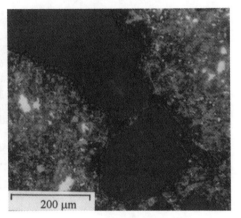

Figure 8. (above left). Mortar containing 4% SO₃, cured at 20°C throughout after 200 days.

Figure 9. (above right). Mortar containing 4% SO₃, heat cured at 90°C after 200 days.

Figure 10. (left). Mortar containing 3% SO₃, heat cured at 90°C after 200 days.

<u>Microscopy</u>

The specimens were observed in the backscattered electron mode in the SEM. Figures 6 and 7 compare the microstructure of the mortar containing 4% SO_3 one day after curing at 20°C and at 90°C. The degree of hydration is much greater in the heat cured mortar with dense rims of hydration around the anhydrous grains. Despite the extent of hydration considerable porosity remains, mainly due to the presence of a large number of hollow hydration rims or Hadley grains.

Figure 8 shows the microstructure of the 4% SO_3 mortar after 200 days curing. The microstructure is typical for a mortar of this age with well developed hydration rims around the cement grains. The corresponding mortar originally cured at 90°C, has undergone expansion by this age and its microstructure is shown in Figure 9. In this specimen ettringite has recrystallised around the aggregates and in pores and voids throughout the paste.

Figure 10 shows the microstructure of the heat treated mortar containing 3% also at about 200 days. This material did not expanded significantly and shows no signs of recrystallised ettringite.

<u>Microanalyses</u>

Microanalyses were carried out on the C-S-H of the hydration rims and the general undifferentiated product of the high temperature cured specimens. The analyses were conducted in the central region of the one inch section and for statistical accuracy, about 50 points were analysed for each specimen. In the SEM, microanalysis will sample an area of just over a micrometer in each dimension. On this scale intimate mixtures of C-S-H with other hydrate phases cannot be differentiated. To interpret the likely composition of the sulpho-aluminate phases intermixed with the C-S-H, the analyses are plotted as atom ratios of S/Ca against Al/Ca, the lines indicate where mixtures of C-S-H with AFt or AFm respectively would lie.

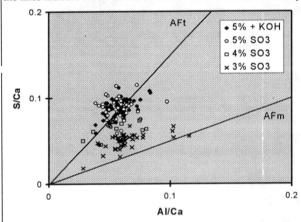

Figure 11.
Microanalyses from the four mortars, heat treated at 90°C after one day.

Figure 11 shows plots of the four mortars one day after heat treatment. XRD indicates that there is no ettringite present in the mortars at this time. Thus it would appear that the sulphate and aluminate detected are variously present: in the pore solution, as AFm phase or absorbed in the C-S-H. The microanalyses indicate that as the SO_3 content of the cements increases, the amount of sulphate incorporated increases. The amount of aluminate incorporated appears to

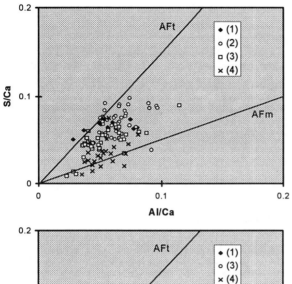

Figure 12.
Changes in microanalyses for the heat cured mortar containing 4% SO_3, throughout the expansion period.

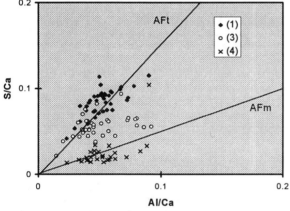

Figure 13.
Changes in microanalyses for the heat cured mortar containing 5% SO_3, throughout the expansion period.

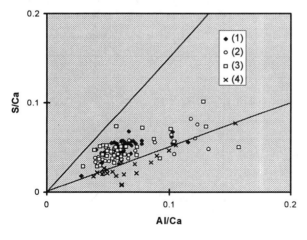

Figure 14.
Changes in microanalyses for the heat cured mortar containing 3% SO_3, throughout the expansion period.

remain roughly constant. The analyses for the mortar containing 5% SO_3 and for that containing 5% SO_3 and KOH are virtually indistinguishable, suggesting that the extra alkali has no significant effect on the incorporation of the sulphate and aluminate.

Figures 12 - 14 show, for each mortar, how the analyses change throughout the expansion period. (The analyses for the mortar containing 5% SO_3 and KOH were virtually identical to those for the mortar containing 5% SO_3 alone so are not included). For the mortar containing 4% SO_3, Figure 12, up to the start of the expansion (2) the analyses remain close to the AFt/C-S-H mixture line, with some increase in the amount of sulphate and, particularly, aluminate intermixed with the gel. However, once expansion starts (3) the analyses fall below this line indicating a loss of sulphate, and to a lesser extent, aluminate, from the gel. By the end of expansion the analyses lie along the line representing mixtures of AFm/C-S-H. A similar pattern is seen for the mortar containing 5% SO_3, Figure 13. In the mortar containing 3% SO_3 the analyses also drop towards the AFm/C-S-H, even though this mortar showed no significant expansion.

DISCUSSION AND CONCLUDING REMARKS

The results presented here clearly confirm previous findings[8], with regard to the pattern of chemical changes which occur in heat cured pastes subsequently stored in water. The XRD results confirm that there is no ettringite present immediately after heat curing, while microanalysis indicates that sulphate and aluminate species are intimately mixed with the C-S-H gel. On subsequent curing in water ettringite forms, probably initially within the C-S-H gel and then recrystallises in cracks and voids and around aggregate particles. At the end of the expansion process the analyses of the C-S-H gel indicate that this is only intermixed with AFm phase. However, it is not possible to identify cause and effect - does the recrystallisation of ettringite cause the expansion or does the expansion of the paste open up spaces into which the ettringite recrystallises.

The central enigma remains the lack of correlation between the amount of ettringite formed and the degree of expansion. In particular the microanalysis data indicate that, from a chemical point of view, the paste in the mortar containing 5% SO_3 alone behaves in an identical fashion to that containing 5% SO_3 and additional KOH; yet the later expands almost three times as much. On the other hand, in the mortar containg 5% SO_3, there is more sulphate intermixed with the C-S-H gel after the heat cure, and more ettringite present at the end of expansion, yet the degree of expansion is less than in the mortar containing only 4% SO_3.

REFERENCES

1. H-M. Sylla, Beton **38**, 449 (1988).
2. D. Heinz and U. Ludwig, Katharine and Bryant Mather International Conference, ACI, Detroit, SP 100-105 (1987)
3. U. Neck, Beton **38**, 488 (1988).
4. I. Odler, S. Abdul-Maula and Z.-Y in Microstructural Development During Hydration of Cement,(Mater. Res. Soc. Symp. Proc.**60**, 1987) pp. 139.
5. C.D. Lawrence, draft papaer, 1993.
6. D. Heinz, U. Ludwig and I. Rudiger, Concrete Precasting Plant Technology **11**, 56 (1989).
7. V. Johansen, N. Thaulow, U.H. Jakobsen and L. Palbøl, RH&H Bulletin No. 47, (1993).
8. K.L. Scrivener and H.F.W. Taylor, Adv. Cem. Res. **5**, 139 (1993).

X-RAY MICROTOMOGRAPHY OF AN ASTM C109 MORTAR EXPOSED TO SULFATE ATTACK

D.P. BENTZ*, NICOS. S. MARTYS*, P. STUTZMAN*, M. S. LEVENSON**, E.J. GARBOCZI*, J. DUNSMUIR+, AND L. M. SCHWARTZ++
* Building and Fire Research Laboratory, Building 226 Room B-350, National Institute of Standards and Technology, Gaithersburg, MD 20899 USA
** Computing and Applied Mathematics Laboratory, Building 101 Room A-337, National Institute of Standards and Technology, Gaithersburg, MD 20899 USA
+ Exxon Research and Engineering Company, Route 22 East, Annandale, NJ 08801 USA
++ Schlumberger-Doll Research, Old Quarry Road, Ridgefield, CT 06877

ABSTRACT

X-ray microtomography can be used to generate three-dimensional 512^3 images of random materials at a resolution of a few micrometers per voxel. This technique has been used to obtain an image of an ASTM C109 mortar sample that had been exposed to a sodium sulfate solution. The three-dimensional image clearly shows sand grains, cement paste, air voids, cracks, and needle-like crystals growing in the air voids. Volume fractions of sand and cement paste determined from the image agree well with the known quantities. Implications for the study of microstructure and proposed uses of X-ray microtomography on cement-based composites are discussed.

INTRODUCTION

The service life and durability of concrete depends on a wide variety of factors [1]. Possible degradation mechanisms include environmental exposure to deleterious compounds of sulfates or chlorides, damage due to frost attack, and spalling as a result of exposure to the high temperature of a fire. The rate of damage can strongly depend on a concrete's transport properties which in turn depend on its microstructure. With knowledge of the three-dimensional concrete microstructure it is possible to determine transport properties that are then used to predict service life. To do this, representative models of the concrete or mortar microstructure are needed.

Two current approaches to representing the microstructure of concrete are 1) constructing ideal models of concrete (e.g., sphere packings modeling the placement of sand in a mortar) and 2) using real two-dimensional images to carry out a variety of studies. For instance, reasonably high resolution two-dimensional images can be made of a mortar or concrete via scanning electron microscopy. Clearly, these two approaches are not mutually exclusive in that data from real images can be used to help guide model building.

In the last decade significant improvements [2] have been made in the development of experimental methods to create three-dimensional images of real microstructures such as sandstone, coal, and biological materials. In particular, it is possible to nondestructively generate maps of X-ray attenuation with about 1 percent accuracy and a resolution of

77

about 1 micron [2]. In this paper, we present results of a study concerning the generation of a three-dimensional image of cement mortar using X-ray microtomography. The image processing techniques are discussed. We find that realistic images of the mortar can be made that preserve the volume fractions of cement paste and sand grains.

CEMENT MORTAR SAMPLE PREPARATION

The material studied was an ASTM C109 mortar [3] with a water/cement ratio of 0.485 and a sand/cement ratio of 2.75. The sand grains ranged between 300-600 μm in diameter. Once the sample was made it was cured for 1 day and then exposed to a 10 percent solution of sodium sulfate for about six weeks. To stop further hydration and sulfate attack at a selected testing time the specimen was potted with an ultra-low viscosity resin using a two-step replacement procedure [4]. This procedure also minimizes the occurance of drying shrinkage cracking. First, the pore soultion is replaced with ethanol and then the ethanol is replaced with resin. The resin was cured at 60 degrees centigrade for 24 hours. A cylindrical sample speciman 3.5mm in diameter is then cored from the original mortar to be scanned. Note, tomography does not require the potting procedure, as specimens could be scanned dry or even wet.

X-RAY MICROTOMOGRAPHY

The three-dimensional x-ray attenuation map of the cement mortar was generated at the National Synchrotron Light Source (NSLS) located at Brookhaven National Laboratory using Exxon's microtomography scanner at beamline X-2B. This instrument is an optical microscope that images scintillation events in a phosphor onto the surface of a 512x512 cooled CCD. X-rays illuminate a cylindrical specimen and radiographic images are acquired at a large number of discrete view angles as the specimen is rotated about its axis. This rotation axis is perpendicular to the x-ray beam and parallel to the CCD pixel columns. At the conclusion of a scan the image data from the CCD are sorted such that the specimen projections for each pixel row are collected together. These collected data are reconstructed to form a single 2-D tomographic slice. Since the synchrotron beam is highly collimated, a fast Direct Fourier inversion algorithm is used to rapidly reconstruct the 512 slices corresponding to each CCD row. Three dimensional volumes are built up by stacking the slices. Beam hardening artifacts are eliminated by using a Silicon monochromator to select a well defined energy.

Under ideal conditions the ultimate spatial resolution of this instrument is comparable to a light microscope and is about 1 micron. For this study the spatial resolution is limited by the specimen size which is about 3.5mm. With a field of view for the mortar specimen of slightly greater than 3.5mm, each pixel represents about 7 microns.

The mortar specimen in this study was scanned at 17 keV and radiographic images were acquired in 0.36 degree increments between 0 and 180 degrees of specimen rotation. Data acquisition and reconstruction were completed in 1.5 hours. This scanning protocol resulted in a reconstructed volume that is more noisy than what is usually achieveable. The protocol for future scans will be modified to reduce noise.

RESULTS-IMAGE PROCESSING TECHNIQUES

Figure 1 is a two-dimensional image of the attenuation coefficients of a plane from the reconstruction of the X-ray tomography data. Note that there are a wide variety of features present in the image. The light areas represent cement paste, the somewhat darker areas are sand grains and the darkest areas are air voids. Inside the air voids are crystalline growths, which most likely are calcium hydroxide or possibly gypsum or ettringite. There is also a crack resulting from the sulfate attack. Notice also the existence of concentric rings, which are an artifact of the tomographical data collection process. Also, there is some noise which may, in part, be due to the local inhomogeneity of the component materials.

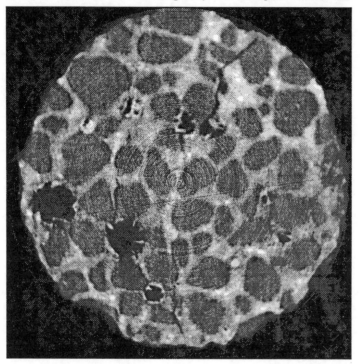

Figure 1: Original two-dimensional image obtained from tomography data set.

For this preliminary study we decided to subdivide the image into three separate phases: sand, cement paste and air voids. Although there are many other features, as described above, these three phases make the largest volumetric contribution and are clearly (at least to the eye) distinguishable.

Once the decision is made to convert the original attenuation coefficients into three phases, a procedure must be developed to decide what voxel (volume element) corresponds to what phase. Due to the noise present in the original image it is not easy to systematically separate the phases. For instance a simple thresholding of the attenuation coefficients, which range in value over two decades, produces an unrealistic image.

In order to produce a more realistic image, a four step image processing procedure was performed. The first step was intended to remove the ringing artifact and some of the local variation from the image. We chose a technique based on the discrete wavelet transform [5]. In many ways the wavelet transform is similar to the Fourier transform. It consists of representing the image as a weighted sum of basis elements. With the Fourier transform, the basis elements are the *sine* and *cosine* functions of different frequencies. The wavelet basis elements also have different frequencies. However, unlike the Fourier transform, a wavelet basis element is nonzero only in a finite region. Therefore, high frequency noise can be removed locally instead of globally as in the case of the Fourier transform. As a result, an advantage of the wavelet transform over the Fourier transform is that it is capable of preserving boundaries better. The particular wavelet technique employed is based on the thresholding idea of Donoho [6].

In figure 2 we show a comparison of attenuation profiles before and after the wavelet processing.

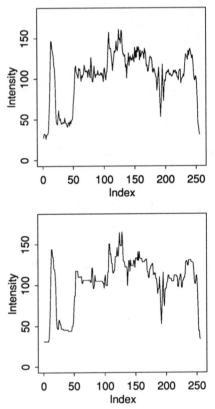

Figure 2: Attenuation profiles of figure 1 along a vertical line before (top) and after (bottom) wavelet filtering.

Clearly plateaus are beginning to form that correspond to the different phases. Also the boundaries are generally preserved. For instance the rapid fluctuation of attenuation coefficients around voxels 190 to 200 has not been diminished. This fluctuation corresponds to the two small air voids in the original image.

Next thresholding was preformed on the image to segment it into the three phases. We found there was still some spurious noise which was cleared up with use of a median filter [7].

The final stage removed the unphysical presence of a band of sand surrounding air voids. This is due to the interpolation of the attenuation coefficient in regions where air voids border the cement paste. To remove the bordering of spurious sand, the regions near the sand-air void boundary were dilated (i.e., sand regions neighboring air-void were converted to air-void). Figure 3 shows the final processed image.

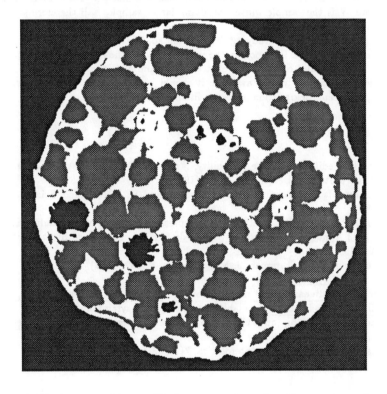

Figure 3: Final processed image. Black corresponds to air voids, grey to sand and white to cement paste.

Once all the images had been processed, we then determined the ratio of sand to cement in the three demensional reconstruction of the mortar and found it was in excellent agreement with the mix design (i.e., 55 volume percent sand, 45 volume percent paste). We are currently studying the sand grain size distribution. Here, there is the difficulty that one must carefully separate touching sand grains. However, our preliminary studies, employing watershed segmentation techniques [8] are encouraging.

CONCLUSIONS-FUTURE RESEARCH

The use of X-ray microtomography to investigate the three-dimensional microstructure of a cement mortar has been demonstrated. For our study, resolution of order 10 μm was sufficient to visualize the main features of a mortar. At higher resolutions other features in the mortar, such as the interfacial zone microstructure at the sand grain/cement paste may become more apparent. Limits to the size of sample studied depends on the resolution needed. Since the CCD used in the data collection had 512 pixels accross, the resolution is W/512 where W is the sample width. Studying larger samples will therefore sacrifice resolution.

Future research includes the studying the size and spacing between sand grains or air voids, the ingress of materials into mortar, the interfacial zone, the flocculation structure of cement grains, fracture, and computation of stress fields and elastic moduli. With realistic three-dimensional images of mortar available it should be possible to better understand transport properties and degradation processes in mortars and concrete.

REFERENCES

[1] J. M. Scanlon, editor, Concrete Durability, American Concrete Institute, P.O. Box 19150, Redford Station, Detroit, MI 48219 (1987).

[2] B.P. Flannery, H. W. Deckman, W. G. Roberge, and K. L. D'Amico, Science, 235, 1439 (1987).

[3] ASTM C109 Standard Test Method for Compressize Strength of Hydraulic Cement Mortar, 1993 Annual Book of ASTM Standards, ASTM, 1916 Race Steet, Philadelphia, PA 19103.

[4] L. J. Struble and P. E. Stutzman, J. Mat. Sci. Let., 8, 632 (1989).

[5] I. Daubechies, Wavelets, SIAM, Philadelphia PA (1992).

[6] D. Donoho, Nonlinear wavelet methods for recovery of signal, densities, and spectra from indirect and noisy data , available through ftp at playfair.stanford.edu.

[7] V. Cantoni, S. Levialdi, and G. Musso, Editors, Image analysis and Processing, (Plenum Press, New York 1986).

[8] J. C. Russ and J. C. Russ, Acta Sterologica, 7(1), 33-4, (1988).

MICROSTRUCTURAL FEATURES OF FREEZE–THAW DETERIORATION OF CONCRETE

TANYA BAKHAREV AND LESLIE J. STRUBLE
University of Illinois, Department of Civil Engineering, Urbana, IL.

ABSTRACT

Microstructural features associated with freeze–thaw deterioration have been studied using scanning electron microscopy and mathematical modelling. Concrete subjected to freeze–thaw cycles in the laboratory and concrete deteriorated in the field were examined in the SEM (BEI). The microcracks produced after a few freeze–thaw cycles are largely vertical and extend deep (several millimeters) into the specimen. With additional cycles horizontal cracks are produced, connecting the vertical cracks to produce spalling. This crack development probably results from localized expansive stresses produced by hydraulic pressure, as described by Powers, plus localized shear stresses produced by differential volume changes.

INTRODUCTION

The mechanisms of frost action on hardened cement paste are discussed in the classical works of Powers and Helmuth[1-2]. Concrete damage was initially attributed to the 9% volume increase when water crystallizes; if this crystallization pressure exceeds the tensile strength of the paste, cracking results. Hydraulic pressure was subsequently proposed to account for the observed migration of the water during freezing. When it was found that frost–resistant concrete contracts on freezing, the concepts of diffusion of gel water to capillary spaces and osmotic mechanisms were added[3]. Investigations of freeze–thaw deterioration have focussed on macroscopic changes, without exploring the process on a microscale. Observation of microstructural characteristics of concrete and cement paste damaged in freezing and thawing cycles using SEM can help to understand the mechanisms of the deterioration. The present study was undertaken to explore microstructural features of freeze–thaw deterioration of both concrete and neat cement paste.

EXPERIMENTAL PROCEDURE

Two concrete samples and several pastes were examined. One concrete was prepared in our laboratory with w/c 0.39 (w/c corrected for aggregate absorption), slump 33 mm, and air content 5.2% (measured on the fresh concrete). This concrete was cured at 100% RH and 23°C for 28 days and had a 28–day compressive strength of 27.6 MPa. The other was a commercial prestressed concrete, similar in composition, slump, and air content, but cured for approximately 12 hours at 70°C prior to detensioning and demolding. This concrete had a specified strength of 48 MPa at 28 days and was 1 year old when subjected to freeze–thaw cycles. The pastes included neat paste, air–entrained paste, and pastes containing silica fume. Two w/c levels were used, 0.4 and 0.5. Three levels of silica fume were used, 0%, 3% and 6% (by mass of cement), and three levels of entrained air, 0%, 4%, and 8% (by volume of paste). All paste samples were cured at 100% RH and 23°C for 28 days.

Samples were exposed to repeated cycles of freezing and thawing using ASTM C–666, Procedure B (freezing in air and thawing in water). The sample size for freeze–thaw exposure was 80 mm x 80 mm x 250 mm for laboratory concrete and 60 mm x 60 mm x 150 mm for cement paste. After 300 or 500 cycles the specimens were cut and polished for microscopical investiga-

83

tion using backscattered electron imaging (BEI). To prevent damage during specimen preparation, a slow, two−step procedure was used to replace pore solution by an ultra−low viscosity acrylic resin (LR White Ted Pella Inc.). Because specimens are not dried, they do not crack due to drying shrinkage during this procedure[4].

RESULTS

Visual examination of the laboratory concrete subjected to 300 cycles did not reveal any signs of deterioration. However, microscopical examination showed vertical cracks extending from the surface to a depth of 200 mm or more (Fig. 1) and spaced 10 to 15 mm apart. There were also horizontal cracks at depths of 5 or 6 mm (Fig. 2). The surface of the concrete subjected to freeze−thaw cycles appeared to have carbonated; EDS analysis of the light gray surface layer in Fig. 1 (approximately 10 μm thick) showed much more calcium and less silicon than in the interior.

Laboratory concrete subjected to 500 cycles showed visible deterioration with map cracking and pop−outs on the surface.

The commercial concrete also did not show any visible signs of deterioration after 300 freeze−thaw cycles. Microscopic examination showed nearly vertical cracks extending 2−3 mm inward from the surface and horizontal cracks at a depth of 200 μm. The vertical cracks were more widely spaced (2−3 times) than in the case of laboratory concrete. The depth of horizontal cracks was also much less. As the water−cement ratios of laboratory concrete and the high strength concrete were the same, the differences are attributed to heat treatment of the commercial concrete.

The sample of neat cement paste without entrained air had surface scaling to a depth of 1−2 mm and damage at the corners. Samples with 4% entrained air had much less scaling than samples without entrained air. The sample with 6% silica fume and no entrained air disintegrated, whereas the sample with 6% silica fume and 8% entrained air showed no visible signs of

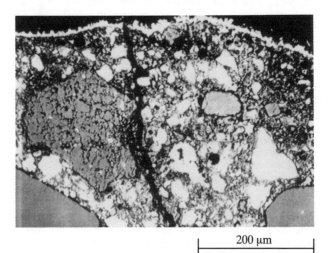

|———— 200 μm ————|

Figure 1. BEI of the surface region of laboratory concrete subjected to 300 freeze−thaw cycles.

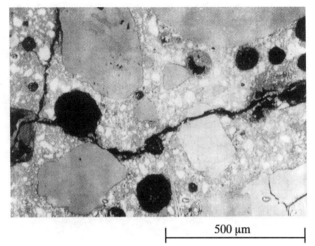

| 500 μm |

Figure 2. BEI of an interior region (5.2 mm from the surface) of laboratory concrete subjected to 300 freeze–thaw cycles.

deterioration. Regardless of whether or not samples showed visible signs of deterioration, their microstructural pattern consisted of vertical cracks extending 5 mm, and sometimes 15 mm, inward from the surface, and horizontal cracks at a depth of 0.5–2 mm, depending on w/c and amounts of silica fume and air (Figure 3). The spacing of vertical cracks in air–entrained samples was 3–6 times larger than in non air–entrained paste. Within the group of air entrained samples, the spacing of vertical cracks was 2–3 times greater with 8% air than with 4% air. The depth of horizontal cracking increased as the amount of air entrainment was increased. The depth of horizontal cracks was greatest for the ordinary portland cement paste and decreased as the amount of silica fume was increased.

DISCUSSION

The microstructural pattern of freeze–thaw deterioration consisted of vertical cracks extending in from the surface and horizontal cracks at some depth from the surface. The main visible feature of freeze–thaw deterioration is spalling. It is reasonable to assume that the spalling results from the connection of vertical cracks by horizontal cracks.

All of the materials had well developed horizontal cracks at different depths from the surface. Hydraulic pressure is thought to be the cause of such cracking. Portland cement concrete tends to freeze and thaw in layers, because the surface freezes while the interior is still warm. If a high degree of saturation exists, considerable expansion occurs in the cement paste accompanied by the expulsion of water from the saturated surface layer into less saturated interior. For this reason, as suggested by Powers[2], hydraulic pressure is greatest at some distance in from the surface, and cracks develop in this region. This mechanism is developed analytically as follows.

Let us assume that heat flow is proportional to the temperature gradient in the surface layer and to the coefficient of heat transfer of concrete. Heat flow is also proportional to the enthalpy of the water–ice transition, the degree of saturation in the surface layer, and the rate of freezing:

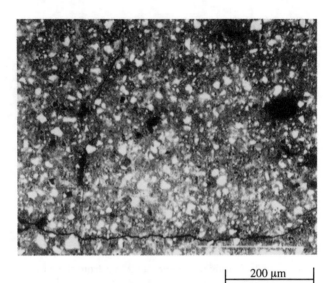

|————— 200 μm —————|

Figure 3. BEI of the surface region of neat paste, w/c 0.4, subjected to 300 freeze–thaw cycles.

$$\frac{Q}{S} = \frac{\lambda(T_f - T_o)}{x + \alpha} = \Delta H c \frac{dx}{dt} \tag{1}$$

where α is the ratio of the heat loss and heat transfer coefficients, x is the depth of freezing, λ is the coefficient of heat transfer, T_f and T_0 are temperatures in the interior and at the surface, ΔH is enthalpy of water–ice transition, c is water content of the concrete, and t is time from beginning of freezing.

From eq. (1) the depth of the frozen layer can be obtained:

$$x = \sqrt{2\lambda(T_f - T_0)\frac{t}{\Delta H c} + \alpha^2} - \alpha \tag{2}$$

As the top layer of the concrete is frozen, water is expelled to the surface and to the interior, causing hydraulic pressure. The flow is proportional to the pressure gradient and the concrete permeability, and the flow gradient is proportional to the pressure and the adsorption coefficient. Because the permeability of the frozen layer is very low, the expelled water tends to move inward. The hydraulic pressure at depth x is:

$$P = \frac{\lambda(T_f - T_0)\Delta V}{(x + \alpha)\Delta H\left(\sqrt{KA} + \sqrt{K_f A_f}\coth\left(x\sqrt{A_f/K_f}\right)\right)} \tag{3}$$

where P is hydraulic pressure, ΔV is the expansion coefficient of water on freezing (1.09), K is the concrete permeability coefficient (K for ordinary concrete and K_f for frozen concrete), and A is a concrete absorption coefficient (A for ordinary concrete and A_f for frozen concrete).

Estimation of hydraulic pressure developed on freezing with a temperature difference of $10°C$ and $K=10^{-10}$ m/s, $K_f=10^{-14}$ m/s, $A=10^{-12}$ m^3/s, $A_f=10^{-16}$ m^3/s using eq. (3) shows that $P=30$ MPa. This is enough to rupture concrete.

The depth at which maximum pressure develops is x, the solution of the following equation:

$$-\sqrt{A_f/K_f} = \frac{\sqrt{A_f K_f}\cosh\left(x\sqrt{A_f/K_f}\right)\sinh\left(x\sqrt{A_f/K_f}\right) - (x+a)K_f}{\sinh^2\left(x\sqrt{A_f K_f}\right)} \tag{4}$$

Another likely mechanism for the horizontal cracks is the development of shear forces at some distance from the surface due to differential volume change of saturated and unsaturated layers. As the surface layer of concrete becomes saturated due to diffusion of water from outside the specimen, the surface layer expands due to crystallization pressure when the concrete freezes. At the same time, interior regions not yet saturated may show contraction due to the movement of water from micropores to macropores[5]. Thus differential volume changes develop, causing shear forces that produce horizontal cracks at some distance from the surface. For both mechanisms, the zone of saturation extends more deeply into the specimen when permeability is greater, so horizontal cracks develop at a greater depth. If the permeability coefficient is very low, horizontal cracks may occur only at very shallow depths or not at all.

Thus it is likely that the depth at which horizontal cracks occurs depends mainly on the depth to which water saturation occurs, which is controlled by the pore structure. We observed horizontal cracks in all specimens, but at various depths, greatest for laboratory concrete and lowest for prestressed concrete and the pastes containing silica fume.

It is possible to estimate the depth of saturation using a linear diffusion model

$$\frac{\partial^2 \Phi}{\partial x^2} - \frac{1}{\gamma^2}\frac{\partial \Phi}{\partial t} = 0 \tag{5}$$

where Φ is the water content, x is the distance from the surface, $\gamma = K^{-2}$ where K is the coefficient of permeability, and t is time. Equation (5) has the following solution for $t>0$:

$$\Phi(x,t) = \frac{1}{t^{0.5}}\exp\left(\frac{-x^2}{4\gamma^2 t}\right) \tag{6}$$

If we normalize $\Phi(x,t)$, using that for $x=0$, $\Phi(0,t)=t^{-0.5}$, then

$$\Phi_n(x,t) = \exp\left(\frac{-x^2}{4\gamma^2 t}\right) \tag{7}$$

Calculations using (7) for known permeability coefficients give values for depth of saturation listed in Table I. (For the experiment with 300 cycles samples were immersed in water approximately 30 days, so $t=2,600,000$ sec.)

Calculated values of the depth of the saturated zone agree reasonably well with the depth of horizontal cracks observed in the SEM. So the hypothesis that differences in saturation cause the horizontal cracks is consistent with the microscopical observations.

The microstructural pattern of freeze–thaw deterioration consisted of vertical cracks extending in from the surface and horizontal cracks at some depth from the surface. The main visible feature of freeze–thaw deterioration is spalling. It is reasonable to assume that the spalling results from the connection of vertical cracks by horizontal cracks.

Table I. Calculated depth of saturation and observed depth of horizontal cracks.

Sample	Coefficient of permeability, K*	Depth of saturation, calculated using eq. (7)	Observed depth of horizontal cracks
High strength concrete	10^{-14}	0.32 mm	0.2 mm
Cement paste with 6% silica fume replacement	$0.3 * 10^{-13}$	0.56 mm	0.5 mm
Laboratory concrete	$5 * 10^{-12}$	7.22 mm	5–6 mm
Neat paste	10^{-13}	1.02 mm	1–1.5 mm

* Coefficients of permeability from ref. 6.

CONCLUSIONS

The microstructural pattern of freeze–thaw deterioration consisted of deep vertical cracks, extending in some cases 200 mm or more, and horizontal cracks at some depth from the surface. This pattern was observed in all tested samples, both air entrained and non–air entrained samples, regardless of whether the sample showed visible signs of deterioration.

Microstructural observations of concretes and cement pastes support the hypothesis of Powers that hydraulic pressure causes horizontal cracking. Also expansion on freezing in the outer, saturated concrete and contraction in the inner, unsaturated concrete may contribute to horizontal cracking.

The depth of horizontal cracking depends on the coefficient of permeability of the material, if there are no other factors that influence saturation.

ACKNOWLEDGEMENT

This work was supported by the University of Illinois Research Board.

REFERENCES

1. T. C. Powers and R. A. Helmuth, Proceedings, Highway Research Board 32, 285–297 (1953).
2. T. C. Powers, Proc., Amer. Concr. Inst. 41, 245–272 (1945).
3. R. A. Helmuth, Proc. 4th Int. Symp. Chem. Cem. 4, 855–869 (1960).
5. G. G. Litvan, Cem. Conc. Res. 6, 351–356 (1976).
4. L. Struble and P. Stutzman, J. Mater. Sci. Lett. 8, 632–634 (1989).
6. R.D. Hooton, in Blended Cements, STP 897, edited by G. Frohnsdorff (American Society for Testing and Materials, Philadelphia, 1986) pp. 128–143.

UNDISPERSED GRANULATED SILICA FUME IN CONCRETE - CHEMICAL SYSTEM AND DURABILITY PROBLEMS

BJÖRN LAGERBLAD * and PEETER UTKIN**
* Swedish Cement and Concrete Research Institute, 100 44 Stockholm Sweden.
**Tallin Technical University, Building Materials Lab, Akadeemia tee 1, 200108 Tallin, Estonia.

ABSTRACT

Granulated condensed silica fume is not easily dispersed. In contact with the pore fluids of the cement paste the granulates rapidly turn into alkali-silica gel nodules. With time this gel absorbs Ca and at least the rim of the AS-nodules transforms into a C-S-H product. The gel by itself does not cause expansion as it forms before the cement paste hardens. However, if exposed to alkali salts the AS-nodules may become further enriched in alkalies, which in turn may trigger an expansive reaction and cracking.

INTRODUCTION

Condensed silica fume (CSF), also known as microsilica, is widely used as an addition to concrete. It is used to improve concrete properties, both in the fresh and hardened state. It works both as a microfiller and a very active pozzolan.

On the market CSF is sold as a granulate and as a slurry. In several cases, when analysing concrete we have observed undispersed silica granulates up to a size of a millimetre. In concrete undispersed CSF will form a lump of amorphous silica and will thus be extremely sensitive to alkali silica reaction (ASR). This report will mainly concentrate on chemical interaction and durability problems. This has a bearing on the mechanism of the alkali-silica reaction.

Condensed silica and dispersion

Silica fume is a by-product from the production of silicon or silicon alloys. When it comes from the electric furnace it is a gaseous suboxide which in contact with air oxidizes and condenses to very small particles of amorphous silica (SiO_2).

Originally, silica fume is very light and thus difficult to handle. While the compact density of CSF is around 2.2 g/cm^3 the bulk density of powder is only around 0.2 g/cm^3. This makes it difficult to handle the material. Thus, the CSF is sold either in a water-based slurry or in a granulated form as small pellets. The slurry is difficult and expensive to handle. Thus financial and technological considerations often make the granulate form preferable.

The CSF can be granulated to different density and sizes. The granules also show a tendency to become denser with time. The different granulated CSF available in Sweden that we have tested have a bulk density from around 0.4 to 0.8 g/cm^3 and a maximum size of around 2 millimetre. Larger density and size makes the granulated CSF more difficult to disperse.

Granulated condensed silica fume (CSF) is not easily dispersed. The granules do not disperse themselves. A series of tests with different mixing sequences show that a certain mixing order is needed. Superplasticizers may help but are not essential. The data shows that mechanical crushing is the most important factor. The fine fraction of the aggregate must be larger than the size of the granule to achieve a good crushing. The tests indicated that the smallest aggregate grain size should be twice that of the CSF granulates. The cement and water should always be added late. The conditions for dispersion is discussed in Lagerblad and Utkin[1].

DURABILITY INVESTIGATIONS

To find out if the undispersed CSF granules may cause durability problems, concrete samples containing different amounts of CSF and different degrees of dispersion were tested by accelerated alkali-silica reaction (ASR) and freeze/thaw (F/T) tests.

In the experiments a whole series of different concretes (Tab I) were prepared. Concrete mixes 1 to 10 (series 1)were prepared in a 250 litre paddle mixer while samples 15 to 20 (series 2) were prepared in a smaller 50 litre paddle mixer. The mixer was kept running while the ingredients were put into it during a short time period. Samples 15 to 20 were deliberately badly mixed. Before testing all the concrete samples were water cured for at least 28 days.

All concretes contain 360 kg of binder (cement + CSF), 1000 kg coarse aggregate (8-16 mm) and 920 kg fine aggregate (0-8 mm). A high alkali (Na_2O eqv = 1,1 wt. %) ASTM type I cement was used in all mixes except nr 18 in which a low alkali (Na_2O eqv < 0.6 wt. %) ASTM type V (sulphate resistant) cement was used. The aggregates were granitoid of glaciofluvial origin. They are not alkali-reactive. An air-entraining agent was used in all mixes. In mixes 9 and 10 a slurry (50 % water) was used. Mixes 5, 6, 15, 16, 19 and 20 contained a lignosulphonate-based plastizicer (Rescon P). The bulk density of the granulated CSF was around 0.7 g/cm^3.

Table I.
Concrete mix design, kg/m^3 and measured properties

Abbreviations; B = bad air void system, A = acceptable air void system, VG = very good air void system. NM = not measured. w/b = water/binder ratio. CSF disp = amount dispersed.

Mix No.	CSF wt. %	Water	w/b	Slump in cm	Air in %	Frost res.	CSF disp.	Mixing in sec.
1	5	169	0.47	5.0	3.2	A	30	60
2	10	180	0.50	2.7	3.1	A	10	60
3	5	169	0.47	5.2	2.6	A	45	90
4	10	180	0.50	2.2	2.8	A	55	90
5	5	173	0.48	4.0	4.3	VG	90	320
6	10	180	0.50	2.4	5.1	VG	99	320
7	5	169	0.47	4.5	3.8	A	50	90
8	10	184	0.51	2.8	2.0	A	60	90
9	5	162	0.45	3.5	4.6	VG	100	90
10	10	162	0.45	3.0	5.0	VG	100	90
15	10	162	0.45	NM	1.8	B	22	45
16	10	162	0.45	NM	1.5	B	10	90
17	5	203	0.56	NM	1.9	B	18	90
18	5	203	0.56	NM	1.4	B	42	90
19	5	148	0.41	NM	1.8	B	69	45
20	5	159	0.44	NM	1.7	B	10	90

The first series (mix 1-10) was mainly done to investigate the effect of different mixing order and mixing time on dispersion. The amount of undispersed CSF granules was obtained by point-counting in thin-sections (25 x 40 mm). To get a reference for dispersion, samples were made from gently mixed CSF (5 and 10 % by weight of cement) and Portland cement. Of this powder 10 mg was squeezed into a pellet and impregnated with water using slight underpressure. The reference showed that 5 wt. % of CSF granulates in the cement paste give slightly less than 4.5 % of the counts. Duplicate thin-sections, however, showed that the mixes were inhomogeneous with areas rich and poor in undispersed granulated CSF. The amount dispersed CSF is therefore only a rough estimate.

The amount of air and air void distribution was determined by point-counting and image analysis in thin sections. The air content in Table I refer to air voids less than 1 mm in diameter. In samples 15 to 20 the air void system was badly developed. The amount of air is small (Tab.I) and there is a lack of very small pores. According to the air void system the concrete can not be regarded as frost resistant. In mixes 1 to 10 the air void system is generally better developed with smaller air voids. The spacing factor is less than 0.25 and the specific surface larger than 25 mm^{-1}. Mixes 5, 6, 9 and 10 have a very good air void system with many very small air voids.

Alkalisilica reaction

The ASR-method used was a modified Nordtest Method (NT BUILD 295). Instead of the mortar prisms (40x40x160 mm) as prescribed in the method, concrete prisms (75x75x400 mm) were used. The prisms were cured in a humid environment for 24 hours, demoulded, and then cured under water for 28 days at 20 °C. The length was measured and the prisms were put in a warm (50 °C) saturated salt (NaCl) solution.

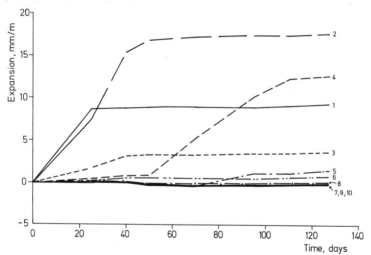

Fig. 1. A SR-Expansion of concrete prisms with CSF.
Samples from mixes 1 to 10. Data of the concrete in Table I.

In the first test series (mixes 1-10) some (mixes 1, 2, 3, 4) but not all the samples with badly dispersed granulated CSF rapidly developed a severe ASR-expansion (Fig. 1). In the second test series (mixes 15-20) with badly mixed concrete only samples 17 and 18 expanded. In the other samples no expansion could be found although they were kept in the salt bath for more than 2 years. Samples 17 and 18 are those with the highest w/b ratio, which indicates that permeability is an important factor, more important than the degree of dispersion. In batch 18 we used a low-alkali cement which normally does not give rise to ASR. This indicates that the expansion is more a function of alkali ingress than the composition of the cement. Once started the expansion was related to the amount of undispersed CSF granulates. Sample 2 with twice as much CSF as sample 1 expanded the double. The CSF in samples 3 and 4 is better dispersed.

Freeze/thaw resistance

Concrete from different mixes was tested using Swedish standard test (SS 13 72 25, frost testing with salt solution). In this test a standard cured cube is taken and cut in half. The sides are sealed with silicon rubber and the surface is covered by a solution containing 3 % of salt (NaCl) by weight. The test sample is put in a freeze/thaw cabinet were the temperature varies between +20 and -18 °C with one cycle per day. The test stipulates 56 cycles and the result is measured in amount of lost material per square meter. More than 1 kg/m^2 is not acceptable.

The test gave very peculiar results (Fig. 2). The concrete mixes that expanded in the ASR test disintegrated very rapidly while others with undispersed granulated CSF behaved surprising well. A normal concrete behaves like samples 6 and 10. However, some of these samples, like nr 4 and 7, suddenly after 56 cycles started to scale very fast. Material analyses showed that the rapid flaking was caused by ASR. The concrete contained expanded CSF nodules and cracks filled with ASR-gel (Fig. 4). Thin section analysis in microscope of sample 5 and 8, that did not scale at all, showed that the surface had microcracks filled with ASR-gel. Presumably this sealed the surface and protected the concrete from salt water ingress and F/T scaling

In the second test series (Fig. 3, sample 15 to 20) all the samples scaled quickly in NaCl solution. The samples with the highest w/b ratio and bad mixing, batches 17 and 18, scaled fastest and developed extensive ASR. In this series we did not observe the phenomena of excellent behaviour prior to a collapse as in the former series. This is probably the result of the poor air void system which give bad frost resistance.

With the surfaces were covered with pure water instead of salt solution nothing happened. When samples were covered by 3% $CaCl_2$-solution, however, concretes with badly dispersed CSF developed ASR. The concrete with low-alkali cement (nr 18) did not develop ASR when exposed to $CaCl_2$. Sample 18 in contrast to 17 (Fig. 3) behave normally when exposed to $CaCl_2$.

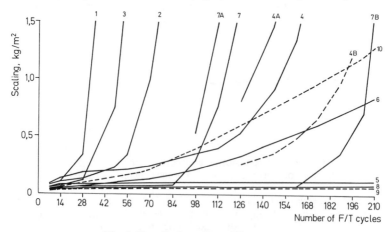

Fig. 2. Results from freeze/thaw tests.
Mixing series nr 1, samples 1 to 10. The numbers give the intermediate value of scaling from two halves of a concrete cube. When the sample is noted A and B they show the value of the individual half.

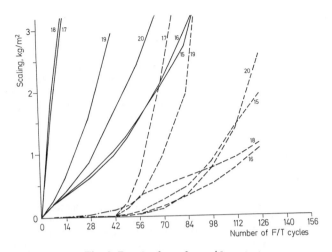

Fig. 3 Results from freeze/thaw tests.
Mixing sequence 2, sample 15-20. One half of the sample was subjected to a normal test as in
those samples in Fig.1. The second sample was first subjected to pure water for 42 cycles. The
pure water was then replaced by a solution with 3 % $CaCl_2$ by weight, and the F/T test continued
up to 156 cycles. The numbers on the top marks the amount of cycles the individual sample was
opposed to and the amount of material that scaled.

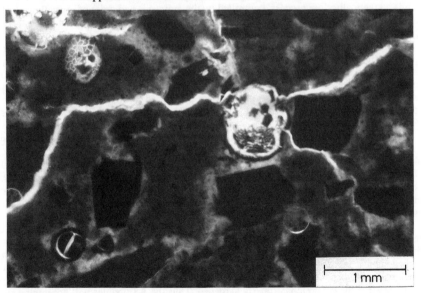

Fig. 4. Microscope photo in thin-section of silica granulates in expanded concrete.
Fluorescence light. Two of the granules have expanded while one small remained intact. A solid
shell (dark as it do not absorb fluorescens dye) can be observed around the bright expanded gel.

Destruction mechanism

After the different tests the material was analysed. Based on the crack pattern it could be concluded that the cracking and rapid disintegration was caused by the formation and expansion of ASR-gel. In thin-sections (Fig. 4) it is possible to observe cracks radiation out from expanding undispersed granules. The crack pattern is, however, uneven. In some parts of the concrete all the granules have expanded and formed cracks that are filled with gel while in other parts the undispersed granules remain intact. In the samples with undispersed granules that did not flake at all i the F/T-test, AS-gel could be found at the surface and filling microcracks in the surface.

In concrete the granules develop a shell. In the expanded and cracked concretes one can notice that this shell remains intact while the interior expands (Fig. 4). When analysed in the scanning electron microscope (SEM) we can only observe a small ball of gel in a shell. This must be due to the high vacuum in the SEM and that the expanding gel, in contrast to the shell, is very rich in water.

The test data indicate that external salt is needed to trigger the expansive reaction. The F/T results with a stagnant period followed by an accelerating indicate that the destruction mechanism is kinetically controlled. We can find that the permeability and homogeneity, apart from the alkali content of the cement, are controlling factors. The shell is probably also an important factor firstly by protecting the interior secondly by mechanically hindering the interior from imbibing water and expand. We will come back to this later after presenting the chemical system.

CHEMICAL SYSTEM

In a normal concrete the AS-gel is formed by the dissolution of silica from aggregates which is a kinetically slow process, controlled by the grain size and order of crystallinity of the quartz. The destruction by ASR is the result of expansion caused by absorption of external water in the AS-gel. The situation in the test samples is different.

To understand the mechanism behind the expansion we must look closer into the chemical systems and the physical properties of the concrete. The chemical composition of the AS-nodules has been determined by energy dispersive X-ray spectrometry (EDS) in SEM. These chemical analysis, however, do not give the amount of water, firstly because the technique do not give it, secondly because the analysis are performed at high vacuum which result in evaporation of water and drying shrinkage of the gel. Thus, in table II and III, the water content is disregard and the compositions are normalized to 100 % based on the weight of oxide.

Repeated analysis showed that with time and in contact with the atmosphere the gel became enriched in calcium and carbonated. Thus the analyses were done on fresh broken surfaces. When possible the analysis were made with a defocused beam as otherwise especially the alkalies evaporated from the loose structure by the energy from the beam. The EDS analysis are calibrated against cement clinker grains and minerals. In all the alkali-silica nodules, former granulates of CSF, we can detect Ca, Na, K and Si. In the nodules subjected to salt we can detect some Cl.

Micro-analysis in scanning electron microscope (SEM) shows that the undispersed CSF granulates are transformed into an AS-gel very rapidly. Table II presents the composition of the undispersed granules at different times after mixing. Already within 24 hours the gel has absorbed significant amounts of Ca and some alkalies. Within 96 hours it is an AS-gel (in the normal case an AS-gel always contain a substantial amount of Ca). This implies that the AS-gel is formed before the concrete is properly hardened. Thus the AS-nodules will not imbibe more water and cause expansion unless the chemical composition of the gel is changed. This is probably the reason why no freeze/thaw scaling can be observed with the samples in pure water. An alkali ingress is needed. Expansion is a matter of timing.

Table II
Typical compositions of undispersed CSF in mortars at different times after mixing.
High alkali cement (1.1 Na_2O equivalent). Granulated CSF = 5 % of cement weight. The mortars are badly mixed with almost all granulates remaining undispersed. Data normalized to 100 %. As water is missing the data only show the relative amounts of the different oxides.

Hours	24	24	96	96	264	264
Pos.	core	shell	core	shell	core	shell
SiO_2	92	87	70	61	70	68
CaO	7	11	11	15	7	29
K_2O	<1	1	17	13	20	3
Na_2O	<1	<1	2	2	3	<1

Table III
Typical compositions of undispersed CSF granulates in well cured concretes.
The first sample is with a high alkali cement (HAC, 1.1 % Na_2O-eqv.). The second analysis is from a concrete with a low alkali cement (LAC, 0.5 Na_2O-eqv). The third concrete (analysis 5 and 6) is from a concrete which has been stored for two years in hot salt bath (ASR-test). The last analysis are from a concrete that expanded in the salt bath. Analysis nr 7 is from an AS-nodule and analysis nr 8 is from AS-gel in crack. Data normalized to 100 %. As water is missing the data only show the relative amounts of the different oxides.

Analys	1	2	3	4	5	6	7	8
Curing	water	water	water	water	salt	salt	salt	salt
Cement	HAC	HAC	LAC	LAC	HAC	HAC	HAC	HAC
%CSF	5	5	5	5	5	5	10	10
Pos.	core	shell	core	shell	core	shell	core	crack
SiO_2	72	50	82	48	67	51	75	59
CaO	3	48	13	52	15	45	16	33
K_2O	21	2	5	<1	2	1	<1	<1
Na_2O	4	<1	1	<1	15	2	13	7

Chemical analysis were done on most of the test samples. Representative analysis are presented in Table III. They all reveal a complicated zonation pattern with a core rich in alkalies and a rim rich in calcium. A similar zonation was recognised and is described by Shyan et al [2]. The outermost rim contain very little, often no, alkalies but a gel with a Ca/Si ratio of 1/1. The product with equal amounts of CaO and SiO_2 is repeated with such a regularity that it must be a distinct product. We can find the same composition in old ASR-gel from damaged structures (Lagerblad & Trägårdh [3]). It is linked to a depletion of portlandite close to the gel. In the high vacuum of the SEM this shell, in contrast to the normal AS-gel, does not shrink. The composition of this gel is reminiscent of and probably is a CSH (I)-gel, the typical product of the pozzolanic reaction. The volume stability, distinct appearance and total lack of alkalies suggest a semicrystalline structure.

From this shell toward the core the relative content of silica increases. At the same time the content of alkalies increases while that of Ca decreases. Sometimes it is possible to observe distinct layers showing jumps in composition like with the outer shell. We have, however, not been able to pin-point specific compositional gaps where the the jump occur. The zoning is a general feature. The exact composition of the gel, however, depends on several factors. We have not been able to identify exact correlations but several compositional trends. They are indicated in Fig. 5.

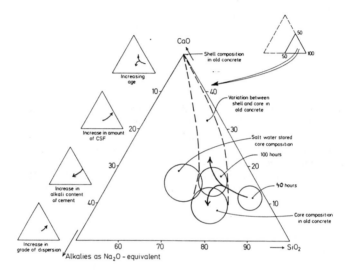

Fig. 5. Schematic triangular diagram
showing the chemistry and the changes of chemistry of the undispersed CSF in concrete.
The data is based on a concrete with a high alkali cement and 5 % undispersed CSF by weight.

DISCUSSION

To understand the composition of the AS nodules we have to look into the chemistry of the young concrete. When cement clinker grains are mixed with water a cement-gel is formed and calcium hydroxide is released to the liquid phase. Within the first hours the pH is increased to more than 13. Successively the alkali ions are released and the portlandite start to crystallize. In the first instance the Si-O-Si bridges are broken under the action of the hydroxide ions in the pore fluid and form silanol groups (Wang & Gillot[4]). This probably happens quickly in the granules as they are porous and contains millions of individual small amorphous balls. Follow this, the calcium and alkalies exchange for the protons of the silanol group and form an alkali-silica gel.

The calcium is, as the data (Tab. I) show, presumably the first to bind as they are more

96

available in the pore solution. Later when the contents of alkalies increase they will be the dominant ions. Much later when the system is stabilized, the alkalies will be replaced by calcium as the Ca-gel is energetically preferable (Wang & Gillot[5]). This will explain why the AS-nodules are zoned with Ca-rich rims. With well dispersed silica fume like in the normal case the gel immediately turns into Ca-gel the pozzolanic product. A comparison can be made with alkali reactive rocks. Corse sand and stones give rise to ASR and expansion while finely ground they are pozzolanes.

When the concrete is subjected to salt solutions, as in the ASR-test and the F/T-test, this increases the content of sodium in the AS-nodules (Table II). From the beginning the AS-nodules are rich in potassium as this is the main alkali ion in the cement we used. The expansion is probably induced by the increased alkali content that allows the gel to imbibe more water. However, not all the concretes with undispersed CSF granulates expanded. Even more than one and a half years of hot salt bath did not result in expansion of most of the concrete mixes although the AS-nodules were enriched in Na. One problem is that it is almost impossible to find out the composition of the gel actually extruded from the AS-nodules as it in contact with the cement-paste has changed composition. However, one would expect that an AS-gel with a composition like that in the samples stored in the salt bath would absorb water and expand.

Presumably an intact C-S-H shell is fairly impermeable to water and resists cracking. In thin-section it is possible to observe strings of expanded AS-nodules which indicates that once cracking has occurred the salt solutions can penetrate and trigger more nodules to expand and continue the crack front. This indicates that the AS-nodules have to be mechanically ruptured to allow the interior to expand. Once the cracking start it will trigger more granulates to expand which will result in the fast detoriation. At a first impression this theory described above is contradicted by the fact that the reaction also occurs with $CaCl_2$ solution during the F/T test. However, as shown by Wang & Gillot[5] excess Ca releases alkalies which allow the remaining gel to be enriched in alkalies.

Acknowledgement

The research has been financed by the Swedish Foundation for Concrete Research, the Swedish Counsil for building Reserch (BFR) and the Swedish National Board for technical and Industrial Development (NUTEK).

REFERENCES

1. B. Lagerblad & P. Utkin " Silica granulates in concrete-dispersion and durability aspects. CBI-report 3:93, 44 p, (1993).

2. A. Shayan, G.W. Quick & C.J. Lancucki,"Morphological, mineralogical and chemical features of steam-cured concretes containing densified silica fume and varios alkali levels", Advances in Cement Research, 5, No 20, 151-162, (1993)

3. B. Lagerblad & J. Trägårdh, " Slowly reacting aggregates in Sweden,"Mechanism and conditions for reactivity in concrete", Proceedings, 9th International conference on alkali-aggregate reactions, London, 570-578, (1992).

4. H. Wang & J.E. Gillot, " Mechanism of alkali-silica reaction and the significans of calcium hydroxide", Cement and Concrete Research, vol 21, 647-654, (1991)

5. H. Wang & J.E. Gillot, " Effect of $Ca(OH)_2$ on the alkali-silica reaction, Magazine of Concrete Research, 43, 156, 215-218 (1991)

EFFECTS OF MICROSTRUCTURE ON FRACTURE BEHAVIOR OF HARDENED CEMENT PASTE

ASIF AHMED AND LESLIE STRUBLE
University of Illinois, Department of Civil Engineering, Urbana, IL.

ABSTRACT

Mechanical properties of any material, including hardened cement paste, are assumed to be controlled by its microstructure. An attempt has been made here to establish a link between bulk fracture parameters of hardened cement paste and its microstructure. Paste microstructure has been varied by changing the initial w/c ratio, curing time and curing temperature, and by addition of chemicals to change the calcium hydroxide morphology. It has been found that, like compressive strength, fracture parameters depend directly on porosity. Contrary to our initial hypothesis, CH morphology was found to have no effect on the fracture parameters.

INTRODUCTION

Hardened cement paste (HCP) is produced by hydration reactions that start as soon as water is mixed with dry portland cement. Microstructure of hardened paste is controlled by various parameters: particle size distribution and composition of cement, morphology of individual hydration products, and composition and age of the paste. In turn, the microstructure determines bulk mechanical properties like strength and toughness. The size scales that are important in the three levels – hydration reactions, microstructure and bulk mechanical properties – differ by orders of magnitude. This wide range of scale is probably why studies of HCP tend to focus on one of the three levels, but very rarely on all three levels at once. From a materials science point of view identifying the microstructural parameters that affect the various bulk mechanical properties is the essential first step in establishing a microstructure–property relationship. The research reported here is an attempt to identify microstructural parameters that affect the fracture behavior of HCP.

MICROSTRUCTURE OF HCP

At the microstructural level HCP is an intimate but inhomogeneous mixture of a variety of crystalline and quasi-crystalline phases and pores of different sizes and shapes. Most of the volume is calcium silicate hydrate (C-S-H), the principal hydration product, which is a highly disordered quasi-crystalline material of variable composition. The molecular structure of C-S-H is layered and provides for a very high internal porosity. The other primary hydration product is calcium hydroxide (CH). In contrast to C-S-H, CH is highly crystalline and has a fixed composition. The crystals appear as thin hexagonal platelets, often layered, typically tens of μm across. With continued hydration they grow massive, lose their hexagonal outline, and encapsulate other regions of the paste. Massive blocks of CH crystals can be easily identified in mature pastes. Unhydrated residues of clinker grains, which are crystalline in nature, are present even in well hydrated systems.

Pores are essential components of HCP microstructure. There are two major sources for these pores: (1) C-S-H, which is inherently porous, containing gel pores with characteristic diameters <10 nm, and (2) remnants of water filled space that are not occupied by the hydration products form capillary pores, typically 10 nm to 10 μm. The total pore volume (i.e.,

99

porosity) of any HCP system is a function of the w/c ratio, degree of hydration, curing conditions, cement grain size, etc. [1].

Empirical relationships between total porosity and compressive strength of HCP are common in the literature [2]. Although these relationships may break down at very low or very high porosities, they fit fairly well at intermediate porosities. However, there are indications that total porosity is not the sole factor that determines strength. Other factors that may be important include the size and shape of pores, the morphology of various constituents, and the nature of the chemical bonding.

Among the solid phases CH crystals are generally massive compared to the rest, but their effect on bulk mechanical properties is not clear. Because of weak cleavage CH crystals have been suspected of weakening the paste [3] although this hypothesis has not been proved. On the contrary, because of their stiffness and size, CH crystals may act as rigid restraints against deformation or as crack deflectors or arrestors [4].

MODIFICATION OF HCP MICROSTRUCTURE

From the discussion above it is clear that two aspects of the microstructure need to be investigated in order to assess the effects of changes in HCP microstructure on its bulk mechanical properties. These are pore volume and size distribution, and the amount and morphology of CH.

Both pore volume and size decrease with continued hydration. The effect of changing pore volume can thus be studied by looking at the same system at different levels of hydration. In addition, pore volume and size can be changed by using different initial w/c ratios. Pore size distribution can also be modified by changing the particle size distribution of the cement or by incorporating an ultrafine inert filler. In both cases there is a risk of changing the other aspects the microstructure.

Chemical admixtures have been shown to affect the morphology of CH crystals in hardened C_3S systems [5]. In C_3S specimens with no admixture, the aspect ratio for CH crystals (length in the c−direction relative to length in the a−direction) varies between 0.5 and 1.5. Addition of sulfate−containing admixtures (e.g., gypsum) reduces the ratio to below 0.5, while addition of chloride−containing admixtures (e.g., $CaCl_2$) increases the aspect ratio to more than 1.5. When organic acids or salts are added the CH crystals tend to be irregular in shape and size.

Curing temperature also affects CH morphology [6]. The size of CH crystals increases with decreasing curing temperature, particularly below 25°C. However, the number of crystals decreases with decreasing temperature. The w/c ratio does not have much effect on the size of CH crystals, but at low w/c ratios CH tends to engulf C_3S particles.

Table 1: Chemicals used to modify CH morphology

Chemical	Identifier	CH Morphology
Ammonium Nitrate (NH_4NO_3)	AN	Elongated along c axis
Calcium Chloride ($CaCl_2.2H_2O$)	CC	Elongated along c axis
Sodium Nitrate ($NaNO_3$)	SN	Elongated along c axis
Maleic Acid (HOCOCH−CHCOOH)	MA	Irregular, different morphologies
Propionic Acid (C_2H_5COOH)	PA	Irregular, different morphologies

EXPERIMENTAL DETAILS

Specimens used in this study were prepared by mixing freshly boiled deionized water with a commercial Type I Portland cement and were cast in the form of notched beams and cylinders. In some specimens reagent grade chemicals were added to alter the size and shape of CH crystals. Table 1 lists the chemicals used in this study and their expected effect on CH morphology.

The notched beams (165 x 35 x 25 mm) were tested in three point bend configuration in an Instron screw driven closed-loop universal testing machine fitted with a fully articulated loading fixture. A crack opening displacement gage mounted across the notch was used to continuously measure the crack mouth opening displacement (CMOD) which was used as the feedback signal to control the crosshead movement. Figure 1 shows a typical load–CMOD curve with variables that were used to compute macroscopic fracture parameters critical stress intensity factor (K_{Ic}) and critical crack tip opening displacement ($CTOD_c$) according to the two parameter fracture model [7].

The compressive strengths (f_c) were obtained by loading cylinders (50 x 25 mm) to fracture under constant crosshead displacement. The fractured cylinders were crushed into 5-10 mm chunks in a mortar and pestle and were preserved in ethanol for microstructural characterization.

Various tests were used to characterize the paste microstructure. The internal pore structure was probed using mercury intrusion porosimetry (MIP) on ethanol exchanged chunks of pastes. Powder X−ray diffraction (XRD) was used to determine relative aspect ratio of CH crystals (length in the c−direction relative to length in the a−direction). Thermal gravimetric analysis was used to measure the degree of hydration and the amount of CH. A scanning electron microscope was used to examine fracture surfaces.

RESULTS AND DISCUSSION

Results from two sets of specimens are reported here. Details are listed in Table 2. The first set was twelve specimens (with prefix "B") cast with varying w/c ratios and tested at various ages. In the second set (with prefix "D"), the CH morphology was modified by adding one of the chemicals listed in Table 1 or by curing at 4°C. It should be noted that specimen B50:57 was the control specimen for the second set.

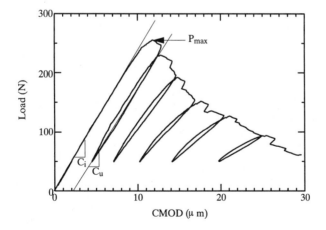

Figure 1: Typical load-CMOD curve

Table 2: Bulk mechanical and microstructural data

Speci-men ID	\multicolumn{4}{c}{Specimen Details}				\multicolumn{3}{c}{Mechanical Data}		\multicolumn{4}{c}{Microstructural Data}
	w/s	Age (days)	Addi-tives[#]	Curing Temp.[*]	f_c (MPa)	K_{Ic} (MPa$\sqrt{m}$)	$CTOD_c$ (µm)	W_{CH} (%)	α (%)	P (%)	$I_{(101)}/I_{(001)}$
B35:01	0.35	1	None	RT	25.7	0.24	1.2	20	46	34	0.81
B35:07	0.35	7	None	RT	58.2	0.36	1.6	26	58	26	0.81
B35:28	0.35	28	None	RT	84.8	0.38	2.1	29	73	18	0.87
B35:57	0.35	57	None	RT	88.1	0.42	2.9	30	76	17	1.01
B50:01	0.50	1	None	RT	9.1	0.10	1.4	23	50	43	0.54
B50:07	0.50	7	None	RT	24.6	0.23	2.2	28	61	36	0.58
B50:28	0.50	28	None	RT	47.7	0.29	2.2	34	79	28	0.63
B50:57	0.50	57	None	RT	60.4	0.29	2.3	34	80	27	0.72
B65:01	0.65	1	None	RT	5.0	0.06	2.6	25	54	50	0.37
B65:07	0.65	7	None	RT	13.6	0.18	2.7	26	61	39	0.55
B65:28	0.65	28	None	RT	31.1	0.24	1.9	33	82	35	0.62
B65:57	0.65	57	None	RT	40.9	0.26	2.8	34	88	33	0.66
DAN:57	0.50	57	AN	RT	53.2	0.32	3.4	39	89	29	0.45
DCC:57	0.50	57	CC	RT	61.5	0.28	3.3	42	96	29	0.41
DSN:57	0.50	57	SN	RT	39.7	0.31	4.2	37	86	31	0.50
DMA:57	0.50	57	MA	RT	54.8	0.26	1.8	37	92	26	0.45
DPA:57	0.50	57	PA	RT	52.4	0.28	3.1	37	92	27	0.54
DRE:57	0.50	57	None	4°C	44.9	0.25	1.8	35	83	29	0.61

[#] Dosage was 1% by weight of the cement.   [*] RT is room temperature (i.e., ~25°C).

Measured parameters characterizing the macroscopic fracture behavior and the microstructure are also presented in Table 2. The gain in compressive strength in series B with continued hydration is shown in Figure 2. As generally expected, f_c increased with increasing hydration time. Specimens with w/c ratio of 0.35 reach a plateau at about 28 days, whereas specimens with higher w/c ratios show upward trends even at 57 days. Figure 3 shows the change in K_{Ic} with hydration time for the same set of specimens. Unlike f_c, for a given w/c ratio K_{Ic} increased rapidly at early ages, then reached a plateau at about 10 days.

Figure 4 combines the data from Figures 2 and 3. There appears to be a bi−linear relationship between f_c and K_{Ic}, independent of the w/c ratio and age. Because f_c is known to depend on porosity, one might expect a relationship between K_{Ic} and porosity (P). Figure 5 does show a negative relationship between K_{Ic} and P (as measured by MIP).

The D series was prepared to assess the effects of changing CH morphology, as characterized using SEM and XRD. Micrographs in Figure 6 show that CH crystals have quite different morphologies in the specimen containing 1% calcium chloride compared to the control specimen. CH morphological differences have been quantified using XRD. A lower ratio of the (101) and (001) peaks corresponds to crystals that are higher in aspect ratio. Results in Table 2 indicate that, despite substantial changes in the amount and morphology of CH, changes in K_{Ic} and P were not large.

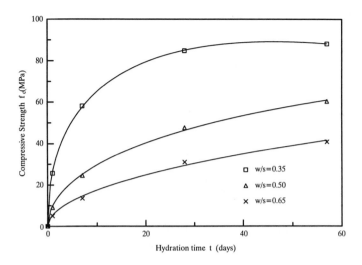

Figure 2: Development of compressive strength with hydration time (series B)

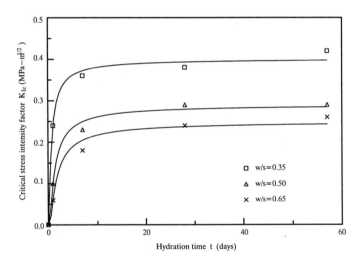

Figure 3: Change in K_{Ic} with hydration time (series B)

CONCLUSIONS

Based on the results presented here, it has been concluded that the macroscopic fracture parameters for hardened cement paste are not affected by changes in the CH crystal morphology. Like compressive strength, the fracture parameters are directly affected by changes in the porosity.

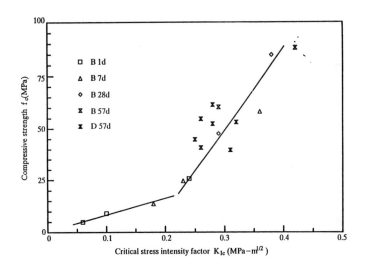

Figure 4: Relationship between K_{Ic} and f_c

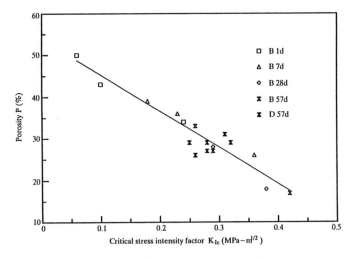

Figure 5: Relationship between K_{Ic} and porosity

ACKNOWLEDGEMENTS

Financial support for this project was provided by the NSF Science and Technology Center for Advanced Cement Based Materials. The support is gratefully acknowledged.

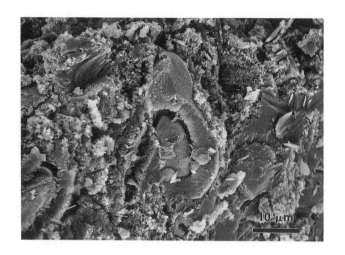

Figure 6(a): CH crystals in control specimen

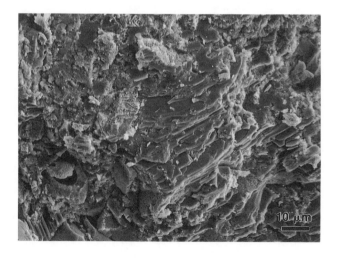

Figure 6(b): CH crystals in specimen containing CaCl$_2$

REFERENCES

1. L.E. Copeland and G.J. Verbeck, Proc. 6th Int. Congr. Chem. Cem. **2**(3) (1974).
2. H.F.W. Taylor, *Cement Chemistry* (Academic Press, San Diego, 1990) 265 pages.
3. R.B. Williamson, Progress in Materials Science **15**, 189−286 (1972).
4. R.L. Berger, Science **175**, 626−29 (1972).
5. R.L. Berger and J.D. McGregor, Cement and Concrete Research **2**, 43−55 (1972).
6. R.L. Berger and J.D. McGregor, Journal of the American Ceramic Society **56**, 73−79 (1973).
7. RILEM TC 89−FMT, Materials and Structures **23**, 457−60 (1990).

IMAGE–BASED CHARACTERIZATION OF FRACTURE SURFACE ROUGHNESS

A.B. ABELL AND D.A. LANGE
University of Illinois, Department of Civil Engineering, Urbana, IL 61801

ABSTRACT

Analysis of images from SEM, optical and confocal microscopes provides insight into the roughness of fracture surfaces of cement–based materials. Several techniques including 3D surface reconstruction and 2–D vertical contour analysis are compared. Roughness parameters and fractal dimensions are computed to describe the degree of tortuosity of the fracture surface. The tortuosity of the surface (i.e. crack path) is an aspect that has not been adequately integrated into fracture mechanics models of cement–based materials. Topographic maps from confocal microscopy reveal orientation of the crack front throughout the fracture process, allowing analysis of mixed–mode character of nominally mode I fracture. These techniques allow quantification of microstructural parameters to be linked to material properties.

INTRODUCTION

Fracture surface characterization is of interest in the relationship of geometry and microstructure of cement–based materials to mechanical properties. Images of surfaces from SEM, optical and confocal microscopes provide qualitative information about the fracture. Images from the confocal microscope provide quantitative data in the form of a topographic map. This data can be analyzed for a roughness number, and profiles and surfaces can be measured with varying "yardsticks" to evaluate a fractal dimension. In addition, this surface data can be used in a mechanics of material model to predict local fracture toughness increases from the tortuosity.

Confocal microscopy has been shown to be a convenient way to obtain three–dimensional descriptions of cement–based materials[1]. A series of optical sections taken at different focal planes construct a digital image in which each pixel is assigned a value that represents the z–level. The surface area can be estimated by the geometric construction of the triangular planes defined by the $x-y-z$ coordinates of each pixel. The surface roughness number (RN) can be determined from the calculated surface area. The coordinates can also be used to determine a fractal dimension for the ruggedness of the surface geometry.

In addition to providing surface data, the confocal image can be "sliced" vertically to obtain profiles in the $x-z$ or $y-z$ plane. These profiles describe the crack front and crack profile tortuosity. A fractal dimension and the crack path deflection angles can be determined.

A mechanics of materials model proposed by K.T. Faber and A.G. Evans[2] predicts fracture toughness increases due to crack deflection around second phase particles. The model predicted the crack deflection and twisting from the particle morphology, aspect ratio, spacing and volume fraction. The confocal image provides an actual description of the crack deflection and twisting angles, and can be used directly in the model to estimate the fracture toughness increases.

IMAGE ACQUISITION

The images were acquired with a confocal laser–scanning microscope with a 2.5X lense at a magnification of 20 to produce a field size of approximately 3.5mm x 3.5mm. The z slice thickness was 10 µm. The specimens used for the images were mortars with 1:1 and 1:2 cement:coarse sand ratios, 1:1 cement:fine sand ratio, and plain pastes with 0%, 5% and 10% replacement with silica fume[3]. Lonestar Type I portland cement, W.R. Grace Force 10,000 silica fume slurry, Ottawa silica sand, and deionized water were the constituents of the pastes and mortars. The mix design and specimen number are listed in Table I. Nine images were taken on each specimen.

Table I. Mix design and specimen number

Name	Spec. #	Water	Cement	Silica fume	Coarse sand	Fine sand
Control paste	L0200	0.4	1.0	0	0	0
1:1 mortar	L0201	0.4	1.0	0	1.0	0
5% s.f. paste	L0202	0.4	0.95	0.05	0	0
10% s.f. paste	L0203	0.4	0.90	0.10	0	0
1:2 mortar	L0204	0.4	1.0	0	2.0	0
Fine mortar	L0205	0.4	1.0	0	0	1.0

Three–point bend test were performed using beam specimens at 28 days. K_{IC} from linear elastic fracture mechanics and Young's modulus were computed, as shown in Table II.

Table II. Young's Modulus and Fracture Toughness (KI_C)

Specimen	E (MPa)	K_{IC} (N m$^{-3/2}$)
Control paste	2.01E+04	4.53E+05
1:1 mortar	2.60E+04	6.25E+05
5% s.f. paste	1.80E+04	4.45E+05
10% s.f. paste	2.10E+04	4.83E+05
1:2 mortar	3.68E+04	7.03E+05
Fine mortar	3.12E+04	7.27E+05

All images were passed through a median noise reduction filter prior to analysis to reduce noise. The median filter created a new image by replacing each pixel with the median z value of the pixel and the 8 pixels surrounding it

ANALYSIS AND DISCUSSION

The resulting images appeared quite noisy with the gold coating, although the exterior surface was properly being recorded. .

Roughness Number

Roughness of the image surfaces is calculated by dividing the estimated area of the surface by the nominal area of the image. For example, a perfectly planar horizontal surface will have an RN value of 1.0.

The characteristic roughness for each specimen was determined by averaging the nine images obtained for that specimen. The average RN for each specimen is listed in Table III.

Table III. Roughness of specimens

Specimen #	Ave. Roughness Number
L0200	3.2236
L0201	3.6958
L0202	3.4178
L0203	3.7087
L0204	4.3258
L0205	4.2479

The roughness of the mortar specimens (L0201, L0204, L0205) was expected to be higher than those without sand. The RN for the pastes (L0200, L0202, L0203) increased with increasing volume of silica fume. The relatively large RN of the 10% s.f. paste (L203) compared to the mortar specimens could be a result of the acquisition noise.

Fractal Dimension

Because the scaling of the image in the z direction was different from the other axes, the images are considered self–affine and use of the Richardson technique for determining the fractal dimension is not appropriate[4]. A modification of the Minkowski method was applied to both the profile and surfaces of the specimen images.

For the profiles, a horizontal structuring element was used. The area of the profile was determined by finding the minimum and maximum z elevation within the element or window width around each pixel. In effect, a box with the width of the window and the height of the elevation difference was constructed around each pixel, and the area marked by the overlapping box boundaries was calculated. Figure 1. This value divided by the window width yields the effective

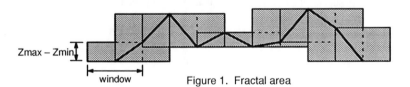

Figure 1. Fractal area

profile length and was plotted as a log against the log of window width to determine the fractal dimension by the equation:

$$D = 1 - h \qquad (1)$$

where h is the slope of the straight line constructed through the data points. The results of the fractal analysis are shown in Table IV.

Table IV. Fractal dimension and line fit

Specimen #	Profile		Surface	
	Fractal Dim.	R^2 (line fit)	Fractal Dim.	R^2 (line fit)
L0200	1.5013	0.9782	2.5327	0.9968
L0201	1.3507	0.9834	2.4074	0.9976
L0202	1.4889	0.9825	2.5185	0.9972
L0203	1.4890	0.9807	2.5194	0.9958
L0204	1.3344	0.9820	2.3720	0.9971
L0205	1.4380	0.9801	2.4678	0.9967

For the surfaces, a circular disc in the horizontal plane was used as the structuring element. The volume of the surface was determined by finding the minimum and maximum z elevation within the circle around each pixel. This elevation difference was assigned to each pixel within the disc. After every pixel was evaluated, the values were summed over the encompassed area. The volume could be considered to be created from overlapping cylinders at each surface point. (Figure 2.) This

Figure 2. Fractal volume (for one row)

volume divided by the circle radius was the effective area and was plotted as a log against the log of circle radius to determine the fractal dimension by the equation:

$$D = 2 - h \tag{2}$$

where h is the slope of the straight line constructed through the data points. The results of the fractal analysis are shown in Table IV.

The results appear to support the fractal principle that D of a surface is 1.0 greater than D of a profile created by a vertical section through the surface.[5]. The variation of the structuring element size was limited by the image size, and only extended one decade in the log–log fractal plot. The fractal dimensions also show little variation, suggesting a single characterizing value.

Fracture Modeling

The fracture mechanics model applied to the confocal microscope image data calculates an average strain energy release rate from the tilted and twisted portions of the crack front. The crack front is deflected by inclusions out of plane by a tilt angle, θ, and the crack projection is bowed around the inclusions by a twist angle, ϕ. (Figure 3.) The tilted crack has Mode I (opening) and Mode II (sliding) contributions to the local stress intensity; while the twisted crack has Mode I and Mode III (tearing) contributions.

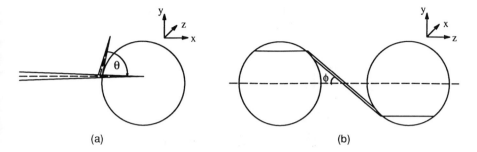

Figure 3. Crack front deflection: (a) tilt, and (b) twist

The strain energy release rate, $\mathcal{G}$, and the local stress intensity factors, $k^t{}_1$ and $k^t{}_2$, for a tilted segment of crack are determined by

$$E\mathcal{G} = k^t{}_1{}^2(1-v^2) + k^t{}_2{}^2(1-v^2) \tag{3}$$

$$k^t{}_1 = K_{11}(\theta)K_I \tag{4a}$$

$$k^t{}_2 = K_{21}(\theta)K_I \tag{4b}$$

where E and v are the Young's modulus and Poisson's ratio of the material, respectively, and K_{11} and K_{21} are angular functions associated with the tilted crack. The solutions for K_{11} and K_{21} are

$$K_{11}(\theta) = \cos^3 (\theta/2) \tag{5a}$$

$$K_{21}(\theta) = \sin (\theta/2) \cos^2 (\theta/2) \tag{5b}$$

The strain energy release rate and the local stress intensity factors, $k^T{}_1$ and $k^T{}_2$, for a twisted segment of crack are determined by

$$E\mathcal{G} = k^T{}_1{}^2(1-v^2) + k^T{}_3{}^2(1+v) \tag{6}$$

$$k^T{}_1 = K_{11}(\phi)k^t{}_1 + K_{12}(\phi)k^t{}_2 \tag{7a}$$

$$k^T{}_2 = K_{31}(\phi)K^t{}_1 + K_{32}(\phi)k^t{}_2 \tag{7b}$$

where the angular functions are determined by resolving the normal and shear stresses of the tilted crack onto the twist plane. These functions are

$$K_{11}(\phi) = \cos^4 (\theta/2) [2v \sin^2\phi + \cos^2 (\theta/2) \cos^2\phi] \tag{8a}$$

$$K_{12}(\phi) = \sin^2 (\theta/2) \cos^2 (\theta/2) [2v \sin^2\phi + 3 \cos^2 (\theta/2) \cos^2\phi] \tag{8b}$$

$$K_{31}(\theta) = \cos^4 (\theta/2) [\sin\phi \cos\phi (\cos^2 (\theta/2) - 2v)] \tag{8a}$$

$$K_{32}(\theta) = \sin^2 (\theta/2) \cos^2 (\theta/2) [\sin\phi \cos\phi (\cos^2 (\theta/2) - 2v)] \tag{8b}$$

111

The angles were determined for each segment from the difference in z elevations and horizontal distance between adjacent pixels. A typical angular distribution is shown in Figure 4. The gaps in

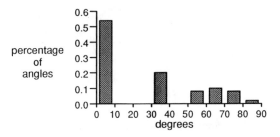

Figure 4. Angle histogram

the histogram result from the fixed z level spacing and the fixed horizontal pixel spacing. Values of 0 degrees were predominant, where values less than 70 degrees were scarce.

The toughening ratio was determined by

$$\text{Toughening ratio} = \frac{K_I}{(\text{average } E\mathcal{G})^{1/2}} \tag{9}$$

where $E\mathcal{G}$ is calculated from equations 3–8 with K_I equal to 1.0. Toughening ratio results are shown in Table V. Comparison of the toughening ratio to the K_{IC} values from testing are shown in Figure 5.

Table V. Ratio of local toughening to fracture toughness

Specimen #	Toughening ratio
L0200	3.3005
L0201	3.6035
L0202	3.3826
L0203	3.3958
L0204	3.7883
L0205	3.6892

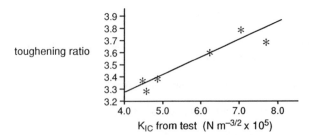

Figure 5. Comparison of toughening ratio and K_{IC}

112

The toughness ratios are generally in the same order as the RN values, but within a narrower range that is probably an effect of the limited angle values. Comparison of the RN values to the toughening ratio is shown in Figure 6.

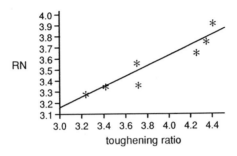

Figure 6. Comparison of RN and toughening ratio

CONCLUSIONS

Roughness number is a useful parameter for characterizing fracture surface ruggedness from confocal microsopy data. A modified Minkowski fractal analysis can also be applied to confocal microscope images to quantify the irregularity of a profile or surface and possibly to quantify a set of similar profiles or surfaces. The roughness of a surface can be shown to relate quite well with the prediction of local toughening increases with respect to measured fracture toughness from a mechanics of materials fracture model.

Use of these characterization techniques for fracture surfaces provides greater insight to the mechanical properties of cement–based materials.

ACKNOWLEDGEMENTS

This research was supported by the NSF Center for Advanced Cement–Based Materials (NSF Grant No. DMR 88808423–01).

REFERENCES

1. D.A. Lange, H.M. Jennings, S.P. Shah, Journal of Material Science, **28**, 3879 (1993).

2. K.T. Faber and A.G. Evans, Acta metall., **31**, 565 (1983).

3. D.A. Lange, H.M. Jennings, and S.P. Shah, Journal of the American Ceramic Society, **76**, 589 (1993).

4. J.C. Russ, Fractal Surfaces, (Plenum Press, New York, 1994).

5. B.B. Mandelbrot, The Fractal Geometry of Nature, (Freeman, New York, 1982).

MICROSTRUCTURAL ASPECTS OF HIGH-PERFORMANCE CEMENT-BASED MATERIALS

W. JIANG AND D.M. ROY
The Pennsylvania State University, Materials Research Laboratory, University Park, PA 16802, U.S.A.

ABSTRACT

The goal of this research is to improve the physical properties, mechanical properties and durability of cement-based materials by controlling chemistry and processing to produce the desired microstructures and properties. An accompanying initiative is to establish the basic scientific understanding relating processing and microstructure with physical and mechanical behavior, for the future development of new, special purpose, high performance cement-based materials. The research involves both experimentation and theory elucidating fundamental strengthening mechanisms for materials such as warm-pressed, MDF, DSP, polymer, and fiber composites. SEM and computer image analysis were used. Of interest are creating process models that serve as a basis for real applications such as in the following areas: High-performance highway concrete, and the immobilization of radioactive waste.

INTRODUCTION

The demand worldwide for high-performance cement-based materials has increased, and predictions are that it will reach a major industrial dimension by the year 2000[1]. Problems encountered during development of high-performance cannot be solved by trial-and-error experiments, therefore it is imperative to understand the microstructure of such products and relate it to the engineering parameters. The key factor in bringing more new or enhanced high-performance cement-based products to fruition is understanding the fundamental microstructure-property-performance relationship. The importance of establishing performance data bases and use criteria is paramount in the battle for survival of the one-third of America's infrastructure which is now classified by the federal government as deteriorated or deteriorating (The Washington Monthly, November, 1991). This research derives from the systematic research effort on cementitious material microstructure in which the Materials Research Laboratory at the Pennsylvania State University has been involved during the last decade.

Although portland cement has been successfully adapted to produce concrete with high strength, it is generally not possible to make products free from defects[2]. There are four major defects.

High porosity — Cement-based material is a porous material. The reaction of cements with water leaves behind pores; the excess porosity non-bound water has detrimental effects on the strength and durability.

Inhomogeneity — This could occur at both micro- and macro- level. The density varies by more than 10 percent from a mean value over dimensional ranges spanning from nanometers to millimeters. The pore size exhibits high gradients or changes in distribution within a specimen. In composites the aggregate/fiber/polymer are difficult to distribute. Honeycombing and segregation of concrete are often encountered due to the lack of consolidation or too much consolidation.

Cracking — A crack or fissure at the surface or in the interior of a solid is a stress concentrating flaw[3]. Cracks of different size exist everywhere in cement-based materials, which could be due to the heat of hydration, hardening and drying shrinkage. The distinguishing mechanical property of these materials is brittleness — that is, catastrophic failure following an almost entirely elastic deformation[4].

Mat. Res. Soc. Symp. Proc. Vol. 370 © 1995 Materials Research Society

Weak bonding between reinforcement and cement matrix — In general, no strong bonding tends to form. The transition zone exhibits some anisotropic gradation properties.

This work attempts to analyze the real image of cement-based materials to establish the critical material properties and process parameters which hold the key to producing macro-defect-free components. This will lead to the development of high-performance cement-based systems on a more logical basis. It is also important to develop microstructural design of cement-based materials in detail, because these affect various stages of the design, especially cement-based composites[5].

EXPERIMENTAL

Materials and specimen preparation

The ordinary portland cement (ASTM type I) was designated as OPC. The chemical composition of materials in the present study was reported in a previous paper[6]. Fine aggregates were ASTM standard sand. In this research, coarse aggregates conformed to the requirements of ASTM C33 (in this case less than 10 mm). The mix proportions of concrete were: cement: sand: aggregate :: 1.0: 2.75: 2.75. The warm-press cell was used to prepare 0.75x1.0x2.75-in bars, 2-in. cubes, and ø0.5, ø0.75, and ø1.0-in. cylindrical specimens. Their mix proportions are presented in Table I. Briefly, the procedure for "mixed" specimens was as follows: weighed amounts of cement, and water were mixed using a small trowel, then transferred to the warm-press cell. The cell was inserted into the press with two heater buffer plates, and kept at 120 °C, 2 hours, pressure at 30 MPa. After demolding, the specimens were cured in the water at room temperature for 4 weeks, then at 70°C for one day defore testing.

Each polymer mortar specimen was cast in a steel mold of dimensions 0.75x1.0x2.75-in. Polymer levels were evaluated at 8% and 12% based on the level of portland cement. The polymer latex was a stryene/butadiene rubber liquid latex emulsion with a pH 10. This emulsion was stabilized with surfactant. The polymer mortars were mixed and cast according to ASTM procedures. Specimens were demolded at 24 hours and cured at constant temperature, 25°C.

MDF cement pastes were prepared by the method described by Birchall *et al.* [7]. The mix proportion is given in Table I. For molding of MDF cement, a C.W. Brabender extruder with a die was used to obtain a sheet (width 5mm and thickness 2.5 mm). The specimens were prepared and dried under room temperature.

SEM and image analysis

Polished sections 3 cm in diameter were prepared after cutting out from the original specimens. The morphology of hydrates was observed by scanning electron microscopy (SEM). The specimens were impregnated with epoxy resin under vacuum at 100°C, to prevent any material pullout during polishing. The SEM micrographs were transmitted via TV camera to a computer image analyzer, details of which are given elsewhere[8].

Table I Mix proportion (by weight gram)

MIX	Warm-pressed	PC	MDF
OPC	100	100	100
Sand	275	275	—
Styrene butadiene	—	8 - 12	—
Polyacrylamide	—	—	10
Glycerin	—	—	0.5
Water/Cement	0.20	0.32 - 0.40	0.18

Porosity and specific surface area measurements

Total porosity — The values of total pore volume can calculated from the differences of bulk and theoretical densities if latter were known; however, due to the complexity of the cement-based system theoretical densities are uncertain. Instead, total porosity of hardened cement paste was measured by the water-replacement method after drying at 100 °C for 4 hours, because it is assumed that free water will be eliminated. Total porosity, ε was defined by the following equation.

$$\varepsilon = \frac{Ws\text{-}Wd}{Vt}$$

Wd and *Ws* are the weights of the dry and water-saturated samples. *Vt* is the total volume of sample. The equation is simplified from Roy and Gouda's original one[9]; here *Vt* is measured accurately.

Apparent porosity — It is defined by the water-replacement method (Archimedes method); the weight of the dry and wet specimen and its weight in water were determined. This terminology adopted by ASTM C-20 has a different meaning from total porosity, because its measurement procedure does not include drying at 100 °C for 4 hours.

Mercury intrusion porosity — This method is based upon the incremental movement of mercury into the porous microstructure, and the results are interpreted in terms of a simple cylindrical uniform cross-section model. This method can characterize open pores of very small diameter (down to 3 nm)

Image analysis — The magnification of (135x) was used, and the surface of each individually analyzed field was equal to 2.67 mm², and all 144 measuring fields covered the total surface of 3.85 cm² of each examined sample. Pore volume was measured on binary (black and white) images, that were taken directly from the monitor of the image analyzer. This optimum microscopic magnification was set up by Konsztowicz and Boutin[8] to characterize refractory materials. This magnification was considered reasonable for mortar samples.

Diffusion and other tests

The steady state diffusion test which was originally set by Hansson and Berke[10] was carried out by an ionAnalyzer ORION EA 920 interfaced to a computer. Shrinkage behavior was studied by an electro-mechanical linear variable differential transformer (LVDT). The displacement of 50x50x350 mm dimension specimens was measured and the signals were also recorded by a computer. The experimental set-up is described in some detail elsewhere[11].

RESULTS AND DISCUSSION

Microstructure: Its influence and evolution

Order of magnitude: Defining the issues — The term "microstructure" is commonly used with different meaning in different disciplines, when it is defined as the description of the individual phase (including pores and dislocations) in terms of shape, size, crystallographic orientation, and position in the material. It provides a background for understanding the macro- and micro-structures observed in cement-based materials. For most utilizations the important structural elements range from the centimeter down to the nanometer scale. This description is made using visual inspection, optical microscopy, scanning electron microscopy, and transmission electron microscopy. Micrography alone may give an excellent nm-mm-scale description, but just one overlooked millimeter- to centimeter-sized pore or rock vein may dramatically affect the materials' properties. In cement-based materials the properties are dependent on the distribution of porosity. Isolated nm-sized regions in the cement hardly affect the strength, but toughness is enormously increased. In Fig. 1 the microstructure of cement-based materials is summed up at three levels: 1) millimeter; 2) micrometer; 3) nanometer. In the past half century, several generations of scientists

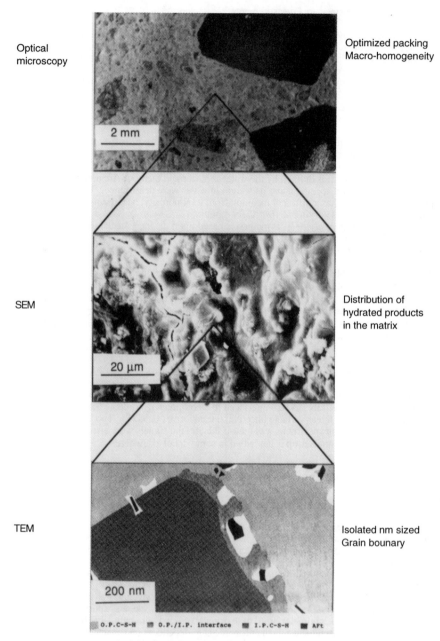

Fig. 1 Cement-based materials: description of the structure from the centimeter scale down to the nanometer scale. TEM schematic diagram is from Richardson and Groves[12].

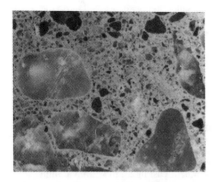

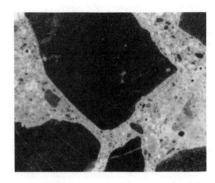

a) Gravel concrete w/c = 0.32 including
superplasticizer

b) Limestone concrete w/c = 0.37 including
fly ash

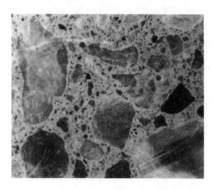

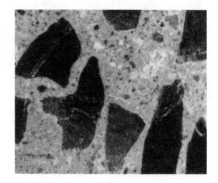

c) Gravel concrete w/c = 0.33 including
slag and silica fume

d) Limestone concrete including fly ash and
silica fume (w/c = 0.36)

Fig. 2 Optimized concrete via particle packing at magnification of (12x)

and engineers have contributed to our understanding at all levels. Understanding microstructure phenomena is crucial in order to develop high performance cement-based materials.

Macro-defect-free mechanism — It is generally accepted that the mechanical properties of porous cement-based materials depend on macro structural defects. Birchall *et al.* (1981) described so-called MDF cements with an absence of large void (pores with diameter greater than 90 μm)[7]. However, this standard for mass concrete is hardly attainable. Neville (1992) pointed out that the issue is not the scale of observation but rather with the scale of application[13]. Parkhouse and Sepangi (1994) recently defined "Macromaterials" which as yet are not given the recognition they deserve[14]. Some highway concrete packing images at the millimeter level (Fig. 2) a give clear picture of the results of processing of optimized concrete via particle packing which was defined by Roy *et al.* (1993)[15]. These materials of samples were from field experiments in the Strategic Highway Research Project (SHRP 201), for which processing is optimized in the laboratory. The 28-day compressive strengths of four samples of concretes all reach 80 MPa. The experiments conducted by SHRP 201 show that the application of packing strategy has its profound effects on concrete properties, and the following approaches work on either the dense mixture of coarse aggregates or cement-sand-aggregate, as well as do the cement-silica fume-fly ash mixtures. It is concluded that both the workability of the fresh materials, and the microstructure development are controlled to a considerable extent by these packing geometric parameters.

Porosity and density

The relation between strength and the total volume of voids is not a unique property of concrete but is found also in other brittle materials in which mixing water leaves behind pores (e.g. plaster). Evidently, if the strength of different materials is expressed as a fraction of the strength at a zero porosity, many studies conformed to the same relation between relative strength and porosity. In concrete practice the optimum air void system improves the resistance to freezing-thawing conditions. Rashed and Williamson (1991) studied microstructure of entrained air at different ages (5 min to 60 days) using small specimens cast in 2 mm x 2mm x 3mm mold[16]. They found in the case of a high w/c (0.49), the air void shells were porous; even after 28 days, some of the spaces adjacent to the air void were empty. On the other hand, when a low w/c (0.29) was used, the air-void shells were solid and sound, the interface was filled with hydration products. The major difference in microstructure between a normal and high-performance cement paste is the greater density of the microstructure in the latter. Elimination of defects in the paste by warm-pressing could be achieved. A warm pressed mortar (2-in cube) made by us using a low w/c = 0.20 compared to w/c = 0.45 is shown in the Fig. 3. Table II gives the values of total pore volume, apparent porosity, porosity by mercury porosimetry, and total area fractions of pores evaluated by the image analysis technique. Three types of mortar samples are listed: warm pressed, with Superplasticizer, and normal casting. The image analysis value is higher than those determined by other techniques, as found also by Konsztowicz and Boutin[8] (interested readers please see their interpretation). Detailed discussion is outside of the scope of this paper, and will be reported elsewhere. Winslow et al. (1994)[17] showed that additional porosity of mortar and concrete occurs in pore sizes larger than the paste's, and they also proposed a computer model of pore structure.

Superplasticizer effect

The advent of superplasticizers has contributed conspicuously to development of high performance concrete. Ozawa et al. suggested that high performance concrete for a certain meaning could been defined as a concrete with high filling capacity[18]. The optimum mix proportion of superplasticizers was clarified for achieving concrete with high filling capacity. From microstructural aspects, this kind of capacity enables concrete to fill all corners of formwork, and realize dense and homogeneous concrete. Fig. 2 (a) concrete is a case using superplasticizer to obtain high performance. Also from Table II, it is shown that when superplasticizer was used, porosity was reduced.

Table II. Porosity of analyzed mortar[#] specimens

| | Porosity (%) | | |
	warm pressed* w/c = 0.20	Superplasticizer** w/c = 0.28	Normal casting w/c = 0.35
Total porosity	13.5 ± 1.7	21.2 ± 1.5	24.8 ± 1.4
Water displccement	10.3 ± 2.3	17.3 ± 1.6	19.2 ± 1.6
Mercury porosimetry	11.6	19.5	20.4
Image analysis	17.0 ± 2.1	25.1 ± 1.8	28.9 ± 2.4

[#]cement:sand = 1:2.75, * 120°C, 2 hours, pressure at 30 MPa; **4% Mighty 150 by weight

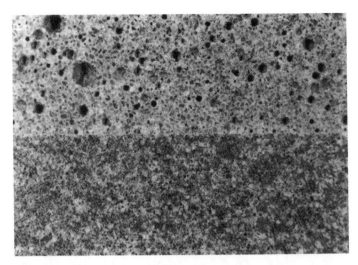

Fig. 3 Comparison between warm-pressed mortar and conventional at magnification of (12x)

Silica fume influences on microstructure

Roy (1989)[19] summed up the effects of silica fume on the microstructural development, and concluded the ultra-fine particle size of silica fume brings the potential of being much more reactive than other supplementary cementing materials. Silica fume particles when properly dispersed fill the interstices of the fresh cement, where they are available to react with the alkali hydroxide and $Ca(OH)_2$ liberated by the hydrating portland cement, forming insoluble C-S-H, and tend to intiate nucleation of C-S-H on the surface. Khayat and Aïtcin (1992)[20] reviewed the beneficial actions of silica fume particles, indicating that it has also been attributed to the great improvement in the microstructure of the hydrated cement paste in the transition zone from the coarse aggregate in concrete. The micrographs near the cemfil fiber in fiber-reinforced MDF cement composites in Fig. 4 and Fig. 5 were an attempt to compare the transition zone of fiber with or without 10% silica fume[21]. Fig. 5 was taken after tensile strength test, and showed crack propagation and fracture surface. The experiments also showed that fiber reinforced composites containing 10% silica fume exhibited high toughness.

Polymer effect on properties related to durability

Mikhailov *et al.* (1992) summed up that the polymer-modified concrete has the following positive effects on microstructure[22]. Firstly, the polymer fills tiny pores and flaws, increases density, improves the cement-aggregate bonding, and reduces stress concentrations. Further, the polymer filler sharply reduces shrinkage, while the modulus of elasticity is greatly enhanced. This permits the use of such concrete in heavily load-bearing and other vital structural members. The Fig. 6 shows that when the polymer latex was used, the shrinkage of the paste was reduced. Fig. 7 illustrates two things: firstly, adding polymer reduces diffusion; secondly, increasing slag content also reduces diffusion. Bakker (1983)[23] explained the beneficial influence of slag, proposing that in cement/slag mixtures, the difference in chemical potentials between the clinker and slag results in the pozzolanic reaction taking place, so the pore structure is refined, and the ease of diffusion and permeation is correspondingly reduced.

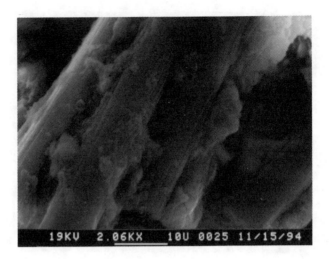

Fig. 4 SEM micrographs near the cemfil fiber-reinforced MDF cement composite

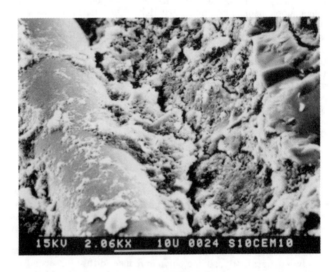

Fig. 5 SEM micrographs near the cemfil fiber-reinforced MDF cement composite containing
10 % silica fume after tensile strength test[21]

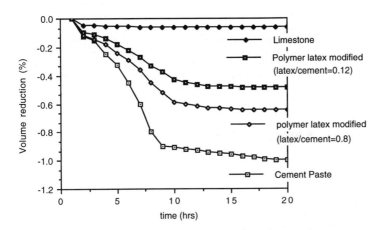

Fig. 6 Shrinkage comparison between the polymer-modified and OPC paste

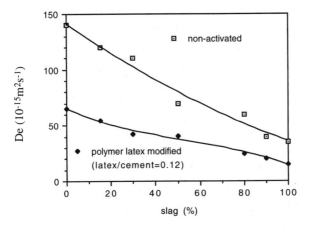

Fig. 7 The effective chloride diffusion coefficient (De) of paste plotted as a function of the content of slag additive

CONCLUSIONS

As discussed above, the following conclusions have been made about the microstructural aspects of high-performance cement based materials.

1. The goal of microstructure research is to understand the mechanisms of formation of such materials in order to create a structure yielding a material with specific physical, mechanical, chemical and durability properties.

2. The information from microstructure should be utilized fully to develop high performance cement-based materials. To create microstructures leading to desirable engineering properties and meaningful codes one must establish a materials science knowledge base.

3. Microstructural aspects must be considered in the design phase of new products as the most common means to solve the problems encountered in high-performance cement-based materials production, yet this information will also be needed for control over materials processing and properties development to approach goals for the next century's materials.

ACKNOWLEDGMENT

This research was partly support by the U.S. National Science Foundations grant MSS-9123239.

REFERENCES

1. Y. Malier, in *High Performance Concrete: from material to structure*, edited by Y. Malier (E & FN Spon, London, 1992) p. xiv.
2. D.M. Roy and R.I.A. Malek, in *Mineral Admixtures in Cement and Concrete*, edited by S.N. Ghosh, S.L. Sarkar and S. Harsh (ABI Book Pvt. Ltd. New Delhi, India, 1993) p. 84.
3. A. Kelly and N.H. Macmillan, *Strong Solids*, 3rd ed. (Clarendon Press, Oxford, 1986) p. 57.
4. R.W. Davidge, Mechanical Behaviour of Ceramics (Cambridge University Press, Cambridge, 1980) p. 1.
5. T.W. Chou, *Microstructural Design of Fiber Composites* (E & FN Spon, London, 1992) p.7.
6. D.M. Roy and W. Jiang, in *SCIENTIFIC BASIS FOR NUCLEAR WASTE MANAGE-MENT XVIII* (Kyoto, JAPAN, on OCT. 23-27, 1994, in progress)
7. J.D. Birchall, A.J. Howard and K.Kendall, Nature, **289**, 388-389 (1981).
8. K.J. Konsztowicz and J. Boutin, J. Am. Ceram. Soc. **76**, 1169-76 (1993).
9. D.M. Roy, G.R. Gouda, and A. Bobrowsky, Cem. Conc. Res., **2**, 367-370 (1972).
10. C.M. Hansson and N.S. Berke, in *Pore Structure and Permeability of Cementitious Materials*, edited by L.R. Roberts and J.P. Skalny (Mater. Res. Soc. Proc. **137**, Pittsburgh, PA, 1989) pp. 253-270.
11. M. Perez, Masters thesis, Pennsylvania State University (1981).
12 I.G. Richardson and G.W. Groves, J. Mater. Sci., **28**, 265-277 (1993).
13. A. Neville, Cem. Concr. Res. **22**, 1067-1076 (1992).
14. J.G. Parkhouse and H.R. Sepangi, in *Building the Future: Innovation in design, materials and construction*, edited by F.K. Garas, G.S.T. Armer, and J.L. Clarke (E &FN Spon, London, 1994) pp. 3-13.
15. D.M. Roy, B.E. Scheetz, and M.R. Silsbee, MRS BULLETIN/MARCH (1993) pp. 45-49.
16. A.I. Rashed and R.B. Williamson, J. Mater. Res. **6**, 2004-12; 2474-82 (1991).
17. D.N. Winslow, M.D. Cohen, D.M. Cohen, D.P. Bentz, K.A. Snyder and E.J. Garboczi, Cem. Concr. Res. **24**, 25-38 (1994).
18. K. Ozawa, K. Maekawa and H. Okamura, in *Admixtures for Concrete: Improvement of Properties*, edited by E. Vazquez (Chapman and Hall, London, 1990) pp. 51-62.
19. D.M. Roy, in *Proc. 3rd Int. Conf. on "The Use of Fly Ash Silica Fume, Slag and Natural Pozzolans in Concrete"* (ACI SP-114, 1989), Vol. I, pp. 117-129.
20. Khayat and Aïtcin,in *Proc. 4th Int. Conf. on "The Use of Fly Ash Silica Fume, Slag and Natural Pozzolans in Concrete"* (ACI SP-132, 1992), Vol. II, pp. 835-872.
21. C. K. Park, Ph. D. thesis, Pennsylvania State University (1993).
22. K.V. Mikhailov, V.V. Paturoev, V.V. and R. Keris (1992). *Polymer Concretes and Their Structural Uses*, Russian Translations Series 91 (A.A. Balkem, Rotterdam, Brookfield), p.1.
23. R.F.M. Bakker, in Proc. 1st Int. Conf. on *"The Use of Fly Ash, Silica Fume, Slag and other Mineral Byproducts in Concrete"* (ACI SP-79, 1983) pp. 588-605.

IMAGE ANALYSIS OF FLY ASH IN THE CHARACTERIZATION OF THE SHAPE OF THE GRAINS

M. BARRIOULET, H. CROS, B. HUSSON, E. RINGOT
LMDC INSA/UPS Génie Civil,complexe scientifique de Rangueil 31077 Toulouse FRANCE

ABSTRACT

Fly ash from power stations is used as concrete additive to improve strength and durability. Surprisingly, studies of ashes of identical mineralogical composition from two different places have reported different results in terms of the rheological properties of the fresh material. The viscosity of the pastes made from these different fly ashes seems to be linked to the proportion of spherical and smooth-shaped grains found in them. A quantitative image analysis was carried out to characterize the shape of the grains of these two ashes from different geographical origins. The main result proves that the higher the glassy particle content of the fly ash, the more the hydraulic matrix is fluid.

INTRODUCTION

Construction engineers had planned to use fly ash from the Le Havre (France) thermal power station, as an additive to concrete. However just before getting down to work, the fly ash from Le Havre became unavailable. They were advised to use fly ash from the Cordemais (France) power station since it produced the same type of silico-aluminous ash (class F) [1] [2]. They soon noticed that the consistency of the Cordemais fly ash based concrete was completly different from the Le Havre type. The former concrete proved to be too dry , so they abandoned the fly ash option and decided on another solution.

The present researchers decided to find out why this phenomenon occured. The mechanism governing the flow of a powder mixed with water depends mainly on the surface forces resulting from the electric state of the environment. The mineralogical nature of the surfaces and the specific area of the powder are the main parameters inducing the rheological behaviour of the paste. But it was observed that, besides the same mineralogical composition (Figure 1), the grain size distribution (Figure 2), the chemical composition (Table 1) and the specific area (Table 2) were nearly the same. But in point of fact two different rheological behaviors were noticed, when the fresh paste flowed through a nozzle such of the Marsh cone type, the Le Havre fly ash based paste being more fluid (Figure 3).

We started a study to define whether the morphology of the grains could be the main cause of the phenomenon. Image analysis was used because a simple microscopic observation cannot determine which one of the two powders contains more glassy grains or smooth-shaped grains which could flow easier than rough-shaped grains. In this paper we present the procedure worked out to obtain binary images of grain boundaries. Subsequently we present the tools we chose from the range of those already known, of shape analysis. Finally we present the limited results of the global investigation that we are carrying out and that is for the present still incomplete.

Table 1: Chemical Analysis

Constituents	Le Havre	Cordemais
SiO_2	54.00	53.00
AlO_3	28.00	29.50
Fe_2O_3	5.75	5.85
CaO	2.20	4.25
MgO	1.00	1.08
Na_2O	0.53	0.43
K_2O	3.15	1.70
equivalent Na_2O	2.60	1.55
SO_3	0.25	0.50
Cl^-	<0,001	<0,01
Loss	0.89	0.92
Free Lime	-	-
Loss of Ignition	4.20	2.70

Table 2: Physical Analysis

Test	Le Havre	Cordemais
Density (t/m^3)	2.22	2.28
Fineness Blaine (cm^2/g)	3810	4475

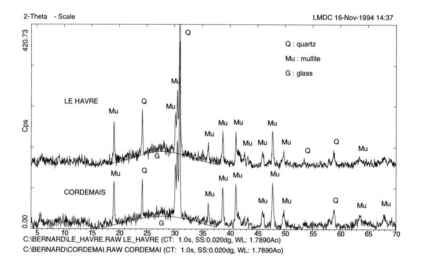

Figure 1: X-ray Diffractograms

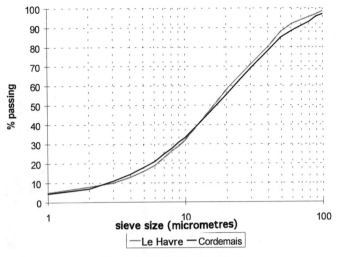

Figure 2: Grain Size Distribution

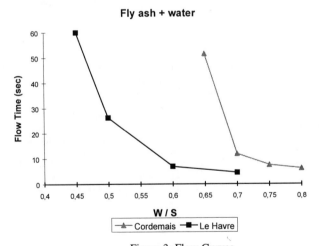

Figure 3: Flow Curves

METHODOLOGY

The procedure to obtain the parameters needed to characterize the shape of fly ash grains can be divided in three parts:
- image acquisition.
- grain boundary extraction.
- shape analysis.

Acquisition

A few drops of a low concentratred blend of water and ash were spread on an optical microscope slide. A super plasticizer was added to the blend so as to avoid contacts between the particles, as we wanted to examine the shape of each grain. Then the preparation was protected by a coverglass. The sample was observed under microscope with a light beam emanating from underneath so as to display the shadow of the particles (Figure 4). A matrix CCD camera was attached to the microscope. The video signal generated by the camera was sampled by a video frame grabber with the electronic card interfaced with a PC. The resolution is 512 by 512 pixels and the scan frequency 10 Mhz, the ratio being: x/y=1.486. The digitized images were transferred to a unix workstation through the local network to be processed and analysed.

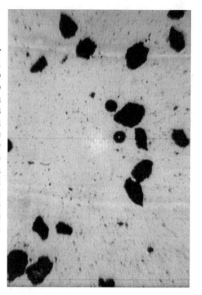

Figure 4: Initial Image

Grain Extraction

Image segmentation was the most important stage in the treatment, because the results of the measurement of the morphological parameters are directly linked to the accuracy of this operation. In histogram based segmentation, the quantity of the information depends on the threshold value, so we prefered an edge detection process based on the morphological gradient rather than an arbitrary process, even for highly contrasted images such as those under study. A combination of mathematical morphological tools allowed the detection of the edge of the objects, which was the largest grey tone variation areas. The major disadvantage of the gradient was the detection of all grey tone variations, consequently use of the gradient based method had to account the areas of direct interest [4]. These areas, i.e grains, are automatically marked by a function based on the minima detection of the grey tone function [4]. When there were several particle contacts, we resorted to a water shade based operation to separate the grain sets.[3]

Study of Shapes

The scope of the study was to differentiate smooth round particles, from the rough and irregular particles. Two aspects were considerated:
- relative roughness (whether smooth or not).
- shape (whether round or not).

For each grain, both aspects were analyzed. In order to test and control the performances of the algorithms implemented to characterize both aspects, we used a typical image (Figure 5) composed of spherical particles, rough-shaped particles and any other particles among the range of different grains contained in the fly ashes that we investigated.

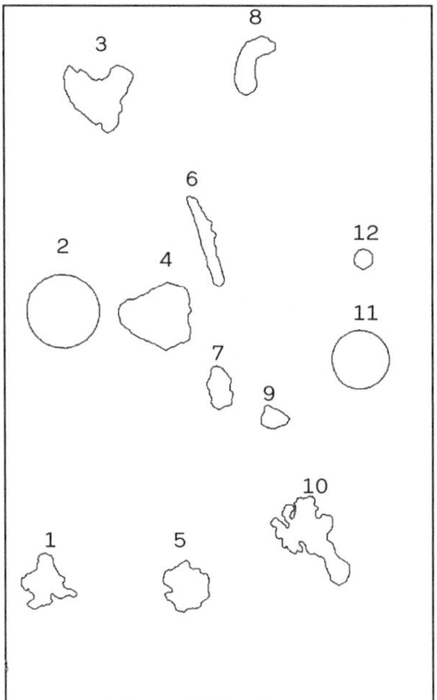

- Roughness

Information about the means of characterizing the roughness of an object are available in the literature, the most obvious in our case seems to be the fractal dimension of the contour. Indeed, Richardson describes the devolopment of the measurement of the profile, which is a function of the size of the step, as a straight line in a log-log diagram, if the material exhibits a self similarity. This means we can observe bumps on the surface of the object whatever the scale. We plotted the fractal curves by using the Minkowsky method. This method is based on a morphological opening operation.The logarithm of adimensional parameter L is the function of the logarithm of the size of the opening operation.

Figure 5: Test Image

$$L = 1 + \frac{P(O^{\lambda B}(X))}{P(X)} \qquad (1)$$

where $O^{\lambda B}(X)$ is set X opened by the structuring element B the size of which is λ, $P(X)$ is the perimeter of X, and X is the boundaries of the object. By means of a linear regression we obtained a straight line where the slope is d such as D=d+ 1 is the fractal dimension of the object:

$$d = \frac{Ln(L)}{Ln(\lambda)} \qquad (2)$$

Thus d is the factor we retained to characterize the roughness of the grains. We can actually confirm that this parameter evaluated up for the particles of the test image, is satisfactory (Figure 6): for a spherical particle, d tends towards zero while it is more as the grain is rougher.

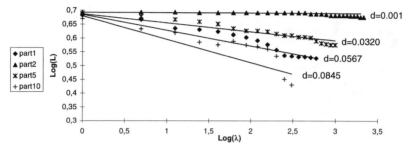

Figure 6: Fractal Curves

- Shape

The literature contains several factors concerning the charaterization of shape [5]. But we felt we had to test them to establish wether they were well-adapted to the sort of particles we were examining. We decided to use three of them:

a) The lengthening index, which was mainly used in our study, the value is between zero and one, but note that the rounder the grain, the closer to one the value is.

$$IA(X) = \frac{Ri}{Re} \qquad (3)$$

where Ri is the maximum inscribed circle radius and Re is the minimum circumscribed circle radius.

b) The ratio of the inscribed circle to the corresponding convex surface, the value is between one and zero.

$$IR_a(X) = \frac{\pi(Ri)^2}{A(C_v(X))} \qquad (4)$$

where $A(Cv(X))$ is the surface of the convex hull of object X.

These factors are stricly for convex particles so in our case, we investigated the convex hull of each grain. $Cv(X)$ is set X closed by a an infinite structuring element.

c) The concavity index in surface and in perimeter which may be useful to get an idea of the concavity of the grains.

$$IC_a(X) = \frac{A(X)}{A(C_v(X))} \qquad (5)$$

$$IC_1(X) = \frac{P(C_v(X))}{P(X)} \qquad (6)$$

where $A(X)$ is the surface of object X. The values of these factors are those for the test image we introduced above (Table 3):

Table 3: Test Image Values

Particle	Area	Perim	Convex Area	Convex Perim	Re	Ri	IA	IRa	ICa	ICl
1	2054	213	3396	184	37	17	0.46	0.27	0.60	0.86
2	5502	243	6029	243	46	38	0.83	0.75	0.91	1.00
3	3750	253	5652	237	49	23	0.47	0.29	0.66	0.94
4	4448	233	5024	227	48	32	0.67	0.64	0.89	0.97
5	2167	178	2550	158	30	21	0.70	0.54	0.85	0.89
6	1547	219	3635	216	58	8	0.14	0.06	0.43	0.99
7	1096	120	1350	118	27	13	0.48	0.39	0.81	0.98
8	1638	167	2542	163	38	12	0.32	0.18	0.64	0.98
9	597	84	675	84	18	11	0.61	0.56	0.88	1.00
10	3854	392	6432	263	60	16	0.27	0.13	0.60	0.67
11	3420	192	3745	192	36	30	0.83	0.75	0.91	1.00
12	395	64	420	64	13	10	0.77	0.75	0.94	1.00

We noted that the value of IA is around 0.8 for spherical particles such as particles 2, 11 and 12. As can be seen, in practice this value is not 1.0, because of the lattice of the image. In fact the inscribed and the circumscribed circles are more like hexagons, when the lattice is hexagonal. The value of IA is a bit less for a rough but round particle, for example particle 1. The value is low for a non-spherical and rough grain such as particles 1 or 10. But the value is definitly lower for a lengthened object like particle 6.

Obviously the IRa factor varies in the same way because it is partly based on the same measurement as IA, the inscribed circle.

We therefore maintained the IA factor as the main parameter to characterize the shape of a grain of fly ash.

Thus, the ICa factor could be considered as an extra item of information about the concavity.

GLOBAL INVESTIGATION

Having demonstrated what kind of tools are available and adapted to the shape characterization of the type of grains we found in the fly ashes we endeavoured to investigate the final step was to use these tools to compare both powders. For the time being the complete study has not been realized, and we hope to present the results of this comparative study in a forthcoming paper. Nevertheless, we are now able to show at least a part of these results. Firstly, we must specify that the grains of the fly ash were sifted into three fractions: 0-40μm, 40-80μm and plus 80μm. In this manner when the dispersed particles were observed the focus plane was easily selected because the size of the objects were not too different. Grains shape was investigated from a hundred images for each section. In this paper only the results of two size ranges are given. These concern the sections of biggest grains; we are currently measuring the last grade of the smallest particles. Nevertheless the results we are editing today are of great interest because the majority of the smallest particles are made up of glassy grain (i.e. spherical). We are therefore convinced that the most of the irregular particles are found in the biggest grains section (>80 μm and 40-80μm).

Shape Factors

We assessed the mean value for both size ranges (Table 4). It can be seen that shape factors IA and IRa are higher in the Le Havre ash Table than in the Cordemais Table. The grains from Le Havre tend to be rounder than those from Cordemais, whatever the size. However there is not a huge difference between them. And one also notes that the difference is smaller for the 40-80μm section.

Table 4: Mean Values

Cordemais										
Size Range	Number	Area	Perim	C Area	C Perim	Re	Ri	IA	IRa	ICa
>80	783	2291	156	2791	152	32	18	0.60	0.51	0.84
40-80	1082	1584	141	16382	129	42	16	0.55	0.45	0.79

Le Havre										
Size Range	Number	Area	Perim	C Area	C Perim	Re	Ri	IA	IRa	ICa
>80	484	3427	200	4121	195	40	25	0.65	0.56	0.84
40-80	1756	1366	121	1645	116	25	14	0.58	0.49	0.83

Furthermore, if we plot the histogram of the distribution lengthening factor, it provides more information about the proportion of glassy particles (Figure 7).

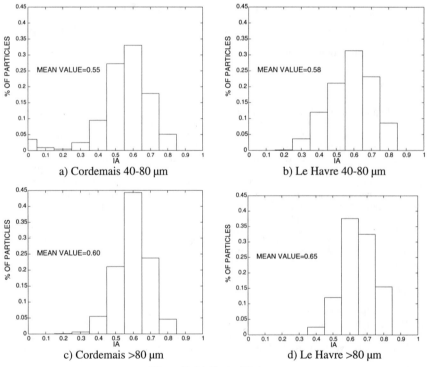

Figure 7: IA Factor Histograms.

In all the histograms the mode value is the same: 0.60, but they are more or less symmetrical. In the plus 80 micron section, the Cordemais histogram is more symmetrical than that of Le

Havre. There are more round particles in the Le Havre plus 80 micron type because the histogram is concentrated more to the right (Figure 7b) than that of Cordemais (Figure 7a). While the Cordemais ash histogram in the 40-80 microns section (Figure 7c) is less symmetrical than that of Le Havre ash (Figure 7d). this means there is a small fraction of lengthened particles which does not flow as efficiently as round particles.

CONCLUSION

We have defined procedures among the range of shape analysis tools which are best adapted to the material we want to be investigated. Thanks to these tools we will be able to characterize each fly ash quantity according to the morphology of the grains it contains. In our case, so far, the partial results prove that the different rheological behaviour of both pastes is linked to the proportion of spherical grains in them. However, only the complete study will confirm this conclusion definitively.

Due to the method we implemented, it is possible to investigate huge field of particles in an objective way. This could not be done manually grain by grain. Nevertheless this method acknowledges certain limits: the smaller the grain is, the harder the grain extraction becomes. Because surface forces are important at this size, it is difficult to separate physically the sets of grains. But is the experimental flocculation state very different from the real flocculation state *in situ*? If not , it may be more logical to examine the shape of the set, instead of the shape of each separated grain. In any case it can be concluded that surface force effects on the rheology are less important than the grain morphology effects, when the grains are of sufficient size.

REFERENCES

1. P. Schiessl, R. Hardtl The Change of Mortars Properties as Result of Fly Ash Processing, 3th Int. Conf. Fly Ash, Silica Fume, Slag, and Natural Pozzolans in Concrete, Trondheim, Norway (ACI, Detroit Michigan, 1989) edited by V. M. Malhotra, pp. 274-294
2. A. Carles-Gibergues and B. Husson Effects of the Use of Sulfitic Fly Ash in mortars and Concretes, 4th Int. Conf. Fly Ash, Silica Fume, Slag and Natural Pozzolans in Concrete, Istambul, Turkey, (CANMET ACI, Detroit, Michigan, 1992) edited by V. M. Malhotra, p. 671.
3. S. Beucher,Ph.D. thesis, centre de morphologie mathématique, école des Mines de Paris, 1990.
4. J.F. Rivest, S. Beucher, J.P. Delhomme Marker-Controlled Picture Segmentation Applied to Electrical Logging Images,(Proceeding SPIE, Vol 1451 Non linear Image Processing II, 1991)
5. M. Costner, J.L. Chermant, Précis d'analyse d'images, (presses du CNRS, Paris, 1989), p. 290.

MICROSTRUCTURE AND MICROCHEMISTRY OF CALCIUM SULFOALUMINATE CEMENT

Ö. ANDAÇ AND F.P. GLASSER
University of Aberdeen, Aberdeen AB9 2UE, Scotland.

ABSTRACT

Calcium sulfoaluminate, $4CaO \cdot 3A\ell_2O_3 \cdot SO_3$, forms the basis of a new class of structural cements. Microstructural examination of commercial clinker is not easy because it is quite friable. However, backscattered imaging of impregnated clinker is successful and reveals an unusual pattern. The clinker has a sponge-like texture connecting better-sintered, denser regions. The microstructure is, however, very similar between regions of differing density. Irregular to equant, well-shaped crystals of $4CaO \cdot 3A\ell_2O_3 \cdot SO_3$ are enclosed in a matrix composed mainly of Ca_2SiO_4 together with a little titanium-rich, ferrite-like phase. The minor phases occluded within ferrite include oxide phases one of which concentrates Pb and Bi; the other, Cr, Fe and Ni. Other minor oxide phases in the matrix include lime, periclase, gehlenite and calcium sulfosilicate and probably bredigite. The crystal chemistry and elemental analyses of the individual phases are presented.

INTRODUCTION

Calcium sulfoaluminate, $4CaO \cdot 3A\ell_2O_3 \cdot SO_3$ or "$C_4A_3\bar{S}$", has long been known to have cementitious properties. Recent experience in China shows that by controlling the clinker mineralogy, a range of dimensionally-stable, cementitious matrices can be formulated. These may replace Portland cement in many constructions or find application in special cements. However, relatively little is known about the mineral composition and microstructure of sulfoaluminate clinkers and of the resulting cement hydration products. Muzhen, et al [1] have presented SEM fractographs of burnt clinkers and claimed that the CaO content of "$C_4A_3\bar{S}$" was variable on "a relatively large scale".

This paper summarises several years of work undertaken in our laboratories on this interesting and novel class of structural materials. Studies of hydration properties will be reported elsewhere [2].

CLINKER COMPOSITION AND MICROSTRUCTURE

Much of the clinker is open-structured and friable, perhaps as a result of the relatively low firing temperatures which are reported; ca 1200°-1300°C. Despite local differences in texture, coherence, colour and apparent bulk density, the clinker is chemically quite uniform on a gram scale: generally within ±0.5 - 0.6% with respect to its major constituents . Table 1 presents an averaged analysis by XRF on a dry weight basis: loss on ignition was variable but generally low, in the range 0.08 - 0.3%. With respect to the minor constituents of Portland cement, sulfoaluminate clinker is very low in alkali but relatively high in TiO_2.

The microstructure of the clinker is shown in Figs. 1-4. Figure 1, at low magnification shows the lace-like open-textured structure of the more friable clinker lumps.

Mat. Res. Soc. Symp. Proc. Vol. 370 © 1995 Materials Research Society

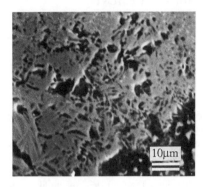

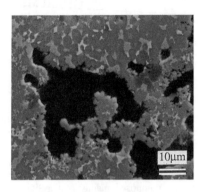

Fig. 1. Backscattered electron micrograph (B.e.m.) of clinker polished section showing lace-like, open textured clinker structures.

Fig. 2. B.e.m. of clinker polished section showing pores surrounded by better-sintered clinker. The cluster of dark grains (right center) is periclase, MgO.

Fig. 3. B.e.m. of clinker polished section. A nest of belite-ferrite core with a little sulfoaluminate, *ca* 30μm, is surrounded by a more sulfoaluminate-rich zone. Note ferrite (bright) and occasional periclase (dark).

Fig. 4. B.e.m. of clinker polished section, showing region of exaggerated grain growth of belite. Partial local melting may have occurred in these zones.

Table I. CHEMISTRY OF SULFO-ALUMINATE CLINKER	
oxide	wt %
SiO$_2$	4.6
TiO$_2$	1.5
Aℓ_2O$_3$	36.5
Fe$_2$O$_3$	2.7
MnO	0.02
MgO	2.5
CaO	42.0
Na$_2$O	0.08
K$_2$O	0.07
P$_2$O$_5$	0.12
SO$_3$	9.6
Total	99.7

This porous structure undoubtedly contributes to the ease of grinding sulfoaluminate clinker and its subsequent reactivity during hydration. Figures 2 and 3 show the microstructure of rather better sintered regions of clinker.

Backscattered electron images at higher magnifications, coupled with analytical microscopy, disclose some of the mineralogical features. The well shaped, equant gray crystals are calcium sulfoaluminate. These are embedded in a matrix consisting mainly of two phases: belite (light gray) and ferrite (bright). Occasional dark gray regions with characteristically indefinite, rather fuzzy margins are also encountered: these are either periclase, MgO, or infrequently, intimate mixtures of free lime and periclase. It is believed that these are relict from former dolomite CaMg(CO$_3$)$_2$ grains. Other evidence of relicts are common but less obvious. Figure 3 shows a region in which a core consisting of belite, ferrite and some sulfoaluminate is surrounded by a sulfoaluminate-rich envelope. The core mineralogy presumably arises from a sulfate-poor region. Figure 4 shows a relatively coarse belite-rich cluster, again believed to be a relict region, perhaps also developed from and around a former sand grain or grain cluster. The extent of local crystal growth and the open margins between grains are interpreted as indicating that partial melting occurred in this zone. Since no remains of former liquid are evident, presumably the liquid either evaporated or drained away to react and/or freeze elsewhere. There are however only occasional indications of melting and, in general, clinker textures and microstructures are indicative of formation by solid state reaction. If liquid did develop at peak clinkering temperatures, it was probably transient and confined to local hot spots and inhomogeneities.

The compositions of the main constituent phases determined from electron microprobe analyses, using a wavelength dispersive system, of clinker polished sections are shown in Table 2. The full analytical data will be presented in a Thesis [3]; to save space, only arithmetic means of the analyses are presented. To facilitate recalculations of mineral compositions, the analytical data have been recalculated to a fixed number of oxygens: e.g. 4 for Ca$_2$SiO$_4$. The standard deviation, δ, is also shown for each data set: footnotes give the total number of analyses.

The Ca$_2$SiO$_4$ contains significant quantities of Aℓ, Ti, P and S. The best chemical balance, to achieve a ratio MO$_4^{4-}$: M^{2+} $\approx$2, necessitates including part of the Aℓ and all of the Ti, P and S as MO$_4$ groups: the remainder of the Aℓ presumably charge balances elsewhere in the structure. It is of course possible that, despite precautions, the analyses arise from phase mixtures: this is discussed subsequently and supplementary calculations are presented.

The majority of belite grains were essentially Mg free. However, a small but significant population of higher Mg belite analyses were found. Electrostatic balance is of course maintained if Mg simply replaces Ca. But the resulting ratio of Ca:Mg in the high Mg suite is close to that of bredigite, as defined from synthetic studies [4,5]. Bredigite, it will be recalled, is very close to Ca$_7$Mg(SiO$_4$)$_4$. Its X-ray powder diffraction pattern is virtually indistinguishable from that of α' Ca$_2$SiO$_4$. However the distinction is important: while α' Ca$_2$SiO$_4$ is reactive with water, bredigite is virtually inert. The presence of bredigite in this clinker is not proven - the grains were too few for accurate statistics and too small for confirmatory electron diffraction - but it is almost certainly present in small amounts.

The analytical totals for calcium sulfoaluminate are quite satisfactory and the general level of substitution is low: some Fe^{3+} replaces $A\ell$ which, together with other minor substituents, brings the atomic population to nearly 6.0 in the M^{3+} site. There is no evidence of significant Ca variation from ideal stoichiometry as was claimed previously [1].

The ferrite-like phase exhibits much the largest standard deviations in analytical totals: moreover, these fall short of 100% when converted to oxides using a conventional ferrite formula, $Ca_4(A\ell,Fe)_2O_{10}$. Difficulties in reconciling the analyses of this phase are discussed subsequently. The ferrite is a notable concentrator of Ti, Mg and Mn.

Table II. Mean Composition of Phases in Calcium Sulfoaluminate Clinker

Ca_2SiO_4[(1)]

	Na	K	Mg	Ca	$A\ell$	Fe	Si	Ti	P	S
	<0.01	<0.01	<0.01	1.96	0.11	0.01	0.77	0.04	0.04	0.08
δ	-	-	-	0.06	0.07	0.00	0.05	0.01	0.01	0.02

$Ca_4A\ell_6O_{13} \cdot SO_3$[(2)]

	Na	K	Mg	Ca	$A\ell$	Fe	Si	Ti	P	S
	0.02	0.03	0.00	4.11	5.60	0.12	0.04	0.02	0.05	1.03
δ	0.04	0.02	0.00	0.04	0.12	0.05	0.02	0.02	0.01	0.05

Ferrite-like phase [(3)]

	Na	K	Mg	Ca	$A\ell$	Fe	Si	Ti	P	S
	0.01	0.03	0.09	3.57	0.99	0.63	0.22	1.39	0.04	0.18
δ	0.01	0.05	0.07	0.17	0.60	0.21	0.10	0.49	0.01	0.12
	0.01	*0.03*	*0.10*	*3.82*	*1.06*	*0.67*	*0.24*	*1.49*	*0.04*	*0.19*

Footnotes
(1) Recalculated to 4 oxygens. 25 analyses; 21 accepted into the above calculation. Two of the rejected analyses could be bredigite, one other was rejected: see discussion.
(2) Recalculated to 16 oxygens. 19 analyses, all accepted.
(3) Analytical totals calculated to 10 oxygens fall short of 100%. The last line, in italics, shows a recalculation to a modified ferrite phase formula with 10.7 oxygens: see discussion.

The rather variable S contents of ferrite arise because of the small size of the ferrite grains, which are close to the analytical limit, *ca* 1-2μm, of the Cameca electron microprobe. Since the ferrite is intergrown with other phases, several of the analyses could be rejected as they almost certainly arose from intimate mixtures of ferrite with calcium sulfoaluminate. A separate series of analyses were undertaken of ferrite obtained by selective dissolution and at higher analytical resolution: these are presented and discussed subsequently.

Four other minor phases were found: lime and periclase, very close to CaO and MgO, respectively, gehlenite, $2CaO \cdot (A\ell,Fe)_2O_3 \cdot SiO_2$, and calcium sulfosilicate, $5CaO \cdot 2SiO_2 \cdot SO_3$. Insufficient analytical data were obtained on gehlenite and calcium sulfosilicate to warrant their inclusion in Table 2 but both are believed to be close to their ideal compositions.

ADDITIONAL DATA AND DISCUSSION

The data sets obtained from microprobe analyses are capable of further treatment. The presence of occasional high-MgO analyses from belite have been noted, and it is suggested that these may arise from the presence of two belitic phases: α' and/or β Ca_2SiO_4 and bredigite, with the later likely to be mistaken for α' Ca_2SiO_4. The Aℓ and S content of belite requires comment. Fig. 5 shows a treatment of analytical data obtained from 23 apparently single-phase analyses, excluding those with significant Mg contents. The diagram is useful to check whether the observed Aℓ and Si contents are due to occluded $C_4A_3\overline{S}$. The trend line at Aℓ/S = 6 shows the expected values if analyses arose from mixtures of the two phases. One high Aℓ analysis lies close to this trend line, suggesting significant inclusion of sulfoaluminate in the analysis signal. This was therefore also rejected. The remaining analyses form a trend line whose ratios, Aℓ/S, lie between 1 and 3. This, coupled with an apparent deficiency of Si in the analyses, suggests that a principal substitution may be replacement of $3(SiO_4)^{4-}$ by $(2[A\ell O_4]^{5-} + SO_4^{2-})$, giving an ideal A$\ell$/S ratio of 2. This may be compared with an observed ratio of ~1.4. Of course, the balance of substitution of the constituents in tetrahedral sites is also influenced by the presence of other tetrahedral units; PO_4 and, possibly, TiO_4 but the significant replacement of Si mainly by sulfate and aluminate is highlighted. The extent of substitution is however believed to be greater than any reported for Portland cement [6]. Two significant points remain to be explored: the influence of Aℓ and S substitution upon belite reactivity and the phase balance between α', β and bredigite-type phases; careful electron diffraction work will be required to establish the structural distinctions.

Dicalcium ferrite and its aluminium-substituted solid solutions form a series of solid solutions with perovskite, $CaTiO_3$, in which complex ordering phenomena occur [7]. These solid solutions are characterised by (i) the presence of oxygen vacancies, necessary to preserve charge balance in some ranges of compositions, (ii) the presence of superstructures based on long-range ordering of oxygen vacancies and (iii) both regular and random interstratifications of structural slabs of perovskite-like and ferrite-like blocks, leading to development of superstructures amongst the regular stacking sequences. It therefore follows

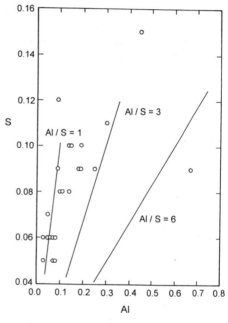

Fig. 5. Aluminium and sulfur analyses from the belite phase in sulfoaluminate clinker. $C_4A_3\overline{S}$ contamination would give analyses on the Aℓ/S trend = 6.0, whereas the majority lie between Aℓ/S = 1.0 and 3.0: the text discusses a possible substitution mechanism with Aℓ/S = 2.0.

that ferrite crystals require to be classified with respect to their ordering sequences and composition by electron diffraction, as individual crystals are typically only $\underline{ca}$ 1-2 μm - too small for conventional X-ray single crystal methods. Since the ferrite is believed to be very unreactive towards water, there has not been incentive to study its detailed crystallography. However, it is noteworthy that a much better fit to the analytical data can be achieved by using a more perovskite-like formula, M^{2+} $(M^{4+}, M^{3+})O_{3-x}$ where x is an appropriate correction factor for oxygen vacancies or, recasting this to a modified ferrite formula, $Ca_4A\ell_{2-2x}Fe_{2-2x}Ti_{4x}O_{10+2x}$ with x ≈ 0.35. The last line of Table 2, in italics, shows a recalculation of the mean analyses in terms of this perovskite-like formula. However the presence of so much S in the analysis still remained a matter of concern since S is not normally a substituent in the perovskite structure [8].

Therefore, the ferrite phase was concentrated by extracting 2g of crushed clinker in sucrose-KOH solution. Two extraction procedures were used: 90g KOH and 90g of sugar in 0.9ℓ water and an extraction time of 9 minutes, or three portions each of 2 minutes duration, of 30g KOH and 30g sugar in 300 mℓ; all extractions were done at 95°C. These extractions were about equal in performance and both gave an apparently pure ferrite phase concentrate, which was subsequently analysed by analytical electron microscopy. This gives much higher analytical resolution in its transmission mode than a microprobe, although its energy dispersive analyser is less accurate.

Table 3 presents the results of 35 analyses, based on an intermediate formula in the ferrite-perovskite series with 10.41 oxygens, (x = 0.205). A conclusion of the data is that sulfur falls to nearly zero, thereby confirming suspicions that it was not an essential constituent. Assuming a perovskite-like environment the structure of the solid would afford 12 and 6 coordinate sites. Probably only Ca is large enough to occupy the twelve-coordinate site and, if other cations are assigned to the 6 coordinate site, the ratio of site occupancies would be 4.30:4.08, sufficiently close to theoretical (1.0) to suggest that this formulation is essentially correct.

Table III. Analysis of Ferrite-like Phase from Selective Dissolution Residues: Atom Ratios

	Na	K	Mg	Ca	Aℓ	Si	Fe	Ti	P	S	Mn	Cr
Mean	0.0	0.0	0.34	4.30	1.23	0.40	0.82	1.18	0.05	0.03	0.02	0.01
δ	0.0	0.0	0.12	0.15	0.31	0.13	0.24	0.51	0.01	0.01	0.01	0.01

One further remark concerns the minor element concentrations associated with ferrite. In the higher resolution of the analytical electron microscopy, anomalously high but very uneven local concentrations of Pb, Cr, Ni, Bi and Fe were found. These seem to be contained in two phases: one, a Pb-Bi phase suspected of being pyrochlore type, ideally $Pb_2Bi_2O_7$. The other phase concentrates Fe, Ni and Cr and was not further characterised. Neither phase could be analysed accurately, but the first probably originates from minerals in the raw meal while the second, comprising Fe-Ni-Cr rich grains, may be derived from grinding and/or crusher and kiln materials. The other phases were not subject to selective dissolution and searched rigorously, so it should not be concluded that these exotic minor phases are associated uniquely with ferrite.

The compatibility relations among the main phases, as revealed by the microstructure and XRD, may be compared with phase equilibria. Data on phase relations within this five component system are scarce, but there is evidence from study of the limiting four component systems that several pairs of constituent phases observed in the clinker are stable. For example, belite and ferrite occur stably in Portland clinker: also, calcium sulfoaluminate and belite are known to be compatible [9,10]. Supplementary unpublished experiments at Aberdeen disclose that ferrite and sulfoaluminate also comprise a stable pair. Taken together, these observations provide tentative support for the microstructural evidence, suggesting that the three-phase assemblage of sulfoaluminate, belite and ferrite is stable at clinkering temperatures, typically 1200° - 1300°C. The upper stability limit of calcium sulfosilicate is about 1310°C [11]; its limited persistence also suggests that clinker firing temperatures were close to this value.

Clinkers formulated to contain belite, ferrite and calcium sulfoaluminate achieve dimensional stability during subsequent hydration. Their mineralogy consists largely of three phases: $4CaO \cdot 3A\ell_2O_3 \cdot SO_3$, belite ($Ca_2SiO_4$) and ferrite. Achieving this phase distribution presents a difficult target for the manufacturer. This is illustrated by comparison with Portland cement, where four principal chemical components: - CaO, $A\ell_2O_3$, Fe_2O_3 and SiO_2 - combine to give four solid phases. However, in sulfoaluminate cement clinker, five components; CaO, $A\ell_2O_3$, Fe_2O_3, SO_3 and SiO_2, must be combined into only three principal solid phases. Moreover, two of the phases, $C_4A_3\overline{S}$ and belite, have only a limited range of compositions. The third, ferrite-like phase is, of necessity, constrained to be a minor constituent of clinker because of its inertness to hydration.

CONCLUSIONS

Calcium sulfoaluminate cements can be used for various purposes. Broadly, these may be defined as primary or secondary: primary use requires that the cement be used as the bonding material with sand and stone to form mortars and concretes. Secondary uses are varied, but include admixtures with Portland cement, where the sulfoaluminate is intended to reduce shrinkage. Perhaps because of this application, as well as its use in strongly expansive cements, cement scientists have become accustomed to treating sulfoaluminate cement products as intrinsically expansive. This need not be so, although careful attention to bulk chemistry, mixing and burning are essential to obtain a desirable phase balance comprising mixtures of belite and $C_4A_3\overline{S}$.

The main clinker phases appear to have attained an equilibrium. Nevertheless, a host of minor disequilibrium features persist. These may be combinations of phases which persist metastably, for example free MgO not taken into solid solution, or in the unequal distribution of some minor components and their association, e.g. of Pb and Bi with each other and possibly also with ferrite; also in the microstructure which shows the presence of xenoliths or relict structures. Microstructures, coupled with microchemical analyses, are an excellent guide to the completeness of reaction and thermal history of the material. The power of modern analytical techniques is such that under favourable conditions, phase detection even at trace level is possible and the overall examination, coupled with a knowledge of the relevant physical chemistry, enables the phase composition and clinkering conditions to be optimised.

ACKNOWLEDGEMENT

The authors thank Dr. R.F. Viles of Fosroc for directing our attention to this interesting problem. The Engineering and Physical Sciences Research Council (UK) provided the Cameca electron microprobe; we are indebted to Drs. A. Coates and E. Lachowski for assistance. One of us (Ö.A.) thanks the Ondokuz Mayis University for leave of absence and a scholarship to enable him to pursue his PhD. studies.

REFERENCES

1. S. Muzhen, D. Junan, W. Zongdao and L. Xioxin, in 9th International Congress on the Chemistry of Cement - New Delhi India 1992 edited and published by the National Council for Cement and Building Materials (India). Volume 2, pp. 94-100.

2. Ö. Andaç and F. P. Glasser (in preparation).

3. Ö. Andaç. PhD Thesis, University of Aberdeen (in preparation).

4. D. Mosely and F.P. Glasser, Cement Concr. Res. **11,** 559-565 (1981).

5. D. Mosely and F.P. Glasser. J. Mater. Sci. **17,** 2736-40 (1982).

6. H.F.W. Taylor, Cement Chemistry (Academic Press, London and San Diego CA. 1990: 2nd printing 1992) p. 475.

7. M.B. Marinho and F.P. Glasser, Cement and Conc. Res. **14,** 360-368 (1984).

8. A.F. Wells, Structural Inorganic Chemistry 5th ed. (Clarendon Press, Oxford 1984) p. 1382.

9. Y.B. Pliego-Cuervo and F.P. Glasser, Cement and Concr. Res. **9,** 573-581 (1978).

10. Y.B. Pliego-Cuervo and F.P. Glasser, Cement and Concr. Res. **9,** 51-56 (1979).

11. Y.B. Pliego-Cuervo and F.P. Glasser, Cement and Concr. Res. **8,** 455-460 (1978).

SEM OBSERVATION OF THE INFLUENCE OF SUCROSE AND KOH ON FORMATION OF SULPHOALUMINATE HYDRATES OF A CEMENT CLINKER DURING THE EARLY HYDRATION

WANG PEIMING, XIA PEIFEN AND CHEN ZHIYUAN
State Key Laboratory of Concrete Materials Research,Tongji University, Shanghai, 200092, China

ABSTRACT

The morphology of early hydrates of a cement clinker was studied by SEM and EDS. The hydrates were formed on thin plates cut from the clinker and immersed in a solution containing $Ca(OH)_2$, $CaSO_4 \cdot 2H_2O$, and varying amounts of sucrose and KOH. The results show the different influences of sucrose and of KOH on the formation of the hydrates, in particular on the crystallization of AFt. The sucrose plays a leading role, and altered the time of primary formation of AFt and its morphology. KOH did not. In $CaSO_4 \cdot 2H_2O$ saturated solution without sucrose, regardless of the presence or absence of $Ca(OH)_2$, the foil - like hydrates, C_xAH_y are formed as the first result of reaction of C3A with water. Then C_xAH_y reacts with $CaSO_4 \cdot 2H_2O$ to form AFt. However, in the presence of sucrose the C3A reacts directly with $CaSO_4 \cdot 2H_2O$ to form AFt. It is supposed that sucrose acts as a catalyst for the formation of AFt crystals. During the early stage of cement hydration, AFt crystals appear to form by both topochemical and by through-solution reaction mechanisms.

INTRODUCTION

In previous experiments[1] we studied by scanning electron microscope (SEM) and energy dispersive spectrometer (EDS) the progress of hydration and the morphology of the hydrates formed of C3A mineral and of a cement clinker. These materials were immersed in 0.2M KOH solutions saturated in $Ca(OH)_2$ and $CaSO_4 \cdot 2H_2O$ and containing varying amounts of sucrose. It was shown that in the presence of sucrose the rate of formation of rod - or prism - form AFt crystals was remarkably accelerated. But sucrose had no effect on the formation and stability of other hydrates of C3A. In that experiment only the amount of sucrose added was varied, but not the amount of KOH. The latter was kept constant at 0.2M. Thus any influence of change of KOH concentration on the hydration was not considered.

In essence, the concentration of KOH in pore solution of cement pastes is highly variable. Stassinopoulos[2] showed that the concentration of KOH after 1 min. hydration for a paste of portland cement of w/c ratio 0.75 reached 0.54M (25.3g K_2O/l), while it was as low as 0.03M (1.47g K_2O/l) for a paste of pozzolanic cement with the same w/c ratio and at the same age. Andersson et al.[3] reported that KOH concentrations varied with the kind of cement used, but their concentrations never exceeded 0.2M. KOH has a remarkable effect on the progress of hydration of cement[4]. Therefore, it is necessary, especially in the case of increasing dry - process production of cement, to study whether an effect of KOH yet exists in the presence of sucrose. The present paper explores the effects mentioned above, with a series of SEM and EDS results. These were obtained on the early-age hydrates formed on polished surface of thin plates prepared from a cement clinker immersed in water solution containing $Ca(OH)_2$, $CaSO_4 \cdot 2H_2O$ and different proportions of KOH and sucrose.

EXPERIMENTAL

Synthesis of cement clinker

The cement clinker to be studied was prepared from pure chemicals, i.e. $CaCO_3$, SiO_2 and Al_2O_3. The powdered materials were proportionally mixed; pressed into a cake of 50 mm in

143

diameter and 10 mm in thickness, burned at 1350°C in a high temperature furnace for 2 h, and then cooled in air. The potential phase composition was C_3S: C_2S: C_3A: $C_4AF=60$: 10: 30: 0. The amount of free CaO was determined as being lower than 0.1%.

Preparation of solutions

Proper amounts of $Ca(OH)_2$ and $CaSO_4 \cdot 2H_2O$ were put into 3 bottles with 1000 ml distilled water, and 2.8g, 5.6g and 11.2g of KOH were respectively added. After maintaining at 20°C for 4 h, 3 supernatant liquids of 0.05M, 0.1M and 0.2M KOH concentration were obtained. They were then transferred to 9 smaller bottles, and then predetermined amounts of sucrose were added into each. To each kind of KOH solution, three concentrations of sucrose were added, i.e. 0.03 %, 0.3 % and 3 %. Thus 9 solutions were obtained for use.

Preparation of samples

The synthesized bulk clinker was cut into small plates of about 5 mm x 5mm area and 3 mm in thickness. These were then ground and polished to obtain a smooth surface. Subsequently, sample plates were immersed into each of the above mentioned 9 solutions for predetermined immersion periods (6 s, 6 min and 1 h) and then taken out and immersed in acetone and ether successively to stop the hydration process. The samples were then sputtered with carbon and studied by SEM and EDS.

RESULTS

Figs. 1 to 9 show typical SEM photographs of hydrated surfaces of the various samples. The hydration product phases were recognized by using EDS data as shown in Table 1. As shown in those figures, the surfaces of all clinker platelets have been etched by hydration after 6 s immersion in the solution saturated in $Ca(OH)_2$ and $CaSO_4 \cdot 2H_2O$ in 6 s, irrespective of the concentrations of KOH and sucrose. The degree of etching increases with the amount of sucrose. The hydrates are readily detectable. Most hydrates appear in the form of needles or prisms (Figs. 1(a), 2(a),..9(a)). Some of them have a perfect crystal forms, as shown in Fig. 3(a). The hydrates with needle or prism forms have been determined to be AFt, according to EDS data.

After 6 min, the degree of etching by hydration, and the morphology of the AFt vary with the hydration conditions. The sample which was immersed in solution with more sucrose is more strongly etched than that after only 6 s. AFt crystals are significantly increased in amount and size (Figs. 1(b), 4(b) ,7(b) and 8(b)). For example, AFt formed in the solution with 0.3% sucrose and 0.2M KOH is about 10 μm in length and 1 μm in diameter (Fig. 8(b)). However, on the sample with more sucrose, AFt crystals almost do not change in amount.

By 1 h, with less sucrose AFt crystals cover almost the entire surface of the samples (Figs. 4(c), 7(c) 1(c)). The amount of AFt crystals increases with the KOH concentration. With an intermediate amount of sucrose, the AFt crystals are longer, but the amount produced is less. With more sucrose, the AFt crystals appear in the form of rounded elongated crystals projecting from common centers (Fig. 6(c), Fig.9(c)), except with 0.05M KOH.

AFm crystals are detectable after 6 min of hydration and show a rosette form. This is observed on the samples immersed in solution with low sucrose levels, irrespective of the concentrations of KOH (Fig. 1(b), Fig. 7(b)). AFm crystals exist only for a very short time, e. g. they appear after 6 min and disappear by 1 h.

Syngenite crystals are observed on all samples with 0.3% sucrose, irrespective of the concentration of KOH. These appear in the form of thin plates (Fig.2(b) and Fig.8(a)).

Of the several hydrates which can be observed, AFt crystals are dominant, as shown in Table 2.

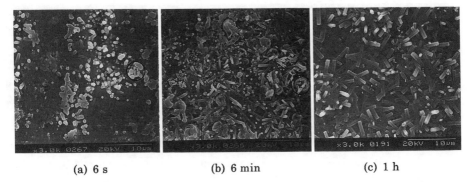

(a) 6 s   (b) 6 min   (c) 1 h

Fig. 1 Morphologies of the surface of clinker platelet immersed in $Ca(OH)_2$ – $CaSO_4 \cdot 2H_2O$ solution with 0.03% sucrose and 0.05 molar KOH

(a) 6 s   (b) 6 min   (c) 1 h

Fig. 2 Morphologies of the surface of clinker platelet immersed in $Ca(OH)_2$ – $CaSO_4 \cdot 2H_2O$ – solution with 0.3% sucrose and 0.05 molar KOH

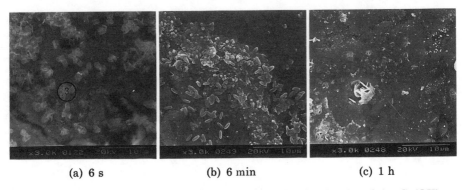

(a) 6 s   (b) 6 min   (c) 1 h

Fig. 3 Morphologies of the surface of clinker platelet immersed in $Ca(OH)_2$ – $CaSO_4 \cdot 2H_2O$ – solution with 3% sucrose and 0.05 molar KOH

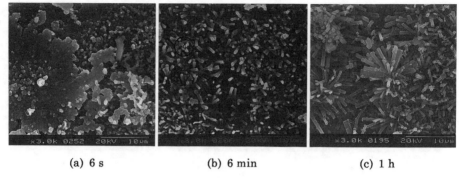

(a) 6 s   (b) 6 min   (c) 1 h

Fig. 4 Morphologies of the surface of clinker platelet immersed in $Ca(OH)_2 -$ $CaSO_4 \cdot 2H_2O$ – solution with 0.03 % sucrose and 0.1 molar KOH

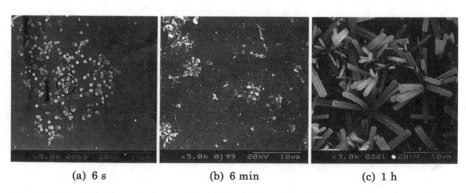

(a) 6 s   (b) 6 min   (c) 1 h

Fig. 5 Morphologies of the surface of clinker platelet immersed in $Ca(OH)_2 -$ $CaSO_4 \cdot 2H_2O$ – solution with 0.3 % sucrose and 0.1 molar KOH

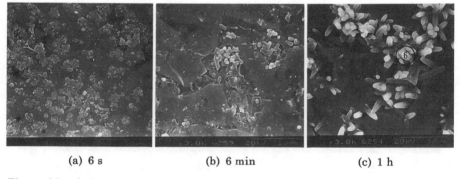

(a) 6 s   (b) 6 min   (c) 1 h

Fig. 6 Morphologies of the surface of clinker platelet immersed in $Ca(OH)_2 -$ $CaSO_4 \cdot 2H_2O$ – solution with 3 % sucrose and 0.1 molar KOH

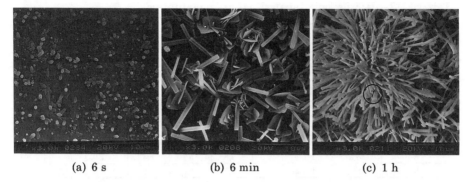

(a) 6 s   (b) 6 min   (c) 1 h

Fig. 7 Morphologies of the surface of clinker platelet immersed in $Ca(OH)_2-$ $CaSO_4 \cdot 2H_2O$ – solution with 0.03 % sucrose and 0.2 molar KOH

(a) 6 s   (b) 6 min   (c) 1 h

Fig. 8 Morphologies of the surface of clinker platelet immersed in $Ca(OH)_2-$ $CaSO_4 \cdot 2H_2O$ – solution with 0.3 % sucrose and 0.2 molar KOH

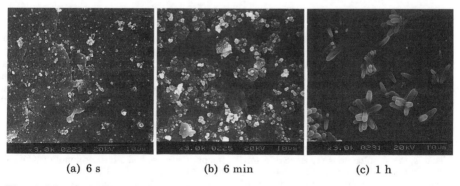

(a) 6 s   (b) 6 min   (c) 1 h

Fig. 9 Morphologies of the surface of clinker platelet immersed in $Ca(OH)_2-$ $CaSO_4 \cdot 2H_2O$ – solution with 3 % sucrose and 0.2 molar KOH

Tab. 1 EDS – data for point analysis (semi – quantitative)

Point No.	Fig. No.	Hydrate form	Element composition (%)					
			O	Al	Si	S	K	Ca
1	1(a)	Needle	4.6	15	2.3	5.4	1.4	71
2	1(b)	Rose	1.9	13	1.9	1.5	0.9	80
3	3(a)	Prism	6.8	19	2.0	1.5	0.8	70
4	4(b)	Prism	7.1	13	3.1	5.8	1.5	69
5	5(a)	Needle	4.1	13	4.5	1.7	4.5	72
6	6(c)	Shuttle	16	12	2.3	12	1.5	56
7	7(c)	Needle	13	8.1	1.6	17	3.4	58
8	8(a)	Plate	2.8	5.7	8.4	6.3	15	62
9	8(c)	Prism	9.7	8.3	1.0	16	1.6	64

Tab. 2 Kind and relative amount of hydrates formed in solution saturated in Ca(OH)$_2$ and CaSO$_4$·2H$_2$O

Fig. No.	Concentration of KOH (M)	Addition of sucrose (%)	Hydration time		
			6 s	6 min	1 h
1		0.03	╳	╳ ◊	╳ ╳
2	0.05	0.30	╳	╳ ※	╳ ╳
3		3.00	╳	╳ ◊	╳
4		0.03	╳	╳ ╳	╳ ╳ ╳
5	0.10	0.30	╳	╳ ※	╳ ╳
6		3.00	╳	╳	╳
7		0.03	╳	╳ ╳ ◊	╳ ╳ ╳ ╳
8	0.20	0.30	╳ ※	╳ ╳	╳ ╳ ╳
9		3.00	╳	╳	╳

╳: AFt crystal; ◊: Rose – like hydrate; ※: Syngenite

DISCUSSION

Influence of sucrose and KOH on formation of AFt

The present results show that both KOH and sucrose influence the formation of AFt on the surface of cement clinker immersed in water solution saturated in Ca(OH)$_2$ and CaSO$_4$·2H$_2$O. However, the influence of the sucrose is much more distinctive than that of KOH. As long as sucrose exists in the solution, AFt crystals are always formed first, no matter what the amount of sucrose and the concentration of KOH are. This is quite different from the phenomenon without sucrose[1] Low sucrose level is associated with formation and growth of AFt crystals. In such cases AFt crystals are greater in amount, and their particles are longer.

The influence of KOH on the AFt formation generally increases with the concentration of KOH. With same sucrose level, the greater the concentration of KOH, the more pronounced the growth of AFt crystals. This trend is evident especially with light addition of sucrose (0.03%). That is, the simultaneous presence of a small amount of sucrose and a large amount of KOH are the best conditions for the AFt crystals being formed. But the effect of KOH is not as restrictive as that of sucrose, which alters the initial age of crystallization of AFt. Without sucrose, AFt crystals are not formed initially, but foil-form hydrates are formed first, regardless of the concentration of KOH, as was shown previously[1] and is seen in Table 3. The previous paper reported results of hydration of clinker in a solution saturated in $Ca(OH)_2$ and $CaSO_4 \cdot 2H_2O$; the latter gives results for a solution saturated in $CaSO_4 \cdot 2H_2O$ without $Ca(OH)_2$ (supplementary experiment). Furthermore, KOH can not fundamentally alter the morphology of AFt crystals, as sucrose can.

Tab.3 Kind and relative amount of hydrates formed in solution saturated in $CaSO_4 \cdot 2H_2O$

Concentration	Addition of	Hydration time	
of KOH (M)	Sucrose (%)	6 s	6 min
0.05	0.00	Δ	✕
	0.03	✕ ◊	✕
0.20	0.00	Δ ※	✕ Δ
	0.03	✕	✕ ◊

✕: AFt crystal; Δ: Foil – like hydrate ※: Syngenite; ◊: Rose – like hydrate

Formation conditions of AFt

As mentioned above, foil - form hydrates are formed first on cement clinker immersed in the 0.2 M KOH solution saturated in $Ca(OH)_2$ and $CaSO_4 \cdot 2H_2O[1]$. The same occurs on the C_3A mineral[1,5]. In a solution saturated in $CaSO_4 \cdot 2H_2O$ only, the first hydrates are still foil - form hydrates (as shown in Table 3), and as reported in the references cited. Without sucrose, AFt crystals can be observed only at a hydration age of 6 min. However, in the presence of sucrose, rod - or prism - form AFt crystals are first formed, and no foil- form hydrates develop. Therefore it is inferred that the C_3A phase of cement clinker reacts first with water and/or $Ca(OH)_2$, to form aluminate hydrate in 0.2M KOH solution saturated in $Ca(OH)_2$ and $CaSO_4 \cdot 2H_2O$ without sucrose:

$$C_3A + aq. \qquad C_xAH_y \qquad\qquad (1)$$

or:

$$C_3A + CH + aq. \qquad C_xAH_y \qquad\qquad (1')$$

Then the aluminate hydrates react with $CaSO_4 \cdot 2H_2O$ to form AFt:

$$C_xAH_y + CaSO_4 \cdot 2H_2O \qquad C_3A \cdot 3CS \cdot H_{32} \qquad\qquad (2)$$

In the presence of sucrose, the C_3A phase reacts directly with $CaSO_4 \cdot 2H_2O$ and water to form AFt:

$$C_3A + CaSO_4 \cdot 2H_2O + aq. \qquad C_3A \cdot 3CS \cdot H_{32} \qquad\qquad (3)$$

Three hypotheses can be suggested as possible reasons for the process given as Eqn. 3.

The first hypothesis may be stated as follows. There must be enough Ca^{2+}, OH^- and SO_4^{2-} ions in the solution to be used for hydration, as indicated by Ludwig et al.[6]. If the concentrations of these ions is high enough, a small quantity of $Al(OH)_4$ ions can lead to crystallizing of AFt. In the presence of sucrose, Ca^{2+} ions react with sucrose to form a complex. The degree of saturation of Ca^{2+} in solution is increased, and hydrolytic decomposition of C_3A is accelerated so as to yield Ca^{2+} and $Al(OH)_4$ ions at once. In addition, the solution was saturated in Ca^{2+}, OH^- and SO_4^{2-} ions before adding sucrose. All these factors provide the necessary conditions for crystallizing AFt.

A second hypothesis may be stated as follows. There must be an appropriate concentration of Ca^{2+} ions. The formation of AFt by hydration of pure C_3A mineral may be different from that in hydration of the cement. With the C_3A phase without sucrose under same conditions, the formation of AFt is not obviously accelerated[5]. Aluminate hydrates are first formed by the hydration of pure C_3A mineral in solution saturated in $Ca(OH)_2$ and $CaSO_4 \cdot 2H_2O$, even though with sucrose[1]. This may be due to a concentration reduced enough in $Ca(OH)_2$ by the effect of hydration of pure C_3A mineral. A low $Ca(OH)_2$ concentration is not conducive to the AFt formation and growth. With cement, although sucrose retards the hydration of C_3S phase, the hydrolytic decomposition of C_3S is faster at beginning with sucrose than without sucrose. In consequence, the Ca^{2+} and Si^{4+} concentrations in the solution are much higher with sucrose[7], allowing rapid saturation and crystallization of AFt. However, if Ca^{2+} ions are in high enough concentration, they may contribute to the formation of aluminate hydrates (monosulfates or their solid solutions), repressing or restricting the formation of AFt. Therefore the concentration of Ca^{2+} ions should be appropriate, i. e. neither too much nor too little.

A third hypothesis may be stated as follows. Sucrose is a catalyst which can accelerate the crystallizing of AFt. This is a conclusion which was developed after the negation of the previous two hypotheses. These previous hypothesis were negated for the following reasons.

1) The hydrolysis of C_3S and C_3A in KOH solution saturated in $Ca(OH)_2$ and $CaSO_4 \cdot 2H_2O$ allows high enough concentrations of Ca^{2+} (super-saturated) and $Al(OH)_4$ to exist, yet no AFt crystals are formed at first[1].

2) In the KOH solution saturated in $CaSO_4 \cdot 2H_2O$ without sucrose, the Ca^{2+} concentration is low enough, but no AFt crystals are formed at first (Table 3). This means that the concentration of Ca^{2+} ions is not the key factor.

3) To observe the influence of concentration of SO_4^{2-} and Al^{3+} ions, some contrasting samples were made, which were immersed in $Ca(OH)_2$ and $CaSO_4 \cdot 2H_2O$ saturated solution, where $Al_2(SO_4)_3 \cdot 18H_2O$ was added. No AFt crystals were observed on these samples at the beginning of hydration.

4) Other theories about sucrose effects on setting of cement can not completely explain the mechanism by which sucrose accelerates formation of AFt. In addition , to observe the influence of $Ca(OH)_2$ concentration on the effect of sucrose, in another experiment a superamount of $Ca(OH)_2$ powder was added in the sucrose solution before the hydration, with a result that AFt crystals were formed on the cement clinker at the beginning.

From these results the conclusion was drawn that AFt crystals are formed first as long as sucrose exists in solution, regardless of the concentrations of Ca^{2+}, OH^-, SO_4^{2-}, and $Al(OH)_4$ ions. This means that sucrose has an obvious accelerating effect on the formation of AFt nuclei and on their growth. Therefore it is supposed that the accelerating effect depends on the catalytic action of sucrose. Addition of sucrose, regardless of the amount added, always shows a catalytic effect leading to the crystallization of AFt at initial age of hydration, and an effect on the amount and morphology of the crystals developed.

In other words, from the experimental results the former two hypotheses are negated and only the last hypothesis is retained, that is, sucrose is a catalyst for the formation of AFt crystals. The catalytic mechanism of sucrose remains to be investigated.

Mechanism of formation of AFt

AFt is observed in unrestricted places in the samples studied, i. e. not only at the locations of the C_3A phase, but also at those of the C_3S phase. Obviously this is the case of "through solution" crystallization, as reported by Mehta[7]. However, many AFt crystals radiate from the locations of the C_3A phases, which may probably be due to" topochemical reaction". It appears that the process of forming AFt can be explained by a several different crystal formation mechanisms even at the early stage of hydration of cement.

CONCLUSIONS

Both KOH and sucrose influence the formation of AFt on the cement clinker immersed in the water solution saturated in $Ca(OH)_2$ and $CaSO_4 \cdot 2H_2O$. The effect of sucrose is much more distinct than that of KOH. AFt crystals are always formed first, as long as sucrose exists in the solution, no matter how low the amount of sucrose and the concentration of KOH are. Small amounts of sucrose are contributory to the formation and growth of AFt crystals. In such cases AFt crystals are more numerous and their particles are longer.

The influence of KOH on AFt formation generally increases with the concentration of KOH. With the same level of sucrose present, the higher the concentration of KOH is, the more favored the growth of AFt crystals is. The trend is especially evident with low levels of sucrose (0.03%). Thus, small amount of sucrose and high concentrations of KOH are the best conditions for the AFt crystals to be formed.

In $CaSO_4 \cdot 2H_2O$ - saturated solution without sucrose, regardless of the presence or absence of $Ca(OH)_2$, the foil- like hydrates, C_xAH_y are formed as the first result of reaction of C_3A and water. Subsequently, C_xAH_y reacts with $CaSO_4 \cdot 2H_2O$ to form AFt. However, in the presence of sucrose the C_3A reacts directly with $CaSO_4 \cdot 2H_2O$ to form AFt. It is supposed that sucrose is a catalyst for the formation of AFt crystals.

The process of formation of AFt can be explained by both through solution and topochemical crystal formation mechanisms even at an early age of hydration of cement.

REFERENCES

1. P.Wang, Proceedings of the 3rd Beijing Int. Sym. Cem. and Conc., Beijing, Vol. 1, pp.429-433 (1993).
2. E. N. Stassinopoulos, Dissertation, TU Clausthal (1982).
3. K. Andersson et al., Cem. Conc. Res. 19, pp. 327-332 (1989).
4. W. Richartz, Zement - Kalk - Gips, 39, pp. 678- 687 (1986).
5. I. Jawed, S. Goto and R. Kondo, Cem. Conc. Res. 6, pp. 441- 454 (1976).
6. U. Ludwig and C. Urrutia, Zement - Kalk - Gips, 42, pp. 431- 436 (1989).
7. P. K. Mehta, Cem. Conc. Res. 6, pp.169 (1976).

GYPSUM-FREE PORTLAND CEMENT PASTES OF
LOW WATER-TO-CEMENT RATIO

FRANTIŠEK ŠKVÁRA
Institute of Chemical Technology, Department of Glass and Ceramics Technology 5,16628
Prague 6, Czech Republic

ABSTRACT

Gypsum-free portland cement is a low porosity hydraulic binder based on finely ground portland cement clinker with addition of synergetic system containing an anion-active surface active agent (usually a sulphonated polyelectrolyte) and an inorganic salt (usually sodium carbonate) for regulation of the hardening process. The properties of GF cement are different from ordinary portland cement; they display, for example, higher strength, better corrosion resistance and thermal stability. These positive differences arise from the different mineralogy and microstructure of the hydration products, for example the absence of portlandite crystals. The main component of the binder product in hardened GF cement pastes is C-S-H (mean C/S ratio 2.7, based on EDAX analysis) intergrown with very fine $Ca(OH)_2$ and highly dispersed C-A-H phases (hexagonal and cubic). The absence of crystalline formations in the GF hardened pastes is responsible for higher mechanical strength. In the Czech Republic, GF cement is produced in the cement works of CEVA Prachovice Inc. (Holderbank group) and is used for special works in the building industry.

INTRODUCTION

Research on gypsum-free portland cements was stimulated by Rebinder and coworkers[1] who substituted calcium ligninsulphonate + K_2CO_3 for gypsum. Their work was followed by that of Brunauer[2]. Our results (plus technological investigations) resulted in the commercial design of gypsum-free portland cements (GF cements). In the Czech Republic GF cement is produced[3,4] on an industrial level in the CEVA Prachovice Inc. cement works (Holderbank group, Switzerland) and has been used for special works in the building industry since 1989.

GF cements may be described as a system of: ground Portland cement clinker (specific surface 300-700 m^2/kg) + anion-active surface active agent with hydroxyl groups (ligninsulphonate. sulphonated lignin, sulphonated polyphenolate) + a hydrolyzable alkali metal salt (carbonate, bicarbonate, silicate). The difference between GF and common portland cement lies in the grinding (absence of gypsum, and/or greater grinding fineness, grinding admixture) and in the set regulator. A particular quality of the GF cements is their ability to set and harden at low and sub-zero temperatures[5]; in addition, they show a rapid strength increase as a result of a small short-time rise in the external temperature. Apart from the above properties, hardened GF cements show a low absorption capacity and high resistance to aggressive media (salts, low and high pH) and recently also resistance to higher temperature (to 1150° C) was reported. The above-described GF cement properties depend, apart from their components, also on the specific surface. The chemical and mineralogical compositions of the clinker affect the properties of GF cements. However, these effects differ from these in the case of Portland cement. The effect of free CaO on initial set and in the development of early strength of GF cements is very marked. Clinkers high in alite but with low free CaO are not suitable for GF cements[6].

This paper is concerned with a study of a GF system of ground portland clinker + sulphonated polyphenolate or ligninsulfonate + alkali metal carbonate - H_2O.

EXPERIMENTAL

In our experiments a clinker of the following composition (wt.%) was used: CaO 66.1, SiO_2 21.4, Al_2O_3 5.6, Fe_2O_3 2.3, MgO 1.85, Na_2O+K_2O 0.85, free CaO 1.0, LOI 0.13. After re-

153

crushing, the clinker was ball milled (100/) in the presence of grinding aids, mostly alkanolamides of dodecylbenzensulphonicacid (Abeson TEA™), the grinding being done to within the specific surface range of 450 - 600 m2/kg. When making pastes, we used Na_2CO_3, Na_2SiO_3 or $NaHCO_3$ and either the anion-active surface active agent sulphonate polyphenolate (sulphonated product of phenol condensation with formaldehyde, Kortan FM™) or commercial ligninsulphonate. This polyphenolate or ligninsulphonate had been used in our experiments, and in some cases a small amount of gypsum (5% by weight). The portland cement (EN standard, CEM I 42.5) originated from the same clinker, and its specific surface was 350 m2/kg. The pastes made ranged in w:c from 0.22 to 0.30, and with PC to as much 0.60. The rheological characteristics of the pastes were studied. Viscosity was measured (using coaxial viscometer) and the nature of the rheological hysteresis loop was examined. The development of strength (mainly in compression) was measured between 1 hour and as much as 180-270 days after paste setting. The test cubes (2 x 2 x 2 cm) were placed for 24 hours in saturated water vapor, then in water at 20°C, and after 28 days in the air at a relative air humidity of 35-45%. Hydration was interrupted at selected intervals by acetone addition and subsequent drying in vacuum. The compositions of hard-set pastes were studied by XRD and TA. The morphology and composition of the hydration products were examined by SEM and EDAX (using ZAF corrections).

RESULTS AND DISCUSSION

Sulphonated polyelectrolyte (ligninsulphonate or polyphenolate) + alkaline carbonate acts as a strong liquefying system in water suspensions of ground clinker. The individual effects of the polyphenolate, ligninsulfonate or carbonate alone in pastes of ground clinker does not come to light since such pastes would set too fast. The correlation between paste viscosity using 500 m2/kg clinker, with varying concentrations of ligninsulphonate + Na_2CO_3 is represented in Fig.1.

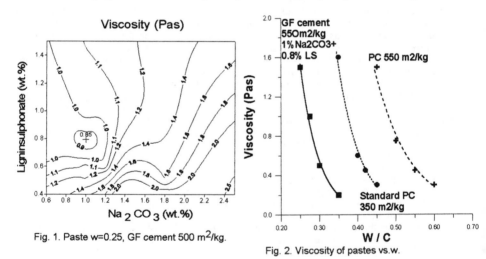

Fig. 1. Paste w=0.25, GF cement 500 m2/kg.

Fig. 2. Viscosity of pastes vs.w.

It is seen that it is possible to prepare pastes having low water-to-cement ratios and good workability (Fig.2). The presence of gypsum in this system is not desirable as it is responsible for very fast loss of workability, and prevents the preparation of pastes of low water-to-cement ratio. The rheological nature of the gypsum-free pastes is in fact Newtonian, with only a minimal yield strength, in contrast to PC pastes.

The GF pastes set very fast after a period of workability, and initial and final time of setting (defined for PC) in fact coincide. They attain 1-8 MPa compressive strength within a short time after setting(Fig.3). Strength rise depends on the nature of the surface active agent, the clinker composition, and the specific surface of cement[7]. Strength increases even at long times, and we

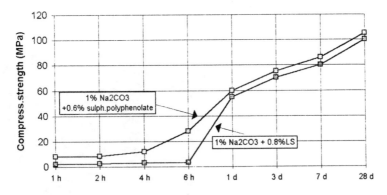

Fig.3. Compressive strength of pastes w=0.25 vs. time, cure temperature + 20°C.

have recorded strength rises even 3-5 years after the cement paste preparation.

Depending on the specific surface and admixtures, the GF cement pastes attain compression strengths of 1 to 9 MPa within 1 to 3 hours. As seen in Fig. 4, the clinker grains are bound by crystalline and amorphous structures.

Fig.4. SEM, GF cement 550 m^2/kg,

0.8% sulphonated polyphenolate

+ 1%Na_2CO_3, w=0.25,

2 hours of hydration, +20° C,

compress. strenght 8 MPa.

   EDAX analysis:

   1. C/S...1, C/A...19

   2. C/S...4, C/A...17

   3. C/S...5, C/A...40

   4. C/S...3.2, C/A...37

The C/S ratio in these structures ranged from 1 to 5, and the C/A ratio from 10 to 50 by EDAX. The XRD and TA showed the presence of C-A-H phases, particularly of the hexagonal C_4AH_n phase. No indication of the presence of either ettringite or monosulphate were found by XRD and TA, and the presence of $C_4A\bar{C}_{0.5}H_{12}$ or $C_4A\bar{C}H_{11}$ is uncertain (in contrast with previously published results[6]. The present results indicate that the initial strength of the GF cements is due both to C-A-H and C-S-H phases. This is how the GF cements differ from the sulphoaluminate or fluoroaluminate cements, in which the initial strengths are primarily due to the ettringite.

After several hours, the microstructure becomes compact, in line with the strength development. The strength development is characterized by a rapid rise after about 6 hours (Fig.3) By 12 to 24 hours the microstructure of the hardened paste changes into one with a distinct gel character (Figs. 5,6). The morphological appearance does not change upon further hydration.

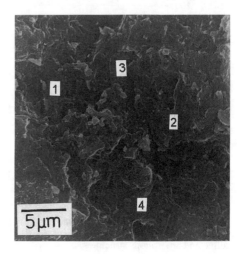

Fig.5. SEM, GF cement 550 m^2/kg,

0.8% LS+ 1%Na$_2$CO$_3$, w= 0.25,

3 years of hydration, +20°C,

moisture regime see "Experimental".

EDAX analysis:

1. C/S...2,5

2. C/S...2,4

3. C/S...2,6

4. C/S...2,3

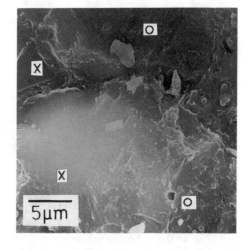

Fig. 6. SEM, GF cement 550 m^2/kg, w=0.35

X.....clinker, O....hydration products

The major morphological part of hardened GF cement pastes is the C-S-H phase corresponding to Type III C-S-H (Diamond[9]). There are no crystalline structures in the compact microstructure, as are typically found in hardened PC in the form of columnar portlandite crystals.

Similar microstructure of hardened cement pastes was found in PRC[10] (pore reduced cement, pressed) and in DSP[11] (portland cement + silica fume + plasticizer) products prepared at low water:cement ratios.

The morphological appearance of hardened GF pastes was not dependent on the specific surface of GF cement used. In most of the EDAX analyses, the C-S-H phase showed C/S ratios of 2 to 3. Sometimes spots of locally increased Ca concentration with C/S values as much as 7 were obtained, yet these did not correspond morphologically to portlandite crystals. Most likely, portlandite is slightly interspersed with the C-S-H phase.

Distinct portlandite crystals found in hardened hydration products invariably had invariably a C/S ratio more than 20-50.

All fracture surfaces of GF pastes analyzed by EDAX (irrespective of GF cement specific surface) were found to contain minor Al, Fe, Na and K. The XRD and TA show that the development of portlandite in the GF takes place later than it does in PC. The portlandite content continued to increase with time, but remained lower than in PC.

The samples of hardened GF cement were found by XRD analysis to contain C_3S and C_2S even after 1-3 years of hydration . The percentage of non-hydrated clinker in polished samples of hardened pastes (GF with specific surface 400-700 m^2/kg) was found by optical microscopy to be up to 10-12% by volume, and in some cases, even more .

The fact of high hydration product density in hardened GF cements follows indirectly from observations of the small effects of aggressive media on hardened pastes and mortars of GF cements. The corrosion processes in hardened GF cements are of a distinctly slower character than in PC cements, as indicated in Fig.7.

Fig. 7. Corrosion of GF cement mortars

(w=0.29) in $(NH_4)_2SO_4$ solution

1.... mortar

2....corrosion layer

(mostly $CaSO_4.2H_2O$, EDAX and XRD

analysis)

The difference in mechanical strength between hardened GF cement and PC and SPC cements results from a shift towards products of greater bonding capacities in GF cements. In the author's opinion the absence of crystalline formations typical of hardened PC and SPC cements in the GF hardened pastes, together with the high density and high degree of dispersion of all hydration products, in addition to the effect of lower water-to-cement ratio, are responsible for the higher mechanical strengths attained in GF cements and also in DSP and similar product with low water-to-cement ratios[12]. In the low porosity pastes crystalline $Ca(OH)_2$ may becomes the weakest part of the structure. As a consequence of steric impediments, pastes with low water-to-cement ratios do not develop the coarsely crystalline structure of $Ca(OH)_2$. *This effect is not believed to depend on the presence or absence of gypsum* in GF cement vs. the other low porosity products; *the processing at low water-to-cement ratio is dominant.*

CONCLUSIONS

1. The system ground clinker-sulphonated polyelectrolyte-alkali salts-H_2O differs qualitatively from the clinker-gypsum-H_2O system. These differences include differences in rheological properties at low water-to-cement ratios, and in the attainment of high short- and long-term strengths.

2. The major constituents of bonding products in hardened GF cement pastes are mostly hydrosilicates interspersed with very fine $Ca(OH)_2$ and highly dispersed hydroaluminates

3. The absence of crystalline formations (typical for hardened PC pastes) is responsible for the higher mechanical strengths attained in GF cements.

REFERENCES

1. O.l.Lukjanova, E.Segalova, P.A.Rebinder, "The influence of hydrophilic plasticizer additions on the properties of concentrated cement suspensions", Koll. Žhurnal 19, 82 (1957).
2. S.Brunauer, Y.Yudenfreund, J. Skalny, and I. Odler "Hardened portland cement pastes at low porosity VII, Summary", Cem. Concr. Res. 3, 279 (1973).
3. F. Škvára , J. Hrazdira , and T. Všetečka, "Method of milling the portland clinker for the production of gypsumless portland cements", US Patent 5,125,976 (30 June 1991).
4. F. Škvara , P. Ďurovec, B. Černovsky , T. Všetečka , J. Hrazdira , and Z. Kadlec, "Mixed gypsumless portland cement", US Patent 5,076,851 (31 December 1991).
5. F. Škvára, K. Kolář, J. Novotný, Z. Zadak , and Z. Bažantová,"The cement for use at low temperatures" in Proc. 7th Intern. Congress on Chem. Cement, Paris 1980, Vol.lll, p.V-57
6. F. Škvára, "The effect of the mineralogical composition of clinker on the properties of gypsum-free portland cements", Ceramics (Prague) 37, 181 (1993)
7. F. Škvára, unpublished results
8. L. G. Špynova , O.L. Ostrovskij, and M.A. Sanickij ,"Concrete for winter conditions, Ed. Visca skola. Lvov 1 (1985).
9. S. Diamond, "Cement paste microstructure-an overview at several levels", in Hydraulic Cement Pastes: Their Structure and Properties pp.2-20 (1976).
10. D. E. Macphee, E. E. Lachowski , A. H. Taylor , and T. J. Brown ,"Microstructural development in pore reduced cement (PRC)", in Advanced Cementitious Systems: Mechanisms and Properties, Mat. Res. Soc. Proc. 245, Boston, MA, pp.303-308 (1991).
11. S. L. Sarkar, Y. Diatta, and P.-C. Aitcin, "Microstructural study of aggregate-hydrated paste interface in very high strength river gravel concretes, in Bonding in Cementitious Composites, Mater.Res.Soc.Proc. 114, Boston, MA, pp. 111-116 (1988).
12. Z. Xu , Y. Deng, X. Wu , M. Tang and J.J. Beaudoin ,"Influence of various hydraulic binders on performance of very low porosity cementitious systems", Cem Concr. Res. 23, p.462 (1993)

THE EVOLUTION OF CEMENTITIOUS MATERIALS THROUGH HISTORY

DAVID BONEN[1], MEHMET A. TASDEMIR[2], AND SHONDEEP L. SARKAR[3]*
[1]Northwestern University, Dept. Civil Engng., 2145 Sheridan RD. Eavanston, IL 60208 U.S.A.,
[2]Istanbul Technical University, Dept. Engng. Materials, 80626 Maslak-Istanbul Turkey,
[3]Sherbrooke University, Dept. Civil Engng., Sherbrooke, QC J1K 2R1, Canada

ABSTRACT

The use of cementitious materials dates back to the beginning of the Epipaleolithic period. Examples for ancient cementitious materials from Israel, Egypt, Turkey, and Italy are numerous.

Prior to Aspdin's patent of portland cement at the first half of the 19th century, cementitious materials were composed of earth, mixture of earth and limestone, calcium sulfates, and slaked lime with and without pozzolans. The latter comprises pozzolanic materials from volcanic and sedimentary origin, crushed burnt clay brick, and dust brick. Frequently, organic fibers were incorporated for reinforcement. This paper describes the evolution of the cementitious materials through time and highlights the durability of ancient cementitious materials as compares to that of portland cement concrete.

Although modern concrete is characterized by its high strength and low permeability, it often faces durability problems. In turn, ancient concretes examined exhibit low strength but have proved to be durable materials. Microstructural examination reveals that the groundmass of the latter has been carbonated and is highly porous. Nevertheless, no specific cracking pattern could be observed. The outstanding performance of ancient concrete structures implies that thermodynamic stability rather that mechanical strength is a key point for a long-term durability.

INTRODUCTION

Anthropologists commonly define a prehistoric society according to its remains of material culture, such as, structures, tool forms, and art. The working hypothesis is that the succeeding society is more developed and sophisticated than the previous one, so its remains. Therefore, the progress and skill of a given society can be evaluated by its construction materials.

Recent progress in the production of macro defect-free (MDF), densified particle (DSP), and fiber reinforced cements testifies to the advanced state of the present holistic knowledge of cement and concrete. Production and fabrication of cementitious materials with compression strengths of hundreds and thousands MPa are attainable. On the other hand, the growing number of concrete structures that display distress and premature deterioration provoke the question to what extent is portland cement concrete durable? Or in other words, is the main hydrated product of portland cement, the C-S-H capable of withstanding the test of time? This question becomes relevant in the light of the outstanding performance of a few ancient structures, some of which are continuously subjected to sequences of drying and

---

* Present Address: S.E.C.A., Inc., Consulting Engineers, 7220 Langtry, Houston, TX 77040

wetting and salt attacks.

Besides an obvious academic interest, comparison of ancient cementitious materials to modern ones provides us with a point of reference as to whether durability is correlated with high-strength and other physical properties, or alternatively, durability is a function of the thermodynamic stability of the final hydrated or alteration products that are formed. It is evident that the remains of ancient materials are the best proof of their durability. We arbitrarily decided to reserve the term ancient cementitious materials to those produced from prehistoric times up to the Byzantine period. Thus, the paper cover these periods only, although some references are made to the Ottoman era.

For the interest of the reader, Table 1 shows the terminology and chronology of old archaeological periods. Attention is drawn to the custom in which old archaeological periods up to about the fifth century B.C.E are denoted according to the material used, i.e., the Stone, Chalcolithic (copper and stone), Bronze, and Iron periods. In addition, although this table was constructed on the basis of a radiometric dating, the chronology presented is a schematic one and serves as a reference rather than a precise delineation of age. Furthermore, in a global view, any archaeological classification is diachronous - periods have different ages at various places. For example, while in the Near East the Bronze period ended at about 1150 B.C.E., in England it lasted up to 600 B.C.E. Finally, because of the abundance of findings, especially form the Epipaleolithic to the Chalcolithic period that were recorded and systematically dated by radiometric ^{14}C methods, the chronological culture of Eretz Israel was adopted in this report. Detailed descriptions of the archaeological remains from these periods are given in references 1-4.

Table 1: Chronology and Terminology of Archeological Paleolithic to Iron Periods[*]

Period	Culture	Age (B.C.E.)
		------520
Iron	I to IIIb	
		------1200/1150
Bronze	Early I to Late III	
		------3200/3000
Chalcolithic	Early to Late	
		------5000/4800
Pottery Neolithic	Yarmukian	
		------6000/5800
Pre-Pottery Neolithic	A, B	
		------8300
Epipaleolithic	Kebaran, Natufian, Harifian	
		------17,000
Upper Paleolithic		
		------42/40,000
Middle Paleolithic	Mousterian, Yabrudian	
		------140/130,000
Lower Paleolithic	Oldowan, Acheulean	
		--------2,3/2,000,000

[*] Compiled from ref. (1, 2, 4)

THE PROGRESSIVE DEVELOPMENT OF BINDERS: PREHISTORIC-NEOLITHIC

During the entire Palaeolithic period-more than two million years ago, people subsisted on hunting and food-gathering. They found refuge from cold and rain in caves, which also provided them with secure storage places. In warm seasons they also encamped in "open stations." Residues of branch shack from almost 2 million years were found in Oldoway in Tanzania. Similar remains pointing at the construction of shacks or tents were found in the Lower through the Upper Palaeolithic in Africa and Europe. In the Upper Palaeolithic in Europe, the lower parts of these shacks were sunk into the alluvial soil. Holes dug in the floors were interpreted as an indication of roofing, almost certainly by using wooden pillars. Similar structures were also found in the Near East. In the succeeding Kebaran culture a greater sophistication was uncovered at Ein Gev site in Israel. The eastern side of a circular shack or hut 5 to 7 m in diameter was cut into the soil, whereas a low wall about 40 cm high constructed form unhewn fieldstone was built on its west side. This tiny wall served as a wind breaker for a stove that was located inside the structure from the north western winds prevailing in this area. In another structure next to the previous one, the presence of piles of stones indicates that the wooden pillars were supported by these stacks of stones.

Evidence for binder use came as early as the Natufian culture. The Natufian structures were still typically circular up to 9 m across. The walls were built from unhewn fieldstone and in two sites, Eynan and Hayonim Cave in Israel, a rendering composed of a mixture of mud-clay and ground limestone was discovered.[2] In the succeeding Harifian culture, limestone plates and unhewn fieldstone were plastered with binder composed of mud-clay and gypsum-bearing marl.

Based on the flint tools used, e.g., hoe for cultivation, axe for cutting etc., it can be safely stated that agriculture goes back to the Natufian age. However, only in the Neolithic (the New Stone Period) people changed their orientation from a food-gathering and hunting society to that of food production: animal domestication and cultivation. Permanent settlements became a widespread phenomenon as is reflected by the large number of Neolithic sites that have been excavated all across Israel and its surrounding. This cultural change was followed by a sociological change from family oriented settlements to multi family or tribal. A notable landmark that occurred at about the middle of this period was the invention of pottery. The discovery that burnt clay changes its characteristics shaped the people's culture and furnished them with inexpensive and sturdy containers and divides the Neolithic to Pre and Post-Pottery periods.

The Pre Pottery A Period is still characterized by circular and oval buildings. But, beside unhewn fieldstone, unburnt mud-clay bricks were fabricated and used as building materials. These mud-clay sun-dried bricks either were laid on a foundation of fieldstone or constituted the entire structure. The floors were made of beaten earth and mud that were smoothly polished, or the polished clay overlaid a layer of pebbles, rubble, and flats fieldstone, which prevented sinking and perhaps helped to keep the moisture out.

Jericho, which is considered as the oldest town on Earth, can be regarded as a key site for the Pre Pottery A period. The town and its fortifications were built about 8000 B.C.E., and covered an area of 25 000 to 40 000 m^2. The houses were slightly sunk into the ground and made of unburnt mud bricks 30 to 40 cm long (which are known as hog backed bricks because of their similarity to loaves of bread). Two to three wooden stairs led to the floor that was made of beaten earth and polished mud clay. The rendering was still made of crushed limestone and clay. Often, the wooden pillars of the doors were also plastered.

The joint communal effort and technical skill of the inhabitants is manifested in the construction of the wall and tower of Jericho some five thousand years before the construction of the first pyramids in Egypt. The wall that was reinforced several times, had an average thickness of 1.7 m and preserved to a height of 3.6 m. Inside the wall, a tower 9 m across its base and 7 m at its top preserved to a height of 8.2 m. Staircase comprising 20 stairs was built inside the tower. Both the wall and the tower were built from fieldstone.

In regard to cementitious materials, the most important discovery at that time was probably the invention of lime that took place in the Pre-Pottery Neolithic B. This period is characterized by two features: (a) the passage from circular or oval structures to orthogonal (rectangular) one, and (b) lime-plastered floors. These lime-plastered floor were widespread in the Levant and Turkey and often were painted in red. In Iran, however, gypsum-plastered floors were built. The lime plaster was composed of crushed limestone, ash, and sand that were burnt over an open fire, recrushed and sieved. Evidently, this pyroprocessed lime had hydraulic characteristics.

A recent study of a lime plastered floor from the village of Yiftahel in Israel was carried out by Ronen et al.[5] The plastered floor, 40-80 mm thick, covered an area over 65 m^2, and weighed about 1.6 tons. According to ^{14}C dating of seeds that were found on it, its absolute age is over 8850±50 years. The floor was placed in two layers, a basal layer about 45 mm thick and an upper finish 5 mm thick. The compressive strength of the former was found to be 34 MPa, and that of the entire floor 45 MPa. X-ray analyses showed that the preponderate constituents of both layers is calcite (that was formed due to a complete carbonation of the original lime), and minor amount of quartz. The greater strength of the upper finishing layer was attributed to it greater density, probably achieved through a mechanical compaction. This floor did not contain aggregate. However, in this as well as other sites in Israel and the northern Levant, mortar floors were discovered.[6,7] Some of these floors weigh 6-7 tons and indicate that some 9000 years ago the inhabitants had acquired a suitable pyrotechnology for the manufacture of a fairly large amount of quicklime that requires furnaces for temperatures of 800 to 900^{0}C, and ability to sustain these temperatures for hours or days. To be noted is the remarkable compressive strength of the floor reported that is comparable to intact modern structural concrete. This could be attained only by a skilled fabrication.

Interestingly, the Pottery Neolithic is a retrogression phase and once again people dwelt in shacks and tents, and probably returned to a seasonal migration lifestyle. This may be attributable to the climatological change at the beginning of the Atlantic period in Europe at the beginning of the sixth millennium B.C.E.

BRICK AND STONE CRAFT EVOLUTION: CHALCOLITHIC-IRON PERIODS

During the Natufian to the Pre Pottery Neolithic, Israel was on the front line of progress. Thereafter, the main Post Neolithic centers were shifted to Mesopotamia, Syria, Anatolia, and Egypt. Surprisingly, the use of lime as a main constituent discontinued for several millennia and only in the Hellenistic period it regained its major role.

In the third millennium B.C.E. the mighty powers of the land of the Tigris and Euphrates in Mesopotamia and the land of the Nile in Egypt had already arisen. The establishment of monarchies in these well-organized empires was followed by massive building activities. As fieldstone masonry never stopped, the beginning of this era is marked by an intensive use of clay for manufacturing and joining dried-mud bricks. This is probably related to the abundance of clay, ease of quarrying, and production.

Plano-convex dried-mud bricks were the common building materials in Mesopotamia. These bricks were placed on their narrow side.[8] In order to control the plastic and drying shrinkage, sand, rubble, crushed burnt bricks, and straw fibers were added. At the Early Bronze I period wooden rectangular brick molds were introduced, which enabled a rapid and smooth placing as is nicely documented in Egyptian wall inscriptions and painting.

An example of stone constructions is the pyramids in Egypt. These megalithic monuments were built from millions of stones that were quarried and shipped hundreds of miles down the Nile. For example, the Great Pyramid (Khufu) at Giza had a basal perimeter of 920 m, and for its original volume of 2.5 million m^3 about 2 300 000 stones were required with an average weight of 2 tons each.[9] Two types of stone were used: the casting stones that were precisely cut from relatively porous lithographic limestone, and hewn or partially dressed filling stones.

Calcium sulfate-based mortar was used as a cementitious material and as a filler for open spaces. Its main constituents were gypsum, anhydrite, calcite, argillaceous limestone, and quartz sand.[10,11,] Ghorab et al. suggested that where the quarried gypsum bed rock was heat treated, it could possess some pozzolanic properties because of the breakdown of clay to amorphous aluminosilicates. It was also proposed that the mechanical integrity of the Pyramids is directly related to the performance of the calcium sulfate binder. Based on microstructural investigation, Regourd et al. concluded that the collapse of the Meidum pyramid built in 2600 B.C.E. was due to the poor quality of the binder. It was composed of argillaceous materials and large gypsum grains. In contrast, other pyramids that were built in 2500 B.C.E. and later, have preserved their integrity because of the transformation of fine anhydrite grains into gypsum.[11]

Dried-mud bricks have low performance as regard to low abrasion resistance, low strength, and low resistance to water. Thus, since the beginning of the second millennium B.C.E., orthostats (smooth slabs of stones) construction was practiced in royal and public structures in Anatolia and in North Syria. Wooden beams and orthostats constituted the frame while the superstructure was generally made of bricks. These buildings, although impressive were not as massive as stone construction that gradually became more widespread. Troy represents an early example of a fine stone masonry that was developed since the mid-second millennium up to its destruction in the late 13th century B.C.E.

A full advantage of stone construction was gained through the passage to ashlar (cut and squared stone) masonry. This technique was probably developed in Phoenician coastal cities. The difficulty in identifying its origin derives in part because it was gradually developed from fieldstone masonry. At the beginning, only the foundations and the corners (which served as columns) were made of ashlar stones, whereas the rest was filled with fieldstone. Often, wooden forms were built for filling the spaces between these columns. In its most advanced form, ashlar masonry appears in the royal centers in Israel at the 10-9th centuries B.C.E and later.[12] The buildings where made of large stones with a specific dressing that characterized each subperiod. Up to the Hellenistic period the stones were not cemented. However, their precise cut and placement were so good that a knife blade could not be inserted between adjacent stones. From the Hellenistic period onward, narrow layers of lime binder were used for cementing adjacent stones. In essence, beside using smaller and lighter stones (or blocks) so that one man can place them , this is the same technique that is used today for manual masonry.

THE CONCRETE ERA

In a broad sense, concrete may be referred to as any substance that is composed of a mixture of solid castable particles that are cemented by a hydraulic binder to form a strong monolith. According to this definition, a concrete technology dates back to the Hellenistic period some 2300 years ago.

As indicated, the production of quick slaked lime dates back to the Pre Pottery B Neolithic period. However, only from the Hellenistic period lime was used for manufacturing structural elements. The fact that hydraulic binder hardens in a moist environment, furnished the masons with a measure to generate water-resistant members and marine structures. A new discovery was that the introduction of pozzolanic materials improves the binder properties. The Greeks used the Santorini Earth as a source of pozzolanic materials, however, the Romans first established the superiority of the lime-pozzolans over the plain lime binder, and used it extensively. They quarried it from Pozzuoli and many other locations. Another important source for pozzolans was ground burnt bricks and ground burnt clay. Additionally, the Romans made use of lightweight aggregates in marine structures, such as, the port of Cosa on the west coast of Italy, and other structural members, for example, the 43 m diameter dome that was built during Hadrian Caesar's time in 128 C.E.[13]

Lime-pozzolanic-based materials (the Roman concrete) became widespread all over the Near East and Europe. Of specific interest is the outstanding performance of Roman concrete structures that are continuously subjected to salt attack. Towards the end of the first century B.C.E., Herod, the great builder-king, constructed a well planned city named Caesaria on the shores of the Mediterranean in Israel. The city was built from 22 to 10 B.C.E. and became one of the most important maritime cities in the eastern Mediterranean. Among the architectural innovations was the building of a port enclosing over than 200 000 m^2 of seawater. It was developed by building a breakwater about 600 m long and 50 m wide made of two external walls and a concrete filling. The outside wall facing the sea waves was made of large stones up to 15 x 3 x 3 m in size that were attached to each other by lead clamps. Following the construction of the walls, the space between them was filled with concrete made of fieldstone, rubble, red clay, and lime.[14] Due to a geological fault, this breakwater submerged in water, and another one from the Byzantine Period was built on it. The port was once again repaired by the Crusaders and is still in service.

Water was brought to the city on two aqueducts that were built in the second century. Figure 1 shows the texture of the superstructure of the western aqueduct. It is composed of local beach rocks about 15 cm in their largest dimensions embedded in thick layers of mortars containing crushed limestone, beach rocks, quartz sand, and pottery. To be noted is that this portion of the aqueduct is located about 150 m from the sea, and despite the continuous spray of salts and sequences of wetting and drying it is still almost intact.

Since the Roman Period up to the 18th century concrete technology developed at a very low pace. It is more likely that the development was from the aesthetic perspective rather than technical properties. Byzantine and Ottoman buildings feature pleasant-looking buildings made of rhythmic and equally spaced layers of red burnt bricks and mortar (Figure 2). The latter which is known as khorasan mortar, is typically composed of 60-70% graded aggregate up to 20 mm in the largest dimension and lime. The aggregate comprises crushed burnt bricks and brick dust, quartz, and other particles from sedimentary or magmatic origin. Fibers, usually chopped straw were introduced at a dosage of up to 3% (Figure 3). Other types of fibers were derived from reed, palm, and even goat hair.

Figure 1: A texture of a Roman concrete from the western aqueduct of Caesaria. (second century B.C.) The aqueduct is located about 150 m form the sea. Coarse fieldstone, crushed burnt bricks, and sand are embedded in thick layers of carbonated lime.

Figure 2: A texture of a khorasan mortar. Byzantine and Ottoman masons used equally spaced layers of burnt bricks (lower) and mortar (upper) for constructing walls. (Yade Kule, 1457 B.C., Istanbul, Turkey).

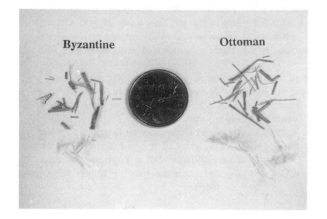

Figure 3: Byzantine and Ottoman fibers composed of chopped straw (upper) and delicate fibers with silky luster (bottom).

Table 2: LOI and Calculated CaCO₃ Contents of Byzantine Mortars.

Sample	Acid Loss wt %	Residue wt% (Aggregate)	LOI °C wt %			Calculated CaCO₃ wt %
			110	500	1050	
1	31.10	68.90	5.38	7.66	23.92	37.0
2	45.36	54.64	5.06	9.50	30.09	46.8
3	29.97	70.03	4.08	7.45	22.38	33.9
4	26.65	73.35	5.29	8.56	20.45	27.0
5	39.56	60.44	5.71	8.82	27.76	43.0
6	36.94	63.06	3.87	8.22	25.51	39.3

Table 2 shows LOI analyses and calculated CaCO₃ content of a few Byzantine mortars from Hagia Sophia (537-563 C.E., Istanbul, Turkey). It is likely that the mixtures were proportioned by volume. Nevertheless, the wide range of aggregate-to-binder ratio implies an absence of strick formulations. That the total of the aggregate and calculated CaCO₃ in Table 2 is slightly greater that 100 may be attributed to the presence of volatile in the aggregate (other than calcareous one since the specimens were acid-treated). It appears, therefore, that most of the original Ca(OH)₂ was carbonated. Similar conclusion is derived from X-ray diffraction and thermal analyses of other Byzantine and Ottoman specimens. Despite a considerable variation in the CaCO₃ content, only minor amounts of Ca(OH)₂ were detected. It should be noted that no compositional difference between the Byzantine and Ottoman mortars could be established. This is in accord with Akman's et al. study illustrating that the two mortars have about the same mechanical properties (e.g. compressive strength up to 8 MPa) that are similar to the corresponding properties of mortars prepared in the laboratory with similar raw materials.[15]

Akman's et al. results deserve further consideration as they indicate that the mechanical properties of these mortars did not degrade over time. It is well accepted that permeability is the single most important factor affecting the durability of portland cement concrete. In turn, in comparison to modern concrete, the khorasan mortar is permeable (K ~ 10⁻⁶ cm/sec) and highly porous (water absorbance 30-45%). SEM analysis in a backscatter mode (Figure 4) shows a porous homogeneous groundmass mainly composed of CaCO₃, and some large residual capillary spaces up to 20 μm. The near interfacial zone between the aggregate and the groundmass is porous as much as the latter. Nevertheless, in contrast to portland cement concrete that has been a few years in service, only a few microcracks were observed in these khorasan mortar. Attention is drawn to the well-preserved bundle of straw fibers (Figure 5) that did not decompose in this alkaline environment over more than 1400 years.

Another observation is related to the absence of C-S-H. Evidently, due to some pozzolanic reaction a limited amount of C-S-H had formed. However, neither visual observation nor X-ray spot analyses carried out by the SEM could confirm the presence of C-S-H in the specimens examined. On the other hand, in other studies C-S-H was positively identified. For example, in another suite of mortars from Hagia Sophia, Livingston[16] noted formation of reaction rims around brick aggregate having a C-S-H composition. Rayment and Pettifer[17] examined 1700 year old lime mortar from Hadrian's Wall in England and noted reaction rims around reacted chert with Ca/Si ranging from about 1 to 1.3. But, on the whole, C-S-H does not seem to have a major role in these mortars.

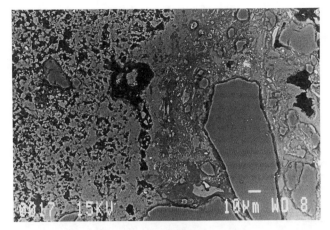

Figure 4: Backscatter SEM micrograph of mortar showing porous groundmass (left) and interfacial zone between the groundmass and the aggregate (burnt brick on the right). The groundmass is mainly composed of $CaCO_3$. (Hagia Sophia, 537-562 C.E., Istanbul, Turkey)

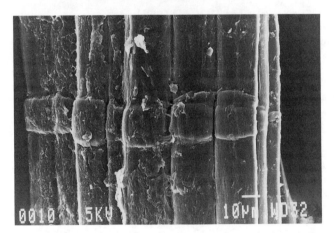

Figure 5: SEM micrograph showing intact bundle of chopped straw fibers in mortar from Hagia Sophia.

At the end of the 18th century and the beginning of the 19th one, a series of clay-bearing limestone combustion experiments had led to the invention of portland cement. This turn point marks the end of $Ca(OH)_2$ domain and the preponderant role of C-S-H as the main binder's constituent in modern concrete.

CONCLUDING REMARKS

Portland cement concrete is distinguishable by its high rate of strength development and superb mechanical properties as compared to lime concrete. However, it appears that the Achilles' heel of portland cement concrete is it susceptibility to environmental attack. In turn, the outstanding performance of the lime concrete is neither attributed to its compressive strength nor to its impermeable nature. Rather the durability of ancient concrete is related to the higher thermodynamic stability of its alteration products.

ACKNOWLEDGEMENT

This study was funded by the NATO Collaborative Research Grant SA.5-2-05 (CRG.931364) 1420/93/JARC-501.

REFERENCES

1. Y. Aharoni, The Archaeology of the Land of Israel from the Prehistoric Beginnings to the End of the First Temple Period, (The Westminster Press, Philadelphia, 1982), p. 344
2. O. Bar-Yosef, in in The Architecture of Ancient Israel from the Prehistoric to the Persian Period, edited by E. Katzenstain, (The Society for Survey of Israel and its Antiquities, Jerusalem, 1987) p. 29 (in Hebrew).
3. M. Avi-Yonah and S. Yeivin with the participation of M. Stekelis, The Antiquities of Israel, (Hakibutz Hameuchad Ltd. Tel Aviv, 1955), p. 343 (in Hebrew).
4. The Archaeology of Ancient Israel,, edited by A. Ben-Tor, (Yale University Press, London, 1992), p. 398.
5. A Ronen, A. Bentur and I. Soroka, Paleorient 17[2], 149 (1991).
6. Y. Goren and P. Goldberg, J. Field Archaeology, 18, 131 (1991).
7. W.D. Kingery, P.B. Vandiver and M. Prickett, J. Field Archaeology, 15, 219 (1988).
8. P. Delougaz, The Oriental Institute of the University of Chicago Studies in Ancient Oriental Civilization, No 7, Part I, 1 (1933).
9. D.H. Campbell and R.L. Folk, Concrete International, August, 28 (1991).
10. H.Y. Ghorab, J. Ragai and A. Antar, Cem. Concr. Res. 16[6] 813 (1986), 17[1] 12 (1987).
11. M. Regourd, J. Kerisel, P. Deletie and B. Haguenauer," Cem. Concr. Res. 18[1] 81 (1988).
12. Y. Shiloh, Foreign Influences on the Masonry of Palestine in the 10th-9th Centuries B.C, , Ph.D. thesis, (The Hebrew University of Jerusalem, 1974) p. 178.
13. T.W. Bremner, T.A. Holm and V.F. Stepanova, in Advances in Cement and Concrete, edited by M.W. Grutzeck and S.L. Sarkar, (Engineering Foundation, ASCE, New York, 1994), pp. 37-51,
14. A. Raban, Kardom, No. 18, Sept. pp. 31-35 (1981) (in Hebrew).
15. M.S. Akman, A. Guner and Hakki Aksoy, in II. Uluslararasi Turk-Islam Bilim Ve Teknoloji Tarihi Kongresi, I.T.U. 1986, pp. 101-111.
16. R.A. Livingston, in Structural Repair and Maintenance of Historical Buildings III, edited by C.A. Brebbia and R.J.B. Rrewer, (Computational Mechanics Publications, 1993), p. 15.
17. D.L. Rayment and K. Pettifer, Mat. Sci. Techn. 3, Dec., 997- (1987).

THE USAGE OF INDUSTRIAL WASTE MATERIALS FOR THE PRODUCTION OF SPECIAL CEMENTS AND BINDERS

HERBERT PÖLLMANN*, JÜRGEN NEUBAUER**, AND HUBERT MOTZET***
*Department of Mineralogy/Geochemistry, University of Halle, Domstr. 5, 06108 Halle
Mineralogical Institute Erlangen ,* Rhein. Kalksteinwerke, Wülfrath, / Germany

By mixing various industrial wastes, as garbage combustion ashes, bottom ashes, fluidized bed ashes, lignite power station ashes, fume purification sulfates and sulfites and lime it is possible to produce cements and binders on the basis of alinite, calcium sulfoaluminate and belite depending on the chemical variety of used wastes.
The fabrication process for these cements was studied by laboratory experiments and the different phases and properties were studied in detail. Alinite cement was already produced on a larger scale in a rotary kiln of 10m length. These cements can be used for application purposes in mining mortars, expansive cements, rapid hardening binders and in landfill technologies. The hydration process and workability can be controlled by using various additives. hus industrial wastes can be a secondary resource for special cement production.

INTRODUCTION

Dealing with the problem of increasing amounts of different wastes problems arise in decreasing amounts of land-fill areas and reduced capacities of primary resources. If landfill technique is not properly managed future problems may occur from leaching out effects of hazardous primary solid phases. For several reasons many contributions came up in the past years to use these materials as potential secondary resources. Depending on the types of waste different economic materials can be applied after processing for reuse. If economic new materials can be produced by using these wastes, besides protecting the environment, new products can be used in industry [4,6,7,8,9,10 - 24].
This paper presents the development of some new materials. Only those wastes which even after specific processing treatments are not usable in any kind should be used for landfill techniques. (Table 1) :

Table 1 : Waste handling and processing treatments for reuse and recycling purposes

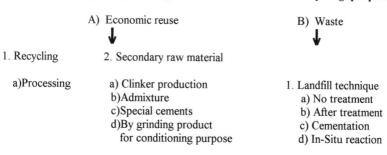

	A) Economic reuse	B) Waste
	↓	↓
1. Recycling	2. Secondary raw material	
a)Processing	a) Clinker production	1. Landfill technique
	b)Admixture	a) No treatment
	c)Special cements	b) After treatment
	d)By grinding product	c) Cementation
	for conditioning purpose	d) In-Situ reaction

The compositions and mineralogical phases of several of the used wastes are given in detail by [7,9,10,12,22]. Sometimes the quality and pureness of the used wastes are even better than those of the equivalent natural primary resources, as there are typical products like the fume purification residues. But normally the wastes used in this study are composed differently with high contents of chlorine, sulfate, sulfite, apart from the "normal" (for instance hard coal ashes) applied secondary resources. Therefore it will be pointed out that some new phases containing chlorine and sulfate are occurring as these described clinker-products. On the other hand a usage in the field of technically specific purposes is not possible in all cases because of the limitations given by the primary composition of the wastes used in that study.

The water contents and temperature of residues influences the formation of different hydrates of the calcium salts. Minor components as fluoride or heavy metals can have a drastic effect on the usage and application of these products. In combination with several other cementitious products it can be used for the preparation of special cements. Those used in that study were of special interest because of their non conformity to typical other residues coming from lignite firing.

The composition of typical flue gas sorption product (sulfate/sulfite) with mineralogically gypsum($CaSO_4 \cdot 2H_2O$), hannebachite($CaSO_3 \cdot 1/2H_2O$), lime(CaO) and calcite($CaCO_3$) as main components is given in Table 2 :

Table 2 : Composition of typical partially oxidized flue gas sorption products

$CaSO_3 \cdot 1/2H_2O$	48. %	SiO_2	3 %	Na_2O	< 1 %
$CaSO_4 \cdot 2H_2O$	39 %	Al_2O_3	2 %	K_2O	< 0.01 %
$CaCO_3$	5 %	Fe_2O_3	1 %	C	0.3 %
CaO	1 %	MgO	<1%	HCl-sol.	1 %

Production of Special Cements and Binders

The combination of several wastes and following mixing process allows to form up a wide range of different chemical compositions which can be transformed, by the burning process, in hydraulic minerals of different amounts. By that process it is possible to get different combinations of hydraulic minerals. The wide variety of applications coming from the processing steps is summarized in Table 4.

The usage of these wastes in the formation of new products by applying high temperature on various waste mixtures can lead to the production of special cements and binders. The most common ones are chlorine containing cements as alinite or sulfate containing cements like calciumsilicosulphate or calciumsulphoaluminate cements. A sulfate-chloride combination at low temperatures is possible too. Depending on the temperature new sulfate and chlorine containing phases are formed. $C_2S \cdot CaSO_4 \cdot CaCl_2$ is hydraulic, but only stable up to approximately 800°C. Above 800°C this phase is replaced by ellestadite or sulfate-silicocarnotite, which are not reacting with water. Ellestadite and sulfate-silicocarnotite can incorporate heavy metals and increase the stability against leaching out effects. For the production of new cementitious materials it is important to control phase formation for hydraulic usage and mineral reservoir formation for heavy metal fixation.

The following table 3 gives an overview on the main phases formed after the burning process of different waste compositions and mixes made for production of new cements and binders:

Table 3 : Typical phases in the clinkers made from waste residues with increased contents in chlorine and sulfate

$Ca_{10}Mg_{1-x}\square_{x/2}[(SiO_4)_{3+x}(AlO_4)_{1-x}/O_2/Cl]$

$0.35 < x < 0.45$

$C_2S \cdot CaSO_4 \cdot CaCl_2$ (< 800°C)

$3C_2S \cdot 3CaSO_4 \cdot CaCl_2$ Chlorellestadite

$2C_2S \cdot CaSO_4$ Sulfate-silicocarnotite

$C_4A_3 \cdot CaSO_4$ Yeliminite

$C_2S \cdot CaCl_2$ $C_{12}A_7 \cdot CaCl_2$

$C_4S_6 \cdot CaCl_2$ C_4AF (A:F var)

$C_9S_6 \cdot CaCl_2$ $\beta\text{-}C_2S$

$CS \cdot CaCl_2$ CaO(free)

The process of forming new products can especially be described by the production of alinite clinker. Yet already a small scale production was made up in a pilot plant (several tons of alinite production). This process can be representative for other mixtures in the field of waste-use in formation of new clinkers and binders. Some others of these phases were used as mineral reservoirs in the fixation process for heavy metals and other substances in the field of immobilization. The well-known system $CaO\text{-}Al_2O_3\text{-}SiO_2$ can only be applied partially because the above mentioned phases replace in some cases the typical calcium silicates and calcium aluminates known from classical cement chemistry.

The realization of formation of new clinkers in a larger scale was proved for alinite as it is given schematically in Fig. 1 with the production steps and later application of alinite instead of ordinary Portland Cement in disposal mixtures.

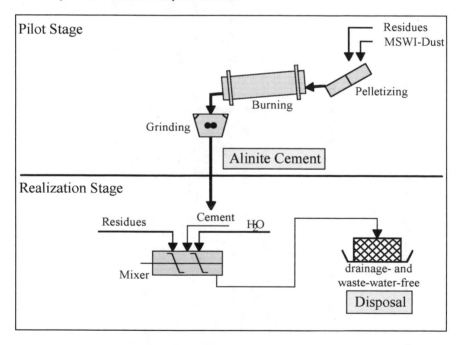

Fig. 1 : Schematic view of the alinite-production and the probable use in the field of solidification of hazardous wastes in landfill technique by partial replacement of Portland cement

Table 4 : Processes, types of residues and probable fields of application

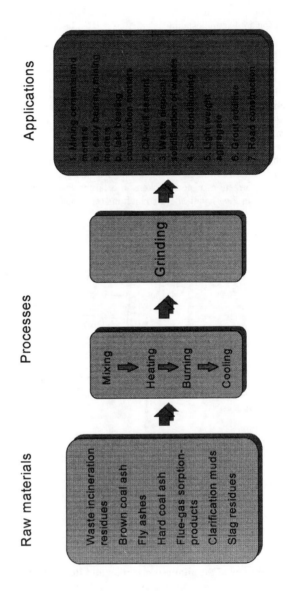

For production of the alinite as given in Fig. 1 the following municipal wastes were used and mixed up according the compositions given in Table 5. Mainly the addition of CaO to the mixtures is important. The alinite can be used for various solidification purposes in disposal sites replacing ordinary Portland cement or it can be used for mining mortars depending on early or late bearing mixtures. Some characteristics of these mortars with little dust formation are given in Fig. 2.

Table 5 : Amount and composition of solid residues from waste incineration plants and of a typical Alinite cement clinker

Constituents	Alinite cement	Flue gas cleaning residue MSWI = Municipal solid waste incineration		
in wt-%		MSWI 1	MSWI 2	MSWI 3
SiO_2	20.2	9.4	9.0	20.2
Fe_2O_3	2.6	1.1	1.1	2.1
Al_2O_3	10.5	4.8	6.4	7.2
Mn_2O_3	0.1	0.1	-	0.1
Ca as CaO	55.0	41.6	25.1	34.7
MgO	1.9	1.3	1.8	2.4
SO_3	3.15	3.4	3.4	4.2
K_2O	0.8	0.9	1.6	3.5
$Na2O$	0.7	0.8	0.7	1.6
Cl	5.6	7.0	-	12.3
CaO free	1.7	32.9	25.1	16.2
LOI 1000°C	1.7	23.4	15.1	14.4
CO_2	0.1	8.5	**	4.2
H_2O	1.1	12.2	**	8.7
Cl+	5.67	8.9	-	14.1
CaO/SiO_2	2.5	4.9	2.8	1.2
$CaCl_2$	-	-	32.5	-
SO_2	-	-	3.1	-

** LOI (CO_2 and H_2O)
+ Chloride content of LOI-free material
+ + Excluding the Ca combined with Cl and is identified as CaO

The main hydration products of alinite seem to be calcium-aluminum chloride hydroxi salts, portlandite and calcium silicate hydrates. Normally some retarders must be added to avoid flash set of alinite mixtures[7]. Small amounts of heavy metals primarily in MSWI-ashes are fixed in the crystal lattice of new formed hydrates as monochloroaluminate and CSHJ-phases. Chloride is fixed in the new hydrate of monochloroaluminate with the general formula $3CaO \cdot Al_2O_3 \cdot (x)Ca(OH)_2 \cdot (y)CaCl_2 \cdot (z)CaCO_3 \cdot nH_2O$.

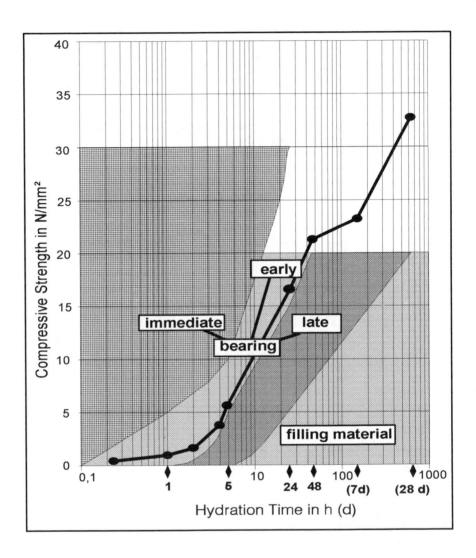

Fig. 2 : Alinite mixtures for underground cements and characteristics of hydration

Besides the formation of alinite clinker the formation of sulphoaluminate clinkers was discussed already in literature [1, 15-20]. The crystal chemistry and properties of sulphoalumi-nate is given by [2,3,5,21]. The raw materials for the formation of these products are normally sulphate containing residues coming from flue gas desulfurization and aluminum-rich wastes from industrial residues. Sulphoaluminates tend to have expansion effects by the formation of ettringite in the course of hydration. It could be proved that in some cases the crystallization of monosulfate can occur and reduce the expansion effect, depending on the

the availability of of sulphate and corresponding solution compositions. The formation of hydrates and corresponding technological effects differ sometimes due to the varying industrial wastes. It is yet not clear what kind of influence is responsible for these different hydration behaviours. Some hydrating phases are given in Figures 3,4,5 and 6 for different mixtures. The micro-structure is obviously influenced on the surface of both mixtures. The CaO-content in Figure 5 leads to the formation of large portlandite-crystals on the surface. In Figure 6 a combination of small lamellar hydrates and primary clinker minerals builds up the main phases. In both cases the primary content of sulphoaluminate was below 15 % of the overall mineralogical composition of the clinker. Besides the use of new clinker products as alinite or sulphoaluminate clinkers it is possible to use waste products as additive. An example could be the addition of flue gas desulfurization products to cement. Laboratory results for 3 different mixtures are given in Table 6.

Especially these results represent a simple decrease in properties by using additives which were not pretreated.

Table 6 : Compressive strength of various mixtures of cements with gypsum ($CaSO_4 \cdot 2H_2O$) and hannebachite ($CaSO_3 \cdot 1/2H_2O$)

Time	cement + 5 % gypsum	cement + 2.5%gypsum + 2.5% Ca-sulfite	cement + 5% Ca-sulfite
1 day	65	58	39
8 days	105	90	67
27 days	110	95	73

Fig. 3: Calciumsilicatehydrate-gel with Portlandite crystal-aggregates on hydrating alinite-cement surface

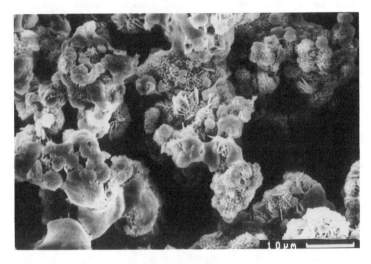

Fig. 4 : Calciumsilicatehydrate-gel with lamellar Portlandite-crystals and roundly shaped
Hydrogarnet at 60°C hydration temperature

Fig. 5 : SEM-micrograph of hydrating surface of clinker made from wastes with large
Portlandite crystals on surface

Fig. 6 : Hydrating clinker made from wastes with lamellar hydrates and unreacted clinker particles with gel-like surface coating

DISCUSSION

The reuse of industrial wastes for the production and application in several fields of special cements and clinkers is possible by selection and treatment of the wastes for their further area of application. The different reactions in thermal and hydraulic behaviour must be studied furtheron, but yet the information available reaches the point, that we can face the waste problem in introducing new techniques for reuse. It is essential that the complex wastes must be treated with appropriate processes and that the chemical and mineralogical compositions must be studied carefully as well as the technological properties. The capability of some of these phases to fix heavy metals and other toxic elements can even be used in the cementitious field for immobilizing purposes. For the future some more basic research can have strong impact for new techniques in these multivariant systems. At the end it is of tremendous effect to discuss as well the discrepancy of the possible usages by obtaining scientific results and comparison with the overall political acceptance of those products.

REFERENCES
1.Cohen; M.D. : Theories of expansion in sulfoaluminate - type expansive cements : schools of thought cement and Concrete Research 13, 809-818, 1983.
2.Depmeier, W. : Aluminate sodalites - a family with strained structures and ferroic phase transitions; Phys. Chem. Minerals 15, 419-426, 1988
3.Depmeier, W.; Yamamoto, A. : Powder profile refinement of a commensurately modulated aluminate sodalite; Materials Science Forum Vol. 79-82. 763-768, 1991
4.Götze, W.; Buschmann, N.; Schroer, D. : Ausbautechnische Anforderungen an Baustoffe im Bergbau; Glückauf 120, Nr. 21, S. 1397-1413, 1984

5.Hanic, F.; Havlica, J.; Kapralik, I; Ambruz, V.; Galikova, L.; und Urbanova, O. : Crystal chemistry and thermodynamics of the sulfate compounds $Ca_4[Al_6O_{12}](SO_4)$ and $Ca_5(SiO4)_2(SO)_4$; Br. Ceram. Trans. J. 85, 52-57, 1986

6.Kurdowski, W. : Expansive cements; 7th International Congress on the Chemistry of Cement, Paris 1, V2/1-V2/11, 1980

7.Motzet,H., Pöllmann,H. & Neubauer, J. : Alinite-cement made from waste incineration residues - from basic research to application - Proc. 16th Int.Con.Chem.Cem., Richmond, 210 - 229, 1994

8.Neubauer, J.; Pöllmann, H. : Hydraulische Sodalite; Kristallographie und Hydratation; Europ. J. Min.,Beihefte, 206, 1992

9.Oberste-Padtberg, R.; Roeder, A.; Motzet, H.; Herbig, J : Alinit aus Müllverbrennungsrückständen; in Faulstich, M. (Hrsg.) : Rückstände aus der Müllverbrennung, EF-Verlag für Energie- und Umwelttechnik, Berlin , S 809-835, 1992

10.Oberste-Padtberg, R.;Neubauer,J.: Laborversuche zur Herstellung von Alinitzement aus Müllverbrennungsrückständen; WLB, Wasser, Luft und Boden, Heft 10ß, S. 63-65, 1989

11.Oberste-Padtberg, R. ; Schweden, K. : Zur Freisetzung von Wasserstoff aus Mörteln mit MVA-Reststoffen; WLB, Wasser Luft und Boden, S. 61-62, 1990

12.Oberste-Padtberg, R.; Roeder, A.; Motzet, H.; Spicker, V. : Alinitzement, ein hydraulisches Bindemittel, hergestellt aus Müllverbrennungsrückständen; ZKG, Zement-Kalk-Gips, 45. Jahrg.; Heft 9, S. 451-455, 1992

13.Okushima, M.; Kondo, R.; Mugurma, H. und Ono, Y. : Development of expansive cement with calcium sulphoaluminious cement clinker; Proc. 5th International Symposium on the Chemistry of Cement, Tokio 4, 419-438, 1968

14.Pöhl,K.:Zur Konstitution und Hydratation deutscher Braunkohlefilteraschen - Dissertation Univ. Leipzig, 1994

15.Pöllmann, H. : Capability of cementitious materials in the immobilization process of hazardous waste materials; Proc.15th Int.Con. Cem.,Micr., Dallas, Texas ,108-126, 1993

16.Pliego-Cuervo,Y.B., Glasser,F.P.: Phase relations and crystal chemistry of apatite and silicocarnotite solid solutions - Cem.Concr.Res. 8, 519-523, 1978

17.Pliego-Cuervo,Y.B., Glasser,F.P. : Role of sulphates in cement clinkering reactions: Phase formation and melting in the system $CaO-Ca_2SiO_4-CaSO_4-K_2SO_4$ - Cem:Concr.Res. 7, 477-482, 1977

18.Pliego-Cuervo,Y.B., Glasser,F.P. : Role of sulphates in cement clinkering: The calcium silico sulphate phase - Cem.Concr.Res.8, 455-460, 1978

19.Pliego-Cuervo,Y.B., Glasser,F.P. : Role of sulphates in cement clinkering:subsolidus phase relations in the system $CaO-Al_2O_3-SiO_2-SO_3$ - Cem:Concr.Res.9, 51-56, 1979a

20.Pliego-Cuervo,Y.B., Glasser,F.P.: Role of sulphates in cement clinkering:Phase formation in the system $CaO-Al_2O_3-Fe_2O_3-SiO_2-CaSO_4-K_2SO_4$ - Cem.Concr.Res.9, 573-581, 1979b

21.Ponomarev, V.I.; Kheiker, D.M. und Belov, N.V. : Crystal Structure of tetracalcium trialuminate; the aluminate analoge of sodalite; Soviet Physics Crystallography 15(5), 799-801, 1971

22.Roeder, A. : Möglichkeiten zur Verwertung von Rauchgasreinigungsrückständen am Beispiel der Herstellung von Alinitzement; in Gutke, K. (Hrsg.) : Symposium : Abfallverbrennung, kein Reizthema mehr ? Abfall und Wirtschaft in Deutschland, Kirsten Gutke Verlag, Köln , S. 133-152, 1992

23Zysk, K.-H.; Schroer, D. : Einsatzmöglichkeiten von REA-Produkten im Steinkohlenbergbau; ZKG, Zement-Kalk-Gips, 46. Jahrg., Heft 5, S. 268-274, 1993

24.Zysk, K.-H.; Roski, H.-J.; Schroer, D. : Anwendungsgerechte Prüfung von Baustoffen bei der DMT; Glückauf 129, 1993,

MINERALOGICAL TRANSFORMATIONS AND MICROSTRUCTURE AFTER DISPOSAL OF CEMENTITIOUS ADVANCED COAL TECHNOLOGY BY-PRODUCTS

G.J. McCARTHY*, R.D. BUTLER**, D.W. BREKKE**, S.D. ADAMEK*, J.A. PARKS*, H.J. FOSTER**, and J. SOLC**
* Department of Chemistry, North Dakota State University, Fargo, ND 58105
** Energy and Environmental Research Center, University of North Dakota, Grand Forks, ND 58202

ABSTRACT

By-products from two advanced coal technologies, Fluidized Bed Combustion (FBC) and Limestone Injection Multistage Burner (LIMB), were found to be cementitious when mixed with water and compacted. However, exposure to natural conditions in test cells resulted in losses of strength and increases in permeability over a period of years. Changes in mineralogy and microstructure with time in recovered core samples have been characterized by powder X-ray diffraction (XRD) and scanning electron microscopy (SEM). Up to 80 wt% of the core materials had converted to crystalline hydrate phases by the end of two years. Ettringite, gypsum and portlandite were the initial hydration products. In the LIMB materials, formation of thaumasite in microfractures, channels and voids was detected after one year, and in the FBC material after two years. Thaumasite formation was accompanied by reductions in gypsum and portlandite; it did not appear to be forming at the expense of ettringite. EDS examination of many ettringite and thaumasite crystals showed that the former always contained some Si and the latter some Al, which is evidence for ettringite-thaumasite solid solution. Thaumasite formation accompanied marked losses in strength and increases in permeability.

INTRODUCTION

In the United States, various laws, such as the 1991 Clean Air Act, require reduced emissions of sulfur dioxide and nitrogen oxides. The advanced coal ("clean coal") technologies necessary to meet these regulations result in solid by-products that differ greatly from conventional combustion fly ash and bottom ash, especially in their high contents of calcium sulfate (or sulfite) and free lime. Utilization or proper disposal of these by-products requires an understanding of their long-term behavior in nature. To begin to obtain such an understanding, by-products from two advanced coal technologies were emplaced in instrumented test cells at three sites in the United States. Results from two sites will be discussed in this paper:

- Limestone Injection Multistage Burner (LIMB), a sorbent duct injection technology combined with low-NO_x multistage burner technology; site near the Ohio Edison Edgewater Plant in Lorain, Ohio;
- Fluid Bed Combustion (FBC) by-product; Midwest Grain Products, Inc. bubbling bed unit; test cell site at the Freeman United Buckheart Mine in central Illinois.

The objectives of the overall project were to: monitor field performance of advanced coal technology by-products; develop a physical-chemical model of by-product behavior; and develop management strategies for by-products [1,2]. The objective of this paper is to summarize the mineralogy of the hydration reactions and microstructure development with time that figure in development of the physical-chemical model. Laboratory modeling of hydration reactions of these materials [3], and some results from earlier stages of the study have been reported [3-7], and the reader is referred here for additional background and detail.

MATERIALS AND METHODS

The test cells were approximately 15 m square by 2-3 m thick. The by-product was conditioned at the plant to 20-30% moisture, hauled to the site, dumped, and spread and compacted by a bulldozer into a series of lifts (layers). These lifts were 10-15 cm thick. Two cells were constructed at the Ohio site, designated North Cell (950 tons of by-product) and South Cell (700 tons). At the South Cell, each lift was sprayed with water through a fire hose and sprinkler system to bring moisture content (mass of water/dry solids * 100) closer to 38%, the ASTM D 698 optimum moisture content for this material. Moisture throughout each lift was apparently not uniform, and caused differential expansive hydration reactions which gave the surface an irregular "heaved" appearance. Approximately 750 tons of FBC by-product was placed in the single Illinois site test cell.

Composite dry and conditioned samples of each by-product were collected in sealed bags at the time of emplacement. Characterization by XRD and SEM/EDS was performed one-two months after sample collection. For the conditioned samples, the phase assemblages and microstructures are those developed by the time of characterization. Core samples were obtained quarterly for the first year, and then annually for years 2,3 and 5 after emplacement. After year 1, all samples were frozen within 24 hours of collection and kept frozen until characterized.

Crystalline phase assemblages were monitored by powder X-ray diffraction (XRD). Diffractograms were collected on Philips diffractometers equipped with variable divergence slits and diffracted beam monochromators and utilizing CuKα radiation. Methods employed for qualitative and semiquantitative crystalline phase analysis have been described in detail in earlier publications [6-9]. Only reactive starting phases (lime, anhydrite) and hydration products (portlandite, gypsum, ettringite), plus quartz, were quantitated. Standards for the Reference Intensity Ratio (RIR) method used for the semiquantitative analyses were pure reagents (lime, anhydrite, portlandite, gypsum, calcite), and synthesized ettringite. No standard for thaumasite was available, so an RIR obtained by calculation from its crystal structure was used as an approximation. A caveat: because the ettringite and thaumasite observed in the hydrated by-products were always solid solutions, and because the RIR of thaumasite is only an estimate, the percentages of phases reported later should be considered as estimates only. There is no way of establishing the accuracy of the method without the availability of standards with the appropriate composition and crystallinity.

Microstructure characterization was done on fracture surfaces using a Noran Instruments ADEM Scanning Electron Microscope (SEM). Samples were carbon coated. The SEM was equipped with a thin window Noran energy dispersive X-ray spectrometer (EDS).

RESULTS AND DISCUSSION

Physical Properties

Physical properties measurements, which constituted a significant portion of the study, are reported and discussed elsewhere [1,2], and only an overview is given here. Moisture contents, initially in the range of 20-30%, increased to as much as 100% or more due to abundant rainfall and infiltration at both the LIMB and FBC cells. Precipitation was on the order of 100 cm/y (40 in./y) at both sites. By year 2, water infiltration saturated the cell and leachate was obtained from lysimeters at the base of the cell. Unconfined compressive strengths measured on cores after two to four quarters were variable, but generally in the range of 3.5-14 MPa (500-2,000 psi), for the LIMB and 1.4-5.5 MPa (200-800 psi) for the FBC material. After the 6-12 months, strengths dropped to about one-tenth of their original values in most samples, although occasionally a stronger sample was found. Average permeability after two-four quarters was in the 10^{-6} to 10^{-7} cm/s range. This average increased to the 10^{-5} cm/s range after two-three years. A principal objective of the mineralogical and microstructural characterization was to provide insights into the marked losses of strength and increases in permeability with time.

Composition and XRD Mineralogy

The bulk chemical composition of the pre-emplacement conditioned by-products that had been sealed in bags on site are given in Table I. The analyses were made five years (with the LIMB sample) or three years (with the FBC sample) after the conditioning took place, so the LOI and moisture are not representative of the original values. Hydration reactions have converted at least some of the moisture (H_2O loss at 110°C) into LOI (loss on ignition at 750°C). Note that LOI also

Table I. Chemical Compositions* (wt%) of By-Products at Emplacement

| | Ohio LIMB By-Product | | | | Illinois FBC By-Product | |
| | North Cell | | South Cell | | | |
	Conditioned	Anhydrous	Conditioned	Anhydrous	Conditioned	Anhydrous
SiO_2	15.9	20.6	14.8	20.3	8.1	13.8
Al_2O_3	7.5	9.7	7.4	10.1	2.2	3.7
Fe_2O_3	8.9	11.5	8.2	11.3	2.6	4.4
CaO	28.3	36.7	25.6	35.1	31.8	54.2
MgO	0.6	0.8	0.5	0.7	0.6	1.0
K_2O	0.7	0.9	0.7	1.0	0.3	0.5
Na_2O	<0.2	<0.2	0.1	0.1	<0.2	<0.2
SO_3	11.3	14.6	11.7	16.0	11.1	18.9
CO_2	3.8	5.0	3.9	5.4	2.0	3.5
H_2O	7.9		11.7		19.2	
LOI	14.3		14.3		23.6	

*Chemical analyses provided by C.M. Lillemoen and D.J. Hassett, UND-Energy and Environmental Research Center. H_2O = loss at 110°C; LOI = loss on ignition at 750°C.

includes CO_2 losses from carbonate decomposition and combustion of char. Because of their different moisture and LOI values, compositions have been recalculated as anhydrous oxides. Both materials are rich in CaO and SO_3, with 74% of the anhydrous oxides of the FBC by-product consisting of these two constituents. The LIMB by-product has nearly twice the amount of the $SiO_2 + Al_2O_3$ components necessary for cementitious reactions, but some of these oxides are present as non-reactive quartz and mullite.

The XRD mineralogy of the original LIMB ash samples received in our laboratory was 8-12% lime (CaO), 15% anhydrite ($CaSO_4$), 4-6% calcite ($CaCO_3$), 2-3% portlandite ($Ca(OH)_2$), and fly ash phases quartz, mullite and hematite. The diffraction patterns showed the features of the usual crystalline phase mineralogy of a low-calcium, bituminous coal-derived, fly ash [8] along with abundant glass. The portlandite, and some or all of the calcite, were the result of atmospheric reactions of moisture and CO_2 with the reactive lime. Mattigod and Rai [10] report the same mineralogy for a sorbent injection technology by-products. The XRD mineralogy of the as-received unconditioned FBC by-product samples was 0-11% lime, 13-19% anhydrite, 5% calcite, 5-24% portlandite, along with quartz and hematite. The lime was highly reactive with moisture in air, and even unconditioned samples hauled to the mine site before collection in moist air were found to have fully hydrated to portlandite. Lime and anhydrite contents from sample to sample were quite variable, suggesting that the FBC unit was not operating under steady-state conditions. The FBC by-product mineralogy was consistent with that reported by other workers [11-13].

These by-products are cementitious because they consist of an intimate mixture of pozzolanic materials and activators that result in self-hardening reactions to form ettringite, gypsum and calcium-silicate-hydrate (C-S-H) when water is added. In the LIMB sorbent injection material, the pozzolan is the fly ash produced by the pulverized coal combustion process, and the activators are residual lime sorbent and anhydrite formed by the SO_2-capture process. In the fluidized bed by-products, the pozzolan is dehydroxylated clay minerals from the coal. The activators are also lime and anhydrite. The FBC system is similar to a lime-gypsum-activated metakaolin.

Approximate average XRD mineralogies of Ohio LIMB core collected after one year from both cells were similar:

gypsum:	5-10%
ettringite	30-40%
calcite	5-15%

All of the lime, portlandite, and most of the anhydrite had reacted. Quartz, mullite and hematite had not reacted. Figure 1a shows a portion of a typical diffractogram. The diffuse scattering maximum ("hump") in the background rises around 25-30° 2θ, which suggests that a significant portion of the material is noncrystalline. There were indications of thaumasite peaks in some of the samples collected as early as the second or third quarter after emplacement. Gypsum showed reduced abundance with time, and was not observed by XRD in some of the older samples. Microstructure characterization discussed below indicated that gypsum is easily dissolved, and reprecipitated, by infiltrating moisture from rainfall. The generation and movement of pore fluids in response to abundant rainfall was a major factor in changes of pore fluid chemistry and in the identity and abundance of secondary minerals over time [1,2].

By the second year, thaumasite was detected in most (but not all) of the Ohio LIMB samples. The mineralogy did not change significantly in the years 2, 3 and 5 core samples. Gypsum became less abundant, or absent more frequently, in the longer exposure samples. The amount of ettringite remained in the 30-50% range. Most samples also contained an estimated 10-40% thaumasite. In some samples, such as that shown in Figure 1b, the estimated amounts of ettringite and thaumasite totaled 85%, with calcite, and unreacted quartz, mullite and hematite making up most of the rest of the sample. The baseline was more nearly flat, suggesting that little residual fly ash glass or C-S-H was present. In samples rich in thaumasite, gypsum was usually absent or only a minor phase.

Even after 5 years exposure, some samples contained no thaumasite (Figure 1c). Such samples always had much greater gypsum contents, suggesting that they may have been protected from much of the latest moisture infiltration. There was a striking correlation between the amount of effort needed for breaking and grinding needed for XRD sample preparation and the thaumasite content. Thaumasite-rich samples were friable, and could be crushed easily. Thaumasite-free samples were tough, and had to be broken up with a hammer.

Semi-quantitative XRD mineralogy of samples Illinois FBC by-product after one year of exposure was:

gypsum:	15-35%
portlandite:	12-20%
ettringite:	20-30%
calcite:	5-10%

This by-product is limited in its ability to form ettringite by low available aluminum (Table I). There is ample gypsum to supply the necessary Ca and SO_4, and the pH is buffered above 12 by the unreacted portlandite. It was only after three years of exposure that thaumasite was definite in some of the XRD diffractograms (Figure 1d). Perhaps because of the lower abundance (Table I) and solubility of their different silica sources, fly ash glass in the Ohio LIMB by-product vs. dehydroxylated clay in the Illinois FBC by-product, thaumasite formed much more slowly in the latter material.

Microstructure

The shape and connectivity of pores in the disposed materials were set by packing density during emplacement, and development of the initial hydration products, ettringite, gypsum and portlandite. Initial bulk density at emplacement at the Ohio site was approximately 1.4 g/cm^3.

LIMB By-Product. Figure 2 shows microstructure development in the LIMB by-product over the first two years of burial.
- Figure 2a: South Cell (to which additional moisture was added between lifts); 3 months; moisture content (M, dry basis) = 41% (increased from 25-30% at emplacement); strength (S) = 8 MPa; permeability (K) = 2×10^{-6} cm/s. This low magnification electron backscatter micrograph shows microcracks, which at higher magnifications revealed deformed ettringite crystals, presumably associated with expansive reactions. XRD analysis of this core gave 30% ettringite (Et), 10% gypsum (Gp), 13% calcite (Cc), a trace of anhydrite (Ah), but no portlandite (Pl).
- Figure 2b: fracture surface of a core from this cell after six months. Et mineralization (26%) covers fly ash grains and fills pore throats. M = 25% (close to the original moisture content); S about 13 MPa; K about 10^{-7} cm/s.

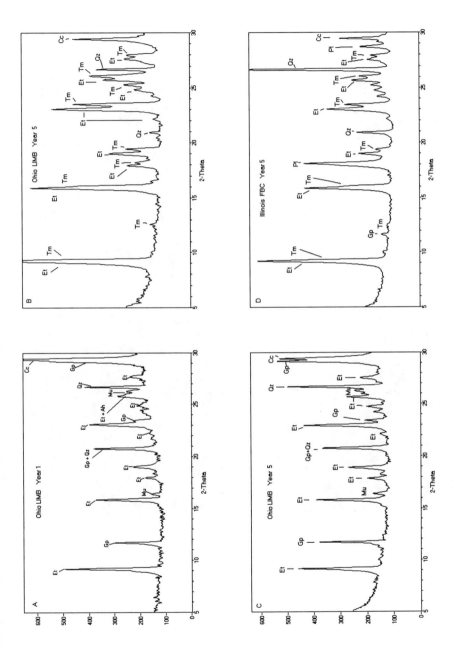

Figure 1. X-ray diffractograms of LIMB and FBC by-product core samples. (Vertical scale in counts/second; 2Θ for CuKα).

183

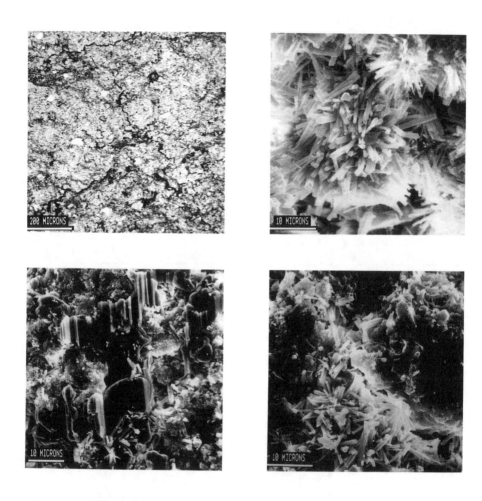

Figure 2.   LIMB by-product core.   (a) top left: macroview, backscatter image, 3 months; (b) top right: fracture surface, 6 months; (c) bottom left: fracture surface, 1 year; (d) bottom right: fracture surface, 2 years.

- Figure 2c: fracture surface; 1 year. M = 63%; S = 5 MPa; K (too fractured to test). Gypsum development, presumably from dissolution and precipitation reactions, is evident. After the first year, dissolution removed most of the gypsum from the upper cell areas. The dissolved calcium sulfate either participated in thaumasite formation, precipitated near the base of the cell, or remained in the leachate. The removal of gypsum from the matrix participates in strength losses and permeability increases.
- Figure 2d: fracture surface; 2 years. M = 78%; S = 12 MPa; K about 10^{-5}. This specimen is typical of ettringite mineralization in pore throats. C-S-H was observed near fly ash grains. The core had moderately high strength, but also relatively high permeability.

Figure 3 shows a series of fracture surface micrographs from LIMB by-product cores after three years of burial.

- Figure 3a: M = 102%; S = 0.4 MPa; K about 10^{-6} cm/s; XRD: Et = 39%, thaumasite (Tm) = 23%, Cc = 9%, no detectable Gp. The specimen is extensively cracked and has little strength.
- Figure 3b: M = 117%; S = 0.2 MPa; K about 10^{-5} cm/s. This micrograph shows an area along a large crack. The matted crystals to the left of the crack are thaumasite, and the crystals to the right are ettringite. EDS analyses indicate that the crystals do not have end member compositions, but are solid solutions. All ettringites contained some Si, and all thaumasites contained some Al.
- Figure 3c: higher magnification view of thaumasite crystals.
- Figure 3d: typical view of partial gypsum dissolution along a microfracture.

Figure 4 shows a series of fracture surface micrographs from LIMB by-product cores after five years of burial.

- Figure 4a: A grain from a depth of 1 m. After 5 years of exposure to the environment, this mm size aggregate mass showed some cracks and voids, but was compact overall. However, the core sample was friable with strength comparable to that of a compacted soil. XRD: 40% Et, 45% Tm, 5% Cc.
- Figure 4b: This higher magnification view of 4a shows porosity development at the expense of the more soluble minerals. Ettringite is found throughout, and thaumasite occurs on surfaces such as that shown at the crosshair.
- Figure 4c: Ettringite and thaumasite are both present in this micrograph. The diffractogram in Figure 1b comes from this region. This core was friable.
- Figure 4d: Core sample from the region which contained no thaumasite and had retained much of its original strength. Ettringite crystals were prominent. The area at the top right of the photograph gave a Ca-Si EDS spectrum, suggesting C-S-H contributes to the strength. The diffractogram in Figure 1c comes from this region.

Throughout the study, EDS analyses of crystal clusters indicated that both ettringite and thaumasite show some mutual solid solution behavior. The spectrometer software is set up to calculate atom fractions of common coal ash elements, and although EDS analyses of fracture surfaces are not expected to be highly accurate, it is instructive to consider the results (normalized to Ca = 6) for typical crystals in Figure 4:

Ettringite, nominally $Ca_6Al_2(SO_4)_3(OH)_{12} \cdot 26H_2O$ $Ca_6Al_{1.45}Si_{0.56}S_{2.80}$
Thaumasite, nominally $Ca_6Si_2(SO_4)_2(CO_3)_2(OH)_{12} \cdot 24H_2O$ $Ca_6Si_{1.92}Al_{0.60}S_{2.25}$.

As noted above, through hundreds of such crystal cluster analyses, no "pure end member" ettringite or thaumasite EDS spectra were observed. Qualitatively, ettringite crystals were identified by their habit, associations and EDS spectra, consisting of Al, Si, S and Ca. The EDS intensity ratio Al:Si varied from 2:1 to 1:2 in crystals identified as "ettringite." Ettringite exhibited the greater range of compositions, while thaumasite compositions were closer to the ideal 6Ca:2Si:2S. For example, in one of the 5 year core samples, a thaumasite analysis gave $Ca_6Si_{2.03}Al_{0.13}S_{2.14}$. Frequently other elements were detected in small amounts, e.g., K and Na (<0.1 per Ca_6), Mg (<0.1), Fe (<0.1), and Cl (<0.2). Additional analysis of the EDS results are underway. One aim is to look for evidence of a miscibility gap between the ettringite and thaumasite solid solution phases.

FBC By-Product. Figure 5 shows four high-magnification fracture surface views of the Illinois FBC by-product after one, two, and three years of exposure to the environment.

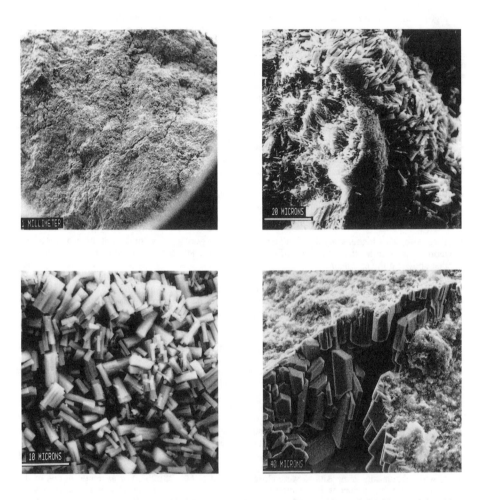

Figure 3.   LIMB by-product core after 3 years; fracture surfaces.   (a) top left; (b) top right; (c) bottom left; (d) bottom right.

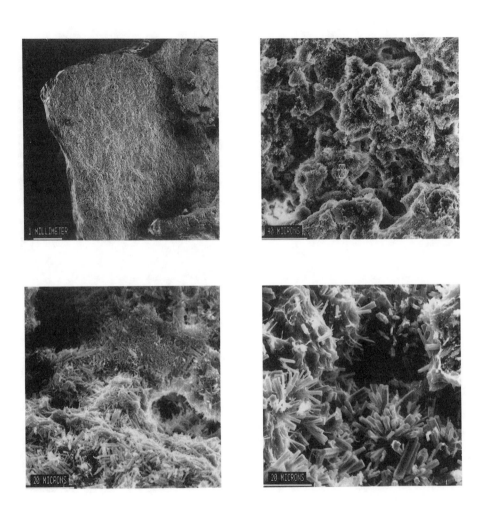

Figure 4.   LIMB by-product core after 5 years; fracture surfaces.   (a) top left; (b) top right; (c) bottom left; (d) bottom right.

Figure 5.   FBC by-product core; fracture surfaces. (a) top left, 1 year; (b) top right, 2 years; (c)
bottom left, 3 years; (d) bottom right, 3 years.

- Figure 5a: 1 year; M = 56%; S = <1 MPa; K about 10^{-5}; XRD: 25% Et, 23% Gp, 16% Pl, 8% Cc. This view of highly mineralized core shows ettringite laths covering grains and portlandite plates in pore throats.
- Figure 5b: 2 years; M = 72%; S = <1 MPa; K about 10^{-5}; XRD: 23% Et, 9% Gp, 17% Pl, 6% Cc. This core had little strength, and the micrograph shows a leached appearance, similar to that seen in the LIMB by-product after several years of exposure.
- Figure 5c: 3 years. XRD: 27% Et, 23% Tm, 5% Pl, <1% Gp, 8% Cc. The diffractogram in Figure 1d comes from this region. EDS spectra suggest an association of calcite and thaumasite in this region. Crystal habits of ettringite and thaumasite were not as well developed as those in the LIMB by-product at the same age.
- Figure 5d: 3 years; XRD: 28% Et, 5% Pl, 4% Gp, 8% Cc; Tm present. Typical view of the condition of the FBC material at the base of the test cell. Leaching is less apparent here after the 3 years. In the bulk XRD, gypsum and portlandite contents are much reduced.

Thaumasite

The appearance of thaumasite in test cell core samples was accompanied by loss of much of the initial strength. If thaumasite formation is not a major cause of strength loss, it is certainly associated with it. This is the first report (that the authors were able to find) of thaumasite formation in advanced coal technology by-products exposed to environmental conditions. However, thaumasite formation is well known as a degradation mechanism in conventional Portland cement concrete, where the combination of cold, wet conditions, sulfate-rich ground water, and carbonation promote attack of the C-S-H matrix and failure of the concrete. A recent report of thaumasite associated with failed concrete in the Canadian Arctic by Hemmings et al. [14] gives an up to date summary of the literature on the association of thaumasite with degradation of conventional concrete.

This study concerned disposal of two advanced coal technology by-products in landfill settings. Although the initial behavior of the material is similar to that of a low-strength concrete, the long term strength declines over the first few years of exposure, and permeability increases. Although the load-bearing strength remains adequate to support equipment, the characteristics of the weathered by-products resemble those of soils more than concrete [1,2]. The results of this study should be noted by those working on utilization of this class of by-products for civil engineering and construction, manufacturing of aggregates, etc. [15]. Initially promising 7-day or 28-day strengths may not be retained on exposure to the environment. In addition, the amount and chemical composition of leachates from by-product landfills are best modeled on the basis of those changing physical structures and altered minerals with which percolating waters react. The source term (solid material) for long-term geochemical modeling is not the assemblage of first-formed hydrated minerals (portlandite, gypsum, ettringite, plus calcite).

Additional research into the nature and role thaumasite in these by-products is recommended. One model developed from observations made to date suggests that thaumasite does not form at the expense of early ettringite. Calcium sulfate mobilized by early-formed gypsum dissolution, available carbonate as calcite, and silicon solubilized into the prevailing high-pH pore solutions combine to form thaumasite. The matrix, perhaps rich in early-formed, strength-producing C-S-H, is attacked by this reaction. The combination of increased porosity from gypsum leaching and matrix attack to form thaumasite leads to the loss of high early strengths.

ACKNOWLEDGEMENTS

This research was supported by the Department of Energy (Morgantown Energy Technology Center), with additional funding and other support from the Ohio Edison Edgewater Station (Ohio Site) and the Illinois Department of Natural Resources (Illinois Site). The prime contractor is Radian Corporation, Austin, Texas; A. Weinberg, Project Manager. NDSU students J.K. Solem-Tishmack, J. Rochocki-Krieg, J.E. Knell, J.A. Bender, and M.C. Oseto, and Dr. J. Longlet are thanked for their participation in the XRD characterization. Chemical analyses were performed by C.M. Lillemoen and D.J. Hassett of the University of North Dakota-Energy and Environmental Research Center.

REFERENCES

1. A. Weinberg, R.H. Petrini and R.D. Butler, Advanced Coal Technology Waste Disposal, draft final report prepared for U.S. Department of Energy, Morgantown Energy Technology Center (1993); A. Weinberg, B.J. Coel and R.D. Butler, Field Study for Disposal of Solid Wastes from Advanced Coal Processes: Ohio LIMB Site Assessment. Final Report, DCN 94-218-044-31, RCN 218-044-03-30, Radian Corporation, Austin Texas, October 1994.
2. R.D. Butler, D.W. Brekke, H.J. Foster J. Solc and G.J. McCarthy, "Mineral Diagenesis in Buried FBC Ash and Effects on Long-Term Physical and Chemical Behavior," in Proc. Seventeenth Biennial Low-rank Fuels Symposium, St. Louis, MO (UND-Energy and Environmental Research Center, Grand Forks, ND, 1993) pp. 515-530.
3. J.K. Solem and G.J. McCarthy, in Advanced Cementitious Systems, (Mater. Res. Soc. Proc. 245, Pittsburgh, PA, 1992) pp. 71-79.
4. J.K. Solem-Tishmack, Use of Coal Conversion Solid Residuals in Solidification-Stabilization Technology. M.S. Thesis, Department of Soil Science, North Dakota State University, Fargo, ND, 173 pp. (1993).
5. J.A. Bender, J.K. Solem, G.J. McCarthy, M.C. Oseto and J.E. Knell, in Adv. X-Ray Anal., Vol. 36, (Plenum Publ. Corp., New York, 1993) pp. 343-353.
6. G.J. McCarthy, J.A. Bender, J.K. Solem and K.E. Eylands, in Proc. Tenth Intern. Ash Use Symp., EPRI TR-101774s (Electric Power Research Institute, Polo Alto, CA, 1993), pp. 58/1-14.
7. G.J. McCarthy and J.K. Solem-Tishmack, in Advances in Cement and Concrete, ed. by M.W. Grutzeck and S.L. Sarkar (Am. Soc. Civil Engineers, New York, NY, 1994) pp. 103-122.
8. G.J. McCarthy, J.K. Solem, O.E. Manz and D.J. Hassett, in Fly Ash and Coal Conversion By-Products: Characterization, Utilization and Disposal VI, (Mater. Res. Soc. Proc. 178, Pittsburgh, PA, 1988) pp. 3-33.
9. G.J. McCarthy, D.M. Johansen, S.J. Steinwand and A. Thedchanamoorthy, in Adv. X-Ray Anal., Vol. 31 (Plenum Publ. Corp., New York, 1988) pp. 331-343; G.J. McCarthy and D.M. Johansen, Powd. Diff. 3, 156-161 (1988); A. Thedchanamoorthy and G.J. McCarthy, in Adv. X-Ray Anal., Vol. 32 (Plenum Publ. Corp., New York, 1989) pp. 565-572; G.J. McCarthy and A. Thedchanamoorthy, in Fly Ash and Coal Conversion By-Products: Characterization, Utilization and Disposal V, (Mater Res. Soc. Proc. 136, Pittsburgh, 1989) pp. 67-76; G.J. McCarthy and J.K. Solem, in Adv. X-Ray Anal., Vol. 34 (Plenum Publ. Corp., New York, 1992) pp. 387-394.
10. S.V. Mattigod and D. Rai, in Advanced Cementitious Systems, (Mater. Res. Soc. Symp. Proc. 245, Pittsburgh, PA, 1992), pp. 65-70.
11. E.E. Berry and E.J. Anthony (1987), in Fly Ash and Coal Conversion By-Products: Characterization, and Disposal III, (Mater. Res. Soc. Proc. 86, Pittsburgh, PA, 1987), pp. 159-170.
12. E.E. Berry, R.T. Hemmings, B.J. Cornelius and E.J. Anthony, in Fly Ash and Coal Conversion By-Products: Characterization, and Disposal III, (Mater. Res. Soc. Proc. 136, Pittsburgh, PA, 1989), pp. 9-22.
13. E.E. Berry, R.T. Hemmings and B.J. Cornelius (1991). Commercialization Potential of AFBC Concrete: Part 2. Mechanistic Basis for Cementing Action, Report GS-7122, Electric Power Research Institute, Palo Alto, CA, Vol. 2 (1991).
14. J.A. Bickley, R.T. Hemmings, R.D. Hooton and J. Balinski, in V. Mohan Malhotra Symposium: Concrete Technology Past, Present, and Future, ACI SP-144, P.K. Mehta, Ed. (Am. Concrete Institute, Detroit, 1994), pp. 159-175.
15. Coal and Synfuels Technology, Vol. 15, No. 42, October 31, 1994, p.1.

PERFORMANCE OF CEMENT-BASED SEAL SYSTEM COMPONENTS IN A WASTE DISPOSAL ENVIRONMENT

PHILIP G. MALONE*, LILLIAN D. WAKELEY*,
J. PETE BURKES*, AND EARL W. MCDANIEL**
*U.S. Army Engineer Waterways Experiment Station, Vicksburg, MS   39180-6199
**Oak Ridge National Laboratory, Oak Ridge, TN   37831

ABSTRACT

A grout based on portland cement, Class F fly ash, and bentonite clay was developed as part of the closure system of shallow subsurface structures for disposal of low-activity radioactive wastes. Heat output, volume change, and compressive strength of the sealing grout were monitored with time, at elevated temperature, and in physical models, to determine if this closure grout could maintain adequate volume stability and other required physical properties in the internal environment of the disposal structure.

To determine if contact with an alkaline liquid waste would cause chemical deterioration of the sealing grout, cured specimens were immersed in a liquid waste simulant containing high concentrations of sodium and aluminum salts. Over a period of 21 days at 60 °C, specimens increased in mass without significant changes in volume. X-ray diffraction of reacted specimens revealed crystallization of sodium aluminum silicate hydrate.

The new phase has an X-ray diffraction pattern similar to that of the commercial synthetic zeolite, Losod. Scanning electron microscopy used with X-ray fluorescence showed that clusters of this phase had formed in grout pores, to increase grout density and decrease its effective porosity. The testing was repeated at 100 °C for 5 days using a simulant containing sodium hydroxide and aluminum nitrate and the results were similar. Physical and chemical tests collectively indicate acceptable performance of this grout as a seal-system component.

INTRODUCTION

Tests of the effect of alkaline wastes on a portland-cement based grout were undertaken as part of a research program on development of sealing grouts for the DOE Hanford Grout Vault Program.[1-3] The grouts were developed to provide stable, cemented layers that fill the void between the top of the wasteform grout and the precast reinforced concrete panels that form the lid of the vault. This material was not designed specifically to serve in contact with the liquid waste. However, to establish what effect such an event would have on the performance of the void-filling grout, samples of cured grout were exposed to simulated liquid wastes under temperature conditions similar to those expected within the vaults.

After initial tests showed that the grout interacted chemically with simulated 106-AN liquid waste (Table I) exposure tests were repeated using a solution containing only sodium hydroxide and aluminum nitrate. The second series of tests demonstrated the causes of reactions that alter grout properties. These investigations allow prediction of the effects on the containment system of free waste liquid within the sealed vault.

METHODS AND MATERIALS

Grout Exposed to Simulated 106-AN Waste

A simulated waste liquid was developed by Oak Ridge National Laboratory to duplicate the chemical characteristics of the contents of Tank 106-AN at the DOE Hanford Reservation. Constituents of this simulated waste are listed in Table I. In preparing the simulant, all of the

Table I
Formulation of Simulated Waste Liquid 106-AN*

Compound	Quantity (gm) required to make 1 liter
$Al(NO_3)_3 \cdot 9H_2O$	158.0
Na_2SO_4	4.40
NaCl	8.77
NaF	0.34
$Ca(NO_3)_2 \cdot 4H_2O$	0.47
$Na_2B_4O_7 \cdot 10H_2O$	0.19
$Cu(NO_3)_2 \cdot 3H_2O$	0.01
$Fe(NO_3)_3 \cdot 9H_2O$	0.08
NaOH (50 mass % aqueous sol.)	188.7
$NaNO_3$	2.30
$NaNO_2$	52.3
Na_2CO_3	40.5
HEDTA	5.29
Na_4EDTA	1.83
Glycolic Acid (70 Mass % aqueous sol.)	4.56
SiO_2 (30 mass % aqueous sol.)	0.04
$Na_3PO_4 \cdot 12H_2O$ (sodium phosphate)	58.9
$Na_3C_6H_5O_7 \cdot 2H_2O$ (sodium citrate)	8.8

* Provided by Roger D. Spence, Oak Ridge National Laboratory.

constituents except for the last two listed were dissolved in 600 ml of distilled water at 40 °C. The sodium phosphate and sodium citrate were dissolved in 100 ml of water and added to the 600-ml batch just prior to dilution to 1000 ml.

Table II gives the formulation for the vault sealing grout developed at the WES and used in this test. Samples of hardened grout were sliced from test cylinders after moist curing at room temperature (23 °C) for 33 days. Each sample was a 10-mm-thick wedge-shaped slice cut from a 100-mm-diameter disk trimmed from the test cylinder. The initial mass of each slice was approximately 10 g. The chemical analyses of the cement, fly ash, and clay are given in Tables III, IV, and V, respectively.

Grout samples were exposed to the simulated wastes by immersing each sample in 25 ml of liquid simulant in separate containers. The sealed containers were maintained at 60 °C, and test samples were removed after 7, 14, and 21 days of exposure. Control samples were maintained in distilled water under identical conditions. At the end of each test period, four containers were opened; two experimental samples and two control samples were removed and washed with distilled water, dried, and weighed prior to examination.

Samples selected for examination using scanning electron microscopy (SEM) were vacuum-dried, fractured, and coated with a 15- to 20-nm-thick layer of gold. The samples were examined using an Hitachi H2500 SEM (Hitachi of America, Mountain View, CA) equipped

Table II
Quantities of Materials for 1 Cubic Meter of Grout

Materials	Weight (kg)
Aggregate and mineral admixture	
Basalt sand	524.5
Bentonite	24.0
Cementitious Materials	
Class H Cement	160.2
Class F Fly Ash (8.3% CaO)	911.4
Admixtures	
High Range Water Reducer (PSP powder)*	2.80
Set Retarder (Placewell LS, liquid)**	523 ml

Water/cementitious materials = 0.32
  * Prochem Technologies, Inc., Denver, CO
** Custom Chemical Inc., Los Angeles, CA

Table III
Chemical Composition of API Class H Cement Used in the Test Grout

Component*	Amount Present (mass %)
SiO_2	23.6
Al_2O_3	2.7
Fe_2O_3	3.5
CaO	61.8
MgO	3.4
SO_3	2.5
Loss on ignition	0.9
Insoluble residue	0.24
Na_2O	0.07
K_2O	0.52
Alkalies-total as Na_2O	0.41
TiO_2	0.19
P_2O_5	0.07
C_3A	2
C_3S	40
C_2S	42
C_4AF	11

* Testing was done using techniques described in ASTM C 114.[4]

Table IV
Chemical Composition of Class F Fly Ash Used in the Test Grout

Component*	Amount Present (mass %)
SiO_2	48.5
Al_2O_3	19.8
Fe_2O	17.6
Sum	85.9
MgO	0.8
SO_3	1.1
Moisture content	0.2
Loss of ignition	3.2
Available alkalies (28-day)	1.1

* Testing was done using techniques described in ASTM C 114.[4]

Table V
Chemical Composition of Bentonite Used in the Test Grout

Component*	Amount Present (Mass %)
SiO_2	77.64
Al_2O_3	16.40
Fe_2O_3	4.34
MgO	1.60
Na_2O	1.17
CaO	1.13
K_2O	0.37
P_2O_5	0.19
TiO_2	0.12
SO_3	0.08
Mn_2O_3	0.06
Total	103.10

* Analyses were done using a Kevex EDX-711 (Fisons Instruments, Valencia, CA).

with a Princeton Gamma Tech energy-dispersive X-ray analyzer (EDXA, Princeton Gamma-Tech, Princeton, NJ).

Samples selected for examination using X-ray diffraction analysis (XRD) were prepared by scraping grout from the outer 1-2 mm of the sample and gently breaking up the grout to separate the sand and paste. The samples were sieved to collect particles that passed a 45-μm screen. Diffractograms were prepared from packed powder samples using Ni-filtered Cu radiation on a Philips PW-1800 XRD unit (Philips Electronic Instruments Co., Mahwah, NJ).

<u>Grout Exposed to Sodium Hydroxide-Aluminum Nitrate Solution</u>

The exposure testing was repeated using a test solution that contained the concentrations of sodium hydroxide and aluminum nitrate shown in Table 1 but with all other constituents omitted. Grout cubes approximately 25 g in mass were exposed to the second test solution by sealing the samples and test solution in alkali-resistant containers and holding the containers at 100 °C for 5 days. Identical control samples were maintained immersed in distilled water under similar conditions. The same XRD and SEM procedures were used with the second sample set.

RESULTS

All experimental grout samples showed the same effects from exposure to either the 106-AN waste simulant or the sodium hydroxide-aluminum nitrate solution. All showed increases in mass. For the 10-g wedges, increases in mass averaged 12.5% over a 7-day period and 14.0% over a 21-day period. The 25-g cubic samples showed an average mass increase of 11.4% over 5 days. In both cases, the controls in distilled water showed minor ($<1\%$) losses in mass. No volume changes were noted for any of the samples.

Examination of the samples with SEM showed extensive alteration with most obvious change occurring as glass spherules in the fly ash were converted to ball-like clusters of crystals. Figures 1 through 3 are SEM photomicrographs. Figure 1 shows the grout prior to exposure. The grout has an open structure with fibrous and reticular networks of C-S-H, and glassy spherules have clean surfaces and little evidence of dissolution. Figure 2 shows a sample after 14 days of exposure to the 106-AN simulated waste. Note the ball-shaped clusters of crystals and the dense matrix. At 21 days, formation of this phase was so extensive that the ball-like clusters were less distinct (Figure 3). Chemical analyses obtained by EDXA showed characteristic lines for sodium, aluminum, and silicon in this newly formed phase. The XRD patterns showed that the phase forming in the exposure testing is a sodium aluminum silicate hydrate with a pattern similar to that of JPDF 31-1270, a commercial zeolite referred to as Losod (Figure 4).

DISCUSSION

The reaction of natural and synthetic glassy silicates with concentrated alkali solutions to form zeolites has been studied extensively.[5] Reactions of alkali solutions with Class C and Class F fly ashes, and with portland cement, have shown that zeolites and C-S-H can form and exist together in the same system.[6] The zeolite that formed in the present study, Losod, ($Na_{12}Al_{12}Si_{12}O_{48}\cdot$ x H_2O) was originally synthesized from reactions of aluminum and sodium-rich solutions containing an organic catalyst, such as bispyrrolidinium.[7] The work described here demonstrates that a low-sodium (Losod) zeolite can form in a glass-alkali reaction.

The formula developed for Losod indicates that the ideal unit structure has equal numbers of atoms of Na, Si, and Al. In the research done in precipitating Losod from solutions with organic Na/Al $\leq$ 1 and Si/Al $\approx$ 1. The fly ash in the test grout is assumed to be the source of silicon for Losod and the sodium and aluminum are contributed by the reacting solution. Sufficient solution to provide aluminum in the amount required to satisfy the requirement that Si/Al $\approx$ 1 provides a concentration of Na that is over 2 times the amount required by the ideal stoichiometry. It is probable that long-range ordering in the glassy fraction of the fly ash is influencing the formation of the new crystalline phase. The proportions of the ions available from the solid phase and reacting solution probably is a secondary factor.

The formation of a zeolite in this system differs markedly from the process in wasteform grouts, where the reactive ions are available in the system initially. In the present study, the samples were not exposed to the reactive ions of the waste until after the cement had hydrated and the grout had achieved measurable strength (near 7.5 MPa).

Figure 1. Fractured surface of grout prior to exposure to waste simulant.

Figure 2. Fractured surface of grout after a 14-day exposure to waste simulant, with crystalline masses after fly ash.

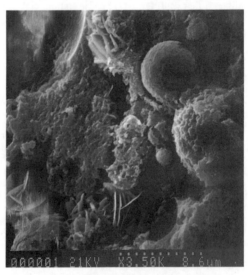

Figure 3. Fractured surface of grout after a 21-day immersion in AN-106 waste simulant, showing dissolution of fly ash spherules.

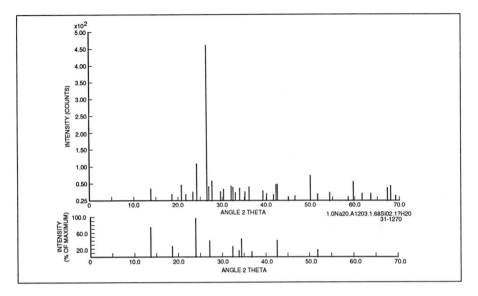

Figure 4. Comparison of XRD patterns for the grout after 21-day exposure to simulated waste (upper pattern) with the published pattern from JPDF 31-1270 (lower pattern).

The reaction that forms Losod in hardened grout appears to proceed with the aluminum-rich, alkali solution selectively dissolving the glass fraction of the fly ash in the hardened paste; then reprecipitating the silica as zeolite in the voids created by the dissolution and in the normal pore space present in the hardened grout. The end product of the reaction is Losod that fills the pore space in the cement paste and creates a solid mass between aggregate grains. The resulting mass appears to be denser and less permeable than the original grout, but the progress of the reaction as a replacement and a pore filler produces no major volume change.

CONCLUSIONS

Observations made on grout sample exposed to aluminum-rich alkali solutions used as waste simulants indicate that the interaction results in the formation of crystalline zeolite phase similar to the commercial zeolite, Losod. The reaction appears to proceed with the dissolution of the glassy portion of the fly ash in the grout. The result of the interaction is the formation of a dense surface layer with reduced porosity. The grout acted as a chemical barrier in that it removed sodium and aluminum from the waste simulant and the properties that make it useful as a physical barrier appear to be improved. The containment provided by the grout layer may be improved, not compromised, by exposure to the wastes.

ACKNOWLEDGEMENTS

This research was authorized under IAG DE-AI05-900R21921 from the DOE Field Office, Oak Ridge. Permission to publish this report was given by the Chief of Engineers. Citation of trade names does not endorse the use of commercial products.

REFERENCES

1. L.D. Wakeley, E.W. McDaniel, J. Voogd, J.J. Ernzen in Fourth CANMET/ACI International Conference on Fly Ash Silica Fume, Slag and Natural Pozzolans in Concrete, Supplementary Papers, (American Concrete Institute, Detroit, MI, 1992), pp. 1-14.

2. L.D. Wakeley and J.J. Ernzen in Advanced Cementitious Systems: Mechanisms and Properties, edited by F. P. Glasser, et al. (Mater. Res. Soc. Proc. 245, Pittsburgh, PA, 1992), pp. 117-122.

3. L.D. Wakeley and J.J. Ernzen, Grout for Closure of the Demonstration Vault at the DOE Hanford Facility, (U.S. Army Engineer Waterways Experiment Station, Vicksburg, MS, 1992, Technical Report SL-92-21), 25 pp.

4. Amer. Soc. for Testing and Materials, Standard Test Methods for Chemical Analysis of Hydraulic Cement, Designation: C 114-88, 1992.

5. R.M. Barrer, Hydrothermal Chemistry of Zeolites, (Academic Press, New York, 1982), p. 230.

6. M.J. Grutzeck, Air Force Office of Scientific Research Report, PSU-AFOSR-91, 1991.

7. W. Sieber and W.M. Meir, Helv. Chem. Acta 57, 1533 (1974).

MICROSTRUCTURAL ASPECTS OF ZEOLITE FORMATION IN ALKALI ACTIVATED CEMENTS CONTAINING HIGH LEVELS OF FLY ASH

A.R. BROUGH, A. KATZ*, T. BAKHAREV, G-.K. SUN, R.J. KIRKPATRICK, L.J. STRUBLE, AND J.F. YOUNG

Center for Advanced Cement Based Materials, University of Illinois at Urbana-Champaign, 204 Ceramics Building, 105 S. Goodwin Ave., Urbana, IL 61801, USA
* Present Address, Department of Civil Engineering, Technion-Israel Institute of Technology, Haifa 32000, Israel

ABSTRACT

Wasteforms made by reaction at elevated temperature of a highly alkaline simulated low-level nuclear waste solution, having high sodium ion concentration, with a cementitious blend high in fly ash have been studied. Significant formation of Na-P1 zeolite (gismondine framework) and of a sodalite occurred. The time evolution of the crystalline phases over the first 28 days is reported for both adiabatic and isothermal curing, and the role of these phases in microstructure development is discussed. The level of carbonate ions in solution was found to have a substantial effect on strength evolution and chemistry.

INTRODUCTION

Cement stabilization has considerable potential for the immobilization of low-level nuclear waste solutions and other hazardous wastes; reactions that occur due to the high pH of the pore solution reduce the mobility of many species of concern [1]. Zeolites have also found considerable application for immobilization of radioactive species; for example, clinoptilolite can selectively absorb species such as cesium ions [2] by ion exchange. *In situ* zeolite formation has been observed previously in alkali activated cement systems with high levels of pozzolanic admixtures [3,4]. Hoyle and Grutzeck have demonstrated the potential for removal of cesium from solution by alkali activated hydration of calcium aluminosilicate glasses [5].

In this paper we consider the phase evolution and microstructure of a cement-stabilized wasteform in which zeolite formation occurs spontaneously. The wasteform contains Na-P1 (gismondine framework) and sodalite zeolites, which may enable immobilization of both cationic and anionic species. The wasteform was generated by reaction at elevated temperature of a simulated alkaline waste solution with a solid blend containing a high proportion of fly ash. The composition of the full solution is given in Table I; sometimes a simplified solution was used (bold type in Table I.) The cementitious solids (dry

Table I: Composition of the full simulated low level waste solution. Components used to make up a simplified solution are shown in **bold type**

Compound	g/l
NaOH	**74.80**
Al(NO$_3$)$_3$.9H$_2$O	**128.00**
Na$_3$PO$_4$.12H$_2$O	**74.40**
NaNO$_2$	**36.90**
Na$_2$CO$_3$	**36.20**
NaNO$_3$	8.58
KCl	1.83
NaCl	2.50
Na$_3$Citrate.2H$_2$O	2.50
Na$_2$B$_4$O$_7$	0.13
Na$_2$SO$_4$	3.89
Ni(NO$_3$)$_2$.6H$_2$O	0.30
Ca(NO$_3$)$_2$.4H$_2$O	0.44
Na$_4$(EDTA).2H$_2$O	1.42
Na$_3$(HEDTA)	5.29
Glycollic Acid	0.65
Mg(NO$_3$)$_2$.6H$_2$O	0.03

Table II: Composition of the dry blend used for solidification of the simulated low level waste solution. A mix ratio of 1kg preblended solids to 1l solution was used.

Material	Amount	Details of Supplier
Cement	20.7%	Type II OPC ex Ash Grove Cement Co., Durkee, OR
Fly Ash	68.3%	Class F (but high Ca) ex Centralia, Ross Sand and Gravel Co., Portland, OR
Attapulgite	11.0%	Attapulgite Clay ex Engelhard Co., Iselin, NJ

blend) was added with a ratio of 1kg / 1 liter of waste solution, and had the composition given in Table II. This blend has been confirmed to give a wasteform meeting specific engineering specifications [6]. The low cement content, for example, is required to limit to acceptable values the heat rise occurring in large monoliths, which would be expected to hydrate under close to adiabatic curing conditions. The oxide composition of the solid phases is given in Table III.

Two major types of phases are produced in the hydration of this material, C-S-H, the binding phase of ordinary Portland cement pastes, and zeolites. The wasteform discussed in this paper loses strength after about 14 days, and we consider some of the compositional and microstructural factors which may cause this loss of strength.

Table III: Oxide analyses by XRF spectroscopy of the raw materials.

Oxide	Cement	Fly Ash	Clay
SiO_2	22.17	46.13	59.71
Al_2O_3	3.24	25.02	9.05
Fe_2O_3	4.24	7.25	3.17
CaO	64.48	8.02	3.13
MgO	1.13	1.81	11.21
K_2O	0.52	0.63	0.83
Na_2O	0.15	4.74	0.07
TiO_2	0.23	4.70	0.42
P_2O_5	0.12	0.42	1.47
MnO	0.06	0.03	0.05
SO_3	2.14	0.12	<0.1
LOI	1.20	0.53	10.64

EXPERIMENTAL

The solids were preblended by initial weighing and hand mixing; the mixture was then placed in a cylindrical container and rotated end-over-end at room temperature overnight. The solution was prepared by addition of the components sequentially to DI water at 50°C and was stored at 45°C in order to prevent precipitation of sodium phosphate. Solids, solution, and mixing implements were then allowed to equilibrate at 45°C in a temperature controlled room. In practice, the waste solution would be at an elevated temperature, and the solids at ambient temperature; all the components were heated to the same temperature in order to facilitate isothermal calorimetry. The liquid was poured into a Hobart mixing bowl at 45°C, and the solid added gradually over 15s, with slow speed stirring. The paddle and bowl was then scraped down, and mixing continued for 2-3 minutes at intermediate speed.

The wasteform was cast into 25mm diameter tubes which were sealed and placed into a custom built programmable oven. Unless otherwise noted, the temperature was raised in a series of ramps to simulate the expected adiabatic temperature profile [6]. After 3 days, when the exothermic hydration was essentially complete, and the temperature had risen to 90°C, the sample tubes were transferred to an isothermal oven for long term curing. 50mm length sections were sawn out of the cylinders for compressive strength measurements, and samples were also studied by powder x-ray diffraction (XRD).

RESULTS

Compressive strength data for batch A of samples, prepared from the solids and full solution given above, are shown for the first three months of hydration in Figure 1. The strength rises to 8.5 MPa at 14 days, but then falls significantly thereafter. Similar results to those of batch A have been observed by us for samples hydrated with the simplified waste solution, Batch B. (See also [7].) On the other hand, samples prepared by researchers at Northwestern University [8], with a lower liquid to solids ratio, continue to gain strength after 14 days; only minor differences have been identified in the physical processes of blending the solids, and preparing the samples, although some of the chemicals used to prepare the solutions were obtained from different suppliers.

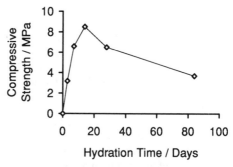

Figure 1: Compressive strength data for samples prepared with the full solution given in Table I and the solids given in Table II, and cured adiabatically for 3 days, and thereafter at 90°C.

An XRD spectrum of the preblended solids is shown in Figure 2(a); peaks due to the crystalline components of the materials can be seen in addition to the broad peak in the region 20-32° 2θ arising from the glassy phase of the fly-ash.

After 6 months of hydration of batch A of the wasteform the major zeolitic product is a sodalite, and some gismondine type zeolite (Na-P1) is also present (Figure 2(b)). The exact

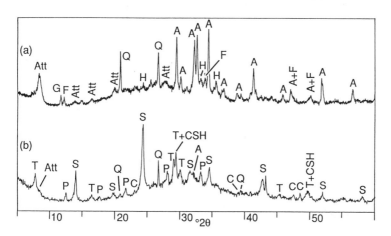

Figure 2: Powder XRD spectra, acquired with CuKα radiation, of (a) the dry blended solids, and (b) the 6 month old wasteform (Batch A, hydrated adiabatically for three days, and thereafter at 90°C). The major peaks are annotated as follows: A = alite, Att = attapulgite, C = calcium carbonate (calcite), F = ferrite, G = gypsum, H = hematite, M = mullite, L = a phase similar to Linde type A zeolite, P = Na-P1 (gismondine) zeolite, Q = quartz, and S = a sodalite. The spectra were acquired under similar conditions, but are not quantitative.

composition of the sodalite has not been determined, and it is possible that the cages are filled with salts, displacing some or all of the zeolitic water. Traces of a chabazite zeolite are also observed. The broad peak from 20-32° 2θ due to the glassy phase of the fly-ash is significantly reduced in intensity, and there is a broad peak due to C-S-H in the region 25-35° 2θ; additionally some tobermorite has formed. The assignments of the calcium silicate hydrate phases were confirmed by the loss of the peaks assigned to these phases upon extraction of the material with salicylic acid in methanol [9]. Partial reaction of the attapulgite clay is also observed.

Samples were also withdrawn and quenched from batch A at an early stage of hydration, in order to follow the evolution of the crystalline phases of this system. Both isothermal hydration at 45°C (for comparison with calorimetry data) and adiabatic hydration were studied; powder XRD spectra of the region from 5 to 20° 2θ were acquired, and selected spectra are presented in Figure 3. In each case, there is initial formation of an AFm phase; at later stages the peaks due to the AFm phase disappear, and simultaneously peaks due to the sodalite appear. These changes occur faster with adiabatic curing, Figure 3(a), than for isothermal curing, Figure 3(b), as would be expected given the considerably higher temperatures for adiabatic curing. Minor peaks due to traces of tobermorite and Na-P1 were found to increase slowly with time, through to 6 months.

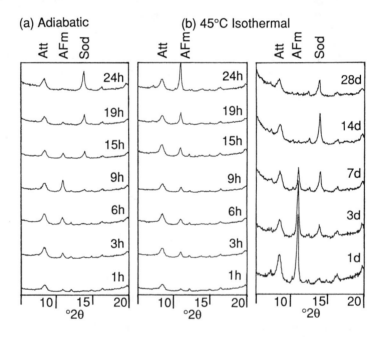

Figure 3: Selected region powder XRD spectra showing the phase evolution of batch A under both (a), adiabatic and (b), isothermal conditions; the spectra were acquired from samples quenched with MeOH and air dried. The spectra are annotated as follows: Att = attapulgite, AFm = an AFm, and Sod = a Sodalite.

Batch B, in which the solid blend was hydrated with the simplified waste solution showed a similar loss of strength to batch A. In addition samples were prepared in which only 4 of the 5 most significant components of the simplified solution were included. These all gave low strengths at 28 days, with one exception: Batch C, in which carbonate was omitted (i.e. the solution contained only NaOH, Al(NO₃)₃.9H₂O, Na₃PO₄.12H₂O, and NaNO₂), gave a very high strength, suggesting that no strength loss had occurred in this material. XRD spectra of batches B and C are shown in Figure 4(a) and (b) respectively; large differences in phase composition occur. Batch B contains some tobermorite, and some amorphous C-S-H as shown by the broad peak from 25-35° 2θ; Batch C appears to contain more amorphous C-S-H, but less crystalline tobermorite. The zeolitic phases also differ; batch B contains predominantly sodalite, while batch C contains predominantly Na-P1 with smaller quantities of sodalite.

The initial hydration at 45°C was studied by isothermal calorimetry, as shown in Figure 5 for batch A. After an initial heat output over the first 5 hours or so, there is a distinct exotherm which occurs at approximately 20 hours, which is the same time at which formation of the majority of the AFm phase occurs (as shown by XRD) under isothermal conditions. Results are also shown for hydration of a sample of batch B (Figure 5(b)), prepared from the simplified waste solution; the overall form of the heat output is similar, but the system is rather less exothermic, and the peak due to AFm formation is delayed from about 20 hours to about 30

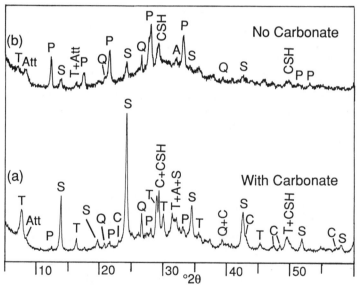

Figure 4: Powder XRD spectra of samples prepared by hydration at 90°C of the standard dry blend with (a), the simplified solution (Batch B; solution contains NaOH, Al(NO₃)₃.9H₂O, Na₃PO₄.12H₂O, NaNO₂ and Na₂CO₃), and (b), with a similar solution from which the carbonate has been omitted (Batch C; solution contains only NaOH, Al(NO₃)₃.9H₂O, Na₃PO₄.12H₂O and NaNO₂). The spectra are annotated as follows: A = alite, Att = attapulgite, CSH = calcium silicate hydrate, C = calcium carbonate (calcite), P = Na-P1 (gismondine) zeolite, Q = quartz, S = a sodalite, and T = tobermorite. The spectra are not quantitative.

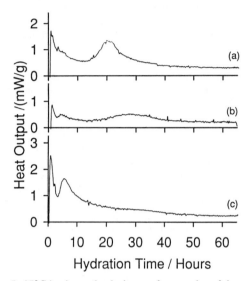

Figure 5: 45°C isothermal calorimetry for samples of the wasteform prepared from the dry blend given in Table II with (a), the full solution given in Table I (Batch A), or (b), the simplified solution (Batch B), or (c) the simplified solution with carbonate omitted (Batch C).

hours. Calorimetry of batch C gave a very different curve, Figure 5(c), with a large exotherm at approximately 6 hours, and only a minor peak at around 20 to 30 hours.

SEM micrographs of a representative fracture surface from a sample of the full wasteform, batch A, are presented in Figure 6. Qualitative EDX analyses indicate the presence of a wide range of products, from C-S-H's of variable composition to materials which also contain high levels of Na and Al, in addition to Ca and Si, and are probably zeolitic.

A number of model systems were also studied by XRD, and by SEM of fracture surfaces: for example, fly ash hydrated with 5M NaOH at 90°C gives a high yield of a zeolite which has an XRD pattern similar to that of the hydro-sodalite $Na_8[AlSiO_4]_6(OH)_2.4H_2O$ [10] (JCPDS file 41-9), and additionally some C-S-H. The microstructure, shown in Figure 7(a) shows quite large angular crystals of the hydro-sodalite intermixed with fly ash spheres which are coated with a thick layer of reaction products.

Fly ash reacted with the simplified waste solution, again at 90°C, gives rather different products, with substantial quantities of chabazite, and of a zeolite similar to Linde type A being formed (as determined by XRD) in addition to some C-S-H; the microstructure (see Figure 7(b)) shows much less of a continuous coating of reaction products upon the fly ash spheres; crystals with distinctive morphologies are observed, which are tentatively assigned as products similar to the zeolites Linde type A (the cubes), and chabazite (the groups of intersecting platelets), by comparison to literature reports [11,12]. A compressive strength of 9.86 MPa is reached at 28 days.

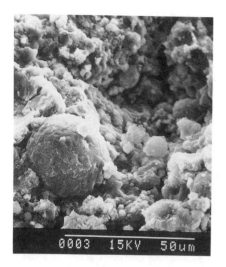

Figure 6: SEM micrographs of a representative fracture surface from a sample of the full wasteform (Batch A), hydrated adiabatically for 3 days, and thereafter for six months at 90°C.

Figure 7: SEM micrographs of representative fracture surfaces; (a) Fly-ash reacted with 5M NaOH at 90°C, (b) Fly ash reacted with the simplified waste solution at 90°C.

DISCUSSION

As noted above, after an initial gain in strength to approximately 14 days, the samples of wasteforms reported in this study lost strength through to 6 months (Figure 1); the major phases observable by XRD during this period of strength loss were amorphous calcium silicate hydrate, crystalline tobermorite, a sodalite which has not yet been fully characterized, and Na-P1 (gismondine) zeolite, in addition to residual starting materials and calcite. In the time period of strength loss, relative to the other phases, tobermorite increased considerably, and Na-P1 zeolite also increased somewhat. Thus it is possible that causes for the loss of strength might include crystallization of amorphous C-S-H to tobermorite, or recrystallization of zeolite to form Na-P1, which has a lower density and higher water content than sodalite.

Our results for progressive dilution of a simplified version of the waste solution [7] show that in the system studied here, the normal hydration of OPC, which would be expected to be significantly accelerated by the high pH, is profoundly retarded; there is no large exotherm in the first few hours. The cement does react at a later stage, however, since the majority of the alite reacts during the first month.

The removal of carbonate from solution was found to give a wasteform (batch C) with high strength at 28 days. This high 28 days strength suggests that no strength loss occurred in this carbonate free system. The isothermal calorimetry also indicates a large exotherm at about 6 hours, which we postulate to arise from the hydration of the OPC in the system. This suggests that in the full wasteform, the hydration of the cement is severely retarded by inclusion of carbonate in the solution, perhaps by precipitation of a layer of calcite, or of a carbonate AFm, around the cement grains. By XRD and SEM, however, significant hydration of the OPC has occurred after 28 days in all of the systems.

The early age XRD spectra of batch A cured adiabatically showed that after initial formation of an AFm type phase in the period from 3-9 hours, the AFm peaks reduced, in the period from 9 - 24 hours, and sodalite peaks grew simultaneously; it is reasonable to suggest that an interconversion reaction takes place. The overall stoichiometry cannot be postulated, since the exact stoichiometry of the AFm, and of the sodalite are both unknown. Olson et al [8] have shown that around the time of this change, in a similar system, the concentration of sodium, aluminum, and silicate in solution fall significantly, as does the bulk permeability of the system.

The time at which the cement reacts will be of considerable importance, since it is likely to determine the composition of products; the cement is high in calcium and silicate, while the fly ash glass will be relatively high in aluminum. (pH is also likely to have an important role [12].) Further work is required to assess the implications of these differences; distinct differences in the composition of the products, however, are observed. The full and simplified wasteforms (batches A and B) have a sodalite (probably Al/Si $\approx$ 1), suggesting a rather lower availability of silicate at the time the zeolite formed, while the carbonate free system (batch C) has an Na-P1 zeolite (Al/Si < 1), suggesting a rather greater availability of silicate at the time of zeolite formation. Since the zeolites form at relatively early age, less than one day, this is consistent with the blocking of the surface of the cement in the full wasteform, which will significantly reduce the availability of silicate species, and which might result in the formation of the zeolite with the higher Al/Si ratio.

While this postulate does not account for the loss of strength beyond about 14 days in some of the systems it does indicate that the systems differ profoundly in chemistry and this may then trigger the change that causes the loss of strength. Note that relatively minor changes in microstructure may be sufficient to cause strength loss in these systems with very high liquids to solids ratios. A number of possible changes may be suggested:

1. There may be a partial recrystallization of amorphous C-S-H to tobermorite, causing a loss of the binding phase, and subsequent loss of strength. (There is no loss of strength associated with tobermorite formation in autoclaved systems, but the wasteform studied here has a much higher water to solids ratio.)

2. As more of the fly ash hydrates, the additional silicate species introduced into the calcium deficient system may result in the recrystallization of sodalite to Na-P1 zeolite.

3. Differences in hydration at early ages may cause differences in the stage and extent of reaction of either the fly ash or the clay.

Microstructurally, an aggressive solution such as 5M NaOH appears to generate significant reaction of the fly ash, as shown in Figure 7(a); a hydro-sodalite with an expected Al/Si ratio of 1 is formed, and there is additional formation of coatings of amorphous material around the fly ash grains. A less aggressive solution, such as the simplified solution described in Table I of this paper, led to less extensive reaction, as shown in Figure 7(b); the zeolites formed were found to have a lower Al/Si ratio, despite the initial presence of aluminum in the solution, and there seem to be considerably less amorphous products. In both cases, the zeolites can be clearly identified morphologically, and have been confirmed by qualitative EDX analyses to be sodium aluminosilicates, perhaps with some calcium substitution.

The microstructure of the full wasteform (Figure 6) is much more complex, though EDX analyses show areas in the gel which have a composition approaching that expected for C-S-H, or for a sodalite. The various phases appear to be coated with a layer of C-S-H, which partially masks their distinctive morphology. Additionally, the crystals are much smaller, and hence harder to identify.

Olson et al [8] have shown that the permeability of a system similar to that studied here falls to a value similar to that for an OPC paste with much lower water content; during the hydration process there is a significant drop in permeability which appears to occur at the same time as zeolite formation.

CONCLUSIONS

Wasteforms made by reaction of a cementitious blend high in fly ash with a simulated highly alkaline waste stream under adiabatic conditions exhibit rather unusual chemistry; significant zeolite formation occurs, in addition to the more normal amorphous and crystalline calcium silicate hydrates. These wasteforms are sometimes found to lose strength after approximately 14 days; the cause of this strength loss has not been determined.

The chemistry of the wasteforms has been found to be quite sensitive to the initial solution composition. In particular, it has been found that the identity of the zeolites formed, the form of the early age isothermal heat output, and the quantity and crystallinity of the calcium silicate hydrates produced are strongly dependent upon the level of carbonate ions in the solution. On the basis of isothermal calorimetry, and by comparison with the results from dilution experiments [7] it is postulated that carbonate ions are responsible for severely retarding the normal hydration of the OPC in these systems.

Further work is required to confirm the role of the carbonate ions, to ascertain what causes the loss of strength in these systems, and to determine if the permeability is adversely affected by the process causing the strength loss.

ACKNOWLEDGMENTS

This work was funded by grant MJGSVV-097600 from the US Dept. of Energy. We thank the anonymous reviewer for helpful comments, R. Olson, P. Tennis and H. Jennings for useful discussions, and the Ross Sand and Gravel Co., the Ash Grove Cement Co. and the Engelhard Co., for the supply of materials..

REFERENCES

1 F.P. Glasser and D.E. Macphee, Mater. Res. Soc., Bull., **18** (3), 66 (1993)
2 L.L. Ames, Jr., Amer. Min., **45**, 689 (1960)
3. S. Kaushal, D.M. Roy, P.H. Licastro and C.A. Langton, in <u>Fly Ash and Coal Convertion By-Products: Characterization and Disposal II</u>, edited by G.J. McCarthy, F.P. Glasser and D.M. Roy, (Mater. Res. Soc. Proc. **65**, Pittsburgh, PA, 1986), p311.
4 J.L. LaRosa, S. Kwan and M.W. Grutzeck, J. Amer. Ceram. Soc., **74**, 1574 (1992)
5 S.L. Hoyle and M.W. Grutzeck, J. Amer. Ceram. Soc., **72**, 1938 (1989)
6 R.O. Lokken and P.F.C. Martin, Report HGTP-91-0303-01: R.O. Lokken, P.F.C Martin and S.E. Palmer, Report HGTP-93-0302-01, Pacific Northwest Laboratory, Richland, WA
7 A. Katz, A.R. Brough, T. Bakharev, R.J. Kirkpatrick, L.J. Struble and J.F. Young, in <u>Microstructure of Cement Based Systems / Bonding and Interfaces in Cementitious Materials</u>, edited by S. Diamond et al (Mater. Res. Soc. Proc., Pittsburgh, PA, 1995) in press
8 R.A. Olson et al, to be submitted to J. Hazardous Waste
9 W.A. Gutteridge, Cem. Concr. Res., **9**, 319 (1979)
10 J. Felsche and S. Luger, Ber. Bursenges. Phys. Chem., **90,** 731-736 (1986)
11 J. Ciric, Science, **155**, 689 (1967)
12 U.Barth-Wirsching and H. Holler, Eur. J. Mineral., **1**, 489 (1989)

LLW SOLIDIFICATION IN CEMENT - EFFECT OF DILUTION

A. KATZ*, A.R. BROUGH, T. BAKHAREV, R.J. KIRKPATRICK,
L.J. STRUBLE, AND J.F. YOUNG

Center for Advanced Cement Based Materials, University of Illinois at Urbana-Champaign,
204 Ceramics Building, 105 S. Goodwin Ave., Urbana, IL 61801, USA
* Present Address, Department of Civil Engineering, Technion-Israel Institute of
Technology, Haifa 32000, Israel

ABSTRACT

A simulated Low Level nuclear Waste (LLW) solution was tested for long term solidification in a cement-based matrix. The waste is characterized by high pH and high concentrations of sodium, aluminum, nitrate, nitrite, phosphate and carbonate. The effect of diluting the waste with additional water was studied. The cementitious matrix was composed of cement, fly-ash and clay (21%, 68% and 11% respectively) with high solution to solid ratio (1 liter / 1 kg.). Mixes were prepared at 45°C and cured at 90°C for 28 days.

Maximum 28 day compressive strengths and early age heat evolution were achieved by diluting the LLW solution to approximately 67% of its original concentration. More dilution led to a lower heat evolution and compressive strength. No dilution was found to give lower compressive strength, and a heat evolution that was delayed, and lower in intensity. XRD spectra showed formation of zeolites and tobermorite at the higher concentrations (67, 85, and 100% of the concentration of the undiluted simulated LLW), with a change from Na-P1 zeolite for 67% of the undiluted concentration to a sodalite at 100%. SEM observations showed a porous system for the low and high dilution rates but a less porous one for an intermediate level of dilution.

INTRODUCTION

Solidification in a cement matrix has been found to be an effective method for the disposal of low level nuclear waste (LLW) [1,2,3]. Many of the LLW's consist of solutions, characterized by high concentrations of sodium, aluminum, nitrate, nitrite, phosphate and carbonate ions. In this paper the solidification of a highly alkaline waste is considered. Cementitious systems containing high levels of fly-ash or granulated blast furnace slag have been found to be suitable for this purpose [4,5,6]; replacing much of the Portland cement with less reactive mineral admixtures reduces the adiabatic heat output.

EXPERIMENTAL PROCEDURE

The low level waste was simulated by 5 species as shown in Table 1. A series of experiments were performed in which this solution was diluted to concentrations of 85%, 67%, 33% and 5.5% of the tabulated concentrations. The solubility of the sodium phosphate is strongly affected by temperature (its solubility in water ranges from 15g/l at 0°C to 1570g/l at

Table I: Composition of the simulated
waste before dilution.

Compound	Amount, g/l	M
NaOH	74.85	1.87
Al(NO$_3$)$_3$.9H$_2$O	128.0	0.34
Na$_3$PO$_4$.12H$_2$O	74.4	0.20
NaNO$_2$	36.9	0.54
Na$_2$CO$_3$	36.2	0.34

Table II: Oxide composition
(% weight) of dry materials.

Oxide	Cement	Fly Ash	Clay
SiO$_2$	22.17	46.13	59.71
Al$_2$O$_3$	3.24	25.02	9.05
Fe$_2$O$_3$	4.24	7.25	3.17
CaO	64.48	8.02	3.13
MgO	1.13	1.81	11.21
K$_2$O	0.52	0.63	0.83
Na$_2$O	0.15	4.74	0.07
TiO$_2$	0.23	4.70	0.42
P$_2$O$_5$	0.12	0.42	1.47
MnO	0.06	0.03	0.05
SO$_3$	2.14	0.12	<0.1
LOI	1.20	0.53	10.64

70°C), so the solution was kept at 45°C in all cases to prevent precipitation of sodium phosphate.

The solid materials were a type I/II Portland cement (Ash Grove, Durkee, OR), a type F fly-ash (Centralia, Ross Sand and Gravel Co., Portland, OR) and an attapulgite clay (Engelhard, Iselin, NJ). The oxide composition of the solid materials is presented in Table II. Proportions by weight of 21% cement, 68% fly ash, and 11% clay were blended by end over end rotation in a cylindrical container for approximately 12 hours. A mixing ratio of 1 liter of waste solution to 1 kg of solids was utilized; the clay acted as a mixing aid, and prevented the slurry from segregation at this high solution/solid ratio. As is discussed below, however, there were cases in which some of the attapulgite reacted with the solution.

The solution was mixed in a Hobart mixer for 30 seconds at low speed, then the solid blend was added and mixing was continued for an additional 2-3 minutes at medium speed. A temperature of 45°C was used in order to mimic the effects of radiolytic heating. Immediately after mixing, samples where placed in the cells of an isothermal calorimeter (Digital Site Systems, Pittsburgh, PA). Considerable care was taken to ensure that the materials, calorimeter and mixing equipment were all maintained at the same temperature prior to mixing, in order to minimize the time taken for the sample temperatures to equilibrate when transferred to the calorimeter. Slurry was also poured into small tubes having a diameter of 25 mm, which were sealed and stored in a programmable oven for simulated adiabatic curing to simulate the temperature rise of a large volume casting. The adiabatic curing regime was developed earlier by Lokken et al. [7] for similar mixes with the undiluted solution but it was also used for the diluted mixes in order to maintain similar conditions for all the samples.

For compressive strength measurements, the samples were cooled to room temperature, demolded and sectioned to produce a cylinder of 25 mm diameter and 50 mm length, which was tested 24 hours later. Three samples were tested for each data point, and the mean value is reported; the extimated population coefficient of variation (C of V) for the individual cylinders was generally < 10% (3-5% in most cases); in the case of the dilution to 5.5%, which gave a very low strength, the C of V was 20%. Additional batches of wasteform gave similar results. The remaining parts of the samples were used to acquire x-ray powder diffraction (XRD) spectra, and to perform scanning electron microscopy (SEM) of fracture surfaces.

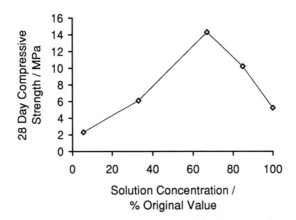

Fig. 1: Graph showing mean 28 day compressive strength results for pastes
hydrated adiabatically with simulated LLW solution which had
concentrations in the range of 5.5% to 100% of the values given in Table I.

RESULTS

Compressive strength

The 28 day compressive strengths of mixes of different concentrations are presented in
Figure 1. It can be seen that increasing the concentration of the simulated LLW from 5.5% to
33% of the values given in Table I has only a modest effect on the compressive strength.
Increasing the concentration to 67% of the tabulated values produces a more substantial
increase in compressive strength (from approximately 2 MPa up to 14.3 MPa for an increase of
the simulated LLW concentration from 5.5% to 67%.) However, further increase of the
concentration led to a reduction in the compressive strength (10.2 MPa and 5.0 MPa for LLW
concentrations of 85% and 100% of the tabulated values, respectively).

The density of the undiluted waste solution is 1.2kg/l, and after allowing for the mass of
dissolved salts, and for the release of waters of crystallization upon dissolution, this indicates a
water content of 950g/l, which is similar to the 1000g/l expected for pure water. Thus the
water to cement ratio increases only very slightly upon dilution of the waste solution.
Nevertheless, the dissolved salts will have a high water demand, and so the effective water to
cement ratio will be increased somewhat upon dilution; the consequences of this change are
considered to be small, however, in comparison with the very large changes observed in the
hydration behavior.

Isothermal calorimetry

For mixes with low and moderate solution concentration, the shape of the curve of the rate
of heat development (Figure 2) is quite similar to typical curves presented in the literature [8],

(see for example the line for hydration with LLW solution having concentrations of 33% of the values given in Table I). An initial peak occurs at around 1 hour, and is attributed to heat of wetting and dissolution; this peak is partly obscured due to the initial equilibration of the sample temperature from the temperature of the materials and mixer to that of the calorimeter. A major peak representing the hydration of the cement can be seen (at about 5 hours for LLW with concentrations of 33% of the tabulated values), having a shoulder on its descending branch (at about 10 hours). Based upon well-established trends in the published literature, the peak and shoulder are attributed to hydration of the C_3S and C_3A phases in the cement.

Increasing the concentration of the simulated LLW solution from 5.5% to 67% of the values given in Table I (Figure 2(a) - (c)) results in an increase in the rate of heat development and of 28-day strength. These results are consistent with an increase in the conventional hydration of the Portland cement component as the concentration of NaOH increases. NaOH is a well known accelerator and because it is the

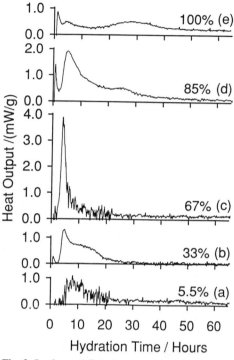

Fig. 2: Isothermal Calorimetry curves for pastes hydrated at 45°C with simulated LLW solution which has concentrations from 5.5% to 100% of the values given in Table I.

species of highest concentration its influence probably dominates in these dilute solutions. However, as the concentration of the LLW solution is increased above about 67% (relative to the values given in Table I) calorimetry shows that conventional hydration of the cement component rapidly diminishes (Figure 2 (d) and (e)). Studies with the undiluted solution with the cementitious blend [9, 10] have shown that aluminate and phosphate are rapidly lost from solution. We conjecture that an amorphous coating containing one or both of these species may be the cause of retardation. Studies by Brough et al. [11] indicate that carbonate ions may also affect hydration. In other words the accelerating tendencies of NaOH no longer dominate. An additional peak which appears with the more concentrated solutions at about 25-30 hours after mixing, indicates a radical change in hydration chemistry. XRD spectra of a closely related system hydrated under the same conditions[11] indicate that during this time AFm is formed. Later it is converted to a zeolite. This conversion to the denser crystalline products is consistent with the observed lower strengths, and is discussed in more detail in the next section.

Fig. 3: SEM micrographs of paste prepared with solution concentrations, relative to the values given in Table I, of (a) 5.5%, (b) 67%, and (c) 100% at the age of 28 days.

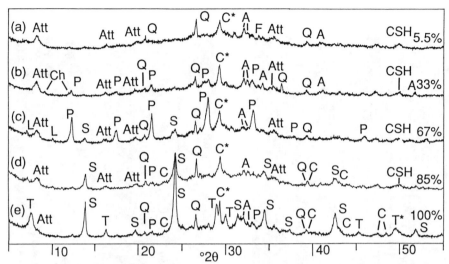

Figure. 4: Powder XRD spectra of one month old samples hydrated adiabatically with solution of the concentration given, relative to the concentrations given in Table I, acquired with Cu$_{K\alpha}$ radiation. The spectra are annotated as follows: A = alite, Att = attapulgite, CSH = calcium silicate hydrate, C = calcium carbonate (calcite), C* = overlapping CSH and / or calcite, Ch = chabazite zeolite, L = zeolite similar to Linde type A, P = Na-Pl (gismondine) zeolite, Q = quartz, S = a sodalite, T = tobermorite, T* = tobermorite + CSH. The spectra are not quantitative.

<u>Microstructure and Composition</u>

The microstructure of the various pastes was examined using SEM. Figure 3 presents typical fields from fractured surfaces of pastes hydrated with simulated LLW solutions which have concentrations of 5.5%, 67% and 100% of the original composition. For the paste of 67%, shown in Fig. 3(b), a denser and more continuous C-S-H field is seen with less voids. Unlike the other two samples individual particles of the cementitious blend can no longer be distinguished. These observations correlate in a broad way with the changes in compressive strength results. The very porous structure of the paste with the most dilute solution (5.5%) corresponds to the low compressive strength, while the pastes of 67% and 100% correspond to high and intermediate strength, respectively.

XRD spectra for materials hydrated with a range of solution concentrations are shown in Fig. 4. The alite from the cement has reacted significantly; at the highest LLW concentration (100% of the values given in Table I) some crystalline 11Å tobermorite is formed. The broad amorphous peak from approximately 25-35° 2θ can be attributed to C-S-H; rather less of this material appears to be present at the highest solution concentration. These phases were confirmed as calcium silicate hydrates by salicylic acid extraction.

Zeolite formation is favored in the samples that have the high NaOH concentration. Sodalite is favored at highest concentration and Na-Pl (gismondine) at intermediate levels. These zeolites have been reported in similar systems [10,11]. Linde Type A and chabazite were identified as minor components. Although only two peaks characterized these latter phases, a positive identification is based on comparison with related systems where these two phases have been observed in greater abundance. Only in the most dilute solution (5.5% of the concentrations given in Table I) were no zeolites detected.

The attapulgite clay appears not to react significantly with the diluted wastes, but in the paste hydrated with undiluted LLW solution the clay is partially reacted at 1 month. Small quantities of quartz, originating from the fly ash and clay, are also observed in all of the spectra. Calcite, possibly arising from the carbonate species in solution is observed in some or all of the samples.

SUMMARY AND CONCLUSIONS

The effect of dilution of a simulated Low Level nuclear Waste (LLW) for the purpose of solidification in a cement matrix was studied. Compressive tests at the age of 28 days showed an increase in the compressive strength of the pastes as the LLW solution concentration increases from 5.5% to approximately 67% of the values given in Table I. However, further increase in the solution concentration led to a reduction in the compressive strength. These changes in strength are consistent with changes in the extent of conventional hydration of the cement, as measured by isothermal calorimetry, SEM, and XRD.

SEM observations on fractured surfaces showed a relatively dense structure of the paste prepared with simulated LLW solution having concentrations of 67% of the values tabulated. Pastes of 100% and 5.5% concentration were more porous, which agrees with the findings from compressive strength measurements.

Isothermal calorimetry at 45°C showed an increase in the heat development as the solution concentration increased up from 5.5% to approximately 67% of the values given in Table I. For higher concentrations the normal hydration of the cement appears to be profoundly retarded. X-ray powder diffraction shows formation of zeolitic phases in the pastes for concentrations of approximately 33% of the values given in Table I and higher; Na-P1(gismondine framework) was formed at intermediate concentrations, and a sodalite at the highest concentrations tested.

ACKNOWLEDGMENTS

The research was funded by grant MJGSVV-097600 from the US Dept. of Energy. We thank the Ross Sand and Gravel Co., Engelhard Co., and the Ash Grove Cement Co. for the supply of materials.

REFERENCES

1. C.D. Hills, C.J. Sollars and R. Perry, Cement and Concrete Research, **23**, 196 (1993).
2. H.L. Benny, Concrete International, **12** (7), 14-18 (1990).
3. W. Jiang, X. Wu and D.M. Roy, in <u>Scientific Basis for Waste Management XVI</u>, edited by C.G. Interrante and R.T. Pabalan, (Mater. Res. Soc. Proc. **294**, Pittsburgh, PA, 1993), pp. 255-260.

4. D.M. Roy and M.R. Silsbee, in Advanced Cementitious Systems: Mechanisms and Properties, edited by F.P. Glasser et al. (Mater. Res. Soc. Proc. **245**, Pittsburgh, PA, 1992), pp. 153-164.
5. E. Douglas and J. Brandstetr, Cement and Concrete Research, **20**, 746 (1990).
6. W. Xuequan, Y. Sheng, S. Xiaodong, T. Mingshu and Y. Liji, Cement and Concrete Research, **21**, 16 (1991).
7. R.O. Lokken and P.F.C. Martin, Report HGTP-91-0303-01; R.O. Lokken, P.F.C Martin and S.E. Palmer, Report HGTP-93-0302-01, Pacific Northwest Laboratory, Richland, WA
8. S. Mindess and J.F. Young, Concrete, (Prentice-Hall, Inc. Englewood Cliffs, New-Jersey, 1981) p.85.
9. S. Diamond and S. Ong, Presented at American Ceramic Society Spring Meeting, Indianapolis, 1993 (unpublished).
10. R.A. Olson et al, to be submitted to Journal of Hazardous Waste.
11. A.R. Brough, T. Bakharev, R.J. Kirkpatrick, L.J. Struble, and J.F. Young, in Microstructure of Cement Based Systems / Bonding and Interfaces in Cementitious Materials, edited by S. Diamond et al (Mater. Res. Soc. Proc., Pittsburgh, PA, 1995) in press.

PORE SIZE DISTRIBUTIONS IN HARDENED CEMENT PASTE BY SEM IMAGE ANALYSIS

SIDNEY DIAMOND AND MARK E. LEEMAN
School of Civil Engineering, Purdue University, West Lafayette, IN 47907

ABSTRACT

Technical requirements for determining the size distribution of capillary pores in hardened cement paste by SEM image analysis are discussed. Results of such measurements are reported for a set of hardened cement pastes of w:c ratio 0.40 and 0.25, and of ages ranging from 1 to 28 days. Pore size distributions based on conventional mercury intrusion porisimetry are presented for the same pastes. Estimates of pore diameters by mercury intrusion are two orders of magnitude smaller than the sizes revealed by the image analysis. Diameters of air voids are even more drastically underestimated by mercury intrusion. Typical micrographs are provided to illustrate the physical reality of the image analysis results, and the technical reasons underlying the conventional misinterpretation of MIP results for hydrated cements are reviewed.

INTRODUCTION

The development of backscatter electron image (BEI) scanning electron microscopy using split quadrant detectors on plane polished surface specimens has opened up a variety of new potentials for investigations of hardened cement paste (hcp). When coupled with appropriate image analysis systems and programs, facets of hcp microstructure not previously accessible to quantitative analysis may now be examined. In the present study we have developed a method for investigating and establishing pore size distribution measurements in hcp using this technology.

A primary motivation for this endeavor has been dissatisfaction with the general uncritical acceptance of pore size distributions assessed by the now standard mercury intrusion porosimetry (MIP) method. It has been considered by the first-named author for many years that MIP badly underestimates the sizes of the pores that are visible in scanning electron microscopy, but quantitative comparisons have not been possible.

A start on this problem was provided by Lange et al.[1] as part of a more general investigation of porosity in cement systems, but these authors did not appropriately assess the comparative pore volumes being measured.

In the present work we have developed an image analysis procedure for the measurement of the sizes of capillary pores in hcp without air voids, and an additional procedure for separately determining the size distribution of spherical air bubbles in air-entrained pastes. Image analysis pore size distributions were then obtained for several series of pastes, and the results compared with MIP pore size distributions obtained on the same pastes.

INSTRUMENTATION

The image analyses described in this work were carried out using an Akashi Beam Technology 55A SEM operated at 15 keV, and equipped with a GW 30-A backscatter detector. The SEM was coupled to a Princeton Gamma Tech Imagist image analysis system. To obtain an optimized image, the BEI signal was adjusted for contrast and brightness at the image analysis system, then digitized and stored as a 512 x 400 pixel image file. The settings were adjusted so as to provide sharp, clear boundaries between solids and pore space, without regard to loss of distinction between the various solid components. Settings were kept constant for the various fields analyzed within any given specimen.

Mat. Res. Soc. Symp. Proc. Vol. 370 © 1995 Materials Research Society

Separate portions of the same specimens were oven dried and subjected to MIP analysis using a Micromeritics Autopore 9220 porosimeter with a maximum pressuring capacity of 414 MPa (60,000 psi). A contact angle of 130° was assumed.

SPECIMENS PREPARED

The cement paste specimens were all prepared from a single Type I portland cement of normal chemical characteristics. The pastes were mixed with de-aired water using an evacuated chamber to prevent the inclusion of air voids, at w:c ratios of 0.40 and 0.25. After demolding at 24 hours, separate portions of each paste were hydrated in saturated calcium hydroxide solution for an additional 6 days or an additional 27 days, thus providing specimens 1, 7, and 28 days old. At the conclusion of the designated period of hydration the pastes were oven dried at 105° C until constant weight was attained.

Specimens for image analysis were prepared by cutting a thin slice from the middle third of each paste specimen as cast, embedding it with a low viscosity epoxy resin while under evacuation, and curing at 70° C The surface to be examined was ground and polished using successively finer diamond paste grades, with the final polish accomplished using 1 μm diameter diamond paste. The prepared surfaces were then sputter coated with palladium to prevent charging in the SEM.

Companion specimens for MIP were broken off also from the middle third of each paste specimen, in such a way that all surfaces of the fragments to be intruded were fracture surfaces.

DEVELOPMENT OF PROCEDURES FOR IMAGE ANALYSIS PORE SIZE DISTRIBUTION MEASUREMENTS

A number of considerations needed to be addressed in developing an appropriate method for measuring pore size distributions by image analysis. These involved such factors as obtaining an appropriate binary segmentation of the BEI to include the pores and exclude the solids, setting the lower limit of the pores that can be tallied, rigorously defining what is meant by "size" of an irregular pore, and deciding on the method of conversion of two dimensional image data to an acceptable approximation of three dimensional reality. Finally, an appropriate way of assessing the three dimensional size distribution so as to permit direct comparison with MIP results needed to be developed.

Binary Segmentation

The gray scale histogram of BEI images normally obtained with cement paste do not show a clearly separate peak for the dark (epoxy-filled) pores. Thus the specific gray level used for the binary segmentation of the pores from the solid features must be arbitrary. In this work the gray level was selected for each image by the operator after repeated observations of the results of trials at various candidate threshold levels. A single operator, the second-named author, performed all of the segmentations, thus avoiding any operator bias. Diligence and extreme care to maintain consistency between images greatly reduced quantitative variation.

Size Limits

The capillary pores in hcp are seen in backscatter SEM observations in sizes up to about 10 μm in largest dimension. There is thus no difficulty concerning the upper limit of size to be tallied. The lower limit that can be assessed at a given magnification is a direct function of pixel size, since some minimum number of pixels must be present in each pore feature to be recognized and geometrically analyzed. In the program used, this minimum is 7 pixels. The magnification that can be used reliably is limited by the resolution of the system. In our case the pore size distribution was evaluated at two different standard magnifications, 400x for the larger pores and 1200x for the smaller. The minimum size of a 7-pixel pore feature accessible at 1200x is approximately 0.8 μm in diameter, thus fixing the lower limit of pore size in our evaluation.

Definition of "Pore Size", Three-Dimensional Conversion, and Assembling the Pore Size Distribution

In this work we adopted the following operational definitions:

1. The area of the feature on the polished surface was assessed by multiplying the number of pixels contained in the feature by the area per pixel.

2. The diameter of a circle of area equal in area to that of the feature was calculated. This constituted the operational definition of "pore size".

3. To establish the three dimensional volume to be associated with each pore recognized, it was assumed that the fraction of the area of the whole field occupied by the pore feature is equal to the fraction of the volume of the whole specimen occupied by the pore. Note that this is equivalent to assuming that the three dimensional pore is a right cylinder extending normal to the plane of observation. In a sense this is consistent with the cylinder pore model assumed in interpreting MIP measurements. Note that the assumption is not that any individual pore actually passes through the entire thickness of the specimen; we only assume that at any section a set of pores exists equivalent to those tallied on the surface being analyzed.

4. To establish the contribution of each pore tallied to the cumulative pore size distribution assessed in volume per unit weight (cm³/g) as normally done in MIP, the fractional volume assigned to the pore is divided by the unit weight of the paste (separately measured).

With this operational definition it is possible to list the individual pore features in order of decreasing size, and then to compile a cumulative pore volume vs. pore diameter distribution, consistent with the cumulative distributions normally derived from MIP results. The image analysis distribution is truncated at 0.8 μm, with the limit set by image resolution. Conventional MIP distributions are truncated at ca. 0.003 μm, the limiting factor being the pressuring capacity of the MIP instrument.

PORE SIZE DISTRIBUTIONS OBTAINED BY IMAGE ANALYSIS

A typical BEI micrograph obtained at 1200x magnification is shown in Fig. 1, for a w:c 0.40 specimen hydrated for 7 days. The pore structure is seen as part of the groundmass of fine solid particles (gray) and interconnected pores (black), with distinct solid grains (phenograins) of several types suspended in it. Some of the larger pores appear to be space within hollow shell hydration grains, but most correspond to partly filled in spaces between the original cement grains - the "capillary porosity" as originally defined by Powers [2]. Pores of sizes up to perhaps 8 μm are visible in this particular field. This is commensurate with what might be expected of spaces between cement particles of the usual mean diameter for modern cements (ca. 15 μm). Note the absence of air voids in this vacuum-mixed paste.

As might be expected, SEM images of a corresponding paste of w:c 0.25 at 7 days show a rather less porous groundmass structure than that of the w:c 0.40 paste shown in Fig. 1. Within a given w:c series, pastes at 1 day show visibly more porous structures than those at 7 days. Differences between 7 day and 28 day old pastes are less obvious

Fig. 2 provides the results of image-analysis pore size distribution carried out in the manner previously outlined for w:c 0.40 pastes hydrated for 1 day and for 28 days. For the 1-day old

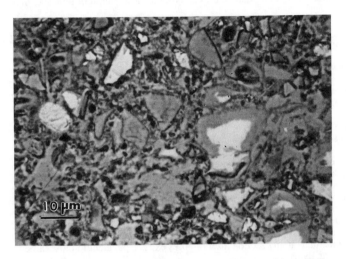

Fig. 1. Representative backscatter electron image of 7 day old w:c 0.40 paste
obtained at an original magnification of 1200x.

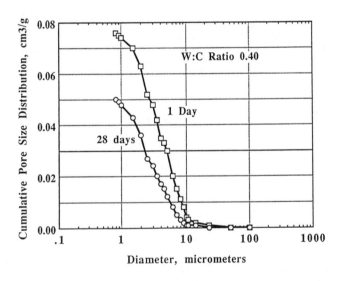

Fig. 2. Image analysis pore size distributions for w:c 0.40 paste hydrated for 1 day
and for 28 days.

paste, a little less than 0.08 cm³/g of pore space has been tallied, most of it in diameters between 10 µm and 1 µm. Note that the tally is truncated at 0.8 µm, as previously discussed.

By 28 days the total pore space tallied in the accessible size range has been reduced to ca. 0.05 cm³/g, and the distribution has shifted slightly (but only slightly) to smaller sizes, almost no pore volume in pores > 9 µm being recorded.

Corresponding results for w:c 0.25 pastes are shown in Fig. 3. At 1 day the total volume of pores tallied is a little over 0.04 cm³/g, much less than the 0.08 cm³/g for the w:c 0.40 paste at the same age. The distribution of sizes within the range tallied is not very different, however. Hydration to 28 days in this case reduces the volume tallied to about 0.025 cm³/g, and shifts the distribution markedly. At 28 days at w:c 0.25 very little pore volume is found for pores > 3µm in diameter.

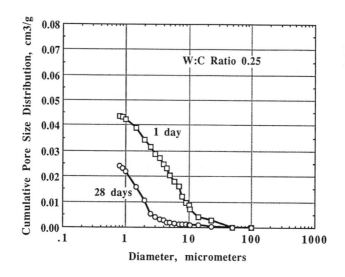

Fig. 3. Image analysis pore size distributions for w:c 0.25 pastes hydrated for 1 day and for 28 days.

PORE SIZE DISTRIBUTIONS OBTAINED BY MIP

Particle size distributions obtained on the same pastes by the conventional MIP technique and assumptions give very different results from those obtained by image analysis.

Fig. 4 compares MIP and image analysis pore size distributions for the 1-day old w:c 0.40 paste. Note that the vertical scale has been changed from that of Fig. 2, with the diameter axis extended to incorporate results two orders of magnitude smaller than shown in Fig. 2. This is necessary because the nominal limiting diameter for the MIP tally is of the order of 0.003 µm.

It is apparent that MIP for this paste records a total intruded porosity almost twice that tallied in the image analysis method. However, it is also evident that essentially all of the porosity tallied in MIP is represented as occurring in sizes less than 0.2 µm, much smaller than even the *lower* limit of 0.8 µm for the image analysis results. The size range indicated (<0.2 µm) is obviously very much smaller than the typical sizes of the pores visible in the BEI micrograph of Fig. 1, for example.

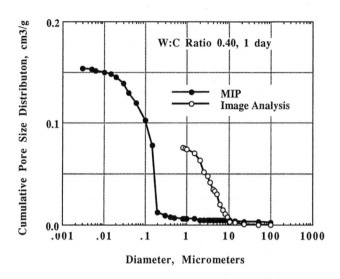

Fig. 4. Comparison of image analysis and MIP pore size distributions for w:c 0.40 paste at 1 day.

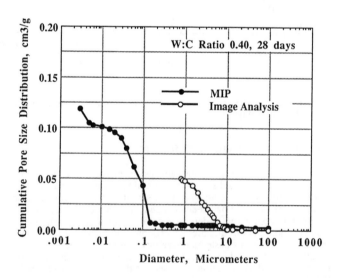

Fig. 5. Comparison of image analysis and MIP pore size distributions for w:c 0.40 paste at 28 days.

Similar results are found in Fig. 5, in which MIP results for the w:c 0.40 paste at 28 days are compared with the image analysis pore size distribution for the same paste. Essentially all of the MIP tally consists of pores smaller than an apparent threshold diameter of 0.15 μm, in contrast to the image analysis tally of pore space in the 0.8 μm to 10μm size range.

It is obvious that MIP and image analysis results for the same paste do not agree, even approximately. Before assessing the substance of the disagreement, mention should be made of the results of further work on pastes with entrained air voids, to be published in detail elsewhere.

EFFECTS OF AIR ENTRAINMENT

A series of pastes were prepared from the same cement used to make the pastes described earlier. Rather than mixing in vacuum to exclude air voids, in this series deliberate air entrainment was sought by incorporating a heavy dose of a standard vinsol resin air entraining agent. The actual mixing was carried out using the ASTM C 305 standard mixing procedure.

As indicated in Fig. 6, the effort made to incorporate air voids was successful. Fig. 6 is a BEI micrograph of a 7 day old w:c 0.40 paste taken at a reduced magnification of 50x,so as to provide a representation of the various sizes of the air voids present. Obvious air voids ranging in sizes up to several hundred μm are visible.

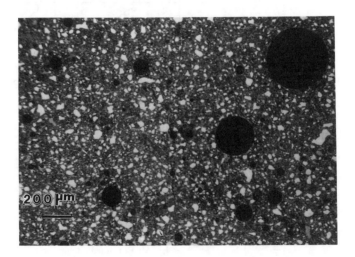

Fig. 6. Representative backscatter electron image of 7 day old w:c 0.40 air-entrained paste obtained at an original magnification of 50x.

The occurrence of large spherical air voids in the pastes necessitates some modification in the image analysis procedure outlined previously. In addition to the previously used magnifications of 1200x and 400x, imaging at 50x was used to help assess the sizes of the air voids. Further, the assumptions previously described for calculating diameters and volumes from the areas of irregular pores depicted at the two dimensional surface are not appropriate for spherical air voids. An apparent diameter can be measured directly, but the apparent diameter of a given air void in a two dimensional slice is always less than its true diameter. A standard statistical procedure (the Johnson-Saltykov method [3]) is available for handling this situation, and was applied to tallying the sizes and volumes of the air voids recognized in image analysis.

In Fig. 7 we plot image analysis results for 28 day old w:c 0.40 pastes with and without entrained air. The results for the air entrained paste combines the Johnson-Saltykov treatment for air voids and the previously described procedure for capillary voids.

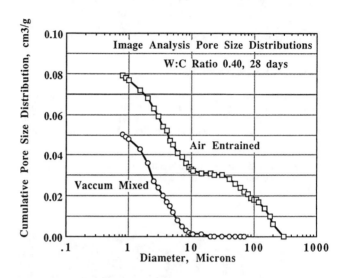

Fig. 7. Comparison of image analysis results for 28 day old w:c 0.40 pastes with and without entrained air.

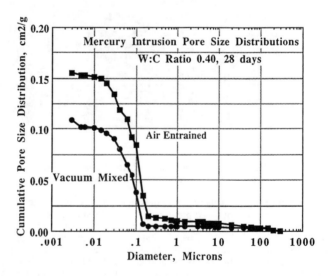

Fig. 8. Comparison of MIP results for 28 day old w:c 0.40 pastes with and without entrained air.

In Fig. 7 it is seen that the entrained air voids detected range in size between about 300 μm and about 10 μm, and that this population of air voids is in addition to the capillary voids tallied at below ca 10μm. The sizes of the air voids are roughly commensurate with their sizes as seen in the various SEM micrographs examined. It is also noted that incorporation of air does not seem to affect the size distribution of the capillary pores tallied below 10 μm, which is virtually identical to that recorded in the vacuum mixed paste.

A very different picture is presented by the comparison of MIP pore size distributions for the same pastes, which are given in Fig. 8.

In Fig. 8 it is seen that the presence of the air voids are detected by the incremental pore volume recorded in MIP. However, nearly all of the incremental volume is recorded not in the range of air void sizes actually present, but rather in sizes below the threshold diameter, here about 0.2 μm. Thus the air voids are tallied by MIP at diameters approximately three orders of magnitude smaller than their true diameters.

DISCUSSION

It is rare in technical work that two methods of measuring the same parameter produce results as discordant as those reported here for image analysis and MIP pore size distributions measured on the same pastes. Obviously image analysis and MIP do not measure the same thing.

The root of the difficulty lies in the nature of MIP when applied to a pore structure like that characteristic of hardened cement paste. Approximately 25 years ago Winslow and Diamond [4] pointed out the existence of a threshold diameter in hardened cement pastes - a minimum diameter of the pores which are geometrically continuous throughout all regions. The concept is equivalent to that of a percolation diameter in modern usage. It was documented that mercury does not flow into most areas of the paste until pressure equivalent to that associated with the threshold diameter is reached. Only then can most of the large pores within the bulk of the paste start to be filled.

The present results confirm this concept, in what the writers consider a convincing manner. When a pore size distribution of paste with entrained air voids does not record these voids until pressure equivalent to a thresholds diameter of ca 0.2 μm is reached, it is obvious that any reliance on the fidelity of the MIP result as a measure of the distribution of the actual sizes of the pores present is lost.

MIP pore size distributions appear to be useful only in providing a comparative measure of the percolation diameters for different pastes, and for an estimate of the total pore volume open to the percolating pore system. The size distribution portrayed is not correct, either for entrained air voids or for capillary pores present in all hcp.

On the other hand, the sizes tallied in image analysis are quite consistent with the sizes of the pores actually visible in SEM. Actually, SEMis not required to visualize the air voids and to some extent the coarser pores. They are visible in optical microscopy and the air voids are routinely tallied in standardized air content measurements for hardened concrete.

We are left with a difficult situation. MIP is well established in the literature, and it is a rapid, reproducible, fully automated technique that covers almost the whole range of possible pore sizes. In contrast, image analysis requires expensive equipment, and a much higher level of skill, is slow and tedious in the extreme, and is limited to only the upper part of the size range known to be present. The lower limit observed in the present work, 0.8 μm, can be improved upon with equipment of better image resolution than that used here. Nevertheless, image analysis is not likely to ever permit the whole range of sizes of pores in hcp to be assessed.

CONCLUSIONS

1. Capillary pores in the size range from ca. 1 to ca. 10 μm are documented by image analysis pore size distribution measurements to constitute a significant part of the pore system of hardened cement pastes.

2. MIP measurements on the same pastes grossly underestimate the content of μm-sized pores and record almost all porosity as though it existed in sizes below ca. 0.1 μm, i.e.. below the percolating diameter of the bulk pore system. Even deliberately introduced air voids of visible sizes up to several hundred μm are tallied only at sizes below the percolating diameter.

3. Image analysis pore size distributions are limited by present BEI resolution to pores of diameters greater than about 0.8 μm. However, analysis appears to provide a reasonably accurate assessment of the actual sizes of the pores tallied, including both capillary pores and air voids. In contrast, MIP pore size distributions measurements grossly underestimate the sizes of pores tallied and do not provide even a rough approximation to the actual size distribution.

ACKNOWLEDGMENTS

This paper is a contribution from the Purdue University component of the NSF Science and Technology Center for Advanced Cement Based Materials. The continued support of this Center by the National Science Foundation is gratefully acknowledged.
The data cited are derived from the M.S.C.E. thesis of the second-named author [5].

REFERENCES

1. D. A. Lange, H. M. Jennings, and S. P. Shah, Cem. Concr. Res. 24, 841 (1994).

2. T. C. Powers, Research Department Bull. 90, Portland Cement Assn., Skokie, IL, 28 pp. (1958).

3. S. Johnson, in Quantitative Microscopy, R. T. DeHoff and F. N. Rhines, 422 pp., McGraw Hill, New York (1968).

4. D. N. Winslow and S. Diamond, J. Materials 5, 564 (1970).

5. M. E. Leeman, M.S.C.E Thesis, Purdue University, 81 pp. (1994)

ESTIMATING THE CAPILLARY POROSITY OF CEMENT PASTE BY FLUORESCENCE MICROSCOPY AND IMAGE ANALYSIS

ULLA H. JAKOBSEN, VAGN JOHANSEN & NIELS THAULOW
G.M. Idom Consult A/S, Bredevej 2, 2830-Virum, Denmark

ABSTRACT

A semi-automatic fluorescence image analysis method to estimate the capillary porosity or the water/cement (w/c) ratio of concrete has been investigated. The method is compared to the usual manual method which also uses thin sections examined in fluorescent light. The two methods correlate very well in the w/c range of 0.35 to 0.5. For higher w/c ratios the semi-automatic method is more sensitive and gives more reliable results. Semi-automatic analyses of standards of different compositions showed that the green tone intensity varied depending on the amount of mineral admixtures in the sample. The fluorescent intensity of the cement paste decreases in the following sequence: OPC, OPC + fly ash and OPC + limestone fines. Analyses of various standards proved that the homogeneity of the paste can be reflected by the standard deviation of the measurement. Finally, the investigation of the semi-automatic method showed that the intensity of the fluorescence light of the paste decreases with the length of illumination and that the intensity of the light bulb changes from bulb to bulb. These observations proved that it is important to carry out the fluorescent w/c analyses, manually as well as semi-automatically, under standardized conditions.

INTRODUCTION

The capillary porosity or the water/cement (w/c) ratio of the cement paste in concrete has for many years been estimated manually using fluorescent impregnated thin sections by assessing the intensity of the green colour emitted from the cement paste when illuminated by UV light[1]. The intensity of the green fluorescence colour of the paste is manually compared to standards with different w/c ratios, normally in the range of 0.35 to 0.60.

This method is generally a quick and reliable method. In recent years, however, new concrete types with for example very low or high w/c ratios and high amounts of additives and admixtures, have shown that the human eye sometimes fails to distinguish intensity differences, especially when the green tone of the cement paste is high.

Attempts have been made to use image analyses to quantify the water to cement ratio. Mayfield[2] used fluorescent impregnated polished samples while Elsen et al.[3] and Wirgot and Van Cauwelaert[4] used thin sections.

In order to differentiate between the various extreme green tones of the cement paste a semi-automatic colour image analysis method has been developed. A number of concrete standards with various w/c ratios and different compositions have been analyzed semi-automatically. From the green tone values of the standards, various calibration curves covering the w/c range of 0.3 to 1.0 have been estimated. These calibration curves can be used to estimate the w/c ratio when analyzing concrete samples with unknown water to cement ratios.

METHODS

Sample Preparation

Both the manual and the semi-automatic estimation of the w/c ratio of the cement paste has been undertaken microscopically on fluorescent thin sections.

A thin section is made by vacuum impregnating a piece (10 x 35 x 40 mm) of mortar or concrete with an epoxy resin containing a yellow fluorescent dye[1]. Subsequently, the impregnated pieces are mounted on a glass ground and polished to a thickness of 0.020 mm, and finally covered by a cover glass. After preparation the thickness of the thin section is checked by looking at the interference colours of quartz in double polarized light. The correct thickness of the section is obtained when the colour of the quartz is white to black.

The vacuum impregnation of concrete samples with epoxy causes all voids, cavities and cracks, as well as the capillary porosity of the cement paste, to be filled with the fluorescent epoxy. By transmitting light from a 100 W halogen source and using a BG12 blue filter and a K510 or K530 yellow filter in the optical polarizing microscope, the fluorescent epoxy in the various porosities will emit yellow light which makes identification easy. The impregnation of the capillary pores of the cement paste causes a dense paste with low w/c ratio to appear darker green while a more porous cement paste with a high w/c ratio appears lighter green.

It is important to use a fixed concentration of fluorescent dye in the epoxy. Furthermore, the thin section should be located in a fully impregnated part of the sample. Experiments[5] with samples of different w/c ratios (0.35, 0.35-0.40 and 0.50) have shown that the thin section has to be placed as close to the impregnated surface as possible since the impregnation depth of the samples is less than 1.5 mm.

Manual Versus Semi-Automatic Analyses

Both methods require an initial microscopic quality check of the preparation of the thin section and an examination of the condition of the concrete. This check is performed both in ordinary polarized and in double polarized light. When looking at the overall condition of the concrete, areas which can cause errors in the w/c estimation are located and marked so that they can be avoided during the w/c estimation. Areas of for example carbonation and $Ca(OH)_2$ depletion should be avoided during w/c estimation as these processes markedly change the capillary porosity and therefore the green tone of the paste. Also areas which have a high content of air voids or cracks should be avoided as the high amount of fluorescent dye in these areas will overexpose the cement paste around.

When estimating the w/c ratio by the *manual method* the intensity of the green colour of the cement paste is compared, by the operator, to known mortar standards made with w/c ratios in the range of 0.35 to 0.60. These standards are normally made with ordinary Portland cement without any mineral admixtures. The operator can either randomly estimate the w/c ratio in 10 different fields covering the whole thin section, and then calculate the average w/c of the thin section or the operator can scan through the thin section and based on experience estimated the w/c ratio. Both ways give for a trained operator a precision of about ±0.02.

The *semi-automatic image analyzing method*, which is built on the procedure from the manual method, has been developed in order to make the method faster and more operator independent.

Furthermore, the semi-automatic method expands the range in which the w/c can be determined to both very high and very low w/c.

The semi-automatic image analysis is performed with the help of the soft-ware program Pippin[6] and a special application the "Green Histogram" developed for the purpose of estimating the green tone variation of cement paste. The thin section is placed in the microscope and the fluorescent light mode turned on. A CCD camera mounted on the microscope scans the image. The image is shown on a video monitor and grabbed by a framegrabber in a computer coupled to the system (Figure 1).

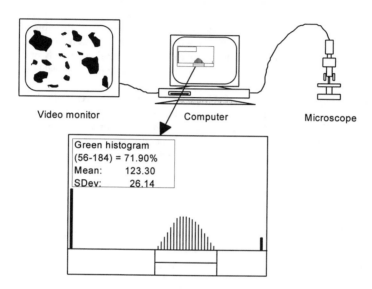

Video monitor   Computer   Microscope

Green histogram
(56-184) = 71.90%
Mean: 123.30
SDev: 26.14

Figure 1 Sketch showing the setup of the semi-automatic image analysis system. The enlarged figure shows an example of the green histogram. The left and right black bar indicates the volume percentage of aggregates (black) and air (yellow), respectively.

The image is then analyzed in the three phases, aggregates, air and cement paste. The operator adjusts the threshold to mark all aggregates black. The porosities (voids and cracks) appear bright yellow and the cement paste appears with shades of green. When the operator is satisfied with the adjustment of the threshold a green tone histogram can be displayed. The histogram represents values from 0 (black) to 248 (yellow). Between the black and yellow a histogram showing the variation of the green tone of the paste appears (Figure 1). From the histogram the amount of paste in vol%, the average green tone and standard deviation is determined.

The final setting (threshold) of the image can be saved in a config.sys file which can be loaded every time the program is started up. Furthermore the setting can be used with the application "repeat marking" on every new image field which is analyzed. This application makes the threshold setting uniform from time to time and the semi-automatic estimation of the green tone value very quick. Depending on the inhomogeneities of the cement paste, 10 to 30 different measurements should be taken covering the whole thin section in order to get a representative value for the w/c ratio. The w/c ratio of the sample is then determined from the average green tone and a suitable calibration curve established for similar concrete compositions.

RESULTS

Repeatability and Reproducibility

In order to estimate the repeatability and the reproducibility of the fluorescent method (manual as well as semi-automatic) various standard samples has been analyzed. It is known that the intensity of the fluorescent dye decreases during time and exposure to light. It is therefore of fundamental importance to prepare new standards relatively often and keep the fluorescent thin sections in a dark cupboard. It is difficult to see with the naked eye the decreasing intensity of the fluorescent dye, however, using the semi-automatic image analyzing method it is possible to quantify the decrease.

Checking the *repeatability* the same field was analyzed over and over again (Figure 2). Within the first 18 minutes and continuous illumination between measurements (2 minute intervals) the green tone intensity steadily decreased, however, with a marked drop the first 2 minutes. When switching off the light for 2 hours the green tone intensity nearly recovered its start value. This value could be reproduced when the light was switched off between measurements (1 hours intervals). If, however, the same field was continuously illuminated again the intensity quickly droped to the value it is before the light was switched off, and then steadily decreased with time (Figure 2).

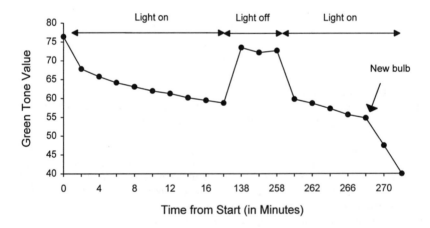

Figure 2 Extent of fading of the intensity of the green tone depending on whether the light is on all time or switched of between measurement. Note jump in time scale.

In continuation of the experiment, the halogen bulb was changed to a new one and the measurements continued with 2 minute light intervals. Surprisingly, the intensity of the new bulb was much lower than the old one and the green tone value was drastically lowered to a new level. Recalculating the green tone values to w/c ratios the apparent w/c ratios actually decrease from

0.33 to 0.30 in the 2 first minutes and then from 0.30 to 0.27 in the next 16 minutes. The final apparent w/c ratio after having changed the bulb was 0.20.

The *reproducibility* of the semi-automatic method was checked by analyzing two standard thin sections (w/c = 0.35 and w/c = 0.55), 3 and 2 times, respectively. The green tone value was each time measured in thirty areas covering the whole thin section. The results of the analysis are seen in Figures 3 and 4. A marked difference in the average green tone value was seen between the first (84.7, black) and second (69.9, stippled) analysis of the 0.35 standard, but not between second and thirth (71.6, blank) analysis.

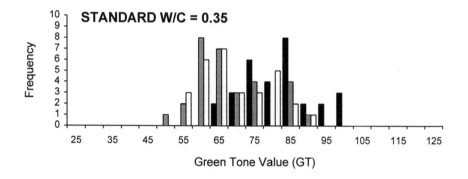

■ w/c=0.36 GT=84.7 std=9.6 ▨ w/c=0.31 GT=69.9 std=10.0 □ w/c=0.31 GT=71.6 std=10.1

Figure 3 Histogram showing the three analyses of w/c standard 0.35. Note scale.

The light intensity of the fluorescence seems to have faded between the two first analyses which can be explained by a very long illumination time of the sample after the first analysis. Between second and third analysis the standard was stored in a closed cupboard for a week.

The two measurements of the 0.55 sample (Figure 4) were performed within a one week period and as for the 0.35 standard the thin section was stored in a cupboard between analysis. The two analyses showed average green tone values which were very close (139 and 134.6)(Figure 4). The standard deviation of the analyses is from all 5 analysis about 10 which compared to normal field concretes is considered relatively low. It is believed that the standard deviation is an expression of the homogeneity of the cement paste. When calculating from the green tone values the w/c ratio of the analyses it is seen that the precision of the w/c estimation of semi-automatic method is ± 0.01. This value, however, dependings on the treatment of the fluorescent section, as discussed above.

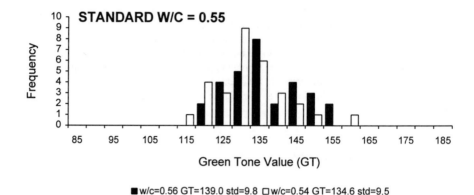

■ w/c=0.56 GT=139.0 std=9.8   □ w/c=0.54 GT=134.6 std=9.5

Figure 4 Histogram showing the two analyses of w/c standard 0.55. Note scale.

Standard Calibration Curves

Several standard samples of different composition and w/c ratios have been prepared (Table 1) and the green tone variation of the cement paste measured on fluorescent impregnated thin sections (Note: in the following designates w/c for the concrete mixture water/(cement+ fly ash) where applicable).

Table 1 w/c ratios of standards used in the investigation. Abbreviations explained in text.

Cement paste	Mortar		Concrete		
PC	OPC	OPC+FA	OPC	OPC+FA	OPC+LF
0.33	0.35	0.40	0.64	0.60	0.69
0.40	0.40	0.79	0.74	0.90	0.74
0.50	0.45	0.99	0.85	1.00	0.83
0.60	0.50				0.94
0.70	0.55				1.04
	0.60				1.14
	0.70				

The standards comprise, pure cement paste (PC), concrete containing (1) ordinary Portland cement (OPC), (2) OPC + fly ash (FA), and (3) OPC + limestone fines (LF), and two sets of mortar standards containing (1) OPC, and (2) OPC + FA. The standards were made within the w/c range of 0.33 to 1.2 (Table 1).

Among the various set of standards a relatively good and positive correlation was found between the measured green tone values and the actual w/c ratio (Figures 5, 6 and 7).

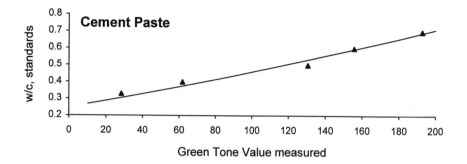

Figure 5 Correlation between green tone value and w/c ratio in cement paste standards.

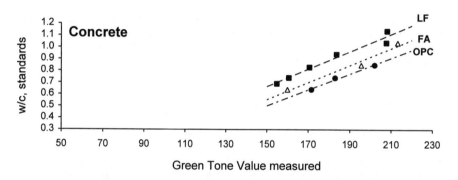

Figure 6 Correlation between green tone value and w/c in concrete sample containing ordinary Portland cement (OPC), OPC + fly ash (FA) and OPC+limestone fines (LF).

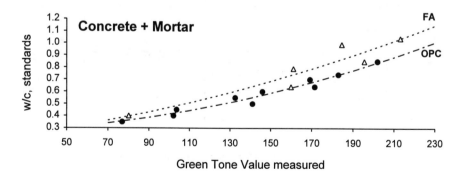

Figure 7 Correlation between green tone value and w/c in concrete and mortar standards containing ordinary Portland Cement (OPC) and OPC + fly ash (FA).

Due to the homogeneous microstructure of the paste and the absence of aggregates in the cement paste standards, the image handling is very simple and difficulties with setting e.g. the right threshold (black markings of aggregates) can be neglected.

The easy image handling of pure cement paste samples is reflected by a relatively low standard deviation of only 6 of each of the green tone measurements. Figure 5 shows that the cement paste standards lie on an almost linear curve with a correlation coefficient of 0.98.

Compared to the cement paste standards, the standard deviation of both the concrete and the mortar samples is a bit higher (about 10), which seems to be due to difficulties in setting the threshold value of the aggregate phase (black). From Figure 6 it can be seen that the green tone intensity of the OPC standard for a given w/c ratio is higher than for the other standards containing FA or LF at the same w/c value.

At a w/c ratio of e.g. 0.8 the intensity of the green tone of the concrete standards decreases from about 200 in the OPC standard to 188 in the FA standard, and 172 in the LF standard.

Combining the data from both the OPC and FA concrete and mortar standards (Figure 7) it is seen that the relationship between the w/c ratio and the green tone value is actually not linear but curved. The curves for the two standards is seen to approach each other in the low w/c end at about 0.3. This pattern fits well with the general observation that it is difficult to impregnate the concrete samples when the w/c ratio is lower than 0.3[5]. In low w/c samples the fluorescence intensity will therefore be very low and independent of the composition.

Correlation between Manual and Semi-Automatic w/c Estimation

Analyzing several field concrete samples, first by estimating the w/c ratio by the manual method and then by the semi-automatic method, two different trends occurred as shown in Figure 8. The first trend represents samples with an automatic measured w/c value of less than 0.50, the second a trend above 0.50. Below 0.50 the correlation between manual and semi-automatic estimation of the w/c ratio is close to 1:1 with a high correlation coefficient of 0.85.

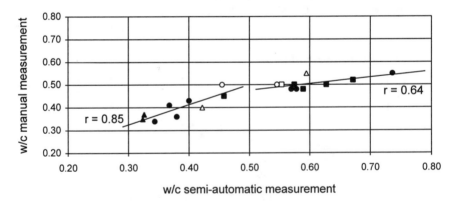

Figure 8 The correlation between manual and semi-automatic w/c measurement of field concretes.

The correlation is somewhat poorer (r = 0.64) above w/c 0.50. This means that below 0.50 the human eye and the camera perceive the same green tones. However, above 0.50 the camera is much more sensitive to variation in the green tone than the eye which seems to have problems differentiating between the high intensity green tones of the high w/c samples.

DISCUSSION AND CONCLUSION

The two methods for analyzing the w/c ratio of cement based samples have both advantages and disadvantages. Common to both is that they are able to estimate the w/c ratios in a relatively fast and reliable way. However, the results of both methods are highly dependent on the quality of the preparation of the thin sections.

Manual Method

The advantages of the manual method are that the w/c ratio can be given, by a trained operator, within an accuracy of ±0.02 in a matter of seconds. The major disadvantage of the method is that the w/c estimation is dependent upon the operator's experience. Another disadvantage is that it is limited to samples with a w/c between 0.3 and 0.5. The manual method does not give a figure for the homogeneity of the cement paste, it can, however, give a range of the w/c ratio as seen in the sample.

Semi-Automatic Method

The advantages of the method are that it gives relatively quick results, the calculated standard deviation can be used as an expression of the homogeneity of the paste, and finally it can measure a wider range of w/c ratios than the manual method. The precision of the method is, under normal storing conditions of the thin sections, ± 0.01. The method has also the disadvantage of being operator dependent, however, only to a certain extent. When first a threshold value is set and stored it can be used again and again by the same or by different operators. This minimizes errors when measuring the green tone.

By using the semi-automatic method the degree of the fluorescent fading has been quantified. As it was shown in Figure 2, the intensity of the light decreases very quickly during time. It is therefore important that every measurement, manual as well as semi-automatic, is taken exactly accordingly to the same procedure. When the operator is aware of this, the measurements are repetitive, however if not, the w/c analysis can be encumbered with uncertainties. Another phenomenon which was observed during the use of the semi-automatic method was that the light intensity of the halogen bulb was different from bulb to bulb. This problem can, however, be solved by making new calibration curves of the standards every time a bulb is shifted or by purchasing a microscope which has the facility of fixing the intensity of the light.

Finally, the semi-automatic analysis showed that the green tone intensity of the paste varies according to the composition of the concrete, e.g. when containing FA and LF the intensity decreases. This phenomenon was earlier observed in our work on microstructure and by

Laugesen[7] who pointed out that the amount of sand in the concrete changes the green tone intensity of the paste. This means that today when making concrete of varying compositions, it is important to have several different standard sets of similar compositions as the field standards when estimating the w/c ratio, manually as well as semi-automatically.

ACKNOWLEDGEMENT

Per Just Andersen is greatly thanked for initiating the semi-automatic work, Lars Palbøl for constructing figures and Anni Harris for correcting the English.

REFERENCES

1. N. Thaulow, A.D. Jensen, S. Chatterji, P. Christensen, H. Gudmundson, Nordisk Betong, 2 - 4, 51-52. (1982)
2. B. Mayfield, Magazine of Concrete Research, 42, 150, 45-49 (1990)
3. J. Elsen, T. Aarre, D. Quenar, V. Smolej, Concrete Across Borders, Int. Conf., Odense Denmark, Proc. Vol II, 449-466 (1994)
4. S. Wirgot, and F. Van Cauwelaert, DeHayes & Stark (eds), Petrography of Cementitious Materials, ASTM STP 1215, 91- 106 (1994)
5. V. Johansen & N. Thaulow, Nato/Rilem Workshop, Saint-Rémy-lès-Chevreuse, in press (1994)
6. Pippin ® by Anacron Consult, Kent Johansen, Denmark (1990)
7. P. Laugesen, Proc. Fourth Euroseminar Microscopy Appl. Building Materials, 15, Sweden, (Abstract)(1993)

DURABILITY AND MICROSTRUCTURE OF PORE REDUCED CEMENTS (PRC)

N.M. GESLIN, D. ISRAEL, E.E. LACHOWSKI AND D.E. MACPHEE.
Department of Chemistry, University of Aberdeen, Meston Walk, Old Aberdeen AB9 2UE.

ABSTRACT

The pore structure characteristics of cementitious materials play an important role in defining their mechanical and durability performance. The mechanical properties of pore reduced cements (PRC) have already been reported [5] as a function of the product density. This paper reports recent data on the durability of PRC and attempts to relate the pore properties of the material to its resistance against deterioration in aggressive environments.

INTRODUCTION

Research into high performance cements is becoming increasingly important. This is partly due to a recognition that essential maintenance of structures made using traditional building materials and methods can be very expensive and also that traditional unreinforced cements and concretes have relatively poor flexural strengths. The relationship between the strength of a material and its porosity has been known for some time. MDF (macro-defect free) and DSP (densified system of homogeneously packed small particles) cements have been developed based on this knowledge and have much improved mechanical performances compared with normal cements. This has been achieved by minimising initial porosities using fillers with further porosity reductions arising due to chemical interactions at the filler-cement interface [1-4].

Reduction of volume porosity not only increases mechanical performance of a Portland cement paste but it also tends to change the nature of the pores and consequently paste durability. Essentially two types of pores exist. At high levels of porosity, interconnected pores will predominate under normal conditions. These have a significant influence on matrix durability, providing tortuous channels through which material transport can occur. Moisture movement and action of chemical species from the environment (e.g. sulfate, chloride, carbon dioxide) can lead to chemical degradation from within the matrix as well as on the external faces of the product. This effect is self-accelerating because degradation usually leads to increased porosity and increased intrusion. At lower volume porosities, the emphasis shifts towards isolated pores which are not connected to the outer boundaries of the product. These are therefore sealed from intrusion and leaching. Thus, the strength <u>and</u> durability performances of cementitious systems can be improved by reduction of porosity.

It is however, the practical achievment of low porosity which has limited the commercial exploitation of high performance cements. While high densities and high strengths have been observed in MDF and DSP products, post-manufacture instability has also been observed due to the nature of the interactions between moisture, the cement and the filler materials. This paper is concerned with the porosity and durability characteristics of pore reduced cements (PRC) where porosity reductions are achieved using mechanical methods as an alternative approach [5,6].

Pore reduced cements (PRC) [5,6]

Pore reduced cements (PRC) arise from the physical compaction, under uniaxial, unidirectional loading, of immature Portland cement pastes with or without blending agents and mixed at normal water/cement ratio (w/c). The technique permits the removal of water excess to hydration requirements at the time of pressing while additionally bringing partially reacted cement grains into intimate contact. The excess water removed is squeezed out of the macropores of the paste as they collapse under the applied load. The w/c can be reduced from an initial value of about 0.3 - 0.35 down to a minimum value of about 0.1 with an accompanying specimen volume reduction of up to 30%. Control using pressing pressure and pre-pressing hydration time can provide a range of product densities and microstructural characteristics. This paper sets out to describe how accessible pore size distributions change as a function of product density and attempts to relate these changes to observations of durability performance of PRC aged in various aggressive solutions.

EXPERIMENTAL

Product preparation and characterisation.

Ordinary Portland cement was mixed with water at a w/c of 0.35, cast in cylindrical perspex moulds (40 mm diameter, 80 mm high) and allowed to set. Moisture loss was prevented using a polythene tent over the filled moulds during setting. At hydration times ranging between 3 and 4 hours, each sample was demoulded and placed in a hydraulic press (Figure 1 (schematic)). Pressure was then applied up to a maximum of 200 MPa while fluid was being expressed from the compressed paste. The sample was then removed and its weight and dimensions were recorded prior to storage in a desiccator conditioned by a saturated solution of copper sulfate (98% relative humidity). After curing for the appropriate period, samples were cut from the pressed cylinders using a diamond saw (South Bay Technology Model 650) for further characterisation

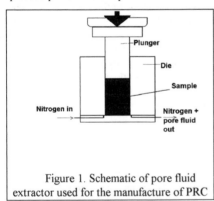

Figure 1. Schematic of pore fluid extractor used for the manufacture of PRC

Electron Microscopy

A scanning electron microscope (ISI-SS40) operating at 20 kV was used to observe the spatial distribution of unreacted grains and hydration products in the PRC pastes. Samples were cut such that portions from various locations in the cylindrical product could be examined to assess paste homogeneity. All samples were polished using silicon carbide (standard F600 and F1200 grit size), diamond paste (6µm) and finally Al_2O_3 (0.3µm) before carbon coating for SEM examination. Typical microstructures are shown in Figures 2 and 3. Image analysis software (Foster-Findlay Associates Ltd.) was used to provide semi-quantitative clinker and pore distribution data.

Figure 2: Backscattered electron image of PRC(2455 kg.m^{-3}) aged 2 months

Figure 3: Backscattered electron image of unpressed Portland cement (1986 kg.m^{-3}) aged 2 months

Pore size distribution (PSD)

The assessment of pore size distribution presents certain difficulties. Firstly, the range of pore sizes may be broad in normal cement pastes and one technique is unlikely to provide complete characterisation. In addition to the image analysis approach, the techniques of N$_2$ adsorption and mercury intrusion porosimetry (MIP) have been used and complement each other with respect to coverage of the pore size ranges. MIP is applicable predominantly to macroporosity (air bubbles entrained during the mixing process) while N$_2$ adsorption is more suitable for characterisation of the gel- or micro-porosity. However, results must be considered carefully as artefacts of the techniques themselves may give misleading information. These measurements were directed at PRC samples prepared over a range of pressing pressures and aged for 5 days at approximately 98% relative humidity.

(a) Nitrogen adsorption

Samples (0.5 - 1.0 g), obtained from the centres of cast or pressed cylinders, were each lightly ground and placed in a vacuum desiccator for about 24 hours before being introduced to the sample cell of a Quantasorb gas adsorption system. The samples were stabilised in a stream of adsorbate (N$_2$) for 24 hours before desorption measurements were taken at progressively lower N$_2$ partial pressures (0.3 < p/p$_0$ < 1.0), pressures being adjusted using helium as the diluting gas. All measurements were made at room temperature (23 ± 2°C) and pore size distributions (Figures 4 and 5) were calculated using a method similar to that of Orr and Dalla Valle [7].

(b) Mercury Intrusion Porosmetry (MIP).

Samples were obtained and prepared as above except that this time, the sample was degassed in the MIP apparatus (Micromeritics Pore Sizer 9300). The intrusion of mercury into the specimen was recorded over the pressure range 0.02 - 200 MPa at a constant temperature (23 ± 2ºC). Pore entry diameter is inversely related to intrusion pressure so that the level of mercury intrusion can be related to apparent pore size. These data are presented in Figure 5.

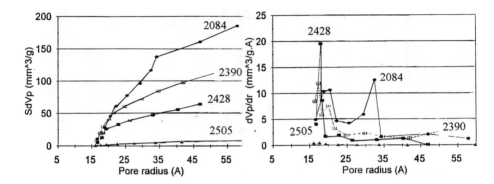

Figure 4: Cumulative (a) and incremental (b) nitrogen desorption isotherms for PRC as a function of density.

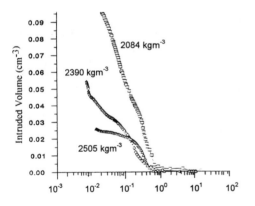

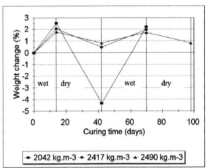

Figure 5: MIP data for PRC as a function of density

Figure 6: Effect of a wet/dry cycle on the weight of PRC samples

Dimensional changes under wet/dry curing

PRC samples were alternately stored under doubly distilled water and in a dry environment on a repeating 28 day cycle. Periodically, their weights and dimensions were recorded and variations in weight are presented in Figure 6.

Resistance to chemical attack

Storage in sulfuric, hydrochloric and acetic acids:

Cylinders were prestored in either moist air or under water for 28 days before being cut horizontally into two halves and immersed completely. Visual inspections and weight recordings were made every month, whereby samples were washed and loosely bound material was brushed away from cylinder surfaces prior to weighing. The solutions were replaced every week. The concentration of the sulfuric, hydrochloric and acetic acids were 0.203 M, 0.274M and 0.05M, respectively, with initial pH's of 0.9, 0.5, 3, respectively. The effects of exposure are summarized in Figure 7.

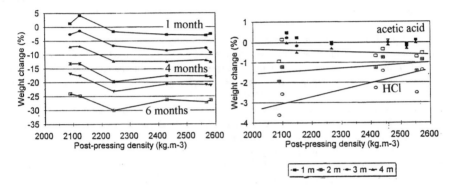

Figure 7: Effect of sulfuric acid (a), hydrochloric (solid lines) and acetic acid (b) exposure on unpressed (<2100 kg.m^{-3}) and PRC products (up to 2600 kg.m^{-3}). Acid concentrations given in text.

DISCUSSION OF RESULTS

Microstructure

Electron microscopy reveals the extent of reaction observed in PRC at the two extremes of density. The high density sample (Figure 2) shows a compacted paste consisting largely of unreacted cement grains in a matrix of hydration product, the latter in relatively small quantities. In contrast, the low density paste (Figure 3) shows a higher degree of reaction but also a higher degree of porosity due to the reduced level of compaction. Image analyses of the micrographs above indicates the concentration of unreacted cement to be around 20% for the low density cement (1986 kg.m^{-3}) compared with 35 - 45% for the pressed product (2455 kg.m^{-3}). The respective porosities around 12% and 5%. There seems to be little variation in these values over a section of the sample indicating reasonable paste homogeneity in both cases.

Porosimetry data reflect the expected trends. Figure 4 shows a shift in the pore size distribution to lower pore radii and a reduction in the volume of desorbed adsorbate respectively, as density increases, indicating that interconnected gel porosity is decreased by pressing. This confirms earlier observations. The MIP data (Figure 5) also indicate a narrowing of the macropore size distribution and a reduction in accessible pore volume as the product density is increased. These observations are likely to explain the lower susceptibility to dimensional variations of pressed samples exposed to wet/dry curing (Figure 6).

Product Durability

The relationship between interconnected porosity and durability can now be tested by observation of the performance of PRC products with different densities (and porosity characteristics) in aggressive solutions. Several exposure media were used to address specific interactions and deterioration was monitored by periodic visual inspection and sample weighing.

H_2SO_4: By 14 days, all of the samples were deteriorating heavily. Circular layers around the cylinders had formed and were scaling off spontaneously. X-ray diffraction (XRD) analyses of degradation products identified predominantly gypsum. The weight changes over this study are shown in Figure 7. At one month, the unpressed samples had increased in weight, whereas the pressed ones have already lost weight. There is a steady weight loss from all of the samples at later ages. The initial weight gain, which was observed with the unpressed samples, has also been reported by Torii and Kawamura [9] and by Metha [10] during similar studies. It is proposed that for low density samples (*ie* unpressed), the decomposition products are retained in the relatively open pore structure, accounting for the weight increase in the first month. Higher density samples do not show this behaviour, presumably due to the more closed microstructure, although the deterioration of all samples is significant (up to 30% at 6 months), regardless of whether pressed or not, or how the specimens were cured prior to exposure.

A preliminary electron microscopic examination of a PRC cylinder exposed for 6 months reveals a number of features (Figure 8). A layer, consisting mainly of gypsum is clearly observed in Figure 8 (a) to have formed on the cement surface. Embedded in this layer are silica-rich grains, probably silica gel (Figure 8 (b)). On the surface of the cement itself is a reaction zone

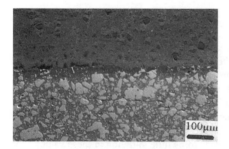

Figure 8 (a): Low magnification image of PRC (2530 kg.m^{-3}) subjected to sulfuric acid attack

Figure 8 (b): Higher magnification image of the same sample showing degraded alite grains in the depleted zone.

approximately 40 - 50 μm deep where the C-S-H is depleted in calcium, (evident as a reduction in the brightness of the image) but only slightly enriched in sulfur. There is considerably greater leaching of calcium from alite grains. They appear dark in the micrographs. The silica gel particles in the outer layer are therefore alite relicts. The apparently greater susceptibility of alite to sulfuric acid attack as compared with that of the C-S-H may explain the enhanced weight loss of the PRC, which is richer in unreacted alite than the unpressed samples.

HCl: Again, the pH of the solution increased (to nearly 12 in the first week) and a yellow-green gel-like product formed on all samples. Analyses of degradation products showed the presence of ettringite, calcite, an amorphous material and Friedel's salt, this phase also reported by Chandra [11] following a similar study. It appears from the weight data (Figure 7) that the pressed products are more resistant to attack than the low density samples. All samples lose weight after about a month but the rate of weight loss is highest in the unpressed samples.

CH₃COOH: There were no visible signs of sample deterioration but a film of calcite was observed on the surface of the solution. The sample weights in this acid (Figure 7) appear to remain nearly constant, within experimental error, over the four months of the test although there is some suggestion of an initial weight gain in lower density samples. Such an observation was not reported by Mehta during similar experiments on concrete using 5% acetic acid solution [10] nor by Nischer [12] with acetic acid solution (pH 3.5) [12]. However, in the present work, the pressed samples are again least sensitive to the acid solution.

It is difficult to make a suitable comparison between the performances of PRC in the different media owing to the variation in concentrations and chemical influences of the different acids. For example, the higher the solubility of the calcium salt of the acid used, the higher is the expected deterioration so that hydrochloric and acetic acids would be expected to be most damaging. In practice, the sulfuric acid appears to be most aggressive. At this stage the mechanism for this enhanced degradation has not been investigated thoroughly but it may be influenced by the expansive formation of gypsum in the dense surface layers of the pastes.

CONCLUSIONS

Pore reduced cements represent a range of cementiteous products having a densely packed microstructure. This is consistent with the good mechanical performance of the material, already reported [5,6]. The degree of cement hydration reduces with increased pressing because much of the mixing water is removed during manufacture so that the amount of unreacted cement remaining is highest in the highest density products. The dense packing of cement grains and the precipitation of products from on-going hydration reactions encourages the development of isolated porosity. Electron microscopy and porosimetry studies confirm that accessible porosity decreases with density and it has been shown that the pore size distribution narrows and shifts to lower pore sizes as post-pressing density increases. The influence of wet/dry cycling on dimensional stability of PRC is also reduced in the higher density samples. These factors therefore promote the belief that high density PRC will be more resistant to chemical degradation than normal OPC pastes.

The durability studies have largely confimed the better performance of pressed cements over normal cement pastes in all but the sulfuric acid media. At this stage, it is proposed that the degradation is due to solubilisation and transportation of calcium to outer regions of the paste where a layer of gypsum is formed causing spalling and exposure of fresh surface for attack. This deterioration mechanism will be the subject of on-going investigation but these data indicate that a high density cement may not be beneficial in certain circumstances.

ACKNOWLEDGEMENTS

The authors wish to thank Blue Circle Industries plc who supplied Portland cement and the financial support of the Engineering and Physical Sciences Research Council.

REFERENCES

[1]   K. Kendall, A.J. Howard and J.D. Birchall, 'The relation between porosity, microstructure and strength, and the approach to advanced cement-based materials', Phil. Trans. Roy. Soc. (London), A310, 139-53, (1983).

[2]   L. Hjorth, 'Development and application of high density cement based materials', Phil. Trans. Roy. Soc. (London), A310, 167-73, (1983).

[3]   L. Struble, E. Garboczi and J. Clifton, 'Durability of High-Performance Cement-Based Materials', Mat. Res. Soc. Symp. Proc., 245, 329-340, (1992).

[4]   J.F. Young, 'Macro-Defect-Free Cement: A Review', Mat. Res. Soc. Symp. Proc., 179, 101-121, (1991).

[5]   D.E. Macphee, 'PRC - Pore Reduced Cement: High Density Pastes following Fluid Extraction', Advances in Cement Research, 3, (12), 135-142, (1990).

[6]   D.E. Macphee, E.E. Lachowski, A.H. Taylor and T.J. Brown, 'Microstructural Development in Pore Reduced Cement', Mat. Res. Soc. Symp. Proc., 245, 303-308, (1992).

[7]   C. Orr and J.M. Dalla Valle, 'Fine Particle Measurement', p271, (Macmillan, New York, 1959)

[8]   F. M. Lea, 'The Chemistry of Cement and Concrete (3rd Ed.)' , (Arnold, London , 1970), 727pp.

[9]   K. Torii and M. Kawamura, 'Effects of fly ash and silica fume on the resistance of mortar to sulfuric acid and sulfate attack'. Cem. Concr. Res., 24, (2), 361-70, (1994)

[10]   P.K. Metha, 'Studies on chemical resistance of low water/cement ratio concretes'., Cem. Concr. Res. 15, 969-78, (1985).

[11]   S. Chandra, 'Hydrochloric acid attack on cement mortar - an analytical study'. Cem. Concr. Res. 18, 193-203, (1988).

[12]   P. Nischer, Zwischenbericht Strassenforschung Nr. 793: 'Hochleistungsbeton für den Strassen- und Brückenbau', B 883/2, Widerstand gegen chemischen Angriff., (1993)

PORE STRUCTURE AND MOISTURE PROPERTIES OF CEMENT-BASED SYSTEMS FROM WATER VAPOUR SORPTION ISOTHERMS

VERONIQUE BAROGHEL-BOUNY AND T. CHAUSSADENT
Laboratoire Central des Ponts et Chaussées, 58 bd Lefebvre, F-75732 Paris Cedex 15, France

ABSTRACT

This paper presents results obtained for hardened cement pastes and concretes, ordinary and very-high-performance materials, concerning pore structure and moisture properties.

Because of the inadequacy of classical pore investigation methods (mercury intrusion, nitrogen adsorption) to provide the exact pore network characteristics of cement-based systems, notably in the case of very-high-performance materials, a more pertinent type of experiment was carried out: water vapour desorption and adsorption measurements.

From these experiments, isotherms were determined at hygrometrical equilibrium and at constant temperature (T = 23 °C) for every desorption/adsorption path. The analysis of these curves gives textural parameters in the mesopore range, not accessible by most of other methods. Thus, from the adsorption curve, the specific surface area accessible to water molecules was calculated by the B.E.T. method. The statistical thickness of the water film adsorbed on the solid surface was also determined. And finally, application of the B.J.H. method on the desorption branch provides relevant porosity values and pore size distribution of the hardened materials, and especially intrinsic characteristics of C-S-H hydrates.

This makes it possible to point out the similarities and the differences between the materials, regarding microstructure and moisture properties, and to quantify the influence of mix parameters such as water-cement ratio or silica fume content.

INTRODUCTION

The behaviour of reinforced or prestressed concrete structures, especially long-term effects, is today at the centre of engineers' concerns. The aim is to understand the physical phenomena that occur during the life of concrete structures, both in order to predict this behaviour and to improve it, by optimizing material mixes, as in high performance concretes. And this unavoidably requires knowledge of microstructure characteristics, especially the pore structure parameters and moisture properties of the material.

These characteristics are indeed expressed, at the macroscopic level, by transfer coefficients (permeability, diffusivity) for gas, liquid or ions, having a direct influence on mechanical behaviour and on durability. In particular, water movements, in liquid or gaseous phase, in the pore network, due to the natural drying of concrete, generate delayed strains (shrinkage, creep) and possible cracking. Furthermore, carbonation resistance, ionic diffusion (for example of chloride ions, responsible for rebar corrosion), freeze-thaw resistance, and also such pathological reactions, as alkali-silica and sulphate attack which damage concrete, all depend on the texture and moisture distribution of the material.

Classical pore investigation methods (mercury intrusion, nitrogen adsorption) provide incomplete descriptions or are sometimes incapable of yielding the exact pore network characteristics of cement-based materials, in particular in the case of very-high-performance (V.H.P.) materials (mixes with low water-cement ratio W/C and containing silica fume and superplasticizer). Limited access to the fine calcium silicate hydrate (C-S-H) gel porosity is the main reason for the insufficiencies of these methods. Moreover, the necessary drying and outgassing, prior to the test, make interpreting the results particularly difficult [1].

Given these difficulties, it was chosen here to characterize pore structure of hardened cement pastes (hcp) and concretes from water vapour desorption and adsorption measurements. These experiments also have the advantage of allowing the determination of moisture properties relevant to understanding and prediction of mechanical behaviour, delayed strains and concrete durability.

The results obtained were completed by those of other methods. A particular aim was to highlight similarities and differences between ordinary and V.H.P. mixes.

EXPERIMENTALS

Materials

Three cement pastes and three concretes were tested. The six mixes were divided into two series, depending on the cement used: series 1 (with French OPC from La Frette) and series 2 (with French OPC from Le Teil). The cement from Le Teil was richer in C_3S (57.28 % vs 39.29 %), and had a lower C_3A content (3.03 % vs 9.73 %) making it possible to have a low water-cement ratio, than the OPC from La Frette. The mix compositions are given in Table I.

The cement pastes "C" and "CO" and the concretes "B" and "BO" are ordinary materials.

The cement paste "CH" and the concrete "BH" are V.H.P. materials. The silica fume came from Laudun (France) and the superplasticizer was a formaldehyde-naphtalene sulfonate copolymer with the trade name Lomar D. The dose of superplasticizer was 1.8 % by mass. At the age of 28 days, the V.H.P. concrete "BH" had a measured compressive strength of 115 MPa, which should be compared to the 49 MPa obtained for the ordinary concrete "BO".

TABLE I. Composition of mixes

Series	Mix	W/C	SF/C	Aggregate/C
1	C	0.45	-	-
2	CO	0.34	-	-
2	CH	0.19	0.10	-
1	B	0.45	-	4.56
2	BO	0.48	-	5.48
2	BH	0.26	0.10	4.55

Water Vapour Desorption and Adsorption Experiments

Water vapour desorption and adsorption experiments were performed at T = 23 °C on the hardened materials. There are not many such measurements on cement-based materials. Some of the first results on hcp were published by Feldman in 1968 [2].

The experiments consisted here in putting specimens into sealed desiccators, where the relative humidity (R.H.) was controlled by satured salt solutions, and submitting them to a first step-by-step water desorption followed by a step-by-step adsorption and possibly a second desorption. Each step lasted until hygrometrical equilibrium was reached and the corresponding water content (ratio of the mass of water contained in the specimen to the mass of the "dry" specimen) was determined by gravimetry.

The specimens were discs about 3 millimeters thick with a diameter of about 10 centimeters for hardened concretes, and crushed ones ($0.8 < d \le 1$ mm) for hcp. The slices were wet sawed from cylinders (after curing for at least five months without water exchange, with the samples wrapped in adhesive aluminium paper) and stored at 100 % R.H. before testing. Given the long duration of both cement hydration processes and water sorption experiments, the specimens were a year and a half old at the beginning of these tests, so hydration of the cement during the test can be assumed negligible (this was verified by measuring the degree of hydration of the cement in the hardened materials at different ages [1]).

The experiments began with *desorption* to study the case of real concrete structures, which undergo drying from their virgin state. Moreover, starting sorption tests with the *adsorption* curve would have entailed first subjecting the specimens to some more-or-less well controlled desiccation process, in thermodynamic conditions different from the isotherms, to reach a "dry" state.

For the first desorption, different sets of specimens were used to obtain the whole experimental curve. In the other cases (adsorption and second desorption), the same specimens covered the whole of the isotherms.

The starting point for adsorption was choosen at different equilibrium R.H. to study the materials' hysteresis behaviour in different regions of the isotherm. In particular, hygrometrical cycles were performed in the high R.H. range, since real ambient average R.H. conditions are 70 % for bridges in France and neighbouring countries.

The lowest R.H. used was R.H. ≈ 3 %. It was therefore possible to define a "dry" reference state for these water sorption tests: equilibrium state at R.H. ≈ 3 % and T = 23 °C, avoiding any more or less thermodyamically well controlled desiccation process, used to calculate the water content and all the textural parameters determined from these experiments.

To obtain correct desorption isotherms, having an intrinsic meaning, it is necessary to take into account the kinetics of the slow water desorption processes. It was therefore essential to wait a long time before equilibrium was reached at each R.H. step (between about five months to one year, with these specimens, depending on the R.H. range, [1]). Water contents were assumed to be at equilibrium state when the mass variation was less than 1 mg for at least three weeks.

Other Techniques

The hydration degree of the cement (α) was measured for every hardened material by direct methods such as microscope image analysis or X-ray diffraction and by indirect methods based on the determination of chemically bound water amounts such as thermogravimetry or water sorption measurements.

Mercury intrusion porosimetry (MIP) experiments were performed with a *Carlo Erba 2000 WS* porosimeter. Since the maximum pressure applied was 200 MPa, this technique allowed the determination of pore size distribution and total porosity for pore radii r_p with 37 Å < r_p < 60 μm.

Nitrogen sorption results on hcp were obtained with a *Micromeritics* automatic apparatus, at T = 77 K. As in the case of MIP, the hardened materials were tested after vacuum drying and outgassing.

Thermoporometry [3] was also used on saturated specimens of hcp "C" with a Differential Scanning Calorimetry apparatus.

For every experiment, the hardened materials were tested at the age of a year and a half.

All of these methods are more fully described in [1].

EXPERIMENTAL WATER VAPOUR SORPTION ISOTHERMS

Water vapour desorption and adsorption isotherms (i.e. "Equilibrium Water Content versus Relative Humidity" curves) obtained from the experiments described above, and for the six mixes, are given in Figure 1. All of the graphs are plotted to the same scale, and the water contents are expressed (in %) per unit mass of "dry" (i.e. at equililbrium state at R.H. ≈ 3 %) hcp, contained in the material (from mix proportions), to make it possible to compare the results between the different mixes, and especially between the hcp and the concretes.

Each isotherm point is the average of experimental values obtained on at least three specimens, but these experimental values were very similar, in particular for crushed specimens, displaying their homogeneity and a very good reproducibility.

The results obtained for V.H.P. materials are of interest to people working on the mechanical and durability behaviour of these materials. It can be seen that a low water-cement ratio and the use of silica fume, inducing intense self-desiccation (H.R. ≤ 80 %) and a very narrow pore network [1], modify the moisture properties of the hardened material. In the high R.H. range, much lower water contents were measured than for ordinary materials. For cycles running between 50 % and 90 % R.H., the curves of V.H.P. materials show very small water content variations. Such materials are fairly insensitive to hygrometrical variations in a wide range of R.H. They will have small desiccation shrinkage strains in this range.

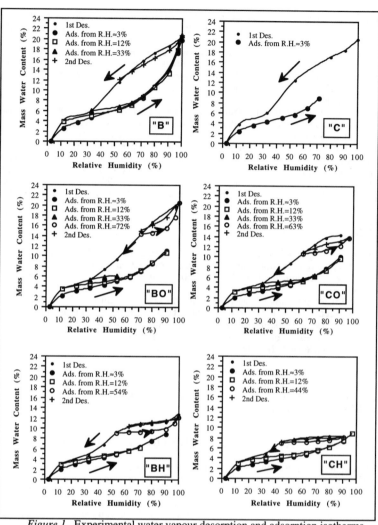

Figure 1. Experimental water vapour desorption and adsorption isotherms obtained at T = 23 °C on the hardened materials.

PORE STRUCTURE CHARACTERIZATION

Pore structure characterization of the hardened materials is possible from these experimental water sorption isotherms assuming the coexistence of an "adsorbed" water phase (multilayer film characterized by a thickness t, a function of relative humidity) and a "capillary" water phase in the pores of the material, assumed to be cylindrical, according to the "B.J.H. model" proposed by Barrett, Joyner and Halenda [4]. In this model, the space where capillary condensation occurs is characterized by the Kelvin radius r_k, given by the Kelvin-Laplace equation as a function of

248

relative humidity and corresponding, at a given R.H., to the pore radius r_p minus the adsorbed film thickness t. The equations and their relative hypothesis are given in [5] and [1].

"Adsorbed" Water Analysis: Determination of the Specific Surface Area by the "B.E.T. Method"

The "adsorbed" water phase can be analysed by applying an adsorption theory like the "B.E.T. theory" proposed by Brunauer, Emmett and Teller in 1938 [6]. This theory is the most widely used and certainly the most relevant, in the case where only surface adsorption in multimolecular layers, without capillary condensation, occurs in the pore network for relative humidities between 5 and 40 % ([7], [8], [9]).

Applying the "B.E.T. method", explained in [9], [5] and [1], to the *adsorption* branch of the experimental isotherms for R.H. ≤ 33 % gives the results reported in Table II, where the specific surface areas accessible to water molecules ($S_{sB.E.T.}$) are expressed in m^2 per gram of "dry" (i.e. at R.H. ≈ 3 %) hcp contained in the material.

TABLE II. Specific surface area ($S_{sB.E.T.}$) determined by the "B.E.T. method" in the low R.H. range of the experimental water vapour adsorption isotherms

Series	Mix	$S_{sB.E.T.}$ (in m^2 per g of "dry" hcp)
1	C	123
2	CO	83
2	CH	85
1	B	122
2	BO	109
2	BH	92

The results show that the surface areas determined for hcp "C" and concrete "B", prepared from the same cement with the same water-cement ratio and having the same degree of hydration of the cement ($\alpha \approx 1$), are identical. This means that concrete and hcp textures, in the micro and the mesopore range explored by water sorption processes, are quite similar. The presence of aggregates, and an interface zone, do not affect the C-S-H surface development.

The calcium silicate hydrate (C-S-H) gel is a finely divided and poorly crystallized product having a high specific surface area. It represents at least 50 % of an ordinary hcp. These hydrates are of the greatest importance in cement-based systems. They account for their compressive strength. They indeed constitute what might be called the "glue" between the different particles of the microstructure. It is the C-S-H gel that contributes the bulk of the large internal surface in the hardened material.

The degree of hydration at a given age was very different between mixes. It is this parameter and the pozzolanic reaction (forming C-S-H from silica fume and portlandite) that in this case determine the amounts of C-S-H formed and, in consequence, the surface area, explaining different values of specific surface area found for the different mixes.

Plotting the specific surface area determined by the "B.E.T. method" versus the degree of hydration of the *ordinary* (i.e. without silica fume) hardened materials, reveals a very good correlation between these two parameters (Fig. 2). The specific surface area determined by water adsorption therefore appears to be a good indicator of the surface developed by C-S-H and consequently of the quantity of outer C-S-H products present in the microstructure.

Considering now the results obtained on the *V.H.P.* materials, it can be seen, by comparison to the straight correlation line of ordinary materials, that it is possible to determine the additional surface area and so the amounts of C-S-H formed by the pozzolanic reaction (Fig. 2). As a result of the additional amount of C-S-H formed by the pozzolanic reaction, there is a gain of specific surface area of 55 $m^2.g^{-1}$ for "CH" and 46 $m^2.g^{-1}$ for "BH". This means that the surface area measured on this last material is twice what one would expect to measure on an ordinary material (without silica fume) with the same degree of hydration as "BH". In other words, regarding the value of their specific surface area, V.H.P. hardened materials contain an amount of C-S-H equivalent to ordinary hardened materials having a degree of hydration of cement around 0.8. Thus, the hardened cement pastes "CO" and "CH" contain equivalent amounts of C-S-H, although "CH" ($\alpha \approx 0.44$) contains much more unhydrated cement particles than "CO" ($\alpha \approx 0.76$).

The values calculated here are not absolute, but intended merely to illustrate the kind of calculations possible for the evaluation of C-S-H quantities from water sorption measurements. A larger number of mixes would allow a more precise quantification.

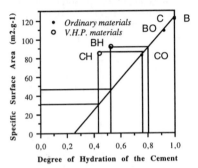

Figure 2. Correlation between the specific surface area accessible to water molecules determined by the "B.E.T. method" and the degree of hydration of the cement of the hardened materials.

"Adsorbed" Water Analysis: Determination of the Statistical Thickness of the Water Film

The average statistical thickness (t) of the water film adsorbed on the solid surface of the pores can be determined as a function of relative humidity, from the experimental water *adsorption* isotherm and with the B.E.T. specific surface area, using the following relationship:

$$t(R.H.) = V_{ads}.(R.H.) / S_{s\,B.E.T.} \qquad (1)$$

But this relationship cannot be used when extensive capillary condensation occurs, as will be the case, at least above a specific relative humidity, with porous solids.

However, it is possible to obtain the statistical thickness (t) as a function of relative humidity for the whole range of R.H. using "t curves" from the literature which provide the statistical thickness (t) of the multimolecular water layer adsorbed on the *non-porous* adsorbent, for which the adsorbate/adsorbent interactions are equivalent, from an energy point of view, to water/hcp interactions. Following the development of the "t method" of de Boer [10] to determine pore size distribution in the micropore region, a number of "t curves" are available in the literature for nitrogen adsorption. Concerning water adsorption, the first experimental "t curves" were published by Hagymassy *et al.* in 1969 [11]. Researchers who have worked on water adsorption since then generally use these curves or computed average "t curves" from these experimental data [12].

Comparison between the experimental t = t(R.H.) relationships calculated with (1) for the hardened materials, illustrated for example by "CO" for ordinary materials and by "BH" for V.H.P. ones in Figure 3, and the "t curve" obtained on non-porous adsorbents proposed by Hagymassy *et al.* [11], for a B.E.T. constant range including our experimental values [1], shows a very good agreement below R.H. = 63 %. In this range, the increase of R.H. is expressed by an increase of the water content in the material that is statistically equivalent to the thickness increase of a layer of water molecules that would be adsorbed on the solid surface as on a non-porous adsorbent.

Above R.H. = 63 %, the experimental quantities adsorbed on *ordinary* mixes are larger than would be found on non-porous adsorbents. This indicates extensive capillary condensation, which causes pore filling in volume (even partial) with the liquid phase in the large pore space situated around r_p = 30 Å (according to B.J.H. calculation).

For *V.H.P.* materials, the behaviour is different. The large pore volume that fills from R.H. = 63 %, i.e. around $r_p = 30$ Å, for ordinary materials, is obviously lacking (or is at least reduced) in the pore size distribution of V.H.P. materials. Water adsorption is restricted, in these materials, by the pore size, and so the experimental curve, in the high R.H. range, is lower than the "t curve" obtained on non-porous adsorbents.

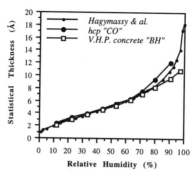

Figure 3. Comparison between the experimental t = t(R.H.) curves obtained with the ordinary hcp "CO" and with the hardened V.H.P. concrete "BH", and the "t curve" from Hagymassy *et al.*

Considering the good agreement between all the experimental results and the "t curve" from Hagymassy *et al.*, in the low R.H. range, it can be claimed that this identifies an intrinsic property of the materials in the form of a *"universal"* adsorption isotherm, valid for the different hcp and concretes in this range. This result is very useful for moisture transfer and shrinkage modelling.

Moreover, it will be possible to use the "t curve" proposed by Hagymassy *et al.* to characterize the adsorbed water in the whole range of R.H. for the determination of the pore size distribution of the hardened materials. It is shown that water can be adsorbed on six layers at most, given that the average thickness of an adsorbed monomolecular water layer is $\mathcal{T}_m = 3$ Å.

Coexistence of "Adsorbed" Water and "Capillary" Water: Determination of Pore Size Distribution by the "B.J.H. Method"

With desorption and adsorption experiments, the method most often used to calculate the pore size distribution of a material in the mesopore range (i.e. pores having an opening between 20 and 500 Å, according to I.U.P.A.C. recommendations [9]), is the "B.J.H. method" proposed by Barrett, Joyner and Halenda in 1951 [4]. However, this kind of analysis, classical with nitrogen at T = 77 K, has not often been used with water, because of the greater experimental difficulties.

This method has been applied here to determine the pore size distributions of the hardened materials in the mesopore range from the experimental water vapour sorption isotherms.

It is an iterative method where computation is based on the step-by-step analysis of the desorption branch (equations are given in [13], [8] and [9]), assuming that:
- this curve represents the capillary condensation (or evaporation) equilibrium, i.e. the pore space filling (or emptying) in volume by liquid water, with the existence of meniscii having the radii given by the Kelvin-Laplace equation,
- the pores are cylindrical with radius r_p,
- the pore walls are covered with a multimolecular water layer, of which the thickness (t) depends on the relative humidity.

The results of the study of pore size distribution can be represented by a curve called: $\Delta V_p/\Delta r_p = f(r_p)$. The curves obtained for the hcp and the concretes are given in Figure 4.

Since water sorption measurements allow pore structure analysis in the mesopore range, a characterization of C-S-H hydrates is possible. It is found that the essential part of the C-S-H pore volume corresponds to $r_p \leq 50$ Å (i.e. R.H. ≤ 76 %). The graphs show a main porosity mode, in which the maximum is identically situated for all mixes with this method at an average radius of r_p = 16.5 Å. This value is in agreement with the results obtained with nitrogen sorption measurements at T = 77 K, for the hcp (maximum situated for all mixes at $r_p \approx 17$ Å) [1]. They are also not very different from those obtained by thermoporometry on the hcp "C" (maximum situated at $r_p \approx 22$ Å, given the calculation modelling is different) [1].

It may also be noted that there is a large pore volume, corresponding to $20 \leq r_p \leq 50$ Å, in the pore size distribution of ordinary materials that is lacking for the V.H.P. materials, which have a very narrow distribution. Exactly the same phenomenon as in the case of the analysis of the adsorption curve is found here.

Identical characteristics have been obtained for concretes and for hcp. This means that the method is sufficiently precise to allow a textural description at the scale of nanometers, and so textural investigation of C-S-H hydrates, from concrete specimens. This point is of interest because it is often concrete mixes, and not hcp mixes, that are to be characterized, and there are very few physico-chemical techniques that allow a direct analysis of concrete.

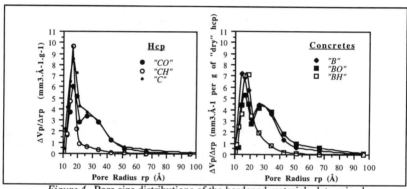

Figure 4. Pore size distributions of the hardened materials determined by the B.J.H. method from experimental water vapour sorption isotherms.

Porosity corresponding to different ranges

From water sorption measurements, it is possible to develop a pertinent evaluation of the *porosity* (ratio of pore volume to total volume of the specimen) of a material. In spite of the long duration of the experiments (at least two years to obtain a complete desorption/adsorption cycle), the relation of water to concrete behaviour and durability parameters and the origin of the porosity of cement-based materials make the water molecule the ideal probe molecule, one that cannot be replaced by other molecules, easier to use but bigger, such as nitrogen, argon or oxygen. For instance, with nitrogen, though the localization of the maximum of pore volume on the pore size distribution is the same, the pore volumes are under-estimated. In particular, it is found in hcp that the pore volume corresponding to the range $10 \leq r_p \leq 20$ Å decreases with W/C+SF, and, notably for the V.H.P. "CH", the volume measured was extremely small compared to those of the other hcp and not in correlation with the respective amounts of C-S-H present in the materials. This is due to the larger size and the lower affinity of nitrogen molecules compared to water, and to the certainly unsuitable drying and outgassing used in the case of nitrogen for all of these materials [1].

The porosity of the hardened materials, accessible to water molecules, stated in %, can be quantified from water sorption tests. This parameter is calculated from the water content at R.H. $\approx$ 100 %, given by the first desorption branch, and from the density of the "dry" materials determined by water absorption and mercury intrusion [1]; it is designated $n_1(H_2O)$ in Table III.

More particularly, the C-S-H porosity $n_2(H_2O)$, essentially concerning pores with $r_p \leq 50$ Å, was quantified. This parameter was thus determined from the water content at R.H. $\approx 76\%$, given by the first desorption branch (Table III).

TABLE III. Porosity relative to C-S-H hydrates determined from experimental water vapour desorption isotherms

Series	Mix	$V (h=1)$ $(mm^3.g^{-1})$	$n_1(H_2O)$ (%)	$V (h=0.76)$ $(mm^3.g^{-1})$	$n_2(H_2O)$ (%) $r_p \leq 50$ Å
1	C	204.97	35.05	162.60	27.9
2	CO	142.28	28.74	131.95	26.7
2	CH	83.89	19.21	82.20	18.8
1	B	42.30	9.60	34.40	7.8
2	BO	38.08	9.22	28.65	6.9
2	BH	27.23	6.89	22.92	5.8

The results show that the ordinary hcp "C" and "CO" (without silica fume) give the same result already provided by the Powers and Brownyard model [14]: the C-S-H gel porosity, whatever the W/C ratio and the degree of hydration of the cement of the hcp, has the intrinsic value of 28 %.

On the other hand, it can be seen that for the V.H.P. hcp "CH", the C-S-H gel porosity is lower; its value is about 19 %. This indicates that, for V.H.P. materials, a lower value than the one accepted since Powers *et al.* would have to be used. So these results show differences in pore network and C-S-H texture between ordinary and V.H.P. mixes. This difference between ordinary and V.H.P. materials can be explained by morphological differences between C-S-H formed in these two different kinds of materials. It can indeed be seen by microscope investigations [1] that, in the ordinary materials, some outer C-S-H products, fibrous, rather well developed (called "type I" C-S-H) and reticulated or honeycomb ones ("type II"), with characteristic pore sizes corresponding to around $20 \leq r_p \leq 50$ Å (Fig. 4), are found. On the other hand, in the V.H.P. materials, the C-S-H are exclusively very dense, amorphous, and featureless ("types III" and "IV"); they are essentially inner products, with characteristic pore sizes corresponding to $r_p \leq 20$ Å. This is due to there being little water and space available in this kind of microstucture, and also certainly to the specific chemical composition of the liquid phase caused by the presence of silica fume.

To complete the description, MIP measurements were performed on the hardened materials. All results and their complete analysis are given in [1]. This technique makes it possible to quantify the macro- and a part of the mesoporosity of the materials, more precisely for pore radii r_p with 37 Å $< r_p < 60$ μm. The porosity $n(Hg)$ then obtained for each mix is given in Table IV. It may be noted that for V.H.P. materials, because of their low water-cement ratio and the use of silica fume, inducing a very narrow pore network, the porosity accessible by mercury intrusion is very small.

TABLE IV. Total porosity determined from experimental water vapour desorption isotherms and mercury intrusion porosimetry

Series	Mix	$V (h=0.72)$ $(mm^3.g^{-1})$	$n_3(H_2O)$ (%) $r_p \leq 40$ Å	$n(Hg)$ (%) 40 Å $< r_p < 60$ μm	n_{total} (%) $r_p < 60$ μm
1	C	153.60	26.3	13.1	39.4
2	CO	124.04	25.1	5.2	30.3
2	CH	81.71	18.7	1.6	20.3
1	B	32.56	7.4	7.3	14.7
2	BO	26.38	6.4	5.8	12.2
2	BH	22.61	5.7	2.5	8.2

Finally, it was then possible to determine, for each hardened material, the *total porosity* n_{total} (in %) corresponding to the very wide range of pore sizes 10 Å $< r_p < 60$ μm. This was done by adding the porosity $n(Hg)$ corresponding to 40 Å $< r_p < 60$ μm (MIP values) and the porosity

$n_3(H_2O)$ corresponding to $r_p \leq 40$ Å (determined from the water content at R.H. ≈ 72 % given by the first water vapour desorption branch). All the values are reported in Table IV.

For example, a total porosity close to 40 % was found for the conventional hcp "C", and only half that (20 %) for the V.H.P. paste "CH". The same was true of the concretes, in which the porosity of "BH" was about half the porosity of "B".

For "CH", the share of porosity corresponding to 40 Å $< r_p < 60$ µm is less than 10 % of that corresponding to $r_p \leq 40$ Å. For the conventional hcp "C", this ratio is close to 50 %.

All of these features illustrate the compactness and the very fine pore spaces of V.H.P. materials.

CONCLUSIONS

These experimental water vapour desorption/adsorption isotherms could actually be called *"hygro-structural" material identity card*. Analysis of these curves, according to for example the established models B.E.T. and B.J.H., has identified several parameters relevant to texture and moisture characterization. In particular, some intrinsic properties of hardened cement pastes and concretes have been found, such as the description of water adsorption in the low relative humidity range by a "universal" curve, giving the statistical thickness of adsorbed water as a function of relative humidity, or such as the C-S-H hydrates characterization.

Completing the description by other techniques has also shown the specific features of very-high-performance materials. For instance, despite large amounts of residual unhydrated cement in these hardened materials (their cement hydration degree was $\alpha \approx 0.5$), the amounts of C-S-H in their microstructure were equivalent to those found in ordinary materials with $\alpha \approx 0.8$, as a result of the pozzolanic reaction.

The compactness of microstructure of materials of this kind was especially illustrated by their very low total porosity (≈ 50 % of that of an ordinary material), their lower C-S-H gel porosity (≈ 19 %, against 28 % for ordinary mixes), and the very small pore radii of their cement matrix (the bulk of pore volume corresponds to pores with a radius $r_p \leq 25$ Å).

These last results could in particular be very useful to people who work on the mechanical behaviour and durability of very-high-performance concretes.

REFERENCES

1. V. Baroghel-Bouny, PhD thesis, Ecole Nationale des Ponts et Chaussées Paris, 1994.
2. R. F. Feldman, in Proceedings of the "5th International Congress on the Chemistry of Cement" Edited by Cement Association of Japan, Tokyo, Vol. 3, pp. 53-66, 1968.
3. M. Brun, A. Lallemand, J. F. Quinson and C. Eyraud, Thermochimica Acta 21, 59-88, (1977).
4. E.P. Barrett, L.G. Joyner and P.P. Halenda, J. Amer. Chem. Soc. 73, 373-380 (1951).
5. V. Baroghel-Bouny and T. Chaussadent, Bull. de Liaison des LPC 187, 69-75, (1993).
6. S. Brunauer, P.H. Emmett and E.J. Teller, J. Amer. Chem. Soc. 60, 309 (1938).
7. D.M. Young and A.D. Crowell, in Physical adsorption of gases Edited by Butterworths, London, 1962, pp. 137-246.
8. F. Rouquerol, Techniques de l'ingénieur P 3645, 14 p. (1968).
9. I.U.P.A.C. (Recommendations 1984). K.S.W. Sing, D.H. Everett, R.A.W. Haul, L. Moscou, R.A. Pierotti, J. Rouquerol and T. Siemieniewska, Pure and Applied Chemistry 57 (4), 603-619 (1985).
10. J. H. De Boer and B.C. Lippens, Journal of catalysis (USA) 4, 319-323 (1965).
11. J.Jr Hagymassy, S. Brunauer and R.Sh. Mikhail, Journal of Colloïd and Interface Science, 29 (3), 485-491 (1969).
12. R. Badmann, N. Stockhausen and M.J. Setzer, Journal of Colloïd and Interface Science 82 (2), 534-542 (1981).
13. C. Pierce, J. Phys. Chem. 57, 149-152 (1953).
14. T.C. Powers and T.L. Brownyard, Bull. Portland Cem. Association 22, 276-287 (1948).

MICROSTRUCTURE-ELECTRICAL PROPERTY RELATIONSHIPS IN CEMENT-BASED MATERIALS

R.A. OLSON*, G.M. MOSS*, B.J. CHRISTENSEN**, J.D. SHANE*, R.T. COVERDALE***, E.J. GARBOCZI****, H.M. JENNINGS*, and T.O. MASON*
*Center for Advanced Cement-Based Materials, Northwestern University, Evanston, IL
**E. Khashoggi Industries, Santa Barbara, CA
***Master Builders, Cleveland, OH
****National Institute of Standards and Technology, Gaithersburg, MD

ABSTRACT

There has been much recent progress on the application of impedance spectroscopy (IS) to the study of microstructure and transport in cement-based materials. The IS spectrum allows for the precise determination of bulk resistance, which is a measure of the pore phase interconnectivity, and calculation of the relative dielectric constant, which is related to the capillary pore size and distribution. High values of the relative dielectric constant ($\sim 10^5$) observed in cement paste at early hydration times are the direct result of the microstructure inducing dielectric amplification. Solvent exchange and freezing experiments, combined with digital-image-based computer modeling, have confirmed the role of large capillary pores in the dielectric amplification in young pastes.

The conductivities (σ) and relative dielectric constants (ε_r) of ordinary portland cement (OPC) pastes were monitored during cooling and solvent exchange with isopropanol and methanol. Dramatic decreases in σ and ε_r, in some cases over two orders of magnitude, occurred at the initial freezing point of the aqueous phase in the macropores and large capillary pores. The same dramatic decreases in σ and ε_r were observed at the onset of solvent exchange. Both effects provide experimental support for the dielectric amplification mechanism within the microstructure on the μm-scale. A secondary dielectric amplification was observed in the frozen and solvent exchanged pastes, which produced dielectric constants on the order of 10^3. This effect is attributed to amplification on the nm-scale within the layered calcium silicate hydrate (C-S-H) gel microstructure. Additional insight into the variable nature of the C-S-H microstructure was obtained by comparing the dielectric behavior of methanol-exchanged OPC pastes to isopropanol-exchanged OPC pastes.

INTRODUCTION

Impedance spectroscopy is a non-destructive electrical technique that has been used to characterize the microstructure of cement-based materials, providing useful information about the relationships between microstructure, electrical properties, and chemical processes during hydration [1-13]. IS consists of applying a small amplitude alternating current to a cement paste sample and measuring the impedance as a function of

Mat. Res. Soc. Symp. Proc. Vol. 370 © 1995 Materials Research Society

frequency. The impedance measurement can be made rapidly and easily while the sample is hydrating, without drying the sample or altering its microstructure.

Recently, it was discovered that the relative dielectric constant of cement paste can attain high values [1-3, 9]. As shown in Figure 1, the value of ε_r for ordinary portland cement paste increases sharply during the first 10-15 hours of hydration to a maximum near 10^5, then decreases gradually to a value around 10^3. This is surprising, since the component with the highest dielectric constant is the aqueous phase ($\varepsilon_r \approx 80$). Even the C-S-H gel ($\varepsilon_r \approx 10^3$) [9], cannot account for ε_r values of 10^5.

Christensen et al. [9] and Coverdale et al. [11] proposed that the dielectric amplification mechanism in cement paste is based on geometric amplification at the μm-scale, where the interlocking network of C-S-H gel and pore fluid create a structure similar to that of a grain boundary ceramic, shown schematically by the bricklayer model in Figure 2. The interconnecting C-S-H gel layers act as low conductivity, high dielectric constant boundaries, and the capillary pores act as high conductivity, low dielectric constant grains. The current is carried mainly by the pore fluid, so a large capacitance results from thin layers of C-S-H gel within the capillary pore network. Porous materials saturated with an electrolytic solution, having microstructures in some respects similar to that of hardened cement paste, also display high dielectric constants [14, 15].

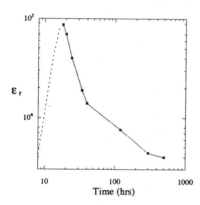

Figure 1: Relative dielectric constant vs. time for a w/c = 0.4 OPC paste [9].

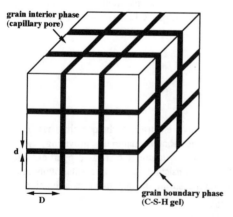

Figure 2: Schematic diagram of the bricklayer model.

This amplification mechanism is demonstrated in the following example. A composite parallel plate capacitor with a high conductivity material inserted between its plates (Figure 3) has a capacitance that is defined by:

$$C = \varepsilon_{eff}\varepsilon_0 (A/D) = \varepsilon_s \varepsilon_0 (A/(D-b)) = \varepsilon_s \varepsilon_0 (A/d) = \varepsilon_s \varepsilon_0 (D/d)(A/D) \quad (1)$$

where ε_{eff} is the effective dielectric constant, ε_s is the dielectric constant of the material in the gap (d), ε_0 is the permittivity of free space, and A is the area of the electrodes. The effective dielectric constant is then:

$$\varepsilon_{eff} = \varepsilon_s (D/d) \quad (2)$$

which can be quite large, depending upon the D/d ratio of the capacitor. This type of dielectric amplification is exhibited by boundary layer capacitors, where conductive grains (D = grain size) are surrounded by insulating grain boundaries (d = grain boundary thickness) [16]. Dielectric amplification in cement paste originates from its microstructure as well, where the dielectric amplification factor (D/d) is attributed to the capillary pore size (D) and the thickness of C-S-H gel layers (d) [9, 11].

The increase in the dielectric constant of cement paste, shown in Figure 1, results from the introduction of C-S-H gel layers into the capillary pores, where they act as small "capacitors". Of significant importance is the fact that the capillary pore network does not need to be disconnected to generate dielectric amplification. This was demonstrated by comparing the dielectric response of two pixel-based computer models of a capillary pore, referred to as the "I" model and the "T" model [11]. In the "I" model (Figure 4a), a capillary pore is disconnected by a "crossbar" of C-S-H gel. The effective dielectric constant, given in Figure 5, decreases as the thickness of the C-S-H "crossbar" increases, very similar to the actual dielectric response of OPC paste given in Figure 1. The "T" model corresponds to the replacement of the C-S-H pixels at the bottom of the "I" by pore fluid pixels (Figure 4b). This model exhibits a similar dielectric response without disconnection, indicating that dielectric amplification can be achieved in young cement pastes when the capillary pore network is percolated.

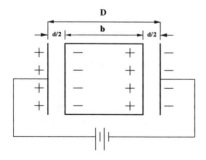

Figure 3: Schematic diagram of a parallel plate capacitor with a dielectric material between the plates.

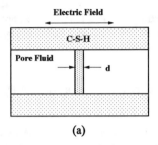
(a)

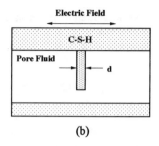
(b)

Figure 4: Pixel-based models of a capillary pore for (a) disconnected case or "I" model and (b) connected case or "T" model.

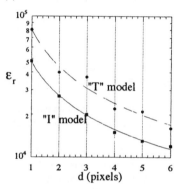

Figure 5: Dielectric constant obtained from the models in Figures 4a and 4b as the thickness (d) of the C-S-H "crossbar" is increased [9, 11].

The decrease in the dielectric constant, displayed in Figure 1, is attributed to hydration. The dielectric amplification factor decreases with growth of the C-S-H layers and reduction of the capillary pore size. Eventually, the dielectric constant of OPC paste reaches a plateau on the order of 10^3, which is still a relatively high value. This suggests the possibility of a secondary dielectric amplification at the nm-scale within the C-S-H gel microstructure, acting on the same mechanism described by equations 1 and 2.

The correlation between the porosity and the high dielectric constant of cement paste imply two methods of testing the validity of this dielectric amplification mechanism:

1) Freezing of Cement Paste

Since the freezing point of pore fluid depends on the size of the pore it occupies, changes in the electrical properties can be attributed to specific pores. Gel pores freeze at temperatures much lower than the freezing point of the larger capillary pores, so when the capillary pores freeze, the conductivity of the C-S-H gel (with unfrozen gel pores) becomes significantly higher than the conductivity of the frozen capillary pore fluid. The µm-scale dielectric amplification should be eliminated when the conductive path is switched from the capillary pore network to the C-S-H gel.

2) Solvent Exchange of Cement Paste

Replacement of the highly conductive pore fluid with an organic solvent makes the porosity highly resistive. Methanol and isopropyl alcohol, low conductivity liquids with relative dielectric constants of 33 and 18, respectively, should drastically reduce dielectric amplification through pore fluid exchange. Additionally, methanol is a smaller, more polar molecule than isopropanol and may exchange a greater degree of porosity, revealing a different level of the microstructure. Knowledge of the nature of solvent exchange could be valuable to a number of current experimental techniques. Additionally, correlating the dielectric behavior with freezing experiments could further substantiate the dielectric amplification mechanism.

EXPERIMENTAL PROCEDURE

Cement paste was prepared by adding de-ionized water to type I portland cement and hand-mixing with a spatula for 1 minute. Water-to-cement ratios of 0.4 and 0.7 were used in freezing, and ratios of 0.35 and 0.7 were used in solvent exchange. The high w/c ratio pastes were rotated for approximately 4 hours before casting to reduce sedimentation and bleeding. Samples were cast into plastic molds and stainless steel electrodes were inserted into the fresh paste. All samples were stored in an airtight chamber above a pool of water to prevent drying.

The freezing apparatus consisted of an insulated stainless steel chamber sealed to a thermoelectric stainless steel plate that was controlled by a variable power source. A thermistor ($10 \text{ k}\Omega \pm 1\%$ at $25°C$) connected to a digital multimeter was used to monitor the temperature inside the chamber. Details on the freezing procedure will be published elsewhere [12].

For solvent exchange, rectangular samples of dimensions 2" x 5/8" x 1/8" were cured for times of 1, 7, 14, and 28 days, then totally submerged in either isopropanol or methanol. Samples were removed from the solvent periodically in order to perform electrical measurements. Solvent baths were replaced frequently to insure a sufficient concentration gradient for diffusion. Details on the solvent exchange procedure will be published elsewhere [13].

A Hewlett-Packard Model 4192A impedance analyzer (Hewlett-Packard, Boulder, CO), with frequency range of 13 MHz to 5 Hz, was used to measure impedance. An impedance curve was generated by collecting 128 data points in logarithmic intervals in a frequency range of 11 MHz to 5 Hz. Data was collected on a personal computer and analyzed using "Equivalent Circuit" [17]. The resistance and capacitance of each sample were determined by fitting a simulated impedance curve to the experimental bulk arc through the origin. This procedure assumes the absence of an offset resistance, supported by recent IS work on OPC paste using a 4-point electrode configuration [18]. Variations in the fitting procedure resulted in less than a 1% difference in the calculated resistance and less than a 20% difference in the calculated capacitance, which was acceptable for this experiment. Impedance data for frequencies above 3 MHz were not used because of the known unreliability of data in this range [9]. High frequency measurements were corrected to account for the residual inductive and capacitive effects of the leads. More details of the measurement and correction processes can be found elsewhere [9].

RESULTS AND DISCUSSION

Figures 6 and 7 are semi-log plots of σ and ε_r vs. temperature in a freezing experiment for a neat OPC paste of w/c ratio 0.4 hydrated for 15 hours [12]. At the initial freezing point of the largest capillary pores ($\sim$ -5°C), σ and ε_r drop sharply by about two orders of magnitude. This effect was observed in several young OPC pastes when the large capillary pores were frozen, where the maximum dielectric constant drop was almost three orders of magnitude. When tap water is frozen, ε_r drops less than one order of magnitude from about 80 at room temperature to about 5. The dielectric constant of the pore solution was shown to be about 80 as well (150 at most) [19], so the drop in the dielectric constant upon freezing must also be around 1 order of magnitude. Since drops of over two orders of magnitude were observed in the dielectric constant of OPC paste at the freezing point of the capillary pores, dielectric amplification must be occurring within the capillary pore network.

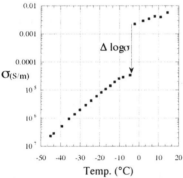

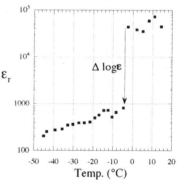

Figure 6: Conductivity vs. temperature for a neat w/c = 0.4 OPC paste hydrated for 15 hours [12].

Figure 7: Relative dielectric constant vs. temperature for a neat w/c = 0.4 OPC paste hydrated for 15 hours [12].

Figure 8 is a plot of the conductivity and relative dielectric constant vs. isopropanol exchange time for an OPC paste of w/c ratio 0.35 hydrated for one day [13]. Both values drop sharply during the first hour as the large capillary pores are exchanged, indicating that dielectric amplification is occurring within the capillary pore network, as demonstrated by freezing experiments. By two hours, the dielectric constant has stabilized, but the conductivity keeps decreasing. Hence, as the smaller capillary pores are exchanged, the dielectric constant of the paste is unaffected. This behavior is also apparent in the freezing experiments. In Figure 7, the dielectric constant does not change much as the smaller capillary pores are frozen between -5 and -40°C. Essentially, when the large capillary pores are exchanged or frozen, the capillary pore network quickly becomes "disconnected", and the small capillary pores become isolated and ineffective.

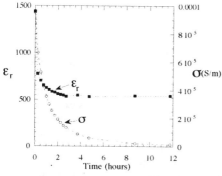

Figure 8: Conductivity and relative dielectric constant vs. isopropanol exchange time for a neat w/c = 0.35 OPC paste hydrated for 24 hours [13].

When the solvent exchange process has essentially stopped and the electrical properties have stabilized, the dielectric constant of a one day old OPC paste of w/c ratio 0.35 is still appreciable at about 500. At this point, the isopropanol has exchanged most of the capillary pores, but has left the C-S-H gel pores intact. Frozen cement paste presents a similar case. As determined by DSC [20], most of the capillary pores are frozen and most of the gel pores are still unfrozen at -40°C. As expected, the dielectric constant of a one day old cement paste cooled to -40°C is also about 500. This secondary dielectric amplification suggests that the C-S-H gel microstructure is geometrically amplified on the nm-scale, in the same manner as described in equations 1 and 2, where the dielectric amplification factor is defined by gel pores (D) and C-S-H layers (d).

The existence of this secondary dielectric amplification is supported by the fact that the dielectric constant of frozen cement paste (-40°C) increases with hydration time and as a function of the volume fraction of C-S-H gel, regardless of w/c ratio, as shown in Figure 9 [12]. This effect is also seen in solvent exchange experiments. Figure 10 is a plot of the relative dielectric constant vs. isopropanol exchange time for three OPC pastes at different hydration times. As shown, the dielectric constant of the solvent-exchanged paste increases with hydration time, correlating with the increase in C-S-H gel [13].

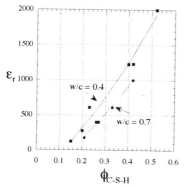

Figure 9: Relative dielectric constant vs. volume fraction of C-S-H gel for OPC paste at -40°C, where capillary pores are frozen and gel pores are unfrozen [12].

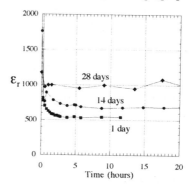

Figure 10: Relative dielectric constant vs. isopropanol exchange time for three neat OPC pastes at hydration times of 1, 14, and 28 days [13].

Isopropanol vs. Methanol Solvent Exchange

In comparison to isopropanol, methanol is composed of smaller, more polar molecules and should exchange a greater amount of gel pores. The difference in electrical properties between the two methods of solvent exchange should reveal information about the C-S-H microstructure. Figure 11 is a plot of the dielectric constant vs. exchange time for similar 1 day old OPC pastes with w/c ratios of 0.35, where one was exchanged with isopropanol and the other with methanol. The sample exchanged with isopropanol displayed secondary dielectric amplification, but the dielectric constant of the methanol-exchanged sample did not, even though the relative dielectric constant of methanol is almost double that of isopropanol. Obviously, the methanol was able to exchange a significant amount of gel porosity and eliminate the secondary dielectric amplification.

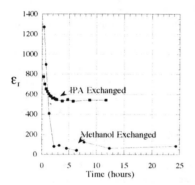

Figure 11: Relative dielectric constant vs. solvent exchange time for similar w/c = 0.35 OPC pastes hydrated for 24 hours, where one was exchanged with methanol and the other with isopropanol [13].

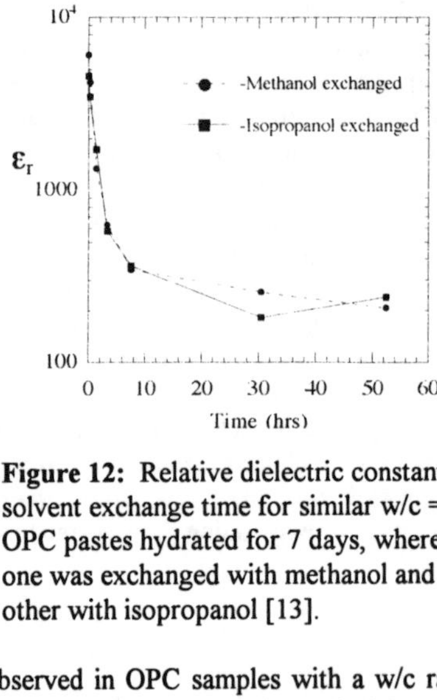

Figure 12: Relative dielectric constant vs solvent exchange time for similar w/c = 0.70 OPC pastes hydrated for 7 days, where one was exchanged with methanol and the other with isopropanol [13].

Interestingly, this difference is not observed in OPC samples with a w/c ratio of 0.7, as shown in Figure 12. The ε_r curves for two OPC pastes exchanged with methanol and isopropanol are identical. The secondary dielectric amplification is present in both samples, suggesting that there is a microstructural difference between C-S-H gel in a w/c ratio 0.35 paste and C-S-H gel in a w/c ratio 0.7 paste, where methanol replacement of the gel pores is allowed in the first case, but prohibited in the latter.

CONCLUSIONS

1) Using impedance spectroscopy to monitor the electrical properties of ordinary portland cement paste during freezing and solvent exchange has provided new information about the microstructure of cement paste.

2) Two modes of dielectric amplification are present in portland cement paste. A primary dielectric amplification occurs at the μm-scale in the capillary pore network, and a secondary dielectric amplification occurs at the nm-scale in the C-S-H gel.

3) The dielectric constant of frozen and isopropanol-exchanged OPC paste increased with hydration, correlating with an increase in the volume fraction of C-S-H gel.

4) Comparing the dielectric behavior of OPC pastes during isopropanol and methanol solvent exchange revealed information about the variable nature of C-S-H gel microstructure. Methanol was able to replace gel pore fluid in an OPC paste with w/c ratio 0.35, but was unable to replace gel pore fluid in an OPC paste with w/c ratio 0.7.

ACKNOWLEDGEMENTS

The authors would like to thank the National Science Foundation's Center for Advanced Cement-Based Materials for financial support under grant number DMR-912002.

REFERENCES

[1] W.J. McCarter, S. Garvin, and N. Bouzid, J. Mater. Sci. Lett. 7 (10), 1056 (1988).

[2] W.J. McCarter and R. Brousseau, Cem. Concr. Res. 20, 891 (1990).

[3] P.R. Camp and S. Bilotta, J. Appl. Phys. 66 (12), 6007 (1989).

[4] C.A. Scuderi, T.O. Mason, and H.M. Jennings, J. Mater. Sci. 26, 349 (1991).

[5] K. Brantervik and G.A. Niklasson, Cem. Concr. Res. 21, 496 (1991).

[6] P. Gu, Z. Xu, P. Xie, and J.J. Beaudoin, Cem. Concr. Res. 23 (4), 531 (1993).

[7] Z. Xu, P. Gu, P. Xie, and J.J. Beaudoin, Cem. Concr. Res. 23 (4), 853 (1993).

[8] Z. Xu, P. Gu, P. Xie, and J.J. Beaudoin, Cem. Con. Res. 23 (5), 1007 (1993).

[9] B.J. Christensen, R.T. Coverdale, R.A. Olson, S.J. Ford, E.J. Garboczi, T.O. Mason, and H.M. Jennings, J. Am. Cer. Soc. 77 (11), 2789 (1994).

[10] R.T. Coverdale, B.J. Christensen, T.O. Mason, H.M. Jennings, E.J. Garboczi, and D.P. Bentz, J. Mater. Sci., in press.

[11] R.T. Coverdale, B.J. Christensen, T.O. Mason, H.M. Jennings, and E.J. Garboczi, J. Mater. Sci. **29**, 4984 (1994).

[12] R.A. Olson, B.J. Christensen, R.T. Coverdale, S.J. Ford, G.M. Moss, E.J. Garboczi, H.M. Jennings, and T.O. Mason, submitted to J. Mater. Sci.

[13] G.M. Moss, B.J. Christensen, E.J. Garboczi, H.M. Jennings, and T.O. Mason, submitted to J. ACBM.

[14] P.N. Sen, Geophysics **46**, 1714 (1981).

[15] F. Brouers, A. Ramsamugh, V.V. Dixit, J. Mater. Sci. **22**, 2759 (1987).

[16] A.J. Moulson and J.M Herbert, Electroceramics, (Chapman and Hall, New York, 1990), p. 261.

[17] B.A. Boukamp, Equivalent Circuit (EQUIVCRT.PAS), University of Twente, Department of Chemical Technology, P.O. Box 217, 7500 AE Enschede, The Netherlands, (1988).

[18] S.J. Ford, B.J. Christensen, R.T. Coverdale, E.J. Garboczi, H.M. Jennings, and T.O. Mason, submitted to J. Mater. Sci.

[19] B.J. Christensen, PhD thesis, Northwestern University, (1993).

[20] D.H. Bager and E.J. Sellevold, Cem. Concr. Res. **16**, 709 (1986).

REVERSIBLE AND IRREVERSIBLE ELECTRICALLY-INDUCED SHAPE CHANGES IN CEMENT-BASED MATERIALS

JIE-FANG LI, LI-JIAN YUAN, AND DWIGHT VIEHLAND
Department of Materials Science and Engineering and The Center for Advanced Cement Based Materials, University of Illinois, Urbana, IL 61801

ABSTRACT

The electromechanical behavior of hardened portland cement paste has been investigated as a function of measurement frequency and DC electrical bias using an inductance technique. Large field-induced shape changes on the order of 10^4 Å have been found in millimeter thick specimens exposed to moisture. Both reversible and irreversible contributions to the total electrically-induced strain were observed. In addition, strong hysteresis effects were found on cycling a large AC field. Dry samples were found to exhibit no field-induced deformations. The mechanism underlying this anomalous behavior is believed to be an electro-osmotically induced swelling of the gel-pore structures.

INTRODUCTION

Nearly 20 years ago, Wittmann[1] observed an electomecanical effect in portland cement[1]. He observed an electromechanical effect in centimeter-sized bars of hardened cement paste. A voltage (~200 mV) was generated when an external load (2000 g) was applied to the specimen, and the converse effect of a bending displacement (~20 mm) was detected upon the application of a voltage (200 V). More recently, Li et al.[2] have investigated the electromechanical behavior of portland cement using an interferometeric technique. Field-induced shape changes (ε) on the order of 100 Å were found in millimeter thick specimens which had been exposed to moisture. Dry specimens were found to exhibit no field-induced deformations. In addition, relatively large longitudonal piezoelectric (d_{33}) coefficients were determined. The mechanism underlying this anomalous behavior was believed to be an electro-osmotically induced swelling of the gel pore structure.

The influence of water diffusion under an electric field on the conduction properties in porous plugs is well-known[3,4], and is generally attributed to electro-osmotic forces. In electro-osmosis, an applied field acts upon ions in a liquid. The ions, then, move under the excitation, carrying along the liquid media. Consequently, it should be possible to establish nonuniform water distributions under AC or DC electrical fields in porous plugs, such as portland cement. This nonuniformity may provide a mechanism by which electrical energy can be converted to a mechanical deformation.

Mat. Res. Soc. Symp. Proc. Vol. 370 © 1995 Materials Research Society

The purpose of this investigation was to study the electromechanical behavior of hardened cement paste. Attention was placed on developing an understanding of the mechanism by which electrical energy can be coupled to a mechanical deformation.

EXPERIMENTAL PROCEDURE

Ordinary portland cement (OPC) was hydrated at water to cement ratio (w/c) of 0.40. The alakali content of the specimens was determined by chemical analysis. The Na_2O and K_2O contents were 0.14 and 0.02 at.%, respectively. Samples were made by casting into 1.25 cm x 2.5 cm cylinders, which were cured in a moist environment (100% relative humidity (r.h.)) for 8 weeks. Cylinders of hardened paste were then cut into disc-shaped specimens having radii of 10 mm and a thickness of 4 mm. The small specimen size was needed to increase the sample impedance for electromechanical property characterization. It was realized that the size of the cement samples was relatively small (in consideration that cement is usually batched in large volumes). However, repeatability of the results was observed within experimental errors. The specimens were polished to assure parallel faces and subsequently electroded with gold.

The electrically-induced strain was characterized by an inductance technique. A rod was placed on top of the specimens, through which an electrical contact was made to the top surface. This rod had attached to it a linear variable differential transformer (LVDT). The sample was electrically excited by a high voltage (HV) power amplifier, in conjunction with a preamplifier and a computer. The preamplifier was a modified Sawyer-Tower circuit. Changes in the shapes of the specimens were induced under the electrical field. As a consequence, the rod placed on top of the specimen was displaced. This displacement was detected inductively by the LVDT. A lock-in amplifier was used to filter random intensity fluctuations from those with the characteristic time constant of the drive, achieving small displacement resolutions. The displacment resolution of the system was ~50 Å. The measurement frequencies used were between 5×10^{-3} and 1 Hz. Both sinuosiodial and step drives were used in this study.

RESULTS

Figure 1 illustrates the electrically-induced strain (ε) for an OPC specimen with a w/c ratio of 0.40. This measurement was made on a virgin sample which had initally contained 14.6 wt.% water. The measurement frequency used was 6.66×10^{-3} Hz. Similar results were obtained from multiple samples. Figure 1 clearly demonstrates the presence of large electrically-induced strains. The magnitude of ε was $\sim 5 \times 10^{-4}$ under 0.5 kV/cm. Previous investigations by Li et al. revealed values of ε of $\sim 10^{-5}$ under similar field strengths, however these studies were performed using a measurement frequency of 10 Hz. The magnitude of the electrically-induced strain is relatively large, especially in consideration of the field strength applied to the specimen. For example, the electrically-induced strain for lead zirconate titanate (PZT) and lead magnesium niobate (PMN) piezoelectric ceramics is ~0.1-0.2% under 30 kV/cm. Clearly, the electrically-induced strains for OPC saturated with water are significant. In consideration that dry samples were found not to exhibit any electrically-induced shape changes in this study, we believe that the shape changes are related to water transport. However, electrically-driven water transport cannot occur due only to a saturation gradient through the specimen, as water is not charged. We believe that ions in the pore solution diffuse under the electrical field and that water transport subsequently occurs by electro-osmosis.

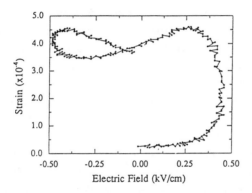

Figure 1. Electrically-induced strain as a function of applied AC electrical field for portland cement. This measurement was made using a sinusoidal drive with a frequency of 6.66x10[-3] Hz.

The sign of the electrically-induced strain was positive, as can be seen by inspection of Figure 1. This clearly shows that the net shape of the specimen expands longitudinally in the direction of the applied field. Equilibrium length-change isotherms on absorption of water have previously been determined for OPC materials by Feldman[5]. These investigations revealed an expansion during absorption of water and a contraction during desorption. The expansion was suggested to develop due to electro-osmotic pressure effects. Electro-osmotic forces may also lead to a redistribution of water internally in the interconnected pore structures under an electrical field, giving rise to local swelling effects. We believe that the electrically-induced strains in OPC occur due to water redistribution between gel pores or between gel pores and interlayer positions. As a consequence, on the average the gel pore structure expands and/or the interlayer spacing increases under an electrical field due to local electro-osmotic pressure effects.

Figure 1 also clearly reveals both irreversible and reversible shape changes on application of an AC electrical field. On starting from the virgin state, a large irreversible (with respect to the time constant of the measurement) change can be seen to occur. The magnitude of this irreversible contribution to the total strain was $\sim 3.5 \times 10^{-4}$. On reversing the field, a switchable (reversible) component of the electrically-induced strain was found. The magnitude of this switchable strain was $\sim 1.5 \times 10^{-4}$. This switchable component of the electrically-induced strain was characterized by the presence of "bow-like" hysteresis loops. Clearly, both irreversible (nonswitchable) and reversible (switchable) shape changes occur on cycling with an AC electrical field from the virgin state. The relaxation time of the irreversible contribution is significantly greater than 150 seconds, which was the time constant of the measurement frequency. After the measurement was completed, a slow drift was found in the sample displacement by observing the digitial readout of the lock-in amplifier. A steady state condition was reached after approximately five minutes.

The electrically-induced strain shown in Figure 1 was measured from the virgin state using a sinusoidal drive. A sequential secondary electrically-induced strain measurement was then performed after ~ 5 minutes using a square-wave drive. The application of a square-wave drive to a sample can more clearly demonstrate the presence of electromechanical relaxations. Figure 2 shows the electrically-induced strain characteristics as a function of time in response to a step function in the electric field of 0.5 kV/cm. This figure unambiguously reveals a gradual build-up of the electrically-induced strain over a long-time period. The electrically-induced strain was observed to be a linear function of time ($t < 50$ seconds) to a first order approximation. The magnitude of the electrically-induced strain was $\sim 3 \times 10^{-4}$, 50 seconds after application of the field. The field was then removed, and the relaxation of the electrically-induced strain measured. The original dimensions of the specimen were not completely recovered after an additional 50 seconds. Rather, a residual strain of $\sim 10^{-4}$ was observed. These results clearly demonstrate the presence of irreversibility in the shape changes, qualitatively similar to that observed in Figure 1. However, if sufficient time were allowed, the orginal shape of the sample would be recovered. The hysteresis arises due to the fact that the time constant of the relaxation (τ_{decay}) of the strain is significantly longer than that of the build-up (τ_{build}) of the strain.

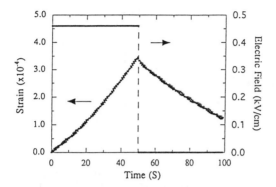

Figure 2. Electrically-induced strain as a function of applied AC electrical field for portland cement. This measurement was made using a square wave pulse.

Electrically-induced strain meaurements would have been made to longer time periods using the square-wave drive, however the system performance characteristics were limited to measurement time constants less than 150 seconds (6.66×10^{-3} Hz). Assuming linearity, the electrically-induced strain can be extrapolated to longer times. The magnitude of the induced strain can be estimated to be $\sim 6 \times 10^{-4}$ at 100 seconds and $\sim 3 \times 10^{-3}$ at 500 seconds. Obviously, saturation effects will eventually reduce the increase in the strain. The extrapolation was performed only to demonstrate the magnitudes of strains which might be induced at ultra-low frequencies ($\omega < 5 \times 10^{-3}$ Hz). These strains are not unrealistic, as the length-changes during absorption of water from the dry state can approach 1% under high partial pressures of water.

Third and fourth sequential electrically-induced strain measurements are shown in Figures 3(a) and (b), respectively. A waiting period of 5 minutes was again used between measurements. The third measurement was made using a sinusoidal drive and the fourth a square-wave pulse. Comparisons of Figures 3(a) and (b) with Figures 1 and 2 will reveal that the magnitude of the irreversible strain decreased with increasing number of cycles. After the third sequential cycling, only the reversible contribution was apparent. The magnitude of this contribution did not change significantly with additional cyclings of the AC electrical field, in the time period investigated. Clearly, the irreversible component of the strain is dominant only in the first several cycles of the electrical field.

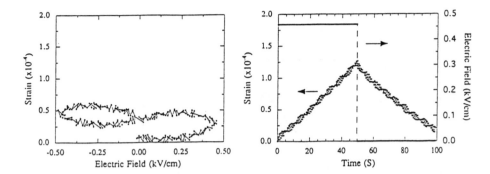

Figure 3. Electrically-induced strain as a function of applied AC electrical field for portland cement on various sequential runs. (a) Third sequential measurement which was made using a sinusoidal drive with a frequency of 6.66×10^{-3} Hz, and (b) fourth sequential measurement which was made using a square wave pulse.

CONCLUSIONS

Large field-induced shape changes have been observed in hardened cement pastes. This electromechanical response was found to be strongly dependent on measurement frequency, DC electrical bias, and exposure to moisture. In addition, strong hystersis effects were found on cycling a large AC electrical field. The induced shape changes are believed to arise due to electro-osmotically-induced swelling of the pore structures under electrical bias.

ACKNOWLEDGEMENTS

This research was supported by the NSF center for the Science and Technology of Advanced Cement Based Materials under contract number DMR-9120002.

REFERENCES

1.   F.H. Wittmann, Cement and Concrete Research, Vol.3 (1973) 601-605.

2.   Jie-Fang Li, Hua Ai, and Dwight Viehland, J. Am. Cer. Soc. (accepted).

3.   R. Hunter, "Zeta Potential in Colloid Science: Principles and Applications", Academic Press, New York (1981).

4.   R. Hunter, "Foundations of Colloid Science, Vol.I", Clarendon Press, Oxford (1989).

5.   R.F. Feldman, Proceedings, Fifth International Symposium on the Chemistry of Cement, Tokyo, 1968, Part III, Vol. III, pp.53-66.

DISTRIBUTION OF PORE SIZES IN WHITE CONCRETE

G. BEST, A. CROSS, H. PEEMOELLER, AND M.M. PINTAR
Physics Department, University of Waterloo, Waterloo, Ontario, Canada, N2L 3G1

ABSTRACT

NMR is used to study the evolution of the discrete distribution of proton magnetization fractions in hydrating synthetic white cement. The results at 100 hours hydration time are modelled using a trimodel pore size distribution. A good correspondence is found between the NMR analysis and SANS results reported in the literature.

INTRODUCTION

During cement paste hydration some water reacts chemically with cement clinker while the remaining water, being liquid-like, resides in pores and voids of the hardening cement paste. The hydrogen atoms on liquid-like water and hydrogen atoms (initially also from water molecules) in the cement's solid structures are in several chemically, structurally, and dynamically distinct environments. In this research we use proton nuclear magnetic resonance (NMR) to study the establishment and evolution of these environments in hardening cement paste. NMR is ideally suited for this purpose because it uses hydrogen nuclei as probes and monitors many details about the evolving structure and dynamics of these environments at the molecular level. All proton spins in a particular environment (repeated throughout the cement paste) are considered to constitute a spin "group" which is characterized by its spin-lattice relaxation time (T_1), its spin-spin relaxation time (T_2), and its magnetization fraction (M_{oi}). The T_1 and T_2 yield dynamical information about the proton bearing molecules in the spin group. While T_2 generally decreases as the molecular environment becomes more rigid, T_1 is normally at a minimum when the molecular dynamical frequencies are equal to the NMR Larmor frequency employed (in our case 40 MHz). The proton magnetization fraction is exactly proportional to the number of protons constituting the spin group, thus it measures the proton group's relative size.

In NMR studies of heterogeneous materials, such as hydrating cement paste, two main analysis protocols are used. The first involves the characterization of the system with a discrete distribution of spin groups; e.g. a sum of delta functions. Different spin groups represent distinct structural components of the cement paste (e.g., C-S-H, CH, other crystallites such as ettringite, and water in several kinds of pores). The resolution of the CH magnetization is remarkable. The C-S-H and other crystallites are resolved relying in part on stoichiometry [1]. The second method involves the fitting of a continuous distribution of different environments (such as water in layers, pores, capillaries, etc.) within the structure of the cement paste. In this case the spin-lattice recovery function is characterized by a continuous distribution of T_1 values. Indications are that both approaches are applicable in hydrating cement paste as the magnetization recovery function in this material has been found to be strongly nonexponential. On the one hand, the nonexponentiality is such that its resolution into several identifiable components by spin-grouping is well justified [1] at hydration times beyond $\sim$ 50 h if the recovery of the total sample magnetization is measured. On the other hand, the cement pore surface has been identified as a fractal object of dimension 2.7, and a distribution [2]

Mat. Res. Soc. Symp. Proc. Vol. 370 © 1995 Materials Research Society

characteristic of this surface has been searched for in the measured proton composite spin-lattice relaxation recovery function. Th fractal dimension has been studied only once [3]. It was found [4] that indeed the proton magnetization recovery curves could be fitted to a "stretched exponential" form, characteristic of fractals if the recovery of the liquid-like water magnetization only is measured; e.g. by selecting a time window where all the solid like magnetization has decayed (at 150 μs).

We propose that a consistent description of the cement system, employing the NMR parameters, is a multi-modal distribution of environments where the position and width of each distribution peak pertains to discrete structural and/or dynamical elements of the system. Due to physical exchange (molecular diffusion) of water molecules between neighbouring pores, and exchange of magnetization between spin groups under each peak of the distribution curve, the distribution of T_1 values narrows (by about 50 h of hydration) into a set of delta-function-like peaks. For the method to work the NMR response has to be characterized with the NMR spin grouping method [5] which utilizes the correlation between spin-spin relaxation times and spin-lattice relaxation times. With this two dimensional (2D) time evolution method the NMR response from the hydrating cement paste (the unresolved composite signal which contains contributions from all spin groups in the material) can be resolved in T_1 and T_2 time domains. This necessitates the 2D decomposition of the total NMR response into a sum of discrete responses. Due to the 2D nature of the information the individual spectroscopic lines of the resolvable spin groups can be reconstructed to yield the magnitudes of the spin groups.

In this report, we discuss briefly the evolution of the discrete distribution of proton magnetization fractions in hydrating synthetic white cement with a water/cement (W/C) ratio of 0.42. These magnetization fractions and their relaxation times are then employed to model a trimodal pore size distribution first proposed by small angle neutron scattering (SANS).

EXPERIMENTAL

The proton NMR relaxation study of the hydration of synthetic white cement (0.42W/C) at 21°C has been undertaken at 40 MHz [6]. The NMR relaxation parameters were studied with a modified SXP Bruker single-coil spectrometer. T_1 was measured using the inversion recovery sequence, 180°-τ-90°, employing typically 30 different τ-values. The NMR spin grouping method is described in detail elsewhere [5].

The composition of the cement clinker is listed in Table I.

TABLE I: Composition of synthetic white cement (wt. %)

Oxide Composition		Principal compound composition	
CaO	65.10	C_3S	46.66
SiO_2	21.39	C_2S	26.13
Al_2O_3	6.32	C_3A	16.75
MgO	3.11	C_4AF	0.02
Na_2O	0	gypsum	7.0
K_2O	0		
SO_3	3.25		
Loss on ignition	1.46		

RESULTS AND DISCUSSION

Evolution of Relaxation Parameters

The magnetization recovery curve in the white cement paste for hydration times less than about 5 hours was fit to a sum of three single-exponentials and for longer hydration times to a sum of four single-exponentials [6]. Figure 1 shows a representative magnetization recovery plot measured at a 8 μs time window on the free induction decay (FID), at a hydration time of 100 hours. Four magnetization components are clearly resolved. Furthermore, these analyses are averaged over 30 time windows on the FID. This improves the reliability. Even more important is that the zero time intercepts are combined into reconstructed FIDs which yield the magnitudes of the fractional magnetizations of spin groups. Such spin grouping analysis of the NMR results at this

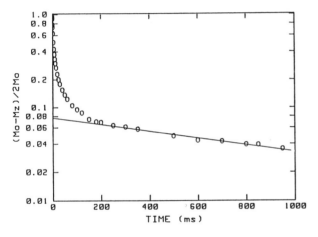

Figure 1a. Recovery curve (solid line) of magnetization with longest $T_1 = 1.2$ s.

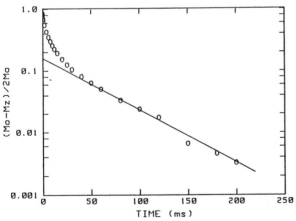

Figure 1b. Plot of the difference between the data in Figure 1a and the recovery curve with $T_1 = 1.2$ s. The solid line represents the recovery curve of magnetization with $T_1 = 50$ ms.

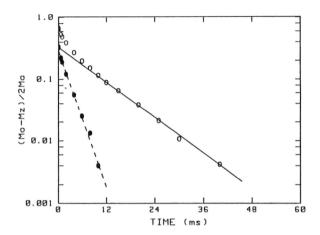

Figure 1c. Open circles represent the difference between the data in Figure 1b and the recovery curve with T_1 = 50 ms. The solid line represents the recovery curve of magnetization with T_1 = 9.5 ms. The filled circles and dashed line represent the recovery curve of magnetization with shortest T_1 = 1.9 ms.

hydrating time gives the following magnetization fractions of the solid matrix [6]:

C-S-H	22%	(23%)
CH	10%	(15%)
crystallites other than CH	19%	(21%)
excess water	48%	(41%)

where the percentages in brackets are magnetization fractions calculated from stoichiometry at full hydration of the studied cement paste. While the cement paste studied at 100 h did not yet reach the exact stoichimetric constitution of full hydration, its CH component of 10% is rather low in comparison with the expected stoichiometric CH magnetization of 15% for this paste.

The excess water magnetization of 48% has been found [6] in three distinct groups of 21%, 5%, and 22%. These were assigned to water in large, medium and small pores, respectively. This assignment is based on the well-known fact that water in restricted geometry with largest surface to volume (S/V) ratio relaxes fastest. This is the consequence of the surface shifting the characteristic frequencies of water molecule motion from its bulk water value of $\sim 10^{13}$ sec^{-1} to a surface water value of $\sim 10^8$ sec^{-1} [7]. On many surfaces the moderated frequencies are very close to the Larmor frequency of protons which results in fast spin-lattice relaxation. Consequently, in small pores, (even without paramagnetic ions) the water relaxation rate is greater than in the bulk phase by a factor of 10 or more. The slowing down effect of water due to the surfaces in cement paste is augmented by the paramagnetic ions attached to these surfaces. This causes an even stronger differentiation between water proton relaxation rates by the S/V ratio of the restricted geometry.

Pore Size Distribution

From scanning electron microscopy of Portland cement paste three types of pores have been identified [8]: (I) micropores with diameters between 10 and 100 nm due to colloidal hydrate gel porosity, (II) micropores with diameters from 100 to 1000 nm which form between crystallites and (III) macropores, or voids, between packed grains, with diameter > 1 μm. Since the proton T_1 of water inside a pore is a function of the pore's surface to volume ratio, it should be possible to relate the distribution of water proton spin group characteristics and the hydrating cement paste pore size distribution. Consider the three quasi liquid-like water proton magnetizations resolved at 100 hours hydration when most of the structure forming has been completed. Renormalizing the water magnetization fractions at this hydration time to 100% we obtain: $M_{0a} = 46\%$ with $T_{1a} = 1.9$ ms; $M_{0b} = 10\%$ with $T_{1b} = 9.5$ ms and $M_{0c} = 44\%$ with $T_{1c} = 50$ ms. We propose that these magnetization fractions are associated with water in pores of type I, II and III, respectively.

For the present discussion we assume that pores are spherical (as was assumed for modelling SANS data) of radius r and contain a fraction b(r) of water (with relaxation time $T_{1surface}$) bound to the pore surface and a fraction [1-b(r)] of bulk-like water (with relaxation time T_{1bulk}). If the exchange between these two fractions is fast on the time scale of the shorter T_1, the average water proton T_1 for a particular pore becomes

$$\frac{1}{T_1(r)} = \frac{b(r)}{T_{1surface}} + \frac{[1 - b(r)]}{T_{1bulk}} . \tag{1}$$

In addition, as long as the pore radii are less than $(D \cdot T_1)^{1/2}$ it is reasonable to expect that water molecules diffuse through a few pores in the time of spin-lattice relaxation and that consequently for pores in close proximity water proton magnetization exchange occurs (including spin diffusion across the solid matrix separating the pores). Then, an effective relaxation time T_{1i} and magnetization M_{0i} for all pores of type i can be calculated by averaging these two observables over an interval of pore sizes employing a pore size distribution function $\rho(r)$:

$$\langle f \rangle = \int_{r_{min,i}}^{r_{max,i}} \rho(r)\, f_i(r)\, dr \tag{2}$$

where $f_i(r)$ is either $1/T_1(r)$ or M(r). The values $r_{max,i}$ and $r_{min,i}$ are the limits on the "i" pore radii. Drawing upon results from small angle neutron scattering experiments [9] we adopt the following distribution function

$$\rho(r) = \begin{cases} C_a r^{-\alpha_a}, & 10\ nm \le r \le 60\ nm,\ with\ \alpha_a = 4.5\ for\ pores\ I, \\ C_b r^{-\alpha_b}, & 60\ nm \le r \le r_x\ with\ \alpha_b = 3.8\ for\ pores\ II, \\ C_c r^{-\alpha_c}, & r_x \le r \le r_{max},\ with\ \alpha_c = 3.8\ for\ pores\ III, \end{cases}$$

$$\tag{3}$$

where r_{max} is the largest pore size. As the NMR experiments yield three distinct relaxing magnetization fractions of water (a, b and c) the following pore size ranges were constructed: a) 10nm to 60 nm b) 60 nm to $r_{max,II} = r_x$ and c) $r_{min,III} = r_{max,II} = r_x$ to r_{max}; where r_x and r_{max} are to be fitted. It is convenient to express b(r) in terms of h, the thickness of the bound water layer:

$$b(r) = \frac{3h}{r} - \frac{3h^2}{r^2} + \frac{h^3}{r^3} \qquad (4)$$

Values of M_{oi}'s and T_{1i}'s were calculated from equation 2 and compared to the five observables (three pairs of T_{1i} and M_{oi}, minus the normalization condition $\Sigma M_{oi} = 1$) using a least squares fitting routine [10] with r_x, r_{max}, and $T_{1surface}$ as adjustable parameters and h and T_{1bulk} as fixed parameters. The best fit was explored for a broad range of h and T_{1bulk} values and the best match between experiment and model was found for h = (6 ± 2) Å; $T_{1surface}$ = (0.3 ± 0.1) ms; T_{1bulk} = (100 ± 30) ms; r_x = (1.0 ± 0.5) μm and r_{max} = (15 ± 4) μm.

The results indicate that only about 2 layers of water molecules on the pore surface are influenced by the surface. This is in keeping with the generally accepted view about the water in hydrated porous systems where paramagnetic ions are attached to the surfaces.

The values of r_x and r_{max} compare well with the upper pore size limits, given in reference 8, for pores of type II and III, respectively. This correspondence between results obtained using such different techniques (SANS and NMR) lends support for the above modelling.

CONCLUSIONS

In summary, it has been shown that the NMR spin-lattice relaxation in hydrating synthetic white cement paste, after 100 h hydration time, are consistent with a trimodal pore size distribution model. The upper cut-off radii of pores belonging to regime II and III, $T_{1surface}$, T_{1bulk}, and h have been found. The NMR analysis is consistent with results obtained using other physical techniques.

Although a number of assumptions have been made in the present approach the methodology is justified for at least two important reasons. One is that it permits the nondestructive and continuous monitoring of the total liquid water component as a fraction of the total during hydration. The other is that it permits the monitoring of the CH growth and of its exact amount, which can be done simultaneously as the liquid water is recorded. Between these two observables the hydration of cement paste and of other cement related materials, can be followed closely. The end of the dormant period, the acceleration of hydration, the secondary hydration, the left over water in the structure, and many other aspects of hydration can be investigated. As NMR characteristics of these two proton structures (liquid H_2O and CH) are very distinct between themselves and from the other structural components, the NMR methodology is at its best.

REFERENCES

1. J.C. MacTavish, L. Miljkovic, M.M. Pintar, R. Blinc, and G. Lahajnar, Cem. Concr. Res. **15**, 367 (1985).
2. W.T. Sobol, I.G. Cameron, M.M. Pintar, and R. Blinc, Phys. Rev. B, **35**, 7299 (1987).
3. A.J. Allen, R.C. Oberthur, D. Pearson, P. Schofield, and C.R. Wilding, Phil. Mag. B, **56**, 263 (1987).
4. R. Blinc, G. Lahajnar, Z. Zumer, and M.M. Pintar, Phys. Rev. B, **38**, 2873 (1988).
5. H Peemoeller and M.M. Pintar, J. Magn. Reson. **41**, 358 (1980).
6. L. Miljkovic, J.C. MacTavish, H. Peemoeller, J.M. Corbett, J. Jian, D.D. Lasic, M.M. Pintar, R. Blinc, G. Lahajnar, and F. Milia, Manuscript to be published in Adv. in Cem. Res.
7. R.G. Bryant and W.M. Shirley, Biophys. J. **32**, 3 (1980).
8. N.M. Alford and D.D. Double in Adsorption at the Gas-Solid and Liquid-Solid Interface, edited by J. Rouquerol and K.S.W. Sing (Elsevier, Amsterdam, 1982), pp. 259-268.
9. A.J. Allen and D. Pearson in Microstructural Characteristics of Material by Non-Microscopic Techniques, N.H. Anderson et al., editors (5th Risco International Symposium on Metallurgy and Materials Science, 1984), pp. 175-180.
10. P.R. Bevington, Data Reduction and Error Analysis for the Physical Sciences (McGraw-Hill Book Company, New York, 1969), Chapter 11.

MULTITECHNIQUE APPROACH TO UNDERSTANDING THE MICROSTRUCTURE OF CEMENT-BASED SYSTEMS

DAVID L. COCKE*, M. YOUSUF A. MOLLAH**, R. K. VEMPATI AND T. R. HESS
*Gill Chair of Analytical Chemistry, Lamar University, Beaumont, TX 77710
**Visiting Professor, Department of Chemistry, Dhaka University, Bangladesh

ABSTRACT

The chemistry of cement, its hydration and the development of microstructure in cement-based systems is extremely complex, and it becomes even more complex in the presence of additives. The elucidation of the mechanisms of these processes is a challenging problem and requires the applications of multiple techniques including the latest microscopic methods. The applications of molecular spectroscopies, surface spectroscopies and microscopies have helped develop models and mechanisms for the retardation of cement setting by Zn, Cd and Pb, the chemical and structural effects of superplasticizers, and the interaction of hydrating cement with aggregates, selective sorbents and fillers. The results of these studies indicated that the inhibition of hydration is controlled by dispersion of various charges present in hyperalkaline solution in cement paste. According to this charge dispersal model, the Ca^{2+} ions from initial hydration form a tightly-bound bilayer with the negatively charged C-S-H surface. Consequent to this intrinsic process, the metalhydroxy or superplasticizer anions immediately surround the bilayer to constitute a trilayer which inhibits further hydration.

INTRODUCTION

The hydration of ordinary Portland cement (OPC) and subsequent development of microstructure involve a series of competing chemical reactions. The mechanisms of these reactions are complex and not yet fully understood. Some of the characterization techniques that have been used for studying the chemistry of cement-based systems include [1]: X-ray photoelectron spectroscopy (XPS), Auger electron spectroscopy (AES), ion scattering spectroscopy (ISS), X-ray diffraction (XRD), scanning electron microscopy (SEM), atomic force microscopy (AFM), energy dispersive spectroscopy (EDS), Fourier transform infrared spectroscopy (FTIR), Raman spectroscopy and ^{29}Si solid state NMR [2-4] employing cross-polarization and magic-angle-spinning techniques. In the present article we are presenting results from our laboratory concerning the application of FTIR, XRD, XPS and SEM/EDS techniques to characterize cement-based systems involving metals, fly ash, and superplasticizer.

CHARACTERIZATION TECHNIQUES

X-ray Photoelectron Spectroscopy (XPS)

The XPS technique is recognized as one of the most important surface sensitive tools. This technique probes the surface core-level electronic states of atoms in the near-surface region (sampling depth ~50 Å) and can provide qualitative chemical state and semi-quantitative information about these surface and near-surface regions [3-5]. In this method, the usual practice is to measure the energy, known as the binding energy (BE), by which the photoelectron had been bound to its parent atom. The binding energies of the electrons are then compared with those of the pure elements or compounds and thus the changes in their chemical environment are reflected in BE shifts. The XPS technique has been successfully used to delineate the: (i) degree of polymerization of silicates in hydrating cement, (ii) effects of carbonation on hydration, (iii) binding and chemical environment of metal pollutants in cement-based systems, (iv) the leaching mechanisms of priority metal pollutants from cement-based solidification/stabilization (S/S)

Mat. Res. Soc. Symp. Proc. Vol. 370 © 1995 Materials Research Society

systems, (v) the influence of dopant metals on the dynamic processes involved during hydration and setting of Portland cement [8-11].

Anhydrous and Hydrated Cement: A typical XPS wide energy scan of dry Portland cement showed that Si, Ca, C, Al and O are the major components present on the surface of cement grains. The BE of Si (2p), Ca (2p), Al (2p) and O (1s) are: 100.8, 347.1, 74.0 and 531.0 eV, respectively. The BE of the corresponding peaks in hydrated cement are: 102.2, 346.8, 74.0 and 531.0 eV, respectively. Two C (1s) photopeaks appear at approximately 285.0 and ~ 289.5 eV in the carbon region. The C (1s) peak corresponding to 285.0 eV is due to adventitious carbon on the surface of carbonate minerals which is commonly formed when the sample is exposed to ambient atmosphere. The other C (1s) peak corresponds to CO_3^{2-} minerals in the matrix. The BE of the Si (2p) electron in the hydrated cement has been shifted to the higher BE by ~ 1.40 eV as compared with the dry clinkers. This shift in Si (2p) is indicative of polymerization of the SiO_4^{4-} units in cement system. The principal products of hydration are ~ 20-25% of $Ca(OH)_2$ and ~ 60-70% of calcium-silicate-hydrate (C-S-H). The formation of C-S-H is accompanied by the polymerization of the SiO_4^{4-} units in cement system. The degree of polymerization of SiO_4^{4-} units during hydration can be ascertained by measuring the change in Si (2p) BE in C-S-H structure.

Effect of Carbonation and Metal Doping: Atmospheric carbon dioxide plays an important role during cement hydration and subsequent development of microstructure in this system. Possible carbonation reactions are shown below:

$$C\text{-}S\text{-}H + CO_2 \rightarrow CaCO_3 + silica + H_2O \quad (1)$$
$$Ca(OH)_2 + CO_2 \rightarrow CaCO_3 + H_2O \quad (2)$$

These reactions convert the C-S-H phase into calcium carbonate and results in the apparent formation of a highly polymerized silica gel with the same morphology as the original hydrate [12,13]. We have investigated the effect of carbonation during hydration of OPC and Zn-doped OPC by XPS technique. The OPC and Zn-OPC (10% by weight of Zn^{2+}) samples were stored in plastic vials and cured in an ambient atmosphere for 28 days at about 85% relative humidity before analyses. Some samples were cured in CO_2-rich atmosphere to see the effect of carbonation. The Zn^{2+} ions were derived from zinc nitrate and a water/cement ratio of 0.35 was maintained to prepare the samples. The XPS spectra of the samples showed that the near-surface region consists of mainly O>C>Ca>Zn (in the Zn-doped samples only) Si > Al > > S (trace amount). The BE for Si (2p), O (1s) and Ca ($2p_{3/2}$) in the OPC (air and CO_2 cured), OPC dry, Zn-OPC (air and CO_2 cured) and silica gel are given in Table 1.

Table I: XPS-Derived Binding Energies for Si(2p), O(1s) and Ca($2p_{3/2}$)

Samples	Si (2p) (eV)	O (1s) (eV)	Ca ($2p_{3/2}$) (eV)
OPC dry clinkers	100.8	531.5	347.1
OPC (air-cured)	102.2	531.6	347.0
OPC (CO_2-cured)	102.2	531.7	347.1
Zn-OPC(air-cured)	101.0	531.6	347.0
Zn-OPC(CO_2-cured)	101.5	531.7	347.1
Pure Silica Gel	104.1	533.9	------
Zn-OPC, leached	102.6	------	------
Pb-OPC(air-cured)	101.2	------	------
Pb-OPC, leached	102.7	------	------

The Si (2p) BE for the OPC dry clinker is ~3.3 eV lower than the Si (2p) BE for a standard silica gel sample (104.1 eV). The higher Si (2p) BE of 104.1 eV in silica gel is indicative of higher degree of polymerization of the SiO_4^{4-} units in this compound since each of the four corners of the SiO_4 tetrahedron is linked by -Si-O-Si- bonds. The silica phase formed by the reaction of C-S-H with the CO_2 (Reaction 1) has the same BE (102.2 eV) for both OPC cured in CO_2 and the OPC cured in air samples. The symmetry of the Si (2p) XPS spectra in these samples and the full-width at half-maximum (fwhm) of these peaks indicate varying chemical environments. It is, therefore, concluded that carbonation promotes polymerization with SiO_4 tetrahedron being reorganized in a different mode as opposed to tridimensional arrangements in silica gel [14].

The Si (2p) binding energy in the air-cured Zn-OPC sample is about 1.0 eV lower than the corresponding peak in air-cured OPC sample. This downward shift in binding energy provides clear evidence for lower degree of polymerization of the silicates in the Zn-doped samples as compared with undoped OPC samples. The presence of Zn^{2+} cations have thus inhibited the polymerization of cement hydration. This observation is consistent with the FTIR results previously reported by us to explain the inhibition of the hydration of OPC in presence of zinc. A charge controlled reaction model (The Charge Dispersal Model) has been proposed to explain the observed inhibition due to the formation of $CaZn_2(OH)_6 \cdot 2H_2O$ on the surface of clinker grains [15]. However, the effect of leaching of the metal-doped samples are amply demonstrated by Zn & Pb-doped cases (Table I). The leaching was carried out according to EPA TCLP tests [16] using acetic acid and sodium acetate buffer (pH=4.85). The Si (2p) BE in these cases are shifted by ~ 1.5 eV due to leaching (see Figure 1). The increase in the BE of the Si(2p) peaks in the leached samples is due to increase in the number of Si-O bonds per silicon atom as a result of enhanced polymerization.

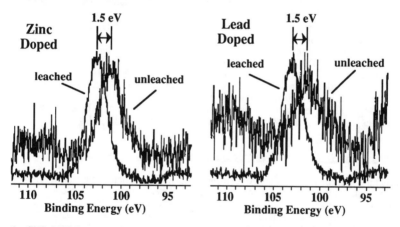

Figure 1: Si(2p) XPS spectra for a zinc-doped and lead-doped Portland Cement before and after leaching. The 1.5 eV BE shift indicates increased polymerization of the Si-O species.

X-Ray Diffraction: The application of XRD in analyzing complex heterogeneous mixture like cement clinkers has not always been successful, although valuable information can be obtained about certain phases [17]. A number of authors have reported XRD examination of the effects of admixtures in cement systems. Since the main product of hydration, C-S-H, is amorphous, most authors have concentrated on analyzing the $Ca(OH)_2$ formed, while others have monitored the change in any of the alite, belite, aluminatesulfate or aluminosilicate phases. Our recent work on the effect of superplasticizer [18] on the hydration of cement has established that superplasticizer inhibits the hydration process, and that a charge-controlled reaction process plays an important role in this system. The XRD peak intensities corresponding to $Ca(OH)_2$ (d=4.90 Å) and the alite phase (d=3.03 Å) were used to follow the hydration process and the resultant microstructure developed. The XRD technique has been proved particularly useful in analyzing cement systems impregnated with superplasticizer and other additives. The XRD analyses of pure and superplasticizer-doped cement indicated that the additive inhibits hydration as demonstrated by much reduced formation of $Ca(OH)_2$ and the lower degree of polymerization of the SiO_4^{4-} anions.

Fourier Transform Infrared Spectroscopy (FTIR)

The FTIR technique has been found to be a powerful tool in monitoring the polymerization of silicate upon hydration of cement and the results may be used to study the effects of carbonation, the addition of metals, superplasticizers and other organics on the hydration process. The principal vibrational band in silicate are: asymmetric stretching vibration (v_3), symmetric stretching vibration (v_1), out-of-plane bending vibration (v_2), and in-plane-bending vibration (v_4) of the SiO_4^{4-} tetrahedra at 940, 810, 525 and 452 cm^{-1}, respectively. Polymerization of the SiO_4^{4-} units upon hydration causes the v_3 SiO_4^{4-} vibration to shift to higher wavenumber, and also the relative changes in the intensities of the v_2 and v_4 vibrations give clear indications of the degree of polymerization of the silicates to produce C-S-H. The water stretching regions at ~3440 cm^{-1} (v_3 and v_1) and 1630 cm^{-1} (v_2 bending) can provide potentially important information to identify the nature of active surface species, the hydroxyl species, the site specificity for the adsorbing species that interact with hydroxyl groups and the nature of the hydroxylated surface compounds formed by metal species like Zn and Cd. The FTIR examination of the hydration of pure cement and Zn-doped cement in CO_2-rich atmosphere have established that CO_2 attacks the C-S-H phase to result a new silica phase having degree of polymerization lower than the original C-S-H phase [19]. Also, the FTIR results confirmed that CO_2 reacts with the metal-bonded and hydrogen-bonded hydroxides, $CaZn_2(OH)_6 \cdot 2H_2O$ [15] and $CaCd(OH)_4$ [20], in Zn- and Cd-doped cement systems, respectively. The FTIR results have confirmed that CO_2 reacts with these metal-bonded and hydrogen-bonded hydroxides, and that C-S-H and $Ca(OH)_2$ competes for reactions with CO_2. The inhibition of silicate polymerization in presence of superplasticizer has also been established by FTIR investigation of superplasticizer-blended cement. These inhibition processes have been explained by a charge controlled reaction mechanisms [21, 22].

Scanning Electron Microscopy & Energy Dispersive Spectroscopy (SEM/EDS)

Structural and chemical analyses of cement based systems containing metal cations using scanning electron microscopy combined with energy dispersive x-ray spectroscopy (SEM-EDS) can provide considerable insight into the microstructure of these systems. The SEM images provide intricate details of the matrix and furnish useful information about morphological and structural changes resulted from treatments, such as leaching or addition of metals. Information about pore-structure and pore-volume can be gleaned from SEM examination of the system. In particular, the SEM images across the leached and unleached front give wealth of information into the structurtal changes caused by such treatments. The effects of Zn^{2+}, Pb^{2+}, Cd^{2+}, Hg^{2+} and Cr^{3+} on the microstructure of cement based systems and the changes upon leaching have been characterized by SEM technique[23-25].

The EDS can probe > 1μm from the surface to provide qualitative and semi-quantitative information of the elemental composition in the bulk of the matrix rather than on the surface. The electron microprobe experiments were carried out to obtain concentration profiles of the constituent elements in the leached and unleached samples. This is illustrated using 10% Cr^{3+} - doped cement samples (Figure 2). On inspecting the spectra, a few changes are apparent. First, the intensity of the chromium signals remain relatively unchanged upon leaching indicating that it is insoluble and is dispersed in the bulk medium.

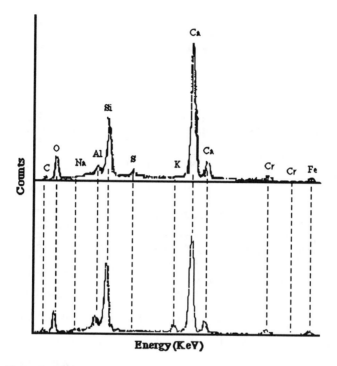

Figure 2: EDS spectra of (a) unleached and (b) leached cement samples containing chromium.

In addition, a calculations based on the normalization of Ca and Si peak areas only show that the average compositions of Ca and Si in the unleached and leached samples are 69.0, 31.0% and 58.0, 42% respectively. This indicates considerable Ca removal although some Ca remains in the calcium silicate[24]. Potassium was detected in the leached sample only which invokes an ion exchange mechanism [26]. Sulfur was present in the unleached sample indicating that it is effectively removed upon leaching. The aluminium and iron signals reamin relatively unchanged. Thus it is abundantly clear that considerable information about metal distribution and fixation can be gleaned from these EDS investigations of the cement-based systems.

CONCLUSION

Cement-based chemical systems have a rich and varied chemistry. Through the application of a combination materials characterization techniques -- molecular spectroscopies, surface analyzing techniques and microscopic methods -- to cement-based systems, valuable information can be obtained to allow a better understand the mechanisms of cement hydration, setting, microstructure development and hazardous waste solidification/stabilization. Results generated by the use of some of these techniques have led us to explain surface precipitation of cement in presence of certain additives and propose a charge controlled reaction mechanism for passivation of cement hydration.

ACKNOWLEDGMENTS

We thank the Gulf Coast Hazardous Substance Research Center, Lamar University-Beaumont, TX, for supporting this work. We acknowledge partial financial support from the Welch Foundation and the Texas Advanced Technology and Research Program and the Texas State Coordinating Board.

REFERENCES

1. D. L. Cocke and M. Y. A. Mollah, in Chemistry and Microstructure of Solidified Waste Forms, edited by R. D. Spence (Lewis publishers, U. S. A.,1992), p.187.
2. J. R. Barnes, A. D. H. Clague, N. J. Clayden, C. M. Dobson, C. J. Hayes, G.W. Groves and S. A. J. Rodger, Mater. Sci. Letter. 4, 1293 (1985).
3. E. Lippmaa, M. Magi, M. Tarmak, W. Wieker and A. Grimmer, Cemt.Concr. Res. 12, 597 (1982).
4. N. J. Clayden, C. N. Dobson, C. A. Hayes and S. A. Rodger, Chem. Soc. Chem. Commun. 1396 (1984).
5. D. Briggs and M. P. Seah, Practical Surface Analysis, (Wiley, New York, 1983).
6. D. L. Cocke, R. K. Vempati and R. L. Loeppert, in Quantitative Methods in Soil Mineralogy, ed. J. Amonnette (Agronomy Soc. of Am. Monograph Ser., 1994), p. 205.
7. M. F. Hochella, Jr., in Spectroscopic Methods in Mineralogy and Geology, edited by F. C. Hawthorn (Reviews in Mineralogy, Min. Soc. Am., Washinton D. C, 1988), p. 573.
8. D. L.Cocke, H. G. McWhinney, D. C. Dufner, B. Horrell and J. D.Ortego, Hazardous Waste and Hazardous Materials. 6, 251 (1989).
9. H. G. McWhinney, D. L.Cocke, K. G. Balke and J. D. Ortego, Cemt. Concr. Res. 20(1), 79 (1990).
10. H. G. McWhinney, M. W. Rowe, J. D. Ortego, D. L. Cocke, and G-S. Yu, Environ. Sci. and Health. A25(5), 463(1990).
11. M. Y. A. Mollah, Y. N. Tsai, T. R. Hess and D. L. Cocke, J. Hazardous Materials. 30, 273 (1992).
12. S. Mindess and J. F. Young, Concretre, (Prentice Hall, N. J., 1981), p. 671.
13. T. G. Baird, A. G. Cairns-Smith and D. S. Snell, J. Colloid Interface Sci. 50, 387 (1975).
14. P. A. Slegers and P. G. Rouxhet, Cemt. Concr. Res. 6, 700 (1976).
15. M. Y. A. Mollah, J. R. Parga and D. L. Cocke, J. Environ. Sci. & Health. 27 A (6), 1503 (1992).
16. Federal register (No. 114, Friday, June 13, 1982), 51, 21672 (1986)
17. H. F. W. Taylor, Cement Chemistry. 1st ed. (Academic Press, New York, 1990).
18. M. Y. A. Mollah, P. Palta, T. R. Hess, R. K. Vempati and D. L. Cocke, Cemt. Concr. Res. (1994, Submitted).
19. M. Y. A. Mollah, T. R. Hess,Y. N. Tsai, and D. L. Cocke, Cemt. Concr. Res. 23, 773 (1993).
20. M. Y. A. Mollah, Y. N. Tsai, T. R. Hess and D. L. Cocke, J. Environ. Sci. & Health. 27 (5), 1213(1992).
21. D. L. Cocke, R. K. Vempati, M. Y. A. Mollah, T. R. Hess and A.K.Chintala, Proceedings of the 207th ACS National Meeting, San Diego, USA, March 13-18. 34(1), 222 (1994).
22. M.Y. A. Mollah, R. K. Vempati and D. L. Cocke, Waste Management, Submitted (1994).
23. D. L. Cocke, M. Y. A. Mollah, J. R. Parga,T. R. Hess and J. D. Ortego, J. Hazardous Materials. 30, 83 (1992).
24. H. G. McWhinney and D. L. Cocke, Waste Management, 13, 117 (1993).
25. M. Y. A. Mollah, T. R. Hess and D. L. Cocke, J. Hazardous Materials. 30, 273 (1992).
26. D. L. Cocke, H. G. McWhinney, D. C. Dufner, B. Horrel and J. D. Ortego, Hazard. Waste Hazard. Mater. 6(3), 252 (1989).

COMPRESSIVE YIELD STRESS OF CEMENT PASTE

KELLY T. MILLER*, WEI SHI*, LESLIE J. STRUBLE**, AND CHARLES F. ZUKOSKI*
*Department of Chemical Engineering
**Department of Civil Engineering
University of Illinois at Urbana-Champaign, Urbana, IL 61801

ABSTRACT

Compressive yield stresses have been measured for pastes ($0.35 \leq$ w/c ≤ 0.50) of portland cement, calcium aluminate cement, and weakly and strongly flocculated alumina ($\phi_o = 0.20$) using the centrifuge sediment height technique. Equilibrium sediment heights are reached quickly, allowing all measurements to be taken during the cement's induction period. The compressive behavior showed little dependence on the compressive history. Compressive yield stress was, however, dependent upon initial volume fraction, decreasing as the initial volume fraction increases. This behavior was observed in both the cements and alumina suspensions, implying that strong dependencies on initial structure may be a general property of the compressive behavior of flocculated suspensions.

INTRODUCTION

Compressive rheology is used to characterize suspension behavior during consolidation (gravitational settling, pressure filtration, centrifugation). As described by Buscall and White[1], consolidation behavior depends on the balance of the various forces acting on the particles: 1) gravitational or centrifugal force, 2) externally applied loads, 3) interparticle forces (which may be either attractive or repulsive), 4) hydrodynamic forces (the drag acting on the particles and fluid as they move past one another), and 5) Brownian forces (which cause submicron particles to move randomly). To get consolidation, the compressive forces must exceed the other forces acting upon the suspension. In stable suspensions, one must overcome the osmotic pressure. In flocculated suspensions, the exact constitutive properties are debatable; one fruitful approach assumes that compaction occurs when the compressional force acting on the suspension exceeds a volume fraction dependent compressive yield stress, $P_y(\phi)$. For pressures acting on the particles less than P_y, the bed stores energy elastically. For an applied load larger than P_y, the bed consolidates irreversibly until it reaches a volume fraction capable of storing the load elastically[1].

The compressive yield stress, $P_y(\phi)$, can be measured using simple centrifuge techniques[1-3]. The approach used here is to compute consolidation pressure, P_y, as a function of volume fraction ϕ from measurements of equilibrium sediment height at a number of centrifuge speeds. In this experiment, the particle bed is subjected to a body force, which increases as the spinning speed is increased. When P_y is exceeded, the particles compact and expel water, reaching an equilibrium height at which the centrifugal load can be accommodated elastically. The $P_y(\phi)$ curve is generated from the sediment heights by calculating the pressure and volume fraction at the bottom of the cake at each spinning speed.

In this paper, we show that this technique can be successfully applied to measure the compressive properties of cement pastes. Measurements are reported for portland cement paste, and calcium aluminate cement paste, and alumina suspensions. The alumina suspensions

285

provided a nonreactive material of known particle size and narrow particle size distribution for validation of the general approach.

EXPERIMENTAL APPROACH

Alumina suspensions were prepared using two commercial α-alumina powders, with average diameters of 0.2 μm and 1.3 μm (Sumitomo AKP-50, AKP-15)[4]. In all experiments, the initial volume fraction, ϕ_o, was 0.20. The particles were suspended in 1 $\underline{M}$ NH_4Cl. The extent of flocculation was adjusted by changing the solution pH using HNO_3 and NH_4OH, using pH 4.0 for weakly flocculated suspensions and pH 9.0 (near the isoelectric point) for strongly flocculated suspensions[4,5].

The portland cement (PC) was a commercial Type I cement (Essroc). The calcium aluminate cement (CAC) was also a commercial cement (Ciment Fondu, Lafarge). The average bulk densities of the cement powders were determined by helium pycnometry to be 3.124 for the PC and 3.193 for the CAC. For both cements, the water to cement ratio (w/c) was varied between 0.35 and 0.50. All pastes were mixed in a paddle mixer (KitchenAid, Model K45SS) mixer for 45 s at low speed (approximately 135 rpm) and 5 min at medium speed (approximately 360 rpm).

The measured parameter, H(g,t), is the height of the suspension (cake) from the bottom of the tube as a function of centrifugal acceleration g and time t. From these measurements we compute $P_y(\phi)$, the compressive yield stress as a function of volume fraction.

Centrifugal force is applied using a Beckman J2-2M/E centrifuge equipped with a swinging bucket rotor with a speed range of approximately 550 to 13,000 rpm. Round bottom tubes, required by rotor design, were filled with sufficient epoxy to provide the flat bottom needed for data analysis. The distance from the center of rotation to the bottom of the suspension in such tubes was 0.1265 m, providing a maximum acceleration ranging from approximately 420 to 234,000 m/s^2. Compressive flow behavior was determined using four centrifuge speeds: 2500, 5500, 8500, and 11,000 rpm. After spinning at the desired speed (equivalent to some centrifugal acceleration g) for some time t, the sample is removed from the centrifuge and H is measured; the sample is then spun further and H measured, repeating until there is no further change in H. The centrifuge speed is then increased and the measurements repeated. H is measured using a steel ruler to a precision of 0.1 mm and an accuracy of $\sim\pm$ 0.3 mm. For cement pastes, H(t) reaches equilibrium very quickly (Fig. 1), allowing all measurements to be taken during the induction period of the cement[6].

The compressive yield stress, $P_y(\phi)$, is computed from measurements of H(g) at various g levels using the approach described by Buscall and White[1,2]. The pressure, P(0), and volume fraction, $\phi(0)$, at the bottom of the tube at each centrifuge speed are computed using the following equations:

$$P(0,g_o) = \Delta\rho g_o \phi_o H_o \left[1 - \frac{<z>}{R} \right]$$

(1)

$$\phi(0,g_o) = \frac{\phi_o H_o \left\{ \left[1 - \frac{<z>}{R} \right] - \frac{g_o}{R}\frac{d<z>}{dg_o} \right\}}{H_\infty \left[1 - \frac{H_\infty}{2R} \right] + g_o \left[1 - \frac{H_\infty}{R} \right]\frac{dH_\infty}{dg_o}}$$

(2)

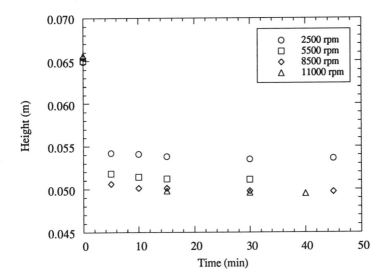

Figure 1. Height change as a function of centrifuge time for portland cement samples
(w/c = 0.45) for samples at four spinning speeds.

$$< z > = \frac{\int_0^{H_\infty} z\,\phi(z)\,dz}{\int_0^{H_\infty} \phi(z)\,dz}$$

(3)

where g_0 is the centrifugal acceleration at the bottom of the cake, H_0 is the initial height of the
suspension, ϕ_0 is the initial solid volume fraction of the suspension, $\Delta\rho$ is the difference in
density between the solid and liquid phases, R is the centrifuge radius, z is the distance from the
bottom of the tube, and H_∞ is the equilibrium height of the consolidated sediment. In general,
Eqs. (1-3) must be solved numerically. A good estimate of the compressive yield behavior is
obtained, however, by assuming a uniform volume fraction profile (i.e., $<z> = 0.5\ H_\infty$)[1].
Measurement of volume fraction profiles indicates that, for cement pastes, this approximation
introduces only a small error in $\phi(0)$.

RESULTS AND DISCUSSION

The $P_y(\phi)$ curves for PC and CAC (Fig. 2) are similar to curves obtained for a strongly
flocculated alumina suspensions (Fig. 3). Furthermore, there appears to be little or no effect of

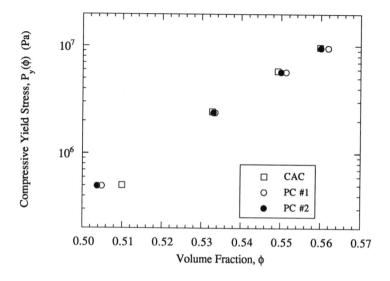

Figure 2. Compressive yield stress versus volume fraction for PC and CAC, intial w/c =0.45. Different symbols for PC indicate repeated experiments and and show reproducibility of data.

compressive history. The $P_y(\phi)$ curves generated from samples tested at four different speeds are very similar to the results computed from four samples each tested at a single speed (Fig. 4). The largest difference is at the lowest centrifuge speed, where estimation of dH_∞/dg_0 is least precise. Moreover, the volume fraction profile in centrifuged samples appears to depend only on the final spinning speed and not on compressive history. Thus, we conclude that the use of sequential increases in g_0 on the same sample (as opposed to a single g_0 per sample) yields accurate $P_y(\phi)$ values for cement pastes, as it does in flocculated alumina suspensions.

The $P_y(\phi)$ behavior also shows a strong dependence on particle size. For alumina suspensions, an increase in average particle size from 0.2 μm to 1.3 μm resulted in a decrease in P_y of about two orders of magnitude (Fig. 3). It remains to be seen whether this effect holds for systems with a wide particle size distribution.

$P_y(\phi)$ also shows a strong dependence on degree of flocculation[3]. For alumina suspensions, going from weakly flocculated to strongly flocculated conditions increased P_y by several orders of magnitude (Fig. 3). In cement pastes, similar effects are expected when dispersant concentration is varied. Based on our studies of shear yield stress[6], we expect that hydration will increase P_y in much the same way as degree of flocculation.

The volume fractions achieved in the lowest experimentally possible g_0 values correspond to low w/c ratios for cement pastes. However, the compressive yield stresses of these suspensions are quite high. For example, the pressure acting on the network at the bottom of 1 m of portland cement paste of w/c ratio = 0.45 (ϕ_0 = 0.42) is approximately 1×10^4 Pa. Extrapolation of the data

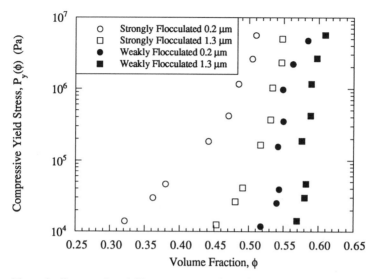

Figure 3. Compressive yield stress versus volume fraction for alumina suspensions of varying particle size and degree of flocculation. Initial volume fraction = 0.20.

in Fig. 2 suggests that the resulting bed would have a volume fraction of $\phi = 0.43$. Thus, there would be little consolidation of a 1m cement column. On the other hand, if dispersants are used, $P_y(\phi)$ will be markedly decreased (note the differences between weakly and strongly flocculated aluminas in Fig. 3). This suggests that one might expect substantial bleeding in meter thick beds of highly dispersed concretes; however, no experiments regarding this have been performed.

Most surprisingly, however, $P_y(\phi)$ shows a dependence on initial volume fraction, ϕ_0. This behavior was observed both for alumina suspensions and for portland cement, indicating that it may be characteristic of flocculated suspensions. As shown in Fig. 5 for portland cement paste, the pressure required to obtain a given volume fraction decreases as ϕ_0 is increased. This dependence means that more consolidation will occur at a given pressure if the initial w/c is lower. This behavior contrasts that expected by Buscall and White[1], who assumed that the compressive yield stress was a unique curve dependent only upon the particle size and strength of flocculation. Our results, however, clearly indicate that while a sample's $P_y(\phi)$ behavior is only weakly dependent on its compressive history, its initial mixing alters the paste microstructure such that higher initial volume fractions result in more easily compacted beds. (Strong dependencies upon initial structure are also seen in the shear rheology of flocculated suspensions.) Understanding the origin and nature of the microstructural alterations which give rise to this behavior requires further work.

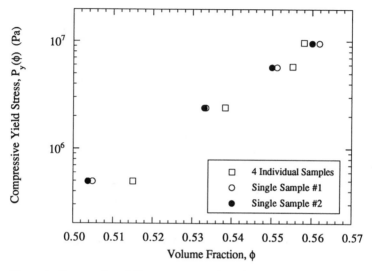

Figure 4. Compressive yield stress versus volume fraction for portland cement samples, w/c =0.45. Two samples were tested sequentially at all four speeds, and are compared with four samples which were tested individually at each centrifuge speed.

CONCLUSIONS

$P_y(\phi)$ is determined through a centrifugal technique where H(g) is measured at various g-levels. For alumina suspensions, PC paste, and CAC paste, the $P_y(\phi)$ behavior shows little or no effect of compressive history for a given initial volume fraction. However, $P_y(\phi)$ depends on ϕ_o, where higher initial volume fractions decrease the P needed to obtain the same ϕ. For a given state of flocculation (particle interaction potential), $P_y(\phi)$ depends on particle size, where smaller particle size increases the P required to obtain the same ϕ. Magnitudes of $P_y(\phi)$ depend on the degree of flocculation, where stronger flocculation increases the P required to obtain the same ϕ.

ACKNOWLEDGEMENTS

This research was supported by Lafarge Fondu International. Additional support was provided by a National Science Foundation Young Investigator Award (Struble) and the U.S. Department of Energy through the Materials Research Laboratory at the University of Illinois (Miller, Zukoski).

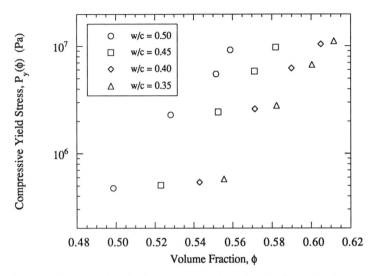

Figure 5. Compressive yield stress versus volume fraction for portland cement pastes at various w/c levels.

REFERENCES

1. R. Buscall and L. R. White, J. Chem. Soc., Faraday Trans. 1 **83**, 873-891 (1992).

2. R. Buscall, Colloids and Surfaces **5**, 269-283 (1982).

3. L. Bergström, C. H. Schilling, and I. A. Aksay, J. Am. Ceram. Soc. **75** (12), 3305-3314 (1992).

4. B. V. Velamakanni and F. F. Lange, J. Am. Ceram. Soc. **74** (1), 166-172 (1991).

5. B. V. Velamakanni, J. C. Chang, F. F. Lange, and D. S. Pearson, Langmuir **6**, 1323-1325 (1990).

6. L. J. Struble and W-G. Lei, J. Adv. Cement-Based Mater. (in press).

PART II

Bonding and Interfaces in Cementitious Materials

FRACTURE SURFACES OF CEMENT PASTES, ROCKS, AND CEMENT/ROCK INTERFACES

S. MINDESS[1], and S. DIAMOND[2]
[1]Dept. of Civil Eng., University of British Columbia, Vancouver, B.C., Canada V6T 1Z4
[2]Dept. of Civil Eng., Purdue University, W. Lafayette, Indiana 47907, USA

ABSTRACT

Stereo pair imaging in a scanning electron microscope was used to document the details of the topography of fracture surfaces at various degrees of magnification. The fracture surfaces had been produced by fracture testing of chevron-notched beam specimens according to ISRM Method I. The surfaces examined included the fracture surfaces of dolomite and andesite rocks, cement pastes with and without silica fume, the fracture surfaces from specimen prepared by casting cement paste on previously fractured rock specimen faces, and the fracture surfaces of mortars prepared using the same rock types and cement pastes. The principal topographic features of the various fracture surfaces are described.

INTRODUCTION

The cement-aggregate interface in concrete has long been the subject of study. Since this has often been assumed to be the "weak link" in concrete, the recent development of very high strength concretes has re-awakened interest in the interface and the related transition zone. Traditionally, methods of studying the interface have included direct measurement of the bond strength between paste and sawn or polished rock surfaces, measurements of the fracture properties of composite specimens using naturally fractured rock surfaces, microstructural studies incorporating SEM observations and measurements of, for example, porosity gradients, and X-ray diffraction studies to try to identify the hydration products in the vicinity of the interface. The work reported here is part of a much larger study to determine the mechanical effects associated with interfaces involving fractured rock surfaces [1-7]. In particular, it describes the use of stereo pair imaging in a scanning electron microscope to document details of the topography of fracture surfaces in cementitious systems at various degrees of magnification. This technique can provide unique information regarding the nature of the fracture process itself. However, stereo pairs have rarely been published for concrete materials. To the knowledge of the authors, the work summarized here [5-7] represents the first systematic use of stereo pair imaging in a scanning electron microscope to describe fracture surfaces in cementitious systems.

Materials and Experimental Methods

The materials and experimental methods are described in detail in Refs. 5-7. Briefly, the rock materials used in these investigations were hard, durable rocks used commercially for crushed rock aggregates in South Africa. They included a dolomite rock from the Olifantsfontein Quarry, an andesite rock from the Eikenhof Quarry and, for the mortar studies, a granite rock from the Jukskei Quarry. The cement used was an ordinary portland cement of

Mat. Res. Soc. Symp. Proc. Vol. 370 © 1995 Materials Research Society

South African origin. The condensed silica fume, also from South Africa, was supplied as a dry powder. A commercial naphthalene sulfonate superplasticizer was used with all of the cement pastes. Initially, the rock specimens were cored to a diameter of 42 mm and a length of 168 mm. Each core was notched at mid-length with a diamond-wheel cut chevron notch, and was subsequently tested in accordance with the International Society for Rock Mechanics (ISRM) Recommendations, Method 1 [8,9]. Similarly, cement pastes (w/c = 0.3, with and without a 15% silica fume replacement by weight of cement) were cast in molds of the same dimensions as the rock cores, and were tested in the same way at an age of about 3 weeks.

In the second phase of the work, cement pastes as described above were cast against the previously fractured rock half-specimens. After curing in limewater for about 3 weeks, these composite rock-cement specimens were tested in accordance with ISRM Method 1. Finally, the andesite and dolomite rocks, and the additional granite rock were crushed in a jaw crusher, and sieved to the desired sand size distribution, with a maximum particle size of 4.8 mm. The fineness modulus values were, respectively, 3.01, 2.96 and 3.39 for the andesite, dolomite, and granite sands. Mortars were then prepared at water:cementitious material ratios of 0.30, with a sand:cementitious material ratio of 2.0. These mortars were cast in molds 42 mm in diameter and 168 mm long; after moist curing, they were tested using the ISRM Method 1.

After the fracture testing, for each type of specimen described above, thin slices bounded by one of the fracture surfaces were cut, and glued to SEM stubs. They were then evacuated and coated for SEM examination in the usual manner. Stereo pairs were taken using a 7° tilt angle difference in the direction of the detector. The magnification generally ranged between 20x and 2000x. The stereo pairs published here are reproduced with the normal spacing between the two images of a given feature of about 50 mm, so that they may be directly viewed using simple stereo viewers without mirrors. Should the reader have difficulty seeing the stereo effect of the pairs as published, we suggest that reproductions be made using a good copier with a photo setting, and that the stereo pair reproductions then be examined at the most comfortable separation distance for the viewer used.

RESULTS AND DISCUSSION

Clearly, it is not possible to present, or to describe in detail, all of the several hundred stereo pairs which were taken in the course of this study. Hence, only a few representative stereo pairs, and the main conclusions, are given here. More detailed analyses of the results are given in Refs. 1, 5-7.

Figure 1 shows the fracture surfaces of representative rock, cement paste and mortar specimens, showing the overall specimen geometry. Fracture is induced in bending at the intersection of the two sawn notches (point A); the failure crack runs progressively to the right across the face of the specimen, broadening out as it goes. Most of the SEM observation were made along a line midway between the two sawn notches.

From an examination of the representative SEM stereo pairs, the following observations can be made:

1. Paste and Rocks (Ref. 5)
The fracture surfaces of both the cement pastes and the rocks are not even approximately planar at a micro-level, nor are they in any sense simple. For the rocks, there are considerable topographic variations, and the fracture surfaces are dominated by local crystal cleavages. The

Fig. 1.   Fracture surfaces of: (a) andesite rock; (b) cement paste; and (c) dolomite mortar.

andesite surfaces were topographically rougher and more complex than the dolomite surfaces. The andesite fracture surface shown in Fig. 2 is very rough.   The outlines of a number of thin sub-parallel plates on edge are visible in the upper central portion of the figure; these are individual rock crystals, about 2 or 3 μm thick.   The uniform-textured grains near the bottom of the field are probably similar crystals exposed on their faces rather than their edges.   On the other hand, the dolomite rock fracture surface (Fig. 3) looks rather different.   The microstructure seems to consist of two distinct classes of units: rounded grains of sizes ranging from a few μm to 20 μm, and darker planar surface patches representing cleaved crystals, presumably of dolomite.

The cement pastes show fracture around, rather than through, individual grains or cluster of grains, and fine cracks intersecting the main failure surface are present.   The silica fume pastes showed considerably less topographic variation than the plain cement pastes.   For example, the fracture surface of the plain cement paste (Fig. 4) is extremely tortuous, showing characteristic "bumps" of hydrated cement grains of the order of about 25 μm projecting from the general level.   There is a wide crack on the failure surface running vertically through the field, and very much narrower cracks, probably due to shrinkage (< 1 μm in width), can be seen in the lower left portion of the field and around several of the hydrated cement grains.   The cement paste with silica fume (Fig. 5) shows a relatively smoother fracture surface, though again multiple fine cracks are evident in several areas.

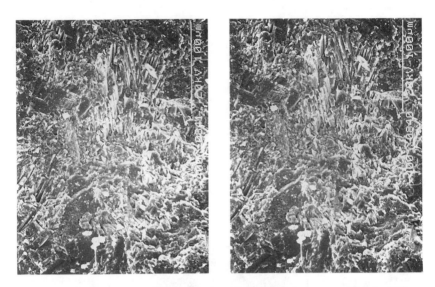

Fig. 2. SEM stereo pair of andesite fracture surface; original magnification 400x.

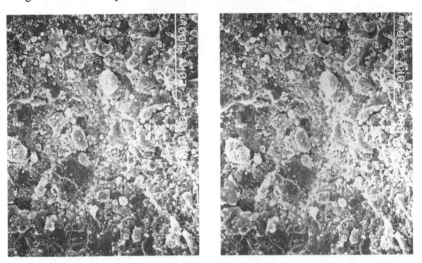

Fig. 3. SEM stereo pair of dolomite fracture surface; original magnification 400x.

2. <u>Paste/Rock Interfaces</u> (Ref. 6)

For paste cast against fractured rock surfaces, the failure crack starts at the notch and either propagates along the original interface, or else enters the paste at an angle before turning and running parallel to the interface, but several mm from it. Silica fume pastes tend to

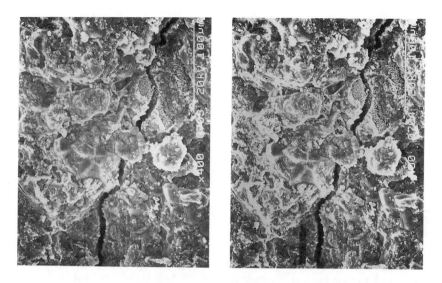

Fig. 4. SEM stereo pair of fracture surface of plain cement paste;
original magnification 400x.

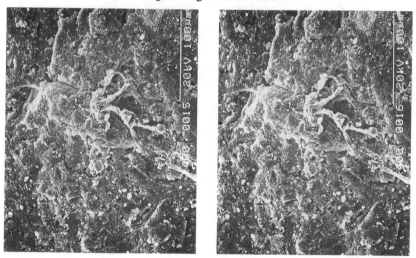

Fig. 5. SEM stereo pair of fracture surface of cement paste with silica fume;
original magnification 400x.

generate a sufficiently strong bond to favour the second mode, especially when cast against the
andesite rock (Fig. 6).

Fig. 6. Fractures occurring with andesite-cement paste composite specimens: (a) andesite-plain cement paste; and (b) andesite-cement paste + silica fume.

In the cleavage failure mode, crack branching may occur on both the paste and the rock sides of the failure crack. In "within paste" failure, the cracks tend to produce branches that travel through the bonded paste layer towards the original interface; these may turn and run parallel to and near the interface for some distance. In both of the above failure modes, crack branching out of the failure plane occurs extensively and constitutes an important energy absorbing mechanism.

Many areas of the rock side of the fracture surface of an andesite - plain cement paste composite specimen (Fig. 7) seem to be covered with an incomplete layer of fine crystallites up to several microns in size, some spherical, some cubical. The fracture surface topography clearly reflects the rock fracture surface. The paste side of the fracture surface of the composite specimens shows topographic features related to the microstructure of the paste (Fig. 8). The surface is relatively featureless at this magnification, except for the small (~1 μm) particles adhering to the surface.

The rock side of a dolomite-plain cement paste composite specimen (Fig. 9) is blocky at this scale, with crystalline micron-sized particles on the surface in some areas. In other areas, the rock surface appears to have been corroded. The paste side of the fracture surface against dolomite (Fig. 10) shows a surface that is, in the lower part, covered with extremely fine spherical particles, about 0.1 μm in diameter.

Qualitative microscopic examination of the paste region near the original paste/rock interfaces gave little visual indication that a transition zone different from the bulk paste well removed from the original interface was generated near the interface in the paste/fractured rock composite specimens (Fig. 11).

3. Mortar Interfaces (Ref. 7)

"High performance" mortars made with silica fume and a heavy dosage of superplasticizer fracture in a generally planar fashion, and primarily through the paste. The fracture surfaces in the corresponding plain paste mortars are much more strongly influenced by the specific characteristics of the individual crushed sand used. All of them are much rougher than those of the silica fume bearing mortars, and this roughness is exhibited over a wide range of size scales. Depending on the specific sand, the fracture surface may either cleave some crystals, or expose previously cleaved crushed sand surfaces, by passing between them and the surrounding interfacial zone paste.

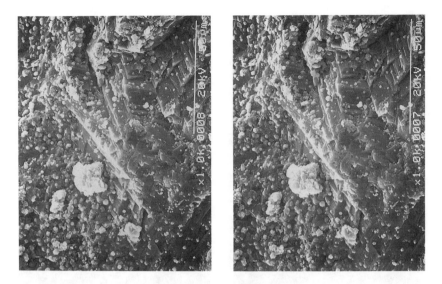

Fig. 7. SEM stereo pair of the andesite side of the fracture surface for an andesite-cement paste composite specimen; original magnification 1000x.

Fig. 8. SEM stereo pair of the paste side of the fracture surface for an andesite-cement paste composite specimen; original magnification 21000x.

For example, the silica fume-bearing mortar made with crushed andesite sand shows a generally flat and featureless fracture surface (Fig. 12). A plain cement mortar made with the

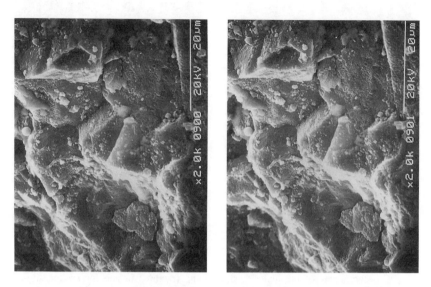

Fig. 9. SEM stereo pair of the dolomite surface; original magnification 2000x.

Fig. 10. SEM stereo pair of the paste side of the fracture surface for a dolomite-cement paste composite; original magnification 2000x.

same sand shows a much more tortuous surface (Fig. 13). Neither fracture surface, however, is much like the blocky fracture surface generated by the original uncrushed rock.

Fig. 11. A portion of the contact zone between an andesite surface and a cement paste containing silica fume; original magnification 300x.

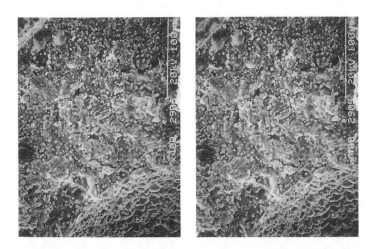

Fig. 12. Stereo pair of an area on the fracture surface of silica fume bearing mortar made with andesite sand; original magnification 400x.

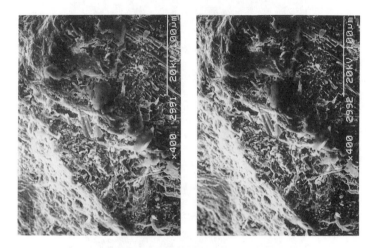

Fig. 13. Stereo pair of the fracture surface of a plain cement mortar made with andesite sand; original magnification 400x.

The silica fume-bearing mortar prepared with crushed dolomite sand shows a generally planar fracture surface (Fig. 14); some cracking may be observed. A plain cement paste mortar made with dolomite sand shows a much rougher fracture surface (Fig. 15).

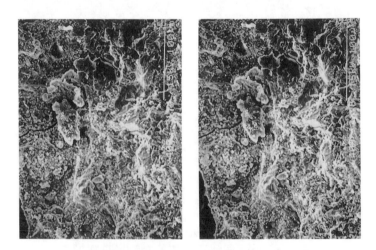

Fig. 14. Stereo pair of the fracture surface of a silica fume bearing mortar made with dolomite sand; original magnification 400x.

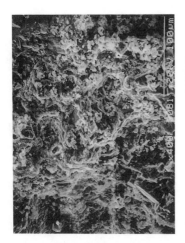

Fig. 15. Stereo pair of the fracture surface of a plain cement paste mortar made
with dolomite sand; original magnification 400x.

The silica fume-bearing mortar made with crushed granite exhibits a generally planar surface, with some exposed cleavage faces of the crushed sand visible (Fig. 16). A plain cement mortar made with the same sand (Fig. 17) is rougher. An area of thin parallel plates, presumably micaceous, is exposed. In general, the texture and geometry of the fracture surface is strongly affected by the nature of the crushed surfaces of the sand.

CONCLUSIONS

1. The fracture surfaces induced in this study reflect specific features dependent on the pre-existing crystal or particle assemblages, and are exhibited at specific size scales. This may be interpreted to mean that the evidence presented here does not support the concept that cement or concrete fracture surfaces may be fractal in nature.
2. The proper evaluation of the nature of the fracture surface produced in a given case requires examination over a wide range of magnifications.
3. This method of examining fracture surfaces, using stereo pair imaging, can reveal useful details of the actual fracture surface. Attention to such details in formulating hypotheses of crack propagation is to be encouraged.

ACKNOWLEDGEMENTS

This paper represents a joint contribution of the Purdue University component of the National Science Foundation Science and Technology Center for Advanced Cement Based

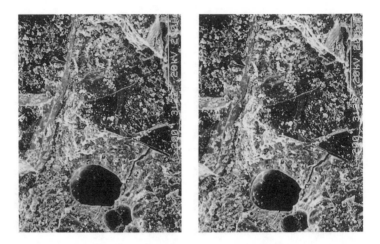

Fig. 16. Stereo pair of the fracture surface of a silica fume bearing mortar made with granite sand; original magnification 400x.

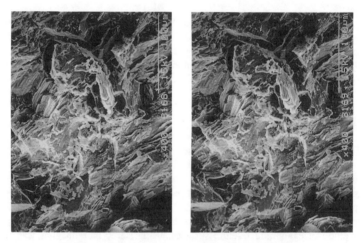

Fig. 17. Stereo pair of a mica exposure on the fracture surface of a plain cement paste mortar made with granite sand; original magnification 400x.

Materials (ACBM), and the University of British Columbia component of the Canadian Network of Centres of Excellence on High-Performance Concrete (now Concrete Canada). The support of the US National Science Foundation and the Natural Sciences and Engineering Research Council of Canada are gratefully acknowledged.

REFERENCES

1. Alexander, M.G., Mindess, S., Diamond, S. and Qu, L., Materiaux et Constructions (RILEM), in press.
2. Mindess, S., Qu, L. and Alexander, M.G., Advances in Cement Research, Vol. 6, No. 23, pp. 103-107, 1994.
3. Diamond, S., Mindess, S., Qu, L. and Alexander, M.G., in Maso, J.C. (ed.), Interfaces in Cementitious Composites, E & FN Spon, London, pp. 13-22, 1992.
4. Qu, L. and Mindess, S., in Proceedings, 3rd Canadian Symposium on Cement and Concrete, Ottawa, pp. 167-176, 1993.
5. Diamond, S. and Mindess, S., Cement and Concrete Research, Vol. 22, No. 1, pp. 67-78, 1992.
6. Mindess, S. and Diamond, S., Cement and Concrete Research, Vol. 22, No. 4, pp. 678-688, 1992.
7. Diamond, S. and Mindess, S., Cement and Concrete Research, Vol. 24, No. 6, pp. 1140-1152, 1994.
8. International Society for Rock Mechanics, International Journal of Rock Mechanics and Mineral Science, Vol. 25, No. 2, pp. 71-96, 1988.
9. Alexander, M.G. and Mindess, S., Cement and Concrete Research, in press.

INFLUENCES OF INTERFACIAL PROPERTIES ON HIGH-PERFORMANCE CONCRETE COMPOSITES

D.M. ROY AND W. JIANG
The Pennsylvania State University, Materials Research Laboratory, University Park, PA 16802 U.S.A.

ABSTRACT

There is a strong motivation to study the interfacial properties of concrete composites because the interfacial region is often the phase where fracture first develops. The aim of this study is to understand phenomena which are unique at high-performance concrete composite interfaces, and how these influence the bulk properties of a concrete composite. Since processes at interfaces must be considered over a range of scales varying from the atomic to the macroscopic, multidisciplinary research approaches are desirable. Model cement/rock (aggregate) and matrix/fiber interaction experiments were carried out. Morphology and microstructure of interfacial regions among mortar/rock, and fiber/matrix were examined utilizing SEM. Computer image analysis performed along a perpendicular to the interface revealed compositional and physical irregularities. The variations in the volume of pores adjacent to interface zones are documented and supported by microscopic observation. The influences of interfacial properties on concrete composite strength and durability are discussed, and influences of fibers on the fracture and fracture resistance behavior are also discussed. Analyses of debonding along interfaces are used to define the role of debonding in fiber-reinforced concrete composites.

INTRODUCTION

Interfaces of concrete as a general term may refer to either external and internal morphologies, or microscopic and macroscropic scales. From environmental and geotechnical points of view, external interfaces are important topics. Mindess defined a number of different types of interfacial bonding which occur in cementitious composites[1]. It is worth noting that the interface between the aggregate (or fiber) and the matrix is an anisotropic transition region exhibiting a gradation of properties. It is the component through which the transfer of stresses between aggregate and matrix occurs. It is must provide adequate chemically and physically stable bonding between the aggregates and the matrix. Its functional requirements vary considerably according to the performance requirements of the concrete composite during its various stages under service conditions. Major strengthening mechanisms which are practically being used are: reduced porosity by low water/cement ratio, absence of macro-defects, and the synthetic composite mechanism[2]. It is generally accepted that the mechanical properties of porous materials depend on macro structural defects. When non-hydrated water leaves behind pores, the porosity directly influences both mechanical and transport properties of cementitious materials. In addition, cementitious materials suffer from shrinkage, creep, and thermal expansion and contraction, which produce cracks that are also detrimental to its performance. For these reasons, concrete is very commonly used in combination with other materials. Most bulk engineering materials are combinations of two or more phases dispersed on a microscopic scale to obtain optimum properties[3]. The composite 'idea' can be related also to the macroscale[4]. This is particularly relevant to civil engineering components (e.g. concrete beam) which may consist of two or more materials combined to give a performance in service which is superior to the properties of the individual materials. During the last decade the composite interface aspects of high-performance concrete have received more attention. There also has been a rapid growth in the use of new fiber reinforced materials in civil engineering applications in the last decade and there is every indication that this will continue[5]. Bonding between them and the cement matrix is an important research topic. This research tries to explore the interfacial zone from a material science point of view.

309

CURRENT RESEARCH (1990—)

Many researchers investigate the interfacial zone by various approaches; however, microscopy, fracture mechanics, and computer simulation are among the dominant current approaches. Diamond and Mindess performed SEM investigations of fracture surfaces using stereo pairs[6]. Pope and Jennings used back-scattered electron microscopy in conjunction with quantitative image analysis to examine the influence of mixing on the cement paste/aggregate bond[7]. Sugama et al. using SEM associated with energy-dispersion X-ray spectrometry (EDX) and X-ray photoelectron spectroscopy (XPS) investigated the interface between zinc phosphate-coated steel fibers and cement paste[8]. Wang et al. using SEM/EDX studied the microstructure of interfacial zones between sulfoluminate cement paste and aggregate[9]. Zhang and Gjørv investigated the microstructure of the interfacial zone between lightweight aggregate and cement paste[10]. Popoola et al. used High Resolution Electron Microscopy (HREM) to observe the interface between the cement grain and PVA matrix in MDF cement[11]. Su et al. studied the interface between polymer-modified cement paste and aggregate[12]. Breton et al. contributed to the understanding of the formation mechanism of the transition zone between rock-cement paste[13]. Sarkar observed an interesting dense mature type ettringite at a paste-aggregate interface from Montreal dating back to the 17th century[14]. Raivio and Sarvaranta used a transition zone program to show that porosity detected by backscattered electron image is higher at the polymer fiber/matrix contact zone than in the bulk matrix[15]. Igarashi and Kawamura studied effects of a size in bundle fibers on the interfacial zone[16]. Stevula et al. studied hydration products at the blastfurnace slag aggregate-cement paste interface[17].

Based on fracture mechanics principles, Li et al. characterized the interfacial properties using pullout tests[18]. Shah and Ouyang reviewed mechanical behavior of the fiber-matrix interface[19]. Ping et al. examined the flat aggregate-portland cement paste interface[20]. Shen et al. introduced a new method of cement-aggregate interfaces which based on a so-called "ideal aggregate"[21] concept. Alexander used two experimental techniques for studying the effects of the interfacial zone between cement and rock, and effects of aging on mechanical properties of the interfacial zone between cement paste and rock[22]. Garboczi and Bentz presented the results of a digital-image-based simulation model of the aggregated-cement paste interfacial zone[23]. Grutzeck et al. used computer simulation of interfacial packing in concrete[24]. Xie and Beaudoin established a relationship between interfacial bond strength and electrical conductivity of the transition zone[25]. Winslow et al. developed a core/shell computer model to examine the percolation characteristics of the interfacial zone pores[26].

From the early development of concrete composite materials, the optimization of the interface between matrix and the reinforcing materials — be it fiber, particle, aggregate, steel bar — has been of major importance; steel bars were roughened to improve mechanical interlocking with concrete. Bonding is a generic term used in the literature. However, from the materials science point of view, bonding must be clarified.

Chemical bonding— A chemical reaction may occur between aggregate (or fiber) and the matrix. Hydrogen bonding and/or hydroxide bridge may play a major role in the bonding of concrete composite. The chemical implications will not be considered quantitatively; it suffices to say that hydrogen bonds may form between fiber and matrix[27].

Physical bonding — The attractive van der Waals forces existing between surfaces is a short-range force. Theoretically, short-range oscillatory forces are in the nanometer regime. In practice, physical bonding, chemical bonding, and hydrogen bonding are hard to discriminate. Zhang and Gjørv use a physical-chemical interaction mechanism to describe the bond between aggregate and cement paste[10].

Mechanical bonding — There are at least two meanings: firstly, bonding related to packing—mechanical interlocking— two rough surfaces may have multiple "hooks" resulting in binding together; secondly is the mechanical bonding as related to a bonding test. The second one is less straightforward, because the bonding is really an assessment of how much load, force or impact an object will be able to withstand and still perform the task for which it was designed. There are various ways in which it can fail; by breaking, bending too much, being crushed, or otherwise exceeding the limits to the amount by which it stretches[28]. No 'bonding' of a material is by any means a simple property like its density or its thermal conductivity. It is a composite property that cannot be measured by a single test or defined by a single formula.

Thermal bonding — The materials (eg. polymer or bituminous) are heated and compressed, which causes partial melting and makes them adhere to aggregate.

Büyüköztürk and Lee pointed out that characterization of interfacial behavior is needed to study the role of interfaces on the global behavior of concrete composites as a basis for the development of high-performance cementitious materials[29]. Parkhouse and Sepangi indicated that there are materials properties existing at macroscopic scales which are relevant to structural performance, safety and economical design which deserve more attention[30]. According to Hull's classification, concretes are *Macrocomposites* (engineering product)[31]. This research is an attempt to combine microscopic and macroscropic aspects. The images explore the precarious balance between two phases in which 'difference' is defined within their intersecting boundaries.

EXPERIMENTAL

Materials

The chemical composition of the materials in the present study is shown in Table I. The ordinary portland cement (ASTM type I) was designated as OPC. Sand is ASTM C 109. The mechanical/physical properties of the limestone aggregate used are shown in Table II.

Specimen preparation

The concrete specimens were standard 4" diameter 8" length cylinders. The limestone (99.9% calcium carbonate) was slabbed, cored and cut into rectangular slices and discs. The warm-press cell used to prepare 2-in. cubes, and ø1.0 in. cylindrical specimens, was described elsewhere[32].

TABLE I Chemical compositions of materials (% by weight))

	Code[#]	SiO_2	Al_2O_3	CaO	Fe_2O_3	MgO	K_2O	Na_2O	SO_3	L.O.I
OPC	I-33	20.30	5.67	60.43	6.23	3.14	0.90	0.36	2.8	2.8
Fly ash	G-07	51.95	27.54	5.60	7.75	1.59	1.43	0.56	1.36	1.83
Slag	G-07	34.70	10.70	39.37	0.41	11.90	0.5	0.25	—	1.34
SF*	G-15	96.00	0.10	0.12	0.10	0.13	0.45	0.10	—	2.57
Min-u-Sil**	B-34	99.70	0.10	0.03	0.01	0.01	—	—	—	0.16
Lime-stone	L-01	0.15	0.14	53.12	0.03	0.02	0.02	0.13	—	44.03

[#] PSU/MRL code, *Silica Fume, **Ground crystalline silica(quartz, median diameter = 5µm).

TABLE II Mechanical/physical properties of limestone

Specific gravity	2.70
Absorption capacity	0.9
Compressive strength, MPa*	120.3
Splitting tensile strength*	7.1
Young's modulus, GPa*	96.4
Poisson's ratio*	0.41

* ø50x80 mm cylindrical specimens were used

A series of modeled interfaces prepared for this research are listed in table III. The limestone -cement paste was performed using cylindrical limestone cores united by cement paste along the diametral plane, ø50x80 mm cylindrical specimens were used as shown in Fig. 1(b). Some samples cored directly from field (highway and building) concrete are described in the figures.

Experimental procedure

Four-point bending and splitting test — The four-point bending is shown in Fig. 1(a). Interface debonding between limestone and cement matrix measured by the splitting test is shown in Fig. 1(b), this type of method has been used for measuring the rock-cement paste bond strength[33]. Test specimens were removed from the molds 24 hours after casting, and cured in water. After 28 days of curing, all specimens were tested. It was expected that the surface conditions of the rock would have an important effect on the bond results, therefore the saw-cut diametral surfaces were sandblasted to increase the roughness.

Image Analysis — Samples were observed under an SEM, then enlarged SEM micrographs were transmitted via TV camera to a computer image analyzer, as indicated in Fig. 2. CCD means charge-couple-device.

Electrical conduction measurements — The electrical conduction cell used in the experiments is illustrated schematically in Fig. 3. An ion Analyzer ORION EA 920 interfaced to a computer was used. Slices of limestone cut from the cylindrical specimens, using a precision saw with diamond tipped blade to 5 mm thick, were glued in a plastic mold, then fresh cement paste with 0.28 w/c ratio was cast in, and mortar was kept 10mm thick . The sample was cured in a moist room at 25°C for 4 hours. The cell was filled with saturated $Ca(OH)_2$ solution, and the sample was located in the middle of the cell. The resistances of the cell with and without the sample and the temperature of solution were measured and recorded. Considering a saturated mortar sample subjected to an external electrical field, Ping et al. used a simplified model containing a combination of three components[34]. Here they are limestone, mortar, and a transition zone. Each has its own electrical conductivity. In this work, a parameter term termed "interfacial excess conductance[34]" derived by Ping et al. was obtained experimentally.

Microscopy — Polished sections 3 cm in diameter were cut out from the modeled interface zone, as indicated in Fig. 4. The sample were impregnated with epoxy resin under vacuum at 100°C, to prevent materials pullout during polishing.

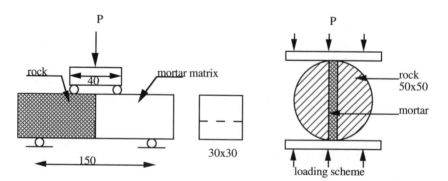

a) fracture energy test by four point slow bending test   b) bond test

(a/d=0.4, a=crack length, d=specimen height; dimensions in mm)

Fig. 1 Four-point bending and splitting test configurations

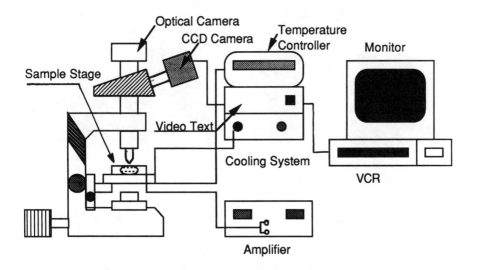

Fig. 2 Configuration of CCD microscope system

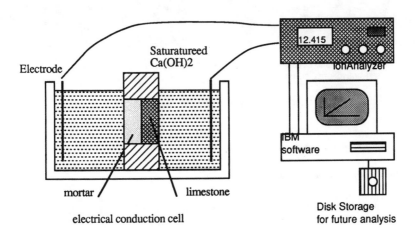

Fig. 3 Schematic diagram of the conduction cell

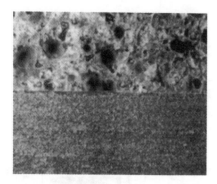

a) interface between warm pressed paste (w/c = 0.2) and mortar (w/c = 0.35)

b) A modeled interface: new mortar (w/c=3.5) cast old mortar (w/c=0.2)

c) A modeled interface between epoxy resin and warm-pressed paste w/c = 0.15

d) Interface between steel bar and alkali-activated slag mortar

Fig. 4 Modeled interface zone investigated at magnification (12x)

RESULTS AND DISCUSSION

Four-point bending and splitting test

Mix proportions and interface toughness values of silica fume (SF) modified mortars are shown in Table III. When SF was added, interface toughness increased. The detailed calculations may be found elsewhere[29]. It is seen that the fracture toughness of mortar/limestone interface when silica fume was added is larger than that of the one without silica fume. The bond strength results are presented in Fig. 5. In each group, 9 specimens were tested. All the standard deviations are less than ±10%. It was noted that the limestone-cement/paste bond strength was lower than the cement matrix. When 5-10% by weight silica fume were added to modify the interface, the interface splitting strength increased. A modeled flat limestone/mortar interface is presented in Fig. 6; and SEM micrographs of the interface show that the limestone surface has low density hydrated products.

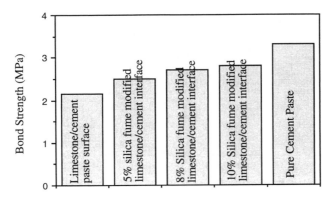

Fig. 5 Bond strength by splitting test

TABLE III Mix proportions and interface toughness (Γ_i) values

	W/(C+SF)	Sand/(C+SF)	σ_c (MPa)	E (GPa)	Γ_i (J/m2)
Control 0%SF	0.44	2.75	46.4	25.8	5.3
5% SF*	0.25	2.75	70.4	33.5	7.5
10% SF*	0.25	2.75	82.9	43.7	9.5

*SF = Silica fume; σ_c = compressive strength; E = Young's modulus.

Fig. 6 Modeled interface between limestone and mortar

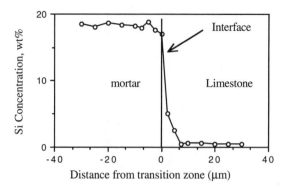

Fig. 7 Auger electron spectroscopy data from a series of points near transition zone

Microstructure and microchemistry of interface

Surface chemistry consideration — Auger electron spectroscopy data from a series of points near the transition zone (3 months age) are shown in the Fig. 7. It seems that the thin film is less than 6~7 μm. Similar experiments conducted with plastic/mortar, steel/mortar, epoxy resin /mortar couples all show that the transition zone is less than 2~3 μm. Barnes *et al.* (1978) detected that the thickness of the combined or "duplex" layer in the transition zone is usually no more than 1-1.5 μm[35], which should be the right boundary condition. Diamond pointed out that this thin duplex is not the "aureole de transition" or the interfacial zone, which typically is 40 or 50 μm thick; rather it constitutes only the thin boundary element in closest contact with the actual aggregate surface[36].

The variation of porosity with distance from aggregate interface by computer image analysis tends to establish a mortar/aggregate transition zone which has a 100 μm boundary condition at

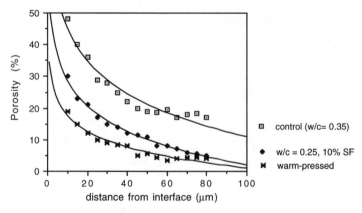

Fig. 8 Variation of porosity of mortar with distance from aggregate interface

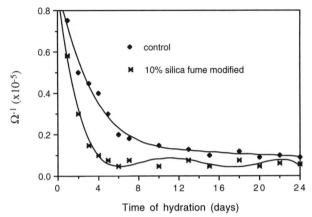

Fig. 9 Interfacials excess conductance vs time

magnifcation (135x). The results are shown in Fig. 8. Comparing the results with those of paste/aggregate done by Scrivener and Gartner (1988)[37], shows similar trends. Apparently, mortar porosity is higher than paste, and the w/c and SF also affect the porosity. Identification of the interfacial phase and defects is important for developing current high-performance concretes and for optimizing processing. Fig. 9 shows that when 10% silica fume is added conductance decreases during the early hydration, as compared the control. It may be evidence that as the silica fume is a much more reactive cementing material.

SUMMARY

The present results show that a better understanding of all interfacial phenomena is needed to establish the complementary roles of surface structure and interface structure at the various levels, and the types of forces operating across thin interfacial zones. In addition, there is a need for more theoretical analysis, including computer simulation.

ACKNOWLEDGMENT

This research was partly supported by the U.S. National Science Foundations grant MSS-9123239.

REFERENCES

1. S. Mindess, in *Bonding in Cementitious Composites*, edited by S. Mindess and S.P. Shah (Mater. Res. Soc. Proc. 114, Pittsburgh, PA, 1988) pp. 3-10.
2. W. Jiang and D.M. Roy, in *High-Performance Concrete*, edited by V.M. Malhotra (American Concrete Institute, SP-149, Detroit, MI, 1994) pp. 753-766.
3. Piggott, M.R., in *Interfacial Phenomena in Composite Materials*, edited by L. Verpoest and F. Jones (Butterworth-Heinemann Ltd, Oxford, UK, 1991) p. 3-8.

4. W.P. Moore Jr., in *Composite Construction in Steel and Concrete*, edited by C. D. Buckner and I.M. Viest (Amercan Society of Civil Engineers, New York, 1988) pp. 1-17; T. Inha, in *Composite Construction in Steel and Concrete II*, edited by W.S. Easterling and W.M.K. Roddis (Amercan Society of Civil Engineers, New York, 1992) pp. 210-225.
5. S. Somayaji, *Civil Engineering Materials* (Prentice Hall, Englewood, Cliffs, NJ, 1995), p. 119.
6. S. Diamond and S. Mindess, Cem. Concr. Res., **22**, 67-68; 678-688 (1992); **24**, 1140-1152 (1994); S. Diamond, S. Mindess, Lie Qu, and M.G. Alexander, in *Interfaces in Cementitious Composites*, edited by J.C. Maso (E &FN Spon, London, 1993) pp. 13-22.
7. A.W. Pope, H.M. Jennings, J. Mater. Sci., **27**, 6452-62 (1992).
8. T. Sugama, N. Carciello, L.E. Kukacka, and G. Gray, J. Mater. Sci., **27**, 2863-72 (1992).
9. Y. Wang, S. Li, Y. Lu and M. Su, in *Proc. 9th Intl. Congress on the Chemistry of Cement*,Vol. V, pp. 184-190 (New Delhi, 1992).
10. M.H. Zhang and O.E. Gjørv, Cem. Concr. Res. **20**, 610-618 (1990); **22**, 47-55 (1992).
11. O.O. Popoola, W.M. Kriven and J.F. Young, Ultramicroscopy, **37**, 318-25 (1991)
12 Z. Su, J.A. Larbi and J.M. Bijen, Cem. Concr. Res. **21**, 242-250; 535-544; 983-990 (1991).
13. D. Breton, A. Carles-Gibergues, G Ballivy and J. Grandet, Cem. Concr. Res. **23**, 335-346 (1993).
14. S.L. Sarkar, Cem. Concr. Res. **22**, 1011-1018 (1992).
15. P. Raivio and L. Sarvaranta, Cem. Concr. Res. **24**, 896-906 (1994).
16. S. Igarashi and M. Kawamura, Cem. Concr. Res. **24**, 695-703 (1994).
17. L. Stevula, J. Madej, J. Kozankova and J. Madejova, Cem. Concr. Res. **24**, 413-423 (1994).
18. Z. Li, B. Mobasher and S.P. Shah, J. Am. Ceram. Soc. **74**, 2156-64 (1991).
19. S.P. Shah and C. Quyang, J. Am. Ceram. Soc., **74**, 2727-38 (1991).
20. X. Ping, J.J. Beaudoin and R. Brousseau, Cem. Concr. Res. **21**, 515-522; 718-726; 999-1005 (1991).
21. Y. Shen, Z. Xu, P. Xie and M. Tang, Cem. Concr. Res. **22**, 612-620; 769-773 (1992).
22. M.G. Alexander, Cem. Concr. Res. **23**, 567-575 (1993); **24**, 1277-1285 (1994).
23. E.J. Garboczi and D.P. Bentz, J. Mater. Res. **6**, 196-201 (1991).
24. M.W. Grutzeck, D. Shi, G. Liu, S. Kwan, J. Mater. Sci. **28** (1993) 3444-3450.
25. X. Ping and J.J. Beaudoin, Cem. Concr. Res. **22**, 47-55; 597-604 (1992).
26. D.N. Winslow, M.D. Cohen, D.M. Cohen, D.P. Bentz, K.A. Snyder and E.J. Garboczi, Cem. Concr. Res. **24**, 25-38 (1994).
27. J.D. Birchall, in *Advances in Cement-Matrix Composite*, edited by D.M. Roy, A.J. Majumdar, S.P. Shah, and J.A. Manson (Proceeding, Symposium L, Materials Research Society, Boston, MA, 1980) pp.25-35.
28. B. Chalmers,*The Structure and Properties of Solids*. (HEYDEN & Son Ltd, London, 1982) p. 87.
29. O. Büyüköztürk and K. Lee, J. Cement and Concrete Composites, **15**, 143-151 (1993).
30. J.G. Parkhouse and H.R. Sepangi, in *Building the Future: Innovation in design, materials and construction*, edited by F.K. Garas, G.S.T. Armer, and J.L. Clarke (E &FN Spon, London, 1994) pp. 3-13.
31. D. Hull, *An Introduction to Composite Materials* (Cambridge University Press, Cambridge, 1981) p. 1.
32. D.M. Roy and W. Jiang, in *Proc.SCIENTIFIC BASIS FOR NUCLEAR WASTE MANAGEMENT XVIII* (Kyoto, JAPAN, on OCT. 23-27, 1994, in progress)
33. P.J.M. Monteiro and W.P. Andrade, Cem. Concr. Res. **17**, 919-926 (1987).
34 X. Ping, J.J. Beaudoin, and R. Brousseau, Cem. Concr. Res. **21**, 515-522 (1991).
35 B.D. Barnes, S. Diamond, W.L. Dolch, Cem. Concr. Res. **8**, 233-244 (1978).
36 S. Diamond, in Proc. 8th ICCC (1986) Vol. 1, p. 122.
37 K. Scrivener and E.M. Gartner, in *Bonding in Cementitious Composites*, edited by S. Mindess and S.P. Shah (Mater. Res. Soc. Proc. **114**, Pittsburgh, PA, 1988) p. 82.

MECHANICAL AND MICROSTRUCTURAL INVESTIGATION OF THE AGGREGATE AND CEMENT INTERFACE

ZONGJIN LI*, SURENDRA P. SHAH**, AND MATTHEW J. AQUINO***
*Hong Kong University of Science and Technology, Department of Civil and Structural Engineering, Clear Water Bay, Kowloon, Hong Kong.
**Northwestern University, NSF ACBM Center, 2145 Sheridan Rd., Evanston, IL 60208.
*** Skidmore, Owings and Morrill (S.O.M.), 224 s. Michigan Ave., Chicago, IL 60604.

ABSTRACT

The understanding of the interface between aggregate and cement is crucial in determining the properties of concrete. Recently, a pushout experimental technique and a theoretical model have been developed to determine the stiffness, strength and surface energy of the interface layer. The validity of these material parameters was verified in the present study by examining the effect of diameter and embedment length of aggregate. In addition, the effect of the pretreatment of aggregate surface and various admixtures was investigated. All pushout tests were performed in a closed-loop manner to obtain the load vs. displacement relationship.

The interfacial zone was further investigated by using backscattered electron imaging and energy dispersive analysis of x-rays (EDAX) to characterize the microstructure of the interface. The relationship between mechanical properties and microstructure of interfacial zone was studied. It was shown that the microstructure of the interface plays a substantial role in the mechanical behavior of the aggregate/cement bond.

INTRODUCTION

The interfacial zone between cement paste and aggregate is an area of particular interest associated with the strength and the durability of concrete. Inevitably, the mechanical and physical properties of the interface should be determined[1][2]. In a study reported earlier[1] a pushout test was developed to obtain the load-slip relationship of the interface. Based on the observed load-slip relationship and a one-dimensional shear-lag model, material parameters to characterize the interface can be determined. These parameters include the stiffness of the interfacial layer, the shear and frictional bond strengths and the interfacial fracture energy. A good correlation was indicated between the values of above mentioned bond parameters and the microstructure of the interface.

The above cited experiments were performed on a single size cylindrical aggregate which was pushed out from a surrounding annulus of a cementitious matrix. If the bond parameters determined from the load vs slip curves are to be considered as valid material parameters then they should be independent of the dimension of the test specimens. One of the goals of the study reported in this paper is to examine the influence of the diameter and the embedment length of the aggregate core on the values of the bond parameters.

Of much importance is to identify the influence of various factors on the properties of the aggregate-cement interface. These factors include the addition of polymer latex, and silica fume to the cement paste mixture as well as pretreatment of the aggregate surface with silica fume plus cement paste.

319

Push-out tests were conducted in a closed-loop manner to enable measurements of load-slip relationship. Results were analyzed using a theoretical model. The study also included the use of backscattered electron image for microstructural investigation of interface.

BRIEF REVIEW OF THE THEORY

To let readers get a clear picture of the method of interpreting interfacial property, a brief summary of the theoretical model used is given below. Fig. 1 represents the schematic figure for the mathematical model. In Fig. 1, L represents the aggregate embedded length. The aggregate is assumed to be elastic with Young's Modulus of E_a and cross-sectional area, A. The bulk matrix is assumed to be rigid except for the interfacial layer which is idealized as an elastoplastic shear layer.

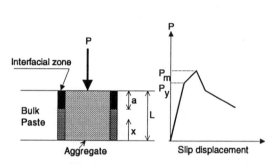

Figure 1   Mathematical model

It is assumed that debonding has occurred over a certain length, a, starting at x=L. Treating the boundary layer as a shear lag and assuming that a constant shear stress is acting at the debonded interface, the following equations can be written,

$$q = \begin{cases} kU(x) & 0 < x < L-a \\ q_f & L-a < x < L \end{cases} \qquad (1)$$

where k is the stiffness per unit length of the interfacial layer for small deformation, q is the shear force per unit length acting on the aggregate, q_f is the frictional shear force per unit length, and U(x) is the aggregate displacement.

Denoting the aggregate push-out force as P, the equilibrium equation and the constitutive relationship for the aggregate can be written as:

$$\frac{dP}{dx} - q = 0 \qquad (2)$$

$$\frac{P}{A} = E_a \frac{dU}{dx} \qquad (3)$$

Introducing eqns. (1) and (3) into eqn. (2), the following differential equation for U can be obtained,

$$U,_{xx} - \omega^2 U = 0 \qquad 0 < x < (L-a)$$

$$U,_{xx} - \frac{q_f}{E_a A} = 0 \qquad (L-a) < x < L \qquad (4)$$

in which, subscript comma "," indicates differentiation. The quantity ω, stiffness parameter, is defined as:

$$\omega = \sqrt{\frac{k}{E_a A}} \tag{5}$$

Equations (4), together with boundary conditions and continuity conditions, constitute a complete set of equations for the determination of $U(x)$. Solving this set of equations the following closed form expression for the slip displacement at the loading end, U^*, is obtained:

$$U^* = U(L) = \frac{P^* - q_f a}{E_a A \omega} \coth(\omega(L-a)) + \frac{P^* - 0.5q_f a}{E_a A} a \tag{6}$$

The relationship between U and P in the elastic stage can be obtained from the above equation by setting a = 0. This leads to a form of

$$\frac{P^*}{U^*} = E_a A \omega \tanh[\omega(L)] \tag{7}$$

The stiffness parameter ω can be determined from the initial slope of the experimental load - slip displacement curves for a pushout test using the above equation[3].

To determine the interfacial yield parameter, q_y (τ_y), the interfacial frictional forces, q_f (τ_f), and the specific energy, Γ, one needs to know the length of debonded crack, a, at the peak load. A method which utilizes the maximum load, P^*_{max}, and the slip displacement corresponding to P^*_{max} was used [3]. The formulae used to calculate τ_f, τ_y and Γ are Eqns. 8, 9 and 10, respectively,

$$2\pi R \tau_f = \frac{\omega P^*_{max}}{a\omega + \sinh(\omega(L - a)) \cosh(\omega(L - a))} \tag{8}$$

$$\tau_y = \tau_f \cosh^2(\omega(L - a)) \tag{9}$$

$$2E_a A \Gamma = \left(\frac{\tau_f}{\omega}\right)^2 [\cosh^4(\omega(L-a)) - \cosh^2(\omega(L-a))] \tag{10}$$

Note that an additional equation (Eqn. 11) is needed to determine the debonding length, a.

$$P^*_{max} \frac{0.5(\omega a)^2 + Cosh^2(\omega(L-a)) + \omega a \; Sinh(\omega(L-a)) Cosh(\omega(L-a))}{\omega a + Sinh(\omega(L-a)) \; Cosh(\omega(L-a))}$$
$$-U^*_{peak}E_{aA}\omega = 0$$

(11)

MECHANICAL PROPERTY MEASUREMENT

Experimental programs

The cement paste used was Type I Portland Cement with a constant water to cement ratio of 0.35 throughout all experiments. The Indiana limestone, cored out of a stone block using diamond edged drill bits, was the aggregate used in all tests. The reference specimens used for comparisons had an aggregate with a diameter of 12 mm and an embedded length of 13 mm and made of plain matrix. For the first purpose of the present study, various values of geometry of the aggregates were used. Three different core diameters of 12 mm, 18 mm, and 28 mm were used in the testing to investigate the influence of the diameter. Plastic cylindrical molds with dimensions $D_a/D_m = 0.23$, where D_a is the diameter of aggregate and D_m is diameter of mold, were used. Four different embedded lengths, 6 mm, 13 mm, 19 mm, and 25 mm were utilized to study the effect of the embedded length.

For the second aim of the present study, in certain tests, additives such as silica fume and latex were used in the cement paste. The influence of the coating of silica-cement slurry on aggregate was also investigated for one group of specimens with silica fume modified matrix.

The aggregate core was first inserted in the mold then the molds were filled with the cement paste matrix and vibrated on the vibrating table. The specimens were allowed to cure for 24 hrs. and then removed from the molds and cured in water at a temperature of $20^0C(68^0F)$ until a testing age of 28 days. After curing, the specimens were sliced, perpendicular to the core, into an appropriate thickness and ground and polished on both top and bottom surfaces before testing.

Experimental setup

The MTS testing machine apparatus set-up is shown in Fig. 2. The specimen is placed on a flat circular plate that was connected to a servohydraulic actuator of a MTS machine through a hollow cylinder. The entire specimen/fixture could move up

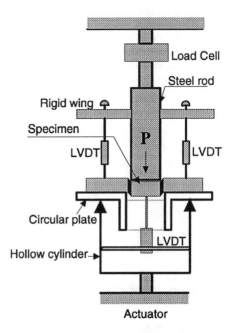

Figure 2   Experiment setup

with the actuator. A steel rod, connected to the load cell, made contact with the top surface of the aggregate. Resistance provided by the upward movement of the loading fixture pushed the cylindrical aggregate core downwards. Two LVDTs, which were fixed between the circular plate and the rigid wings of the steel rod, were used to measure slip displacement at the top of the aggregate relative to the surface of the cement matrix annulus. The average output of the LVDTs was used as a feedback signal to control the servohydraulic system. The rate of slip displacement of the push-out test was about 1 mm/hour. Slip displacement of the bottom of the aggregate was measured by an additional LVDT. Push-out load, slip displacement of aggregate and stroke of the actuator of MTS were recorded using a data acquisition computer.

Experimental results

For a typical pushout test, the four measurements of importance are time, pushout load, average displacement at the top of the aggregate and slip displacement at the bottom of the aggregate. Shown in Fig. 3 are two curves of slip displacement superimposed over the pushout loading curve as a function of time. For the measurement of the bottom displacement, the points at which the curve changes its slope signal the transition of interfacial damage stages. The first portion of the curve represents the elastic deformation of the aggregate-cement interface. The change of the slope, becoming steeper, implies that the deformation rate is increased. This means that the interface becomes less stiff than before and the elastic bond must have been broken. Hence, the end point the first portion of the curve marks the initial debonding and

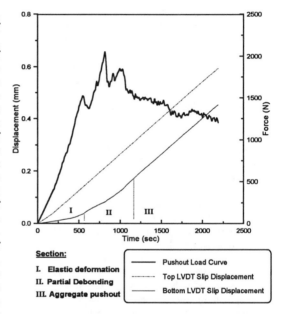

Section:
I. Elastic deformation
II. Partial Debonding
III. Aggregate pushout

——— Pushout Load Curve
---------- Top LVDT Slip Displacement
——— Bottom LVDT Slip Displacement

Figure 3   Pushout force and displacement as a function of time

the second stage of the curve could be called as the partial debonding stage. The point at which the up measured displacement and bottom measured displacement become parallel signals the start of the third stage. The parallelism of the two lines means that the top of the aggregate and the bottom of the aggregate underwent the same amount displacement. This corresponds to complete debonding. It should be noted that complete debonding of the interface occurs after peak load has been achieved as can be seen from the figure by matching the load curve with the second transition point.

A typical pushout load vs. slip displacement curve is shown in Fig. 4. Taken from this graph are the peak load, displacement at peak load and the slope of linear elastic stage for obtaining the interfacial properties ω, τ_f, τ_y, and Γ using the formula derived in the previous section. The results of the pushout test can be arranged into two groups which will be discussed separately. The groups are: (1) examination of effect of surface area of aggregate and (2) examination of effects of variations of cement paste composition.

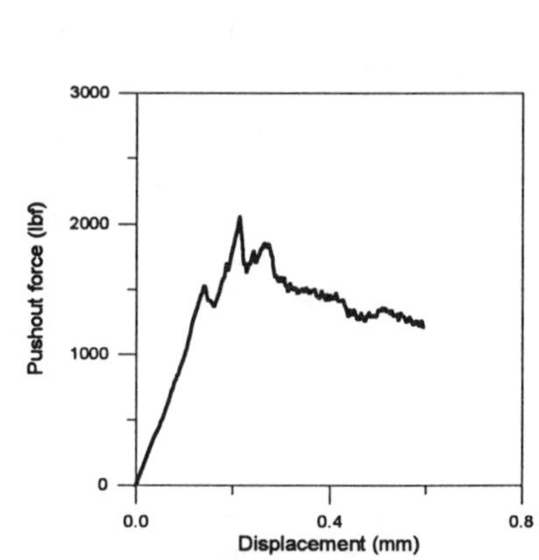

Effect of Aggregate Surface Area

In this set of experiments, the varied surface areas are obtained by either changing aggregate embedded lengths over a range of 6 mm, 13 mm, 19 mm and 25 mm with a fixed diameter or using different diameters of 12 mm, 18 mm and 28 mm with a fixed embedded length. Due to the existence of a transition layer, the interfacial zone between aggregate and cement can be characterized by a bond stiffness parameter, ω. In Fig. 5, two group of interfacial stiffness parameters are plotted versus surface area. One group of results is from varying aggregate embedded length while the other group of results is from varying aggregate diameter. The graph shows that these two group of data remain constant. A similar trend can be seen for the shear and frictional bond strengths. They are more or less on the same level for the aggregates with different surface area. Overall, it seems that these three parameters generally reflect the nature of the aggregate/cement interface properties since all three variables remained constant.

On the other hand, shown in Fig. 6 is the graph of specific surface energy plotted against bond surface area. In this case, specimens with increasing embedded length showed an increase in specific surface energy while specimens with increasing diameters stayed constant with specific surface energy. The observed influence of the embedment length of the aggregate on fracture energy can perhaps be explained by the R-curve concept. According to the energy concept in fracture mechanics, a crack propagates when the rate of change of strain energy release (G) is equal to the rate of energy consumption (R). The rate referred to is with respect to crack extension. For the pushout specimen, crack length is the length of the debonded zone, a, as shown in Fig. 1. For perfectly brittle materials, when G = R, a crack catastrophically propagates. However, in quasi-brittle materials the crack steadily propagates until a second condition is also satisfied

Figure 4 **Pushout force as a function of displacement**

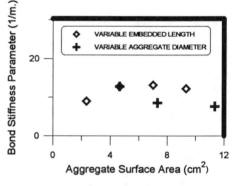

Figure 5 **Stiffness parameter, ω, as a function of surface area**

$$\frac{\partial G}{\partial a} = \frac{\partial R}{\partial a} \qquad (12)$$

Thus, the fracture resistance or R-curve is rising for quasi-brittle crack propagation. It was shown by Ouyang et al.[4] that specimens with different embedment length will have different G curve. Thus the critical point when Eqn. (12) is satisfied is different for different embedment lengths. This may explain why the bond fracture energy calculated assuming a perfectly brittle crack propagation gave different values for aggregates with different embedded lengths.

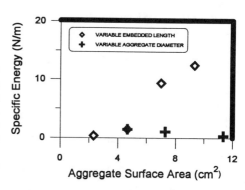

Figure 6   **Specific energy as a function of surface area**

Effects of Variation in Cement Paste Composition

The second part of the experimental pushout tests was to study the effects of additions of admixtures and the coating of silica-cement slurry upon interfacial bond strength. In this study, three admixtures; silica fume, latex and latex with an antifoaming agent were tested and analyzed. In addition, effects of pretreatment of silica-cement slurry on coarse aggregate were also tested and analyzed. Bond parameters were calculated using the theoretical model. The results of these parameters are listed in Table I.

The latex copolymer (emulsion) was added in a latex/cement ratio of 1:2 (by weight), which was recommended by the latex manufacturer. As shown in Table I increases in the shear and frictional bond strengths were about 20% over the reference cement. On the other hand, the increase in specific surface energy of the interfacial bond, was in the range of 50%.

In addition to the latex modified cement matrix, a combination of latex plus an antifoaming agent was added to the cement paste for the study. Again, the latex/cement ratio was 1:2 plus 1%, by weight of cement, of antifoaming agent. Including the antifoaming agent to the cement matrix proved to be very valuable in improving the interfacial bond characteristics. An increase in the specific surface energy was obtained with a magnitude of almost three fold over the reference specimen and with an increase of 80% over the latex modified specimen. The increases in shear and frictional bond strengths were observed to be in the range of 60% over the reference specimen and about 35% over the latex modified specimen as can be seen in Table I.

The third admixture used in this study was silica fume. In the tested silica fume modified matrix, 10% by weight of cement was replaced with silica fume. As shown in Table I, the specific surface energy was almost 5 times greater for the silica fume matrix specimen over the reference specimen. The frictional and shear bond strengths were also over twice as large as compared to the reference specimen.

The precoating of an aggregate was made by painting cement and silica fume-mixed slurry on aggregate, and then exposing the aggregate to air at room temperature for 1 hr. The additional effect of treating the aggregate with slurry of the cement plus silica fume can be observed from the Table I. The surface fracture energy increased about 10 times for the specimens with SF-modified matrix together with cement and SF slurry coating when compared with the reference specimen.

TABLE I   EXPERIMENTAL TEST RESULT

Specimen type	ω (1/m)	τ_y (MPa)	τ_f (MPa)	Γ (N/m)
Reference specimen	12.80	2.28	2.26	1.40
Latex additive matrix	10.71	2.70	2.67	2.08
Latex + antifoamer matrix	14.92	3.70	3.63	3.81
Silica fume additive matrix	13.91	4.79	4.71	6.59
SF matrix + (SF +C) coating	27.36	11.44	10.34	59.02

MICROSTRUCTURAL ANALYSIS

Specimen Preparation for Microscopic Study

Specimens were cast in the same manner as previously discussed using exactly the same materials in order to match the mechanical test results. After curing for 28 days, the samples were sliced, perpendicular to the axis of the cored aggregate, to a thickness of approximately 1 mm using a diamond saw.

The sliced specimens were then placed in ethanol in order to stop hydration until they were ready to be prepared for examination. The specimens were then placed in a vacuum in order to remove all moisture and foreign substances from the surface. While under the vacuum, the specimens were impregnated with a low viscosity epoxy. The samples were then removed from the vacuum and allowed to cure for 24 hours with the addition of a hardening accelerator. The specimens then needed to be sanded and polished for a clear examination. The aggregate/cement interfacial zone was examined on a microscopic level through the use of a scanning electron microscope with an energy dispersive x-ray analysis system. Based on previous work, an image magnification of 400x was selected for the analysis. [1, 2][5]

Using conventional Polaroid micrographs taken from the scanning electron microscope, a quantitative analysis of the aggregate/cement interface was performed. The micrographs were digitized by a video camera attached to a Tracor-Northern 8500 image analysis system. The binary images are constructed by determining a grey level threshold value which appears to segment the porosity from the other hydration product. The final digitized binary images were 512 x 512 pixels which resulted in approximately 0.5 microns per pixel at a resolution of 400x. [1, 2]

From each of the digitized images, the interfacial zone was divided into segmental bands 20μm in width up to 120μm from the aggregate surface. The digitized binary images of each segmental band were analyzed to determine an area percentage of pore space. By plotting the average area percentage versus the distance from the aggregate surface, gradient plots for the aggregate/cement interface were established for each image. The porosity gradient is useful since it can lead to a better understanding of the material behavior of the interfacial bond between cement and aggregate.

Microscopic Aggregate/Cement Interface Evaluation

Figs. 7 to 8 are two SEM photographs of the aggregate/cement interfacial zone. Fig. 7 shows the interface of the reference specimen with a very clear and visible interface. On the other hand, the specimens incorporated with silica fume both in the matrix and surface coating (Fig. 8), the porosity of the interface is almost eliminated. Fig. 9 is a porosity gradient graph taken from the results of the image analysis study of SEM photographs of the aggregate/cement interface. The graph shows an average decrease in porosity for the latex + antifoaming agent specimen, silica fume specimen and silica fume matrix + coating specimen throughout the interface as compared to the reference specimen. These results show a direct correlation to the improved mechanical bond properties obtained from the pushout test results as referred to Table I.

Figure 7 Reference specimen interface   Figure 8   SF + Coating specimen interface

CONCLUSIONS

The effect of aggregate surface area on the interfacial bond properties was studied. The geometry of the cylindrical aggregate was altered by either varying the aggregate embedded length or by varying the aggregate diameter. The interfacial stiffness, shear bond strength and frictional bond strength remained relatively constant since the physical and chemical composition of the interface was not altered. However, the specific surface energy shows an increase for the longer embedded lengths. The

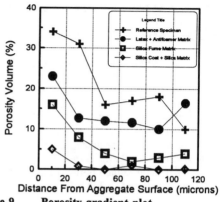

Figure 9   **Porosity gradient plot**

interfacial properties varied with the addition of several chemical admixtures. Substantial improvements in bond properties were obtained by the addition of admixture to the cement matrix. The addition of silica fume showed the largest improvements, especially in the area of bond fracture energy.

The interfacial zone was examined using a scanning electron microscope and the porosity of the interface was quantified using an image analysis system. From the backscattered electron imaging, it can be seen that the addition of chemical admixtures can decrease the porosity of the interfacial zone. The specimens with silica fume showed the most significant improvements in porosity and microstructure of the interface. Comparison of the bond properties and interfacial porosity gradients has shown a direct correlation. It can be seen that the microstructure of the interface plays a substantial role in the mechanical behavior of the aggregate/cement bond.

ACKNOWLEDGEMENT

The financial support from the National Science Foundation Center for Science and Technology of Advanced Cement-Based Materials is greatly acknowledged.

REFERENCES

1.   K. Mitsui, Z.Li, D.A. Lange, and S.P. Shah, ACI Materials Journal, V. 91, No. 1, 30, (1994).

2.   N. McN Alford, . and A.B. Poole, Cement and Concrete Research, Vol. 9, No. 5, 583, (1979).

3.   Z. Li, B. Mobasher, and S. P. Shah, Journal of American Ceramic Society, 74(9), 2156, (1991).

4.   C. Ouyang, A. P.Alvarez, S. P. Shah, accepted for publication, ASCE Journal of Engineering Mechanics, (1994).

5.   K. L. Scrivener, A. Bentur, and P. L.Pratt, Advances in Cement Research, Vol. 1, No. 4, 230 (1988).

CHEMICAL MODELS FOR THE MECHANISMS THAT INFLUENCE THE NATURE AND STRUCTURE OF THE INTERFACIAL TRANSITION ZONE

D. L. COCKE*, M. Y. A. MOLLAH**, T.-C. LIN, R. K. VEMPATI, AND T. R. HESS
*Gill Chair of Analytical Chemistry, Lamar University, Beaumont, TX 77710
** Visiting Professor, Department of Chemistry, Dhaka University, Bangladesh.

ABSTRACT

The chemical nature and microstructure of the interfacial transition zone (ITZ) are influenced by the interfacial mechanisms of hydration which are primarily controlled by surface and near-surface phenomena. Recent works from our laboratory involving superplasticizers, metal anions that retard hydration and interaction with silicate based aggregates have established that the interfacial chemistry is of fundamental importance in understanding the mechanisms of CSH formation, surface chemical compound formation and mineralization of the interfacial transition zone and its interaction with the hydrated cement particles and the aggregate. A surface charge control reaction model that accounts for the importance of calcium and other cations and anions will be outlined and used to discuss the interfacial physical chemistry of the ITZ and its ultimate chemical nature and structure.

INTRODUCTION

Considerable importance has been given to understanding the surface and interfacial chemistry of minerals, soils and sediments. Recently, the interaction of pollutants with these materials as well as with cement in solidification/stabilization (S/S) has spurred increased research on the interfacial region [1-3]. Cement-based systems are widely used to immobilize and stabilize waste materials containing hazardous substances before landfilling. Interfaces in cementitious systems influence hydration reactions, aggregate-cement interactions, action of additives and interaction with wastes. In our interfacial studies of cementitious solidification/ stabilization systems with metal anions, aluminosilicate sources such as fly ash, and superplasticizers results and models have been generated that are of importance to understanding the interfacial interactions in cementitious systems.

STRUCTURE OF SOLID-LIQUID INTERFACE

The adsorption of an ionic species at the mineral surface (solid-liquid interface) is largely controlled by the affinity of the adsorbing species for the surface sites. The adsorption is also dependent on the solution properties as well as the electrostatic interaction at the solid-liquid interface. A number of models for adsorption processes have been developed to clearly understand the intrinsic mechanisms involved in these processes [4-6].

The Bi-layer Model

The adsorption of cationic species on a negatively charged surface is schematically demonstrated in Figure 1. The figure shows that the initial adsorption of cations is accompanied by the adsorption of an equivalent number of opposite-charged ions (counter-ions) so that the system as a whole is electrically neutral. The distribution of counter ions in solution and the formation of an electric double layer is well established phenomenon [5]. The separation of charges in the double layer results in an electrical potential difference across the solid-liquid interface. The maximum potential is at the surface, and it decreases exponentially with the distance from the surface.

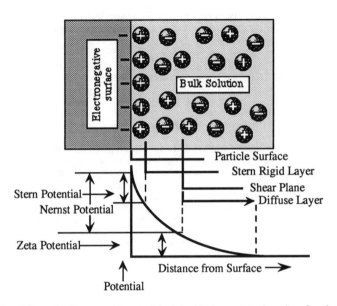

Figure 1.   Schematic diagram of the electrical double layer at a mineral surface in an aqueous
phase and the change in potential with increasing distance from the surface.

The potential determining ions (PDI) consist of ions of which the mineral is composed of,
H^+ or OH^- and collector ions that form insoluble salts with surface mineral species or complexes
on the mineral surfaces [7]. The PDI at the mineral surface will adsorb the counter-ions from the
solution either by Coulombic (e.g. electrostatic) attraction or by chemisorption through covalent
bond formation. The concentrations of the PDI and the net surface charge are dependent on the
pH of the medium which becomes zero when the densities of the positive and negative charges
are equal. The activity of the PDI, defined when the surface charge is zero ($s_0=0$), is called the
"Zero Point of Charge" (zpc) of the mineral [8] which is a characteristic property of the oxide
minerals. When H^+ and OH^- are the potential-determining ions, the zpc is expressed in terms of
pH and it is designated as pH_{zpc}. The pH_{zpc} is an important parameter since it determines the
acidity and basicity of the hydrated oxides which controls the adsorptive potential of a particular
oxide. When $pH > pH_{zpc}$ the surface is negative, and cations will be adsorbed. Conversely,
when $pH < pH_{zpc}$ the surface becomes positive, and anions will be adsorbed. Strong amphoteric
oxides, like α-Fe_2O_3 (zpc=8.5) and γ-Al_2O_3 (zpc=8.5), have zpc near neutral pH; while acidic
oxides, such as SiO_2 (zpc=2.0) have zpc at lower pH values since they are proton donors [9,10].
Some authors have preferred, "Point-of-Zero-Salt-Effect" (pH_{pzse}) instead of pH_{zpc}, since "Zero-
Point-of-Charge" is, in practice, not independent of ionic strength [11,12]. The change of zeta
potential 'ζ' at the mineral-water interface (Figure 1) and its variation with solution conditions is
also important in understanding the adsorption phenomena. Anions will be adsorbed when the
'ζ' of a mineral is positive; conversely, cations will be adsorbed when the 'ζ' is negative [4].
However, the point at which the zeta potential becomes zero is termed as "The Isoelectric Point"
(IEP). The zpc and the IEP of most mineral systems are extremely close to each other and may
even coincide in many cases, except in the case of specifically adsorbing polyvalent and
surfactant ionic species [13]. Another term - "The Point-of-Zero-Net-Proton Charge" (pH_{pnzpc})
is used sometimes, and it corresponds to the pH where the surface is neutral [14].

Models for describing the phenomena at the solid-liquid interface are important when considering the concentration and potential gradients as a function of distance from the solid surface. However, these models are not intended for understanding the critical mechanistic role that concentration and potential gradients play in the surface reactions of solid-liquid systems. We are proposing a charge controlled reaction mechanism in cement-based systems to explain the passivation of the hydration process in OPC through the formation of bi- and tri-layers in the presence of certain metals and superplasticizers.

The Charge-Dispersal Model (CDM)

The adsorption and speciation of toxic metals in cementitious materials is strongly dependent on the aqueous chemistry of the metal cations in a very basic solution, like cement paste (pH = 13± 0.5). The hydration of ordinary Portland cement (OPC) is dominated by the hydration reactions of Ca_3SiO_5 (alite) and Ca_2SiO_4 (belite) to produce $3CaO \cdot 2SiO_2 \cdot 3H_2O$ (C-S-H) and $Ca(OH)_2$ which constitute ~60-70% and ~20-25% respectively of the total products and about 5-15% of other minor phases [15]. The stoichiometry of these reactions clearly indicate that the hydration of the alite and belite phases yield Ca^{2+} ions. Initially, hydration occurs rapidly, forming a thin C-S-H film on the surface of the cement clinkers. The C-S-H membrane is permeable to the inward flow of water molecules and outward flow of smaller ions like Ca^{2+} and OH^- from inside the solid matrix [16]. The excess Ca^{2+} ions thus produced are diffused through the C-S-H gel membrane into the solution of increasing chemical potential and undergo subsequent reactions to produce $Ca(OH)_2$. However, with time an excess of $Ca(OH)_2$ will be precipitated on the fluid side while an excess of silicate ions build on the grain side of the membrane. This intrinsic process will cause an osmotic pressure differential which will rupture the membrane periodically and reform by extruding concentrated silicate solution, thus allowing secondary growth of C-S-H in the accelerated stage of hydration [16]. However, the newly formed C-S-H phase will possess a negative charge because the ZPC of the commonly occurring silicate minerals in the cement system is much below the pH (13±0.5) of cement paste. The most abundant counter ions in cement solution is Ca^{2+}, diffused from inside through the C-S-H membrane. These oppositely charged counter ions will immediately form a layer of positive charge adjacent to the negatively charged surface to constitute what is known as 'Electrical Double Layer'. However, the dispersions of ionic species present in such a system needs to be clearly understood, since this is expected to play a significant role in s/s of metals. For example, FTIR analyses of the Zn [17] and Cd-doped [18] cement systems have established that the inhibition of hydration were due to surface precipitation of $Ca Zn_2 (OH)_6.2H_2O$ and $CaCd(OH)_4$ respectively. These results combined with a recently developed chemical equilibrium model [19] of the zinc-doped cement/water system have established that a 'charged Dispersed' reaction mechanism successfully accounts for the surface precipitation of $Ca Zn_2 (OH)_6.2H_2O$.

The chemical equilibrium model of the early hydration of zinc-doped cement/water (EH-Zn-OPC) system, $Na-K-Ca-Zn-H-SO_4-OH-Zn(OH)_2-Zn(OH)_3-Zn(OH)_4-H_2O$, has been developed for simulating the reactions between the zinc-containing cement porewater and available mineral components to high ionic strengths at 25 °C [19]. The model utilizes the Pitzer's model (ion interaction model) [20] as a basis for calculation of the activity coefficients of ions and neutral species. In addition to the parameters needed for activity coefficient expressions in Pitzer's model, the necessary thermodynamic data have been carefully examined and provided in order to enable quantitative analysis of the chemical equilibria in the EH-Zn-OPC system.

Two porewater compositions of high alkali cement pastes after 60 minutes hydration time were selected from Gartner et al. [21] and Michaux et al. [22] as the initial solution conditions for the model. The EH-Zn-OPC system was then simulated by assuming that the solution were at equilibrium with ZnO. The results of the simulations of two initial solutions equilibrated with ZnO are presented in Table I. The saturation indices of calcium hydroxyzincate are –0.0455 and 0.0073 for case I and II, respectively. It shows that only a slight increase of either pH as well as Ca^{2+} ion concentration changes the saturation condition of calcium hydroxyzincate from undersaturation to supersaturation which favors precipitation.

Table I. Model Results of the Porewater Compositions of High Alkali Cement Pastes in the Presence of ZnO at the First Hour of Hydration.

	Case I[†]		Case II[§]	
Temperature	25°C		25°C	
pH	13.11		13.12	
Activity H_2O	0.9869		0.9861	
Ionic Strength	0.7687		0.8180	
Solubility of ZnO	1.33 mmol/kg		1.42 mmol/kg	
Species	Conc. (mmol/kg)	Activity Coefficient	Conc. (mmol/kg)	Activity Coefficient
H^+	1.423E–10	0.5398	1.406E–10	0.5342
Ca^{2+}	19	0.0738	21.5	0.0707
Na^+	103	0.6419	116	0.6347
K^+	395	0.6433	425	0.6362
SO_4^{2-}	186	0.0964	196	0.0932
OH^-	214.6	0.6016	219.8	0.6002
$Zn(OH)_4^{2-}$	1.182	0.1347	1.266	0.1312
$Zn(OH)_3^-$	1.458E–1	0.4891	1.512E–1	0.4814
$Zn(OH)_2^0$	1.796E–3	1.0000	1.794E–6	1.000
Phase	log SI		log SI	
Gypsum	–0.0309		0.0118	
Portlandite	0.5583		0.6126	
Calcium Zincate	–0.0455		0.0073	
Zinc Oxide	0.0000		0.0000	

[†] Gartner et al. [Ref. 21]. [§] Michaux et al. [Ref. 22].

A general model of the porewater of high alkali cement pastes is defined by considering the average values of pH and the concentration of Na^+, K^+, SO_4^{2-} from case I and II. The zinc-doped cement/water system at different pH or Ca^{2+} ion concentration can be simulated by changing the pH or Ca^{2+} ion concentration in the model. The change of saturation indices for gypsum, portlandite and calcium hydroxyzincate were monitored. The results are shown in Figures 2 and 3 for the different pH case and different Ca^{2+} ion concentration case, respectively. It is observed that a supersaturated solution with respect to calcium hydroxyzincate occurs only when the Ca^{2+} ion concentration is higher than 0.02 (mol/kg) and the pH must be higher than 13. It can, thus, be concluded that the zinc-doped cement porewater is nearly saturated with respect to calcium hydroxyzincate at the early stage of hydration.

According to the solution-precipitation theory [23], it is not likely that calcium hydroxyzincate will be precipitated in the bulk solution of zinc-doped cement porewater, because the corresponding saturation index is not high enough to initiate the nucleation process. However, at such high pH environment the crystalline and amorphous phases of hydrated cement will form a negatively charged surface. The activity of Ca^{2+} ions in the proximity of the charge charged surface of solid phases are then increased due to electrostatic interactions between these opposite charged phases.

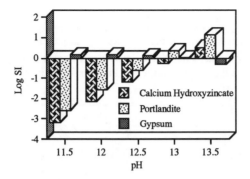

Figure 2. The Saturation Indices of Minerals at Different pH at the Early Stage of Hydration in the Zinc-doped Cement/Water System.

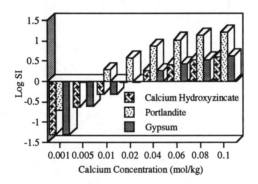

Figure 3. The Saturation Indices of Minerals at Different Ca^{2+} Ion Concentration at the Early Stage of Hydration in the Zinc-doped Cement/Water System.

It has been shown that the dominant anionic species at pH 12 to 13 are $Zn(OH)_3^-$ and $Zn(OH)_4^{2-}$ constituting ~40% and ~60% respectively [3]. At pH>13.0 the $Zn(OH)_4^{2-}$ ions are the dominant species. The diffuse ion cloud in such a system will, therefore, be dominated by OH^-, $Zn(OH)_3^-$ and $Zn(OH)_4^{2-}$ ions. The negatively charged zinc-hydroxy ions are not expected to be adsorbed on the C-S-H surface, since it is negatively charged. However, the negative C-S-H surface is charge compensated by Ca^{2+} ions to form an electrical bi-layer as shown in Fig. 4. The next zone in the solution will consists of a diffuse layers of mainly $Zn(OH)_3^-$ and $Zn(OH)_4^{2-}$ ions and constitute a, "Tri-layer of Diffuse Ions" (The Diffuse Ion Swarm). As shown schematically in Figure 4, the negatively charged CSH surface and the Ca^{2+} counter ions constitute the bi-layer while the CSH surface, the Ca^{2+} ions and zinc hydroxy anions form the tri-layer. The ions constituting the tri-layer are remote from the surface, and are unlikely to undergo any specific interaction with the surface components. These anions are charge dispersed uniformly over the entire C-S-H surface to form $CaZn_2(OH)_6 \cdot H_2O$ by reactions of $Zn(OH)_4^{2-}$, $Zn(OH)_3^-$ and $Zn(OH)_2^0$ with Ca^{2+} ions [17]. The charge-controlled reactions leading to the retardation of cement hydration and subsequent setting are due to competition mainly between the Ca^{2+} ions and metal-hydroxy anions in the system. It is the higher concentration of Ca^{2+} ions that control the formation of calcium hydroxyzincate in this system. In the presence of sufficient Ca^{2+} ions in solution the zinc hydroxy anions $Zn(OH)_3^-$ and $Zn(OH)_4^{2-}$ are transformed into $CaZn_2(OH)_6 \cdot H_2O$ which completely covers the cement grains and thus passivate them from further hydration reactions.

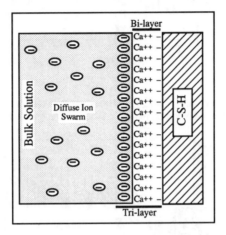

Figure 4:   Schematic diagram of charge dispersed tri-layer model.
⊖ represent negatively charged Zn-hydroxy anions.

Our recent work involving the use of water soluble superplasticizer (sodium lignosulfonate) also support this 'Charge-Dispersal' mechanism to account for the inhibition of the hydration reactions [24]. In this case, the tri-layer is composed of mainly superplasticizer anions. Superplasticizers have multiple attachment sites to cause the molecule to loop along the surface with the 'Train' segments adsorbed on the negative surface, and this affects the Stern layer in a manner reported for polymer adsorption on negative surfaces [25]. Since the adsorption of the 'Train' segments cause a displacement of the specifically adsorbed counter ions the surface charge will be reduced. However, in addition to the electrostatic forces, the adsorption of superplasticizers and surfactant molecules on charged surfaces are also significantly influenced by hydrophobic bonding between CH_2 groups on the adjacent adsorbed superplasticizer units.

CONCLUSIONS

The surface and near-surface phenomena at the mineral-water interface critically control the chemistry and structure of the interfaces in cement-based S/S systems containing metals and superplasticizers. In the present article we have reviewed the fundamental concepts pertinent to mineral-water interface chemistry and developed a "Charge Dispersed Reaction Model" to successfully explain the inhibition of cement hydration due to the formation of calcium hydroxyzincate type compounds on cement clinkers.

ACKNOWLEDGMENTS

We wish to thank the Gulf Coast Hazardous Substance Research Center, Lamar University - Beaumont for financial support. We also acknowledge partial financial support from the Welch

Foundation and the Texas Advanced Technology and Research Program and the Texas State Coordinating Board.

REFERENCES

1. D. L.Cocke, D. L., J.Hazardous Materials. **24,** 231 (1990).
2. D. L. Cocke and M. Y. A. Mollah, in Microstructure of Solidified Waste Forms, edited by R. D. Spence (Lewis publishers, USA, 1993) pp. 187-242.
3. M. Y. A. Mollah, R. K. Vempati, T.-C. Lin and D. L. Cocke, Waste Management, (1994) (in Press).
4. A. Breeuwasma and J. Lyklema, J. Colloid Interface Sci. **43,** 437(1973).
5. J. Lyklema, Chemistry and Industry, Nov. 2, 741(1987).
6. J. Lyklema, in The Scientific Basis of Flocculation, edited by K. J. Ives. (Sijthoff and Noordhoff, The Netherlands, 1978), pp. 3-36.
7. W. Stumm and J. J. Morgan, Aquatic Chemistry, 2nd edition, (Wiley-Interscience publication, N.Y.,1981).
8. G. A. Park, Adv. in Chem. Series **67,** 121 (1967) (edited by R. F. Gould).
9. G. A. Park, Chem. Rev. **65,** 177(1965).
10. J. Parker, W. Zelazny, S. Sampath and W. Harris, Soil Sci. Soc.Am. J. **43,** 668 (1979).
11. M. A. Pyman, J. W. Bowden, and A. M. Posner, Aust. J. Soil Res. **17,**191 (1979).
12. I. Iwasaki and P. L. de Bruyn, J. Phys. Chem. **62,** 594 (1958).
13. J. A. Davis and D. B. Kent, in Reviews in Mineralogy, edited by M. F. Hochella Jr. and A. F.White (published by The Mineralogical Society of America)**23** 177 (1990).
14. P. Schindler, B. Furest, R. Dick and P. Wolf, J. Colloid Interface Science **55,** 469 (1976).
15. F. P. Glasser, in Chemistry and Micro Structure of Solidified Waste Forms, edited by R.D. Spence (Lewis publisher, USA, 1993) pp.1-40.
16. J. Hansen, Science. **3(10),** 48 (1982).
17. M. Y. A. Mollah, J. R. Parga and D. L. Cocke, J. Environ. Sci. Health. A **27(6),** 1503 (1992).
18. M. Y. A. Mollah, Y. N. Tsai, T. R. Hess and D. L. Cocke, J. Environ. Sci. Health. **A27(5),** 1213 (1992).
19. T.-C. Lin, Chemical Modeling of the Zinc-Doped Cement/Water System and Characterization of Calcium Hydroxyzincate, (M.S. Thesis, Lamar University, 1994).
20. K. S. Pitzer, "Theory: Ion Interaction Approach." in Activity Coefficients in Electrolyte Solutions. Edited by R. M. Pytkowicz. (CRC Press, Boca Raton, 1979).
21. E. M.Gartner, F. J. Tang and S. J.Weiss, "Saturation Factors for Calcium Hydroxide and Calcium Sulfates in Fresh Portland Cement Pastes." J. Amer. Cer. Soc. **68(12),** 667-73 (1985).
22. M. Michaux, P. Fletcher and B.Vidick, "Evolution at Early Hydration Times of the Chemical Composition of Liquid Phase of Oil-well Cement Pastes with and without Additives. Part I. Additive Free Cement Pastes." Cemt. and Concr. Res. **19,** 443-56 (1989).
23. W. Stumm, "Precipitation and Nucleation." in Chemistry of the Solid-Water Interface. (John Wiley & Sons, New York, 1992).
24. M. Y. A. Mollah, P. Palta, T. R. Hess, R. K. Vempati and D. L. Cocke, Cemt. Concr. Res. submitted (1994).
25. J. Lyklema, Pure and Appl. Chem. **52,** 1221(1980).

OPTICAL MICROSCOPY AND DIGITAL IMAGE ANALYSIS OF BOND-CRACKS IN CEMENT BASED MATERIALS

ADRI VERVUURT AND JAN G.M. VAN MIER
Stevin Laboratory, Delft University of Technology, P.O. Box 5048, 2600 GA Delft, The Netherlands

ABSTRACT

A new testing method is presented in which the properties of the bond zone between aggregate and matrix in concrete are investigated. The reason for starting the investigation is the improvement of the lattice model, which is developed at the Stevin Laboratory. In the lattice model the material is schematized as a lattice of beam elements. The fracture properties of the individual beam elements are purely brittle, and the use of fracture energy parameters as a material property can be avoided for simulating the non-linear fracture process. In spite of the fact that a locally brittle law is used, the global behaviour shows softening of the material. An improved optical microscopy technique in combination with image processing is adopted for investigating the fracture process during loading.

DEBONDING IN HETEROGENEOUS MATERIALS

It is a known fact that cracking in concrete and other heterogeneous materials like sandstone is initiated along the interface between matrix and aggregate [1]. Even before loading, cracks are present in the specimen as a result of drying shrinkage and due to differential temperatures that develop during the hydratation process. During loading of the specimen more microcracks are formed and after the peak, bond-cracks join and form macrocracks which propagate through the matrix of the material. The load which can be transferred by the specimen decreases. This softening behaviour was explained by Hillerborg and co-workers [2] by a band of microcracks ahead of the crack tip. For numerical calculations with the model according to this theory (Fictitious Crack Model) fracture energy is introduced as a material property. However through the years is has been proven that it is very difficult to arrive at a unique description of fracture energy, see for example [3, 4, 5]. Later a complementary link between softening and the microstructure of concrete was given by the second author of this paper [6]. In the theory of crack face bridging softening is explained by intact ligaments connecting the two faces of the (discontinuous) crack. Experimental evidence for this phenomenon is for example given in [6] and [7].

Lattice models have proven to be capable of simulating this latter mechanism in heterogenuous materials like concrete. In the model the material is assumed to be perfectly brittle at the level of the beams in the lattice. Cracking is simulated by removing in each load step one element of the mesh. Consequently the beams in the lattice have to be chosen rather small as the size of the beams depends also on the microstructure of the material. Less than 2 mm beam length seems appropriate for concrete, however if the elements are chosen smaller more detail can be implemented in the mesh, which improves the amount of simulated details. The application of lattice models for simulating cracking in concrete is for

337

example presented in [5, 8, 4]. Yet the idea of using a lattice model for simulating a continuum is not new. Hrennikoff [9] used a regular triangular truss model for solving problems of elasticity. With increasing computer capacity in the last decades the use of lattice models for simulating non-linear stress-strain behaviour received new interest. A regular square lattice was for example introduced by Herrmann et al. [10]. Schlangen & Van Mier [11] introduced a triangular variant of the Hrennikoff model with beams instead of trusses (Fig. 1a). Next to these regular shaped lattices, also a number of lattices with random length of the beams have been introduced during the last years, see for example [12]-[16]. The random lattice applied in this paper for simulating cracking due to drying shrinkage in concrete was developed by Moukarzel & Herrmann [14]. The random lattice has a square grid as a basis. In each cell a node is selected at random (Fig. 1b). Subsequently the random lattice is defined by connecting always the three nodes which are closest to each other. It is obvious that lattices with a random length of the beams already contain a certain heterogeneity. When a regular lattice is used for simulating cracking in disordered materials, heterogeneity has to be implemented separately, for example by projecting the material structure on top of the lattice and assigning the properties of the beams to the properties of the corresponding phase in the material. For concrete three phase can be distinguised, i.e. aggregate, matrix and interface between matrix and aggregate. Of course the material structure can also be projeted on top of a random lattice. In this case each phase in the material is assumed to be individually heterogeneous as well.

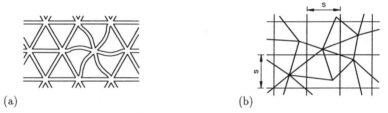

(a)   (b)

Figure 1: Lattice models: regular triangular lattice (a) and random lattice (b).

A simulation of a tensile test on a single edge notched specimen is shown in Fig. 2. The area where cracks are expected is schematized as a regular triangular lattice in which heterogeneity is implemented by projecting a grain structure on top of the lattice. The mesh used for the simulation is plotted on the right side of Fig. 2. The crack opening is calculated as the average crack opening on the left and right side of the specimen. The simulated stress-crack opening diagram is enhanced with crack patterns at four stages in the fracture process. Debonding of the material occurs in the beginning of the fracture process along the interfacial zones between matrix and aggregates. The mechanism of crack face bridging can be seen in the crack patterns belonging to the tail of the stress-crack opening diagram. However when the load-displacement response of the simulation is compared to a similar experiment, a much too brittle response is observed for the simulations [4, 17]. The amount of detail included in the mesh seems an explanation for this [8].

In order to obtain a better understanding of the fracure processes in the interface region, a new testing program has started in which the cracked surface is observed very closely with an optical microscope. Special attention is given to the existence of initial cracks (due to drying shrinkage) and cracks in the interfacial zone. A very simple specimen geometry is chosen in which a few (1, 2 or 3) cylindrical aggregates are embedded in a matrix of cement

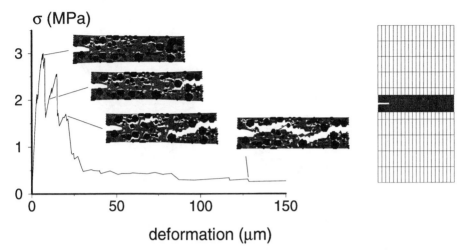

Figure 2: Example of a uniaxial tensile test with the lattice model. The stress-crack opening diagram is shown with the crack patterns at several stages in the fracture process.

paste. The idea of simplifying the complex three dimensional process of interface fracture is not new [18, 19], however the approach of applying an optical microscopy technique for detailed observations of crack growth, and inverse modeling with the lattice mode has not been used to date.

EXPERIMENTS

Materials and Specimens

To investigate the interfacial zone between matrix and aggregate in concrete, a special two dimensional geometry is developed in which cylindrical aggregates are embedded in a matrix of white cement paste, see Fig. 3a. For studying the influence of the density of the bond zone two different types of aggregates are used. Either dense aggregates (Polar White granite from Brasil) or porous aggregates (Bentheimer sandstone from Germany) are placed in the specimen. The cylindrical aggregates with a diameter of 20 mm are positioned perpendicular to the casting direction in a mould of size 210x165x170 mm³. Cement is cast around the aggregates and after one day of hardening the block is demoulded and placed under water. After three days of hardening, specimens with a thickness of 10, 15 or 20 mm are sawn from the centre of the block. The dimensions of the specimens are 125x170 mm and a notch (5x30 mm) is sawn at half width at the top of the specimen. The specimens are subjected to two types of loading. A drying shrinkage load (20 °C and 50% RH) is applied for three days after one week of curing and a mechanical load is applied to determine the residual strength. For reference, half of the specimens are not subjected to a shrinkage load but a mechanical load only.

For scanning the surface of the specimen with a long distance microscope during shrinkage and mechanical loading a special set-up was developed. The microscope will be described

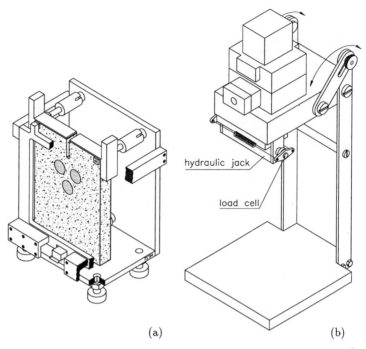

hydraulic jack

load cell

(a)   (b)

Figure 3: Specimen with three embedded aggregates in a loading frame (a) and (b) the loading device. During the test the frame of Fig. (a) is placed in the loading device (b).

in detail in the next section. The set-up consists of a loading frame (Fig. 3a) and a loading device (Fig. 3b). Because of the high accuracy of the microscope, several modifications were made to the loading frames. The frame is roughly adjustable in all directions and three spring supports are positioned at the back of the specimen for keeping the front of the specimen vertical during the test. Friction in the supports was minimized by using ball bearings between specimen and springs (Fig. 3a). After shrinkage a mechanical splitting load is applied similar to [20]. However to avoid vertical forces in the specimen and consequently compressive stresses near the lower support of the specimen, a miniature hydraulic jack was provided through which horizontal loads are applied directly to the specimen (Fig. 3b). An important constraint in the development of the loading device was that the load should be applied symetrically because of the fixed position of the specimen with respect to the microscope. The elastic deformations of the specimen during loading must be limited because the area of interest might move out of the field of view of the microscope. Two miniature load cells with a capacity of 1100 N are placed between the hydraulic jack and the specimen (Fig. 3b).

A conventional technique was used for obtaining a stable deformation controlled test. At each side of the specimen a LVDT is attached across the notch of the specimen to measure the crack mouth opening displacement (CMOD). The average CMOD is used as a feedback signal for the closed-loop servo-controlled system. The loading device is developed such that a complete set of specimen and loading frame can be centered under the hydraulic actuator (a combination of Fig. 3a and 3b). In spite of these precautions, bending stresses will still appear near the lower support of the specimen which should be kept in mind when the crack enters this zone. However normally this will cause no problems as crack growth is monitored in the upper part of the specimen containing the cylindrical aggregates, see Fig. 3.

Optical Microscopy

In addition to the conventional measuring technique described in the previous section, optical measurements are performed in the region between the notch and the aggregates. These crack measurements are carried out with a QUESTAR Remote Measurement System (QM100-RMS), see Fig. 4a. The optical system which was already available in the laboratory has been adapted for automatic scanning the specimen surface during loading [21]. The microscope is combined with a CCD camera (type ICD 46E REV) which provides for images of 756x581 pixels with a grey value varying from 0 (black) to 255 (white). At the moment only the smallest magnification factor (3.65 μm^2/pixel) is used, however five larger factors can (and will) be used in future experiments. Camera and microscope are both fixed on a camera cradle which can be moved in three orthogonal directions with a stepper motor (range 50 mm) for each direction. The specimen is illuminated by two fiber optic arms connected to a light source. Both the motions of the stepper motors and digitizing the

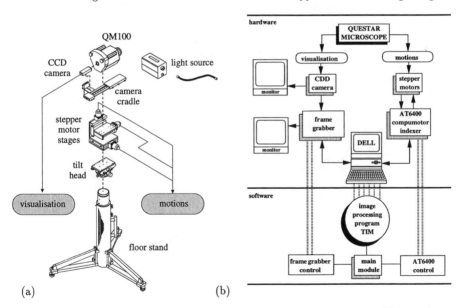

(a)   (b)

Figure 4: Schematic view of the QUESTAR Remote Measurement System (a). The boxes *visualisation* and *motions* refer to the connections with the frame grabber and the four-axis indexer respectively (b).

images is controlled by a DELL personal computer. The stepper motors are connected to a four axis indexer (Compumotor AT6400) and the CCD camera is connected to a variable scan framegrabber. This is schematically shown in Fig. 4b from which it also can be seen that the software for both components (visualisation and motions) is implemented in the image processing program TIM.

With the system described above it is possible to perform real time observations of the fracture process during shrinkage and mechanical loading of the specimen. The system is programmed for automatical scanning a prescribed path. Cracked positions are stored in the computer and the path is adjusted as soon as cracking proceeds.

PRELIMINARY RESULTS

Scanning Initial Cracks

In this section an example will be given in which a lattice is projected on the images of a precracked specimen. Through this procedure a numerical simulation can start at any stage into the fracture process of a specimen. Hopefully this will give more insight in the fracture law parameters of the lattice model. Unfortunately not enough tests have been performed up till now to verify this procedure. In Fig. 5 an overview is given of a precracked specimen containing three Polar White granite aggregates. The overview consists of a coupled set of images retrieved during one scan with the microscope. Area A in Fig. 5 is selected to illustrate the lattice projection and is shown in detail in Fig. 6a. The crack is filtered from

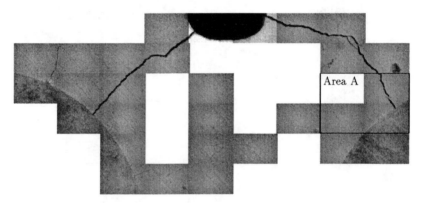

Figure 5: Example of shrinkage cracks in a specimen with three granite aggregates embedded in a cement matrix.

the original images by a threshold value of 128 and a lattice is projected on top of the images (Fig. 6b). Pixels in black represent the crack. Fig. 6c shows the lattice projected on top of the original images. Areas where no images are retrieved, are assumed to be uncracked. Finally beam properties can be changed or a beam can be removed from the mesh when it intersects with the crack. In this example intersecting beams are plotted thicker than the remaining beams, see Fig. 6d. The small bridge between the two crack faces is also expressed in the lattice discretization and contributes to stress transfer, as explained in the first section of the paper. Through inverse analysis we hope to retrieve interface properties.

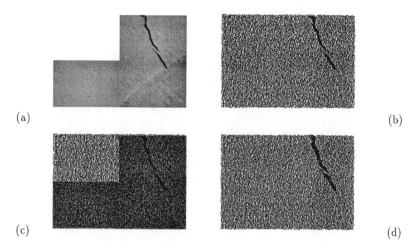

(a) (b)

(c) (d)

Figure 6: Lattice overlay on Area A of the overview in the previous figure.

Simulating Shrinkage Cracking ·

As mentioned no simulations have been performed of a precracked specimen. However it has been attempted to simulate the first stage in the new experiments, i.e. (drying) shrinkage cracking in a specimen with grains embedded in a matrix. No attention has been given to stress-strain behaviour, but only crack growth processes have been evaluated. In the example (Fig. 7) a random lattice with a cell size of s=1 mm is used. The lattice represents a specimen (75x75 mm²) containing 1, 2 or 3 aggregates (Fig 7a, b and c respectively). The specimen is notched (3x10 mm) at half width. The overall stiffness of the matrix beams is 25

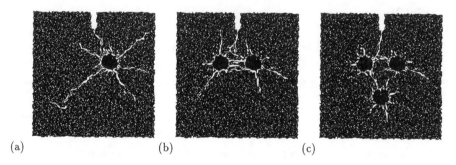

(a) (b) (c)

Figure 7: Example of simulated shrinkage cracking for one (a), two (b) and three (c) aggregates embedded in matrix of cement paste. In each of the patterns shown 500 elements have been removed from the mesh.

GPa, the aggregates have a stiffness of 75 GPa. The strength ratios for the beams are chosen as follows: $f_{t,A}/f_{t,B} = 8$ and $f_{t,A}/f_{t,M} = 4$, where f_t is the tensile strength of the beams in the aggregate (A), matrix (M) and bond between matrix and aggregate (B). Assuming that no shrinkage occurs in the aggregates (compared to the matrix), only a uniform shrinkage

343

load has been applied to the beam elements in the matrix and bond region. Fig. 7 contains the final crack patterns (500 beams removed) of the simulations. It can be seen that mainly radial cracks (perpendicular to the grains) and bond cracks are found in the crack patterns and that the influence of the notch is substantial. Similar behaviour was found in shrinkage experiments with the geometry presented earlier in this paper.

CONCLUSIONS

A new experimental and numerical program is presented in which the bond zone between matrix and aggregate in concrete specimens is studied. The (white) cement paste specimens contain 1, 2 or 3 cylindrical aggregates and are subjected to a drying shrinkage load. The residual strength is determined with a mechanical splitting load. During the tests real time observations are performed using a long distance microscope (QUESTAR QM-100 RMS). Next to the presentation of the new experimental set-up the principle of the numerical lattice simulations is outlined. The analyses will be used for studying the fracture law parameters in the numerical model. Shrinkage cracking is simulated and a lattice is projected on top of a precracked specimen. The results of the simulated crack patterns seem very promising. However attention has to be paid to the stress-strain behaviour of the specimens. This has not been done to date. By simulating precracked specimens we hope to obtain further information about the model parameters.

ACKNOWLEDGEMENTS

The authors greatly acknowledge the expert assistance of Mr. A.S. Elgersma for building up the experimental set-up and carrying out the experiments.

References

[1] S. Mindess. The Application of Fracture Mechanics to Cement and Concrete: A Historical View, *Fracture Mechanics of Concrete*, ed. F.H. Wittmann, Elseviers Science, Amsterdam, (1983), 1-30.

[2] A. Hillerborg, M. Modéer & P.-E. Petersson. Analysis of Crack Formation and Crack Growth in Concrete by Means of Fracture and Finite Elements, *Cem.Concr.Res.*, **6(6)** (1976), 773-782.

[3] G.V. Guinea, J. Planas & M. Elices. Measurement of the Fracture Energy using Three-Point Bend Tests: Part 1-Influence of Experimental Procedures, *Mat.&Struc.*, **25** (1992), 212-218.

[4] E. Schlangen. Experimental and Numerical Analysis of Fracture Processes in Concrete *Ph.D. Dissertation*, Delft University of Technology, (1993).

[5] A. Vervuurt & J.G.M. Van Mier. Experimental and Numerical Analysis of Boundary Effects in Uniaxial Tensile Tests, *Localized Damage III: Computer Aided Assesment and Control*, eds. M.H. Aliabadi et al., Computational Mechanics Publications, Southhampton, (1994), 3-10.

[6] J.G.M. Van Mier. Mode I Fracture of Concrete: Discontinuous Crack Growth and Crack Interface Grain Bridging, *Cem.&Con.Res*, **21(1)** (1991), 1-15.

[7] J.G.M. Van Mier. Crack Face Bridging in Normal, High Strength and Lytag Concrete, in *Fracture Processes in Concrete, Rock and Ceramics*, eds. J.G.M. Van Mier J.G. Rots & A. Bakker, Chapmann & Hall/E&FN Spon, London, 1 (1991), 27-40.

[8] J.G.M. Van Mier, A. Vervuurt & E. Schlangen. Crack Growth Simulations in Concrete and Rock, *Probabilities and Materials. Tests, Models and Applications*, ed. D. Breysse, NATO-ASI Series E Applied Science, Kluwer Academic Publishers, **269** (1994), 377-388.

[9] A. Hrennikoff. Solution of Problems of Elasticity by the Framework Method, *J.Appl.Mech.*, **12** (1941), A169-A175.

[10] H.J. Herrmann, A. Hansen & S. Roux. Fracture of Disordered Lattices in Two Dimensions, *Physical Review B*, **39(1)** (1989), 637-648.

[11] E. Schlangen & J.G.M. Van Mier. Boundary Effects in Mixed Mode I and II Fracture of Concrete, in *Fracture Processes in Concrete, Rock and Ceramics*, eds. J.G.M. Van Mier J.G. Rots & A. Bakker, Chapmann & Hall/E&FN Spon, London, 2 (1991), 705-716.

[12] N.J. Burt & J.W. Dougill. Progressive Failure in a Model Heterogeneous Medium, *J.Engng.Mech.Div. (ASCE)*, **103** (1977), 365-376.

[13] Z.P. Bažant, M.R. Tabbara, M.T. Kazemi & G. Pijaudier-Cabot. Random Particle Model for Fracture of Aggregate or Fiber Composites, *J.Eng.Mech. (ASCE)*, 116 (1990), 1686-1705.

[14] C. Moukarzel & H.J. Herrmann. A Vectorizable Random Lattice, *Preprint HLRZ 1/92, HLRZ-KFA*, Jülich, Germany, (1992).

[15] M.R.A. Van Vliet & J.G.M. Van Mier. Comparison of Lattice Type Fracture Models for Concrete under Biaxial Loading Regimes, *in Proc. IUTAM Symp. on Size-Scale Effects in the Failure Mechanisms of Materials and Structures*, ed. A. Carpinteri, Politecnico di Torino, Torino, Italy, Chapmann & Hall/E&FN Spon London/New York, (1994), (in press).

[16] J.G.M. Van Mier, A. Vervuurt & M.R.A. Van Vliet. Damage Analysis of Brittle Materials: Concrete and Rock, To appear in *Proc. ICM-7 (Mechanics of Materials)*, The Hague (The Netherlands), May 1995.

[17] A. Vervuurt, J.G.M. van Mier & E. Schlangen. Analyses of Anchor Pull-Out in Concrete, *Mat.&Struct.*, **27** (1994), 251-259.

[18] T.T.C. Hsu. Mathematical Analysis of Shrinkage Cracking in a model of Hardened Concrete, *J.Applied Mech.*, (1963), 371-390.

[19] K.M. Lee, O. Buyukozturk, A. Oumera. Fracture Analysis of Mortar-Aggregate Interfaces in Concrete, *J.Eng.Mech.*, **118(10)** (1992), 2031-2047.

[20] H.N. Linsbauer & E.K. Tschegg. Fracture Energy Determination of Concrete with Cube Shaped Specimens, *Zement und Beton*, **31** (1986), 38-40.

[21] A. Vervuurt & J.G.M. Van Mier. An Optical Technique for Surface Crack measurements of Composite Materials, *Recent Advances in Experimental Mechanics*, eds. S. Gomes et al., Balkema Rotterdam, (1994), 437-442.

CHANGES INDUCED BY PVA IN THE CEMENT MICROSTRUCTURE SURROUNDING AGGREGATE

TUN-JEN CHU, JAE-HO KIM, CHEOL PARK, AND RICHARD E. ROBERTSON
Department of Materials Science and Engineering and Macromolecular Science and Engineering
Center, The University of Michigan, 2300 Hayward St, Ann Arbor, MI 48109-2136

ABSTRACT

The presence of poly(vinyl alcohol) (PVA) in portland cement during its hydration was found to induce changes in the microstructure and behavior of the paste that forms around aggregate. The microstructure was studied with scanning electron microscopy in secondary electron, backscattered electron, and EDX mapping modes and with Fourier-transform infrared spectroscopy. The adhesion between cement paste and aggregate was examined with planar aggregate surfaces. With concentrations of the order of 1% by weight of PVA based on the weight of the cement, the deposition of calcium hydroxide on the aggregate surface was found to be diminished and the porous layer that usually surrounds the aggregate to be reduced in thickness. Also, the bond between the aggregate and cement paste was increased enough for the bond to approach the strength of the cement.

INTRODUCTION

The weak link in the mechanical behavior of concrete is the transition zone between the bulk cement paste and the aggregate.[1,2] This zone, which is often found to be approximately 50 μm in thickness, has been shown by scanning electron microscopy to be relatively porous and to consist of a particular sequence of materials.[3] A layer of calcium hydroxide (CH) typically forms on the aggregate, and on this a layer of calcium silicate hydrate (C–S–H) is deposited to form a "duplex" layer.[4] Either depositing on the duplex layer or substituting for it, large calcium hydroxide crystals often form, usually oriented with their c-axes normal to the aggregate surface.[5,6] Ettringite is present in much of the rest of the transition zone.[7] The porosity is greatest near the aggregate and decreases toward the bulk.[8]

A number of measurements have been made of the cement paste-aggregate bond strength. Many of these were conveniently summarized in tabular form recently by Mitsui et al.[9] Though there is some variation between experiments, bond strengths in tension of only 3 MPa have typically been found for limestone aggregate and only 1 MPa for granite aggregate.

The admixing of poly(vinyl alcohol) (PVA) was recently reported to have doubled the pull-out strength and friction of steel fibers pulled from cement paste.[10] The microstructure surrounding the fiber was found to have been considerably altered by the PVA.[11] A ductile, fine-grained interfacial layer had formed around the steel fibers. The formation of this microstructure is suggested to arise from the effect of PVA on the nucleation of CH and C–S–H at the fiber surface and on the presence of polymer around the fibers. In this report, the measurements and observations from a study of the effect of PVA on the transition zone and bond strength between cement paste and aggregate are described

EXPERIMENTAL

The aggregate studied was limestone, which was washed before using. The cement used was type I ordinary portland cement. For the control specimens, the cement was mixed with water in the ratio w/c = 0.35. For the PVA-modified specimens, the cement was mixed with a solution of PVA in water in the same ratio, 0.35, of PVA solution to cement. The PVA solution consisted of 4 wt% PVA in water. The PVA used was Airvol-203 (Air Products and Chemicals, Inc.), which was 87–89% hydrolyzed poly(vinyl acetate) in granular form. Its molar mass was 7,000–13,000 number average, 13,000–23,000 mass average. After mixing, the PVA constituted 1.4 parts per one hundred parts cement or slightly over 1 wt% of the combined mass of cement plus water.

Two types of specimens were made, and for both, the cement was allowed to harden for 28 d at saturated humidity. In the first type of specimen, the aggregate was cut and ground to generate flat surfaces on which the cement/water and cement/PVA-water solution mixtures were cast. After washing the ground surfaces and casting the cement against them, these specimens were sealed in plastic containers and allowed to harden for 28 d. After the 28 d, some of these specimens were sectioned with a diamond saw and examined with a polarizing optical microscope (Olympus BH-2). Other specimens were cleaved with a razor blade inserted along the interface, and a layer of the cement paste within 10 μm of the aggregate interface was collected from each for analysis by a Fourier-transform infrared (FTIR) spectrometer (Mattson Galaxy Series FTIR 3020) that was used in a diffusive reflectance mode.

In the second type of specimen, the washed and dried aggregate was embedded in the cement slurries. After 28 d, the material was sectioned, and pieces were immersed in a low viscosity epoxy resin (Epon 815/Epon V-40 (Shell Chemical Company) mixed in the ratio of 100/43) to allow the specimens to become impregnated with the resin. The resin was hardened at 25°C for 4 days. The impregnated aggregate-cement surfaces were then polished with 12 μm alumina paper and 1 μm alumina powder. The non-polar solvent kerosene was used as a coolant for the final polishing step and to wash the specimens in an ultrasonic cleaner between polishing steps. After coating the dried specimens surfaces with carbon, the specimens were examined with a scanning electron microscope (SEM) (Hitachi S–570) equipped with detectors for secondary and backscattered electrons and energy-dispersive X-ray spectroscopy (EDX).

RESULTS

Cement-Aggregate Bond

To test the bond strength between the cement and aggregate, large limestone rocks was sectioned and ground, and the cement pastes were applied to these surfaces. Cements both with and without

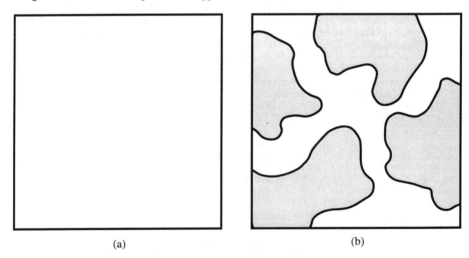

(a)   (b)

Figure 1.   A graphic showing the two types of bond failures between cement paste and a flat aggregate (limestone) surface. (a) Cement paste without PVA: adhesive failure over the whole surface. (b) Cement paste with PVA: adhesive failure over less than 50% of the surface; cohesive failure in the cement phase over the rest.

PVA were examined. The flat aggregate surface removed essentially all of the mechanical locking contribution to bond strength. After allowing the paste to harden in a moisture-saturated atmosphere for 28 d, the specimens were tested.

Cement-aggregate interfaces were cleaved by inserting a razor blade into the aggregate-cement bond at the interface. The bond for specimens without PVA separated very easily without any debris from the opposite surface being left on the two opposing surfaces. On the other hand, the bond for specimens with PVA was not easily broken, and the separation of the two parts occurred by the mixed modes of adhesive failure at the interface and cohesive failure in the cement phase. A schematic representation of these two types of failure is shown in Figure 1.

Transition Zone Morphology

Optical microscopy. After undergoing the hardening reactions for 28 days, the flat aggregate-cement composite specimens, with and without PVA, were sectioned with a diamond saw to expose the aggregate-cement transition zones. The cut surfaces were washed and dried and observed using reflected polarized light.

Specimens without PVA usually had white interfacial bands having a thickness of 50–100 μm around the aggregate. In these specimens, if cracks developed, they usually ran around the aggregate, along the interface.

The morphology was more complex in specimens with PVA. Much of the aggregate surface was in contact with thin (10–15 μm) white bands, though some parts of the surface seem to be in direct contact with bulk cement. In these specimens, if cracks developed, they often ran around the aggregate in the cement phase, though they sometimes ran along the interface.

These two types of morphology are shown in Figure 2.

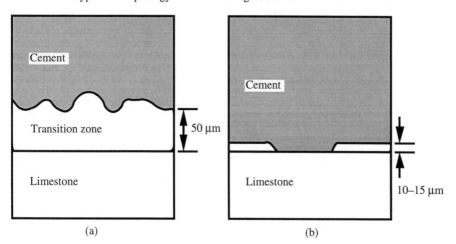

(a)   (b)

Figure 2.   Graphics showing the general size of the transition zone between the cement and aggregate phases: (a) without PVA and (b) with PVA.

Scanning electron microscopy. The specimens used for SEM examination were aggregate-cement mixtures that had been hardened for 28 d and sectioned with a diamond saw to expose the aggregate-cement transition zones. The sectioned specimens were impregnated with epoxy, which was then hardened. The excess hardened epoxy was removed and the surface polished before being coated with carbon.

Scanning electron micrographs of the transition zones at a modest magnification are shown without and with PVA in Figures 3(a) and 3(b), respectively. The limestone aggregate is at the bottom, and the cement is above it in both micrographs. These images were obtained with the microscope operating in backscatter mode. Particularly noteworthy are the thicknesses of the

| 50 μm | | 50 μm |

Figure 3.   SEM micrographs obtained in backscatter mode of cut and polished specimens showing the transition zone between the aggregate and the cement. (a) Cement without PVA. (b) Cement with PVA.

transition zones. In Figure 3(a), without PVA, the transition zone is about 50 μm thick. A more extensive survey of the transition zone showed that its thickness, though often 50 μm, could range up to 100 μm. In Figure 3(b), with PVA, the transition zone is about 15 μm. A more extensive survey showed that the transition zone thickness with PVA ranged from 0 to 30 μm and averaged around 15 μm. Seen also in these micrographs are the bright unreacted cement grains (the brightest of these being tricalcium silicate (C_3S)), calcium hydroxide at the surface of the aggregate (next in brightness), and calcium silicate hydrate (C–S–H) rims surrounding the C_3S.

The transition zones observed with SEM are essentially the same as observed optically and are represented in Figure 2.

Compositional Changes in the Transition Zone

Fourier-transform infrared analysis (FTIR). The cement phases were analyzed by FTIR. Specimens from both the bulk and from within 10 μm of the interface with the aggregate were examined. The interface was locatable because the aggregate had been sectioned and ground flat before the cement paste was packed against it.

In the FTIR diffuse reflectance spectra shown in Figure 4, the sharp peak at 3643 cm⁻¹ arises from calcium hydroxide (CH), the absorption at 1400-1550 cm⁻¹ arises from calcium carbonate (CC), and the peak around 950–1000 cm⁻¹ arises from silicate ions (S). As the silicate becomes hydrated, its absorption moves upward in wavenumber (cm⁻¹). (The limestone aggregate contributed much if not all of the calcium carbonate.) The absorption spectra from the bulk with and without PVA, Figures 4(a) and 4(c), are seen to be very similar, with strong calcium hydroxide and silicate peaks and a modest amount of calcium carbonate. The absorption spectrum from the cement without PVA, 0–10 μm from the aggregate surface, Figure 4(b), is considerably

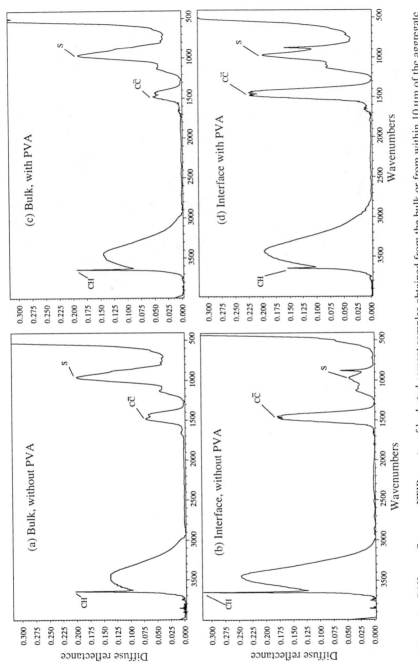

Figure 4. Diffuse reflectance FTIR spectra of hydrated cement samples obtained from the bulk or from within 10 μm of the aggregate interface. (a) Bulk, without PVA. (b) Interface, without PVA. (c) Bulk, with PVA. (d) Interface with PVA.

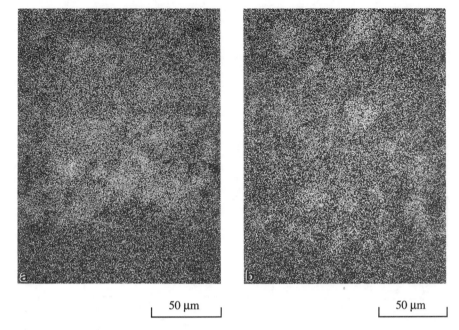

| 50 μm | 50 μm |

Figure 5. Calcium map of the images shown in Figure 3. (a) Cement without PVA. (b) Cement with PVA.

enhanced in calcium hydroxide, especially, and calcium carbonate, but is largely depleted of silicate. In contrast, the absorption spectrum from the cement with PVA, 0–10 μm from the aggregate surface, Figure 4(d), has a concentration of silicate more nearly like that of the bulk, much more calcium carbonate, and perhaps one-third to one-half as much calcium hydroxide as the bulk. The enhanced calcium carbonate in the latter specimen is thought not to have arisen from CO_2 in the air, because each of the specimens was treated the same and were minimally exposed to air, nor from chips of the aggregate left on the cement surface following separation of cement and aggregate, but from an etching of the aggregate.

The ratios of the peak heights for the calcium hydroxide, calcium carbonate, and silicate are given in Table I. (The broad absorption below that of calcium hydroxide is due to water, and this is subtracted from calcium hydroxide peak height.)

Table I. FTIR peak intensity ratios from the cement bulk and interfacial regions, with and without PVA.

Specimen	I_{CH}/I_S	I_{C-C}/I_{CH}	I_{C-C}/I_S
Bulk, without PVA	0.57	0.63	0.36
Bulk, with PVA	0.58	0.47	0.27
Interface, without PVA	4.10	0.90	3.70
Interface, with PVA	0.30	3.80	1.13

EDX analysis. The same images shown in Figures 3 were analyzed for the elements calcium and silicon with an EDX instrument. The results are shown in two forms, as a mapping and as

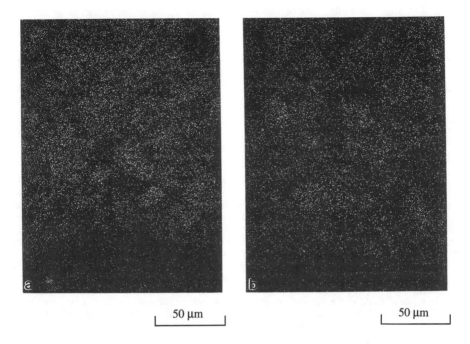

| 50 μm | 50 μm |

Figure 6.   Silicon map of the images shown in Figure 3. (a) Cement without PVA. (b) Cement with PVA.

elemental peak height averaged over small regions.

The calcium maps for the images in Figure 3 are shown without and with PVA in Figure 5, and the corresponding silicon maps are shown in Figure 6. Figure 5(a) shows that without PVA, the density of calcium ion in the transition zone can be comparable to that of the high calcium areas in the bulk, despite the porosity of the transition zone. Though the aggregate is limestone (calcium carbonate), the apparent density of calcium in the transition zone is greater than that in the aggregate. With PVA (Figure 5(b)), the apparent calcium density in the transition zone is about the same as that in the aggregate and in many areas of the bulk. Neither a depletion of calcium in the transition zone nor a concentration of calcium, as seen without PVA, were seen with PVA.

Though the silicon maps are less distinct than those of the calcium, one can see that without PVA (Figure 6(a)), the silicon is depleted in those regions near the aggregate surface having a high calcium concentration. Also elsewhere in the transition zone, the silicon concentration is on average less than most regions of the bulk. With PVA (Figure 6(b)), the silicon concentration in the transition zone is similar to the background silicon concentration of the bulk (excluding the tricalcium silicate grains).

The atomic ratios of calcium to silicon taken from small regions (10 μm by 100 μm) of the cement are given in Table II. (The ratios in Table II are peak height ratios corrected by multiplying these ratios by the factor that gave a Ca/S ratio of three for the tricalcium silicate regions.) The Ca/Si ratios were essentially the same for the bulk, without and with PVA, as would be expected. Without PVA, the Ca/Si ratio at the interface was almost a third higher than that in the bulk, but with PVA, the ratio at the interface was about a third less than that in the bulk.

Table II. Atomic ratios of calcium to silicon from selected regions of the cement phase obtained from EDX.

Specimen	Ca/Si
Bulk, without PVA	3.3_2
Bulk, with PVA	3.3_5
Interface, without PVA	4.3_0
Interface, with PVA	2.4_0

DISCUSSION

The addition of PVA to aggregate-cement mixtures has induced a number of changes. The aggregate-cement failure mode was changed. Judged from the fact that the crack often deviated from the interface into the cement, the bond seems to have approached in strength that of the cement. The morphology and composition of the transition zone were also changed. The zone was decreased in thickness, and the ratio of calcium silicate hydrate (C–S–H), which is presumed to be the source of the silicate in the transition zone, to calcium hydroxide (CH) was increased. Also, there was less tendency for large crystals of calcium hydroxide to deposit on the aggregate surface.

Previous work[12] suggested that the presence of PVA during the hydration of cement may increase the amount of C–S–H while it decreased the amount of CH. The effect was suggested to arise from the effect of PVA on nucleation. An increase in the absolute amount of C–S–H was not able to be confirmed in the present experiments, though an increase in the relative amount of C–S–H in the transition zone seems to have occurred. The FTIR spectra indicated a near doubling of the silicate to calcium hydroxide ratio in the transition zone, and the silicon map in Figure 6(b) indicates a near constant silicon concentration in going from the bulk to the transition zone, in spite of the growing porosity.

The increase in adhesion with PVA may have arisen partly by displacement of CH deposits from the aggregate interface and partly by the etching and involvement of calcium carbonate in the bonding. Though they seem to conform to the surface, the CH deposits seem not to adhere to it very strongly. Without the preponderance of CH at the surface, the C–S–H is able to bond to the surface, and the PVA itself may contribute to the bond. Also, the development of adhesion by the etching of limestone by cement without PVA after 56 d of curing that was reported by Mehta and Monteiro[2] seems to be accelerated by PVA.

The shrinkage of the transition zone may arise from better compaction of the cement grains about the aggregate that is provided by a lubricating (superplasticizing) effect of the PVA. By bringing the cement bulk closer to the aggregate, the concentration of C–S–H in the transition zone and the average density of the transition zone are expected to be increased.

CONCLUSIONS

The bond strength between cement and limestone aggregate can be considerably increased by the addition of 1.4 wt% PVA, based on the weight of the cement. The gain in strength seems to arise from several or all of these three changes. (1) The thickness of the transition layer is significantly reduced, possibly by a superplasticizing effect of the PVA. (2) The calcium hydroxide crystals normally present at the aggregate surface are reduced or eliminated and possibly replaced by calcium silicate hydrate (C–S–H) crystals. (3) The etching of the limestone aggregate by the cement seems to be accelerated with PVA, and a possible role in the adhesion may be played by the dissolved calcium carbonate. In addition, the PVA itself may contribute to the bond.

Acknowledgement

This study was supported by the NSF Center for Science and Technology for Advanced Cement-Based Materials (ACBM), NSF Grant No. DMR-8808432.

References

1. P.K. Mehta and P.J.M. Monteiro, *Concrete: Structure, Properties, and Materials,* 2nd ed. (Prentice Hall, Englewood Cliffs, NJ, 1993), pp 36–41.
2. P.K. Mehta and P.J.M. Monteiro, in *Bonding in Cementitious Composites,* edited by S. Mindess and S.P. Shah (Mater. Res. Soc. Proc. **114**, Pittsburgh, PA, 1988) pp. 65–75.
3. A. Bentur, S. Diamond, and S. Mindess, *J. Mater. Sci.,* **20**, 3610 (1985).
4. B.D. Barnes, S. Diamond, W.L. Dolch, *Cement Concr. Res.,* **8**, 233 (1978).
5. P.J.M. Monteiro, J.C. Maso, and J.P. Ollivier, *Cement Concr. Res.,* **15**, 953 (1985).
6. P.J.M. Monteiro and P.K. Mehta, *Cement Concr. Res.,* **15**, 378 (1985).
7. S. Diamond, in *Microstructural Development During the Hydration of Cement,* edited by L.J. Struble and P.W. Brown (Mater. Res. Soc. Proc. **85**, Pittsburgh, PA, 1987) pp. 21–31.
8. K.L. Scrivener, A. Bentur, and P.L. Pratt, *Adv. Cement Res.,* **1**, 230 (1988).
9. K. Mitsui, Z. Li, D.A. Lange, and S.P. Shah, *ACI Mater. J.,* **91**, 30 (1994).
10. H. Najm, A.E. Naaman, T.-J. Chu, and R.E. Robertson, *Adv. Cement-Based Mater.,* **1**, 115 (1994).
11. T.-J. Chu, R.E. Robertson, H. Najm, and A.E. Naaman, *Adv. Cement-Based Mater.,* **1**, 122 (1994).
12. J. K. Kim, R.E. Robertson, and T.-J. Chu, to be published.

THE INTERFACIAL TRANSITION ZONE AND ITS INFLUENCE ON THE FRACTURE BEHAVIOR OF CONCRETE

DAVIDE ZAMPINI*, HAMLIN M. JENNINGS**, AND SURENDRA P. SHAH*
*Department of Civil Engineering
**Departments of Materials Science and Civil Engineering
Northwestern University
NSF Center for Advanced Cement Based Materials
Evanston, IL 60208, U.S.A.

ABSTRACT

Concrete specimens of constant water-to-cement ratio and varying amounts of gravel aggregate were tested under 3-point bend. The fracture toughness of the composite and surface roughness of the paste are determined. Fracture parameters obtained from the Two Parameter Fracture Model (TPFM) such as the critical stress intensity factor (K_{IC}) and critical effective crack extension (Δa_c) are found to be related to the average surface roughness of the paste. The Interfacial Transition Zone (ITZ) and the bulk paste are distinguished by different values of roughness. The surface roughness of the area adjacent to the aggregate particle is evaluated as a function of the distance from the aggregate surface. In particular, the surface roughness of the paste near the aggregate is greater than that of the paste far from the aggregate, and it decreases with distance from the aggregate. The higher roughness of the paste near the aggregate indicates that aggregate particles, and more specifically the ITZ associated with them, act to toughen the paste in concrete.

INTRODUCTION

Jacques Faran first identified the "aureole de transition"[1], which more recently has been recognized as the Interfacial Transition Zone (ITZ). Most of the studies related to the ITZ tend to fall into two general categories: characterization of microstructure and mechanical properties. Ultimately, the objective is to understand how the microstructure of the ITZ influences mechanical properties in concrete. However, this goal is difficult to achieve since frequently the ITZ is that of model systems in which paste is cast against a flat aggregate[2,3,4,5] or a glass slide[6]. In cases where the microstructure of the ITZ has been characterized in systems representative of actual concrete, the influence on mechanical properties is inferred based on the observed microstructural features[7]. Thus, the findings that relate the ITZ's microstructure to mechanical properties are indirect, and the mechanisms by which aggregate particles and their associated ITZ's contribute to the fracture process in concrete are far from being completely understood.

Microstructural studies of the ITZ have predominantly considered the porosity[7,8,9], the formation and orientation of CH crystals[10,11,12], and the formation and distribution of anhydrous products[8,9,13]. The gradients of these microstructural features, measured with respect to the distance from the aggregate surface, are used to quantify the size of the ITZ. These studies report a thickness of the ITZ between 40µm and 50µm. The porosity and anhydrous natures of ITZ's are quantified through image analysis techniques[8,9,13]. The orientation and distribution of CH crystals have been measured through X-ray diffraction and pole figures[14]. On the other hand, mechanical properties, such as the bond strength between aggregate and paste, have been measured through microhardness tests[15], push-out tests[16], or sandwich specimen tests[2,3,4].

The motivation for the present study is to examine how the aggregate particles interact with the paste to influence the fracture behavior of concrete. In this investigation, it is evident that surface roughness of the paste is a microstructural parameter that can be used to characterize the ITZ. Roughness is used as a direct means of evaluating the influence of the paste-aggregate bond on the fracture behavior of concrete. The paste roughness is related to fracture toughness. In addition, the roughness of the paste is examined as a function of distance from the aggregate surface.

EXPERIMENTAL APPROACH

Experimental Matrix

The primary variable between specimens was the aggregate content. The gravel was sieved so that all of the particles were smaller than 4.75mm and greater than 2.36mm. Four types of specimens were cast, each with a water-to-cement ratio (w/c) of 0.45, and the following aggregate-to-cement ratios (A/C): 0.0-specimen C (control paste), 0.5-specimen AA05, 1.0-specimen AB10, and 2.5-specimen AC25. Each group of specimens consisted of 4 beams. All beams were cured for 28 days in a lime bath prior to mechanical testing. A summary of the experimental matrix is shown in Table I.

Table I. A summary of the parameters considered in the design of the experimental matrix.

Specimen	Aggregate type	Aggregate size (mm)	Water-to-Cement ratio (W/C)	Aggregate- to-Cement ratio (A/C)	Specimen age (days)
C0	none	--	0.45	0.0	28
AA05	gravel	2.36 - 4.75	0.45	0.5	28
AB10	gravel	2.36 - 4.75	0.45	1.0	28
AC25	gravel	2.36 - 4.75	0.45	2.5	28

Mechanical Testing

The specimens that were tested consisted of notched beams having a loaded span (s) to height (h) ratio of 4. Details of the specimen geometry are given in Figure 1. Three-point bend tests were carried out on each of the beams as specified by Jenq and Shah's Two Parameter Fracture Model (TPFM)[17]. A load versus crack mouth opening displacement curve (CMOD) is acquired from the mechanical testing (see Figure 2). Information such as the peak load (P_{max}), the initial compliance (C_i) and unloading compliance (C_u) is extracted from the curve. The values obtained from the graphs are used to derive fracture parameters, such as critical stress intensity factor (K_{IC}) and the critical effective crack extension (Δa_c) based on the TPFM.

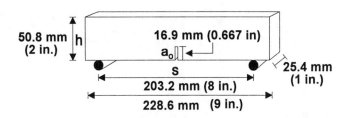

Figure 1. Geometry and dimensions of the beams tested under 3-point bend.

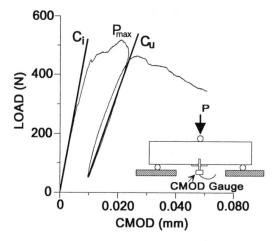

Figure 2.   A typical Load versus CMOD curve obtained from 3-point bend tests. The maximum load (P_{max}), initial compliance (C_i), and unloading compliance (C_u) are obtained from the graph and used as input for the TPFM.

Confocal Microscopy

Fracture surfaces were examined using a Confocal Tandem Scanning Reflected Light Microscope. The fracture surface was optically sectioned to form a two-dimensional digitized topographic image (Figure 3), employing a method previously implemented by Lange et al[18]. Three dimensional information was acquired at a magnification of 600X (Figure 4). The high magnification made it possible to isolate the paste regions and focus directly on the areas surrounding the aggregate particles (i.e. the ITZ). The images consisted of a 512 pixels x 480 pixels picture with a dimension of 146μm x 137μm (pixel size = 0.285μm). The height of the slices of the fracture surface were taken at 4μm intervals, starting from the highest point on the surface and moving progressively towards the deepest point.

Figure 3. A two-dimensional topographic image of the paste fracture surface.

Figure 4. A three-dimensional image of the paste fracture surface.

Roughness Analysis

The roughness of a fracture surface is evaluated by the Roughness Number (RN), which is given by ratio of the apparent actual surface area to nominal projected surface area:

$$RN = \frac{apparent\ actual\ surface\ area}{nominal\ projected\ surface\ area} \qquad (1)$$

The apparent surface area is computed by summing the area of triangles formed by connecting the z-height of adjacent pixel points, as shown in Figure 5. In this study, the roughness of the paste was evaluated in two stages:

Stage 1 - Roughness of the paste was randomly evaluated for 10 different regions of the fracture surface and averaged. No consideration was given to the location of the area and its respective distance from the aggregate particles.

Stage 2 - Roughness was evaluated as a function of distance from the aggregate surface. In this analysis, a computer program, previously developed by Lange et al[19], was modified so that the RN could be evaluated for smaller subregions having a dimension of 146μm x 32μm.

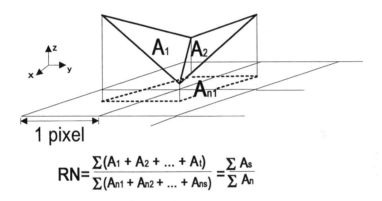

$$RN = \frac{\sum(A_1 + A_2 + ... + A_t)}{\sum(A_{n1} + A_{n2} + ... + A_{ns})} = \frac{\sum A_s}{\sum A_n}$$

Figure 5. A schematic diagram of how Roughness Number (RN) is computed (ref. 19).

RESULTS AND DISCUSSION

Roughness of the Paste

The roughness of a fracture surface is a microstructural parameter that reflects the amount of energy absorbed during crack propagation. A more tortuous surface is indicative of a material having a higher fracture toughness. Also, the roughness of a fracture surface in concrete is the "post-mortem" evidence of how the individual components interact microstructurally to yield the resulting surface fracture. The results from the Stage 1 analysis show that the Roughness Number (RN) of the paste increases with increasing volume of aggregate particles. This is illustrated in Figure 6. Values for the paste Roughness Number, summarized in Table II, ranged from 3.27 for the control specimen (consisting of paste only) to 4.35 for the specimen with the highest aggregate content (A/C = 2.5).

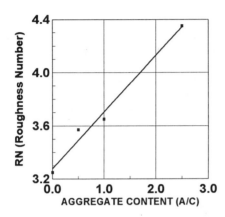

Table II. Values of the paste average RN evaluated for concrete of different aggregate-to-cement ratios (A/C).

A/C	Roughness Number (RN)
0.0	3.27
0.5	3.58
1.0	3.65
2.5	4.35

Figure 6. Roughness Number of the paste plotted as a function of aggregate content.

A more detailed analysis was carried out (Stage 2) to evaluate RN as a function of distance from the aggregate surface. A series of three, adjacent, topographic images were acquired in a sequence moving away from the aggregate surface. Each topographic map was subdivided into smaller areas, or subregions. The Roughness Number of the subregions, spaced at 32μm intervals, was evaluated and plotted against distance from the aggregate. The results, shown in Figure 7, indicate that for the specimens containing aggregate, the RN was greater near the aggregate, and decreased with the distance from the aggregate surface. An interesting trend is observed from the graph: the RN's near the aggregate tend to converge to a value of 4.3, and at distances removed from the aggregate the RN converges to a value of 3.2. The RN of areas far away from the paste-aggregate, interface is similar to the RN of neat paste (control), whereas, the RN at the interface is similar to that obtained for specimens having the highest aggregate content. The latter observation indicates that, on a random basis, the probability of including an interfacial region in the RN evaluation is proportional to the aggregate content.

If the size of the ITZ is defined as the distance from the aggregate surface at which the RN begins to level off, then the measured size is between 96μm and 128μm (see Figure 7). Thus, the size of the ITZ reported in this study is about double the size cited in the literature. Obviously, the size of the ITZ as measured by surface roughness has a different meaning than that deduced from other types of microstructural characterization. Measurement of the ITZ by way of surface roughness is a characterization of both microstructure and the distribution of stress during the fracture process. Roughness, therefore, characterizes the ITZ in terms of a mechanical property, and not simply a microstructural feature. Unlike the other methods of characterizing the ITZ, roughness measures the extent by which the interface influences the bulk paste, and this influence extends beyond the boundaries delineated by changes in microstructural features. More importantly, the aggregate particles, present in concrete, are believed to alter the distribution of stresses in the paste. Thus, roughness is a parameter that is sensitive to the changes in the state of stress induced by the aggregate particles, and possibly by its size, shape, and spatial distribution. Of more significance is the fact that roughness is a means of characterizing the influence that aggregate particles, and their associated ITZ's, have on fracture in a 3-dimensional sense. Surface roughness distinguishes the ITZ from the bulk paste. Since the ITZ is rougher than the bulk paste, its presence represents the extent by which the bulk paste is toughened.

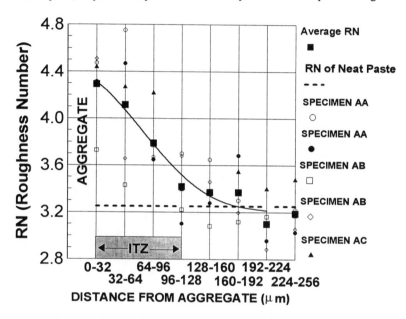

Figure 7. Roughness Number of the paste evaluated as a function of distance from the aggregate surface.

The ITZ and Its Role in Fracture of Concrete

Fracture parameters, such as the critical stress intensity factor (K_{IC}) and critical crack length extension (Δa_c), were evaluated using the Two Parameter Fracture Model (TPFM) and related to RN. The K_{IC} of the composite is related to the roughness of the paste in Figure 8. It is evident from the graph in figure 8 that K_{IC} increases with increasing RN of the paste. The presence of the ITZ increases the average roughness of the paste and, therefore, increases its toughness and the toughness of the composite. This provides new insight into the influence of the aggregate on the fracture process. The aggregate does not simply deflect the cracks and, thereby, create additional surface area, but it also fundamentally alters the fracture path near the interface. The path in the paste is made more tortuous by the close proximity of an aggregate particle. Thus, even though an increase in aggregate particles means less paste, the roughness of that paste is related directly to the toughness of the sample. The presence of stiffer aggregate particles and their associated ITZ's act to modify the state of stress in the paste.

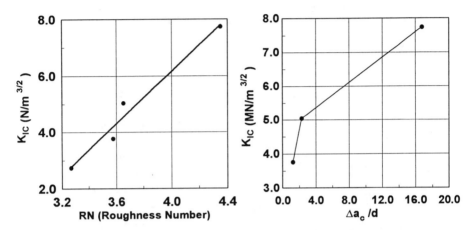

Figure 8. Fracture toughness as measured by K_{IC} and its relationship to the RN of the paste.

Figure 9. Fracture Toughness (K_{IC}) and its relationship to the critical crack extension to particle spacing ratio ($\Delta a_c/d$).

In order to develop a better insight on how the ITZ influences the fracture behavior in concrete, a connection should be made relating fracture parameters to the physical process of crack propagation. The critical crack extension ($\Delta a_c = a_c - a_o$, where a_c is the critical crack length and a_o is the initial notch length) is an approximate value for the extension of a crack prior to its rapid propagation across the specimen. Using this parameter and the surface-to-surface spacing between aggregate particles (d) a relationship between Δa_c and the ITZ can be established. The theoretical average surface-to-surface particle spacing is calculated by assuming a close-pack arrangement of particles having a coordination number of 12. The results of these calculations and of the fracture parameters obtained for each series of tests are summarized in Table 3. The ratio of the critical crack extension to particle spacing ($\Delta a_c/d$) is an estimation of the fraction of the critical crack extension that passes through an ITZ. As shown in Figure 9, the fracture toughness of the concrete specimens increased with increasing $\Delta a_c/d$. From the results of the roughness analysis, it is difficult to establish whether a crack initiates in the bulk paste or the ITZ. However, since the bulk paste is less tough than the ITZ, crack propagation is more prevalent in the bulk paste. As the crack grows and propagates through the bulk paste, a greater fraction of the critical crack length (Δa_c) will be forced into the ITZ as the distance between aggregate particles decreases. In specimens C, AA05, and AB10, the Δa_c values do not differ considerably, however, the propagating crack has to traverse a greater fraction of ITZ's as the volume of aggregate particles increases in each of the respective specimens. The aggregate spacing is the smallest in specimen AC25, therefore, not only is Δa_c attained through a higher degree of interaction with ITZ's but, to some extent, it is subjected to the energy-consuming process of deflection caused by the aggregate particles.

Table III. A summary of the fracture parameters evaluated from the TPFM, and the theoretical aggregate particle surface-to-surface spacing resulting from changes in aggregate volume.

Specimen	K_{IC} (MN/m$^{3/2}$)	Δa_c (mm)	Theoretical Aggregate spacing - d (mm)
C0	274	2.43	--
AA05	376	2.42	2.00
AB10	504	2.57	1.12
AC25	774	6.37	0.38

CONCLUSION

This study shows that microstructure and the fracture toughness of concrete are related. Experiments have been performed with the aim of establishing the influence that aggregate particles have on the surface roughness of the paste portion in concrete. The fracture toughness of concrete is found to be sensitive to the presence of an Interfacial Transition Zone (ITZ). Results indicate that roughness of the paste increases with increasing volume of aggregate, and this is attributed to the increased presence of an ITZ. Near the aggregate, the fracture surface of the cement paste is shown to have a greater roughness than it does at distances greater than 100µm. Thus, the roughening of the paste, due to the ITZ, is an important toughening mechanism in concrete. It is also noted that, in terms of roughness, the size of the ITZ is greater than that defined in terms of microstructural features such as porosity, anhydrous product distribution, and orientation and distribution of CH crystals. This indicates that roughness characterization of the paste also measures the altered state of stress brought about by the presence of aggregate particles and their associated ITZ's. Quantitative roughness analysis of the bulk paste and the ITZ provides a basis for improving our understanding of the fracture mechanisms responsible for the toughening of concrete.

ACKNOWLEDGEMENT

The National Science Foundation (NSF) and its Center for Advanced Cement-Based Materials are greatly acknowledged for the support of this research (Grant DMR 9120002/03).

1. J. Farran, Rev. Mater. Constr. Trav. Publ. **490-491**, 155 (1956); **492**, 191 (1956).

2. L. Struble and S. Mindess, The International Journal of Cement Composites and Lightweight Concrete **5**(4), 79 (1983).

3. O. Buyukozturk and K.M. Lee, Cement and Concrete Composites **15**, 143 (1993).

4. C.A. Langton and D.M. Roy, in Morphology and Microstructure of the Cement Paste/Rock Interfacial regions (7th International Congress on the Chemistry of Cement **3**,Paris, Editions Septima, Paris, France, 1980,1981) pp. VII 127-VII 132.

5. C. Gibergues, J. Grandet, and J.P. Ollivier, in Bond in Concrete, edited by P. Bartos (Applied Science Publishers, London, 1982) p. 24.

6. B.D. Barnes, S. Diamond and W.L. Dolch, Cement and Concrete Research **8**, 233 (1978).

7. K.L. Scrivener and P.L. Pratt in A Preliminary Study of the Microstructure of the Cement/Sand Bond In Mortars (8th International Congress on the Chemistry of Cement, Rio de Janeiro **3**, Abla Grafica e Editora, Rio de Janeiro, Brazil, 1986) pp. 466-471.

8. K.L. Scrivener, A. Bentur and P.L. Pratt, Advances in Cement Research **1**(4), 230 (1988).

9. J.P. Ollivier and M. Massat, in Advances in Cement and Concrete, edited by M.W. Grutzeck and S.L. Sarkar (American Society of Civil Engineers, New York, 1994) p.1994.

10. J.A. Larbi and J.M.J.M. Bijen, Cement and Concrete Research **20**, 461 (1990).

11. P.J.M. Monteiro, J.C. Maso and J.P. Ollivier, Cement and Concrete Research **15**(2), 953 (1985).

12. P.J.M. Monteiro and C.P. Ostertag, Cement and Concrete Research **19**, 987 (1989).

13. K.L. Scrivener and E.M. Gartner in Bonding in Cementitious Composites, edited by S. Mindess and S.P. Shah (Mater. Res. Soc. Proc. **114**, Boston, MA, 1987) pp. 77-85.

14. R.J. Detwiler, P.J.M. Monteiro, H.R.Wenk and Z. Zhong, Cement and Concrete Research **18**, 823 (1988).

15. M. Saito and M. Kawamura, Cement and Concrete Research **16**, 653 (1986).

16. K. Mitsui, Z. Li, D.A. Lange and S.P. Shah, ACI Materials Journal **91**, 30 (1994).

17. Y. Jenq and S.P. Shah, Journal of Engineering Mechanics **111**(10), 1227 (1985).

18. D.A. Lange, H.M. Jennings, S.P. Shah, J. Am. Ceram. Soc. **76**, 589 (1993).

19. D.A. Lange, H.M. Jennings, and S.P. Shah, J. of Mater. Sc. **28**, 3879 (1993).

FURTHER STUDIES ON FRACTURE PROPERTIES
OF PASTE/ROCK INTERFACES

MARK G. ALEXANDER* AND ANDREAS STAMATIOU**
*Department of Civil Engineering, University of Cape Town, 7700 Rondebosch, South Africa
**Department of Civil and Environmental Engineering, University of the Witwatersrand,
   P O WITS, 2050, South Africa

ABSTRACT

The results of fracture tests on rocks, pastes, and paste/rock interfaces are presented. Variables included rock type (andesite, dolomite, and granite), w/c ratio (0.3 - 0.55) and age of test (3 - 91 d). The results illustrate the variability inherent in toughness measurements on naturally fractured rough rock interfaces, but also indicate that interface toughness is not necessarily intrinsically lower than that of pure paste.

REVIEW OF PREVIOUS WORK

Work has been underway for about 4 years on an experimental study of "artificial" rock/paste interfaces. The study arose in an attempt to help explain the aggregate-dependency of various mechanical properties of concrete.[1] Details of experimental methods are given in refs. 2, 3 and the studies are reported in refs. 4, 5.

Experimental Method

A cylindrical bend specimen based on the ISRM Method 1[3] has been selected for paste/rock interface tests. The rock specimen is obtained by coring. This method produces "naturally" fractured rock surfaces against which paste can be cast to produce an interface for testing. Subsequently, the rock surfaces can be ground or polished to remove the effects of macro-texture and re-tested. Figure 1 gives details for the ISRM Method 1 test on chevron-notched cylindrical specimens, which in this work are typically 42 mm in diameter.
   Parameters yielded by the test are :
• Fracture toughness K_{CB} (which can be corrected for non-linearity if required)
• Work of fracture (or fracture energy) R_{CB}
• Elastic modulus E
The usual procedure is to undertake rock and hardened paste tests, followed by interface tests, which involve casting paste against previously fractured rock half-cores to produce composite specimens.

Previous Results

Table 1 summarises previously measured 28 d K_{CB} and R_{CB} (and E) values from tests on paste, rock and composite beams. The w/c ratio used was 0.3, and 3 rock types are represented, all used commercially for crushed rock aggregates in South Africa. The table indicates that paste values may be lower than interface values (at w/c = 0.3), that use of silica fume generally causes embrittlement, that rock values are very much higher than paste values, and (by comparison with other investigators[4]) that use of rough rock surfaces yields significantly higher fracture values than sawn or polished specimens. In fact, rock surface roughness is, in our opinion, probably the single most important factor in the values measured.

TABLE I - Values from previous fracture tests (28 d)

		Measured		
		K_{CB} (MN/m$^{1.5}$)	R_{CB} (J/m^2)	E (GPa)
Paste (w/c=0.3)	OPC	0.54	14.9	24.1
	OPC+SF	0.50	9.1	24.4
Rock	andesite	3.49	147.1	99.7
	dolomite	2.23	59.9	112.6
	granite	2.22	162.6	75.2
Rock/paste interface	andesite/OPC	0.80	24.6	45.6
	andesite/OPC+SF	0.71	14.5	38.2
	dolomite/OPC	0.57	7.5	46.8
	dolomite/OPC+SF	0.64	9.9(65d)	41.7
	granite/OPC	0.63	25.2	-

OPC - Ordinary Portland Cement; SF - silica fume at 15% replacement

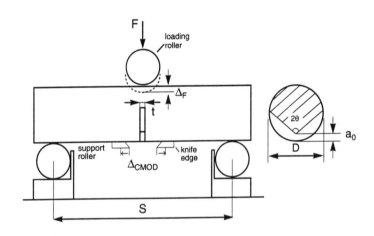

D = diameter of cylindrical specimen a_o = chevron tip distance from bottom fibre
S   = support span, 3.33D   F = load on specimen
2θ = chevron angle, 90° Δ_F = load point displacement (LPD)

FIGURE 1 - Chevron bend specimen for ISRM Method 1

Previous results also indicate significant differences between the rock types :

- In general, andesite composites have superior properties. This rock type also gives substantial improvements in concrete mechanical properties.
- Dolomite produces composites with enhanced stiffness.
- Granite provides considerable post-cracking ductility.

Modes of Failure

Previous studies[5] have shown that, for composite specimens :

- The failure crack starts at the notch and either runs at the original interface (cleavage failure), or else enters the paste at an angle before turning and running parallel to the interface.
- In the cleavage failure mode, crack branching may occur on both the paste and the rock side of the failure crack.
- In "within-paste" failure, the crack produces branches that travel through the bonded paste layer toward the original interface.
- In both failure modes, crack branching out of the failure plane occurs extensively and constitutes an important energy absorbing mechanism.

In the case of within-paste failure, the ISRM test is strictly no longer a valid test of interface toughness; all that can be said is that the interface toughness is greater than the value measured in the test.

PRESENT STUDY : EXPERIMENTAL DETAILS

The work reported here is part of an expanded study to look at the influence of two important variables: w/c ratio of the paste, and age. Test details were as reviewed earlier in this paper.

Paste specimens were cast in steel cylindrical moulds which were placed in a fog room for 20 to 24 hours before stripping. The specimens were then placed in lime water at standard temperature. The rock types used were as before, andesite, dolomite and granite.

An ordinary ASTM Type I Portland cement, at w/c ratios of 0.3, 0.4, 0.5 and 0.55, was used for the paste. Superplasticiser was added at the rate of 2% and 0.5%, by mass of the cement, for w/c = 0.3 and 0.4 respectively. Specimens were tested at 3, 7, 28 and 91 days after casting. To minimise the effects of bleeding and settlement in fresh pastes, especially at high w/c ratios, the moulds, which were sealed at both ends, were slowly rotated about their transverse axis until the cement had set. Despite this, it was initially found that high water/cement ratio specimens (w/c = 0.5 and 0.55) developed bubbles 15 to 20 mm in diameter. If the paste was allowed to stand in the mixer for 2 hours, i.e. until after a degree of initial setting of the cement had taken place, and then briefly remixed before casting, these bubbles no longer formed.

RESULTS AND DISCUSSION

Each "result" quoted is usually the mean value from tests on 3 individual specimens. Coefficients of variation for fracture parameters, from this and previous work, are generally less than 15%, although for R_{CB}, this figure can be higher. In the present work it was found that variability of individual measurements increased at the higher w/c ratios, presumably due to greater inhomogeneity.

Values for rock K_{CB} were 4.30, 2.73, and 2.95 MN/m$^{1.5}$ for andesite, dolomite, and granite respectively. Corresponding rock values for R_{CB} were 155.8, 52.9, and 157.4 J/m^2 respectively.

As in previous work, the two failure modes of interface cleavage and within-paste failure were observed. The average percentage of samples exhibiting cleavage failure over all ages of test amounted to about 30% for the andesite samples, 75% for the dolomite samples, and 55% for the granite samples. This is consistent with previous work, except that the present granite samples showed much improved behaviour, due probably to better quality rock samples from the quarry.

Fracture Toughness K_{CB}

Results for K_{CB} are given in Figures 2 (a) - (d), reflecting the effects of w/c and age for the pure paste, as well as the interface specimens from the three different rock types. General trends reveal the following :

- The fracture toughness values of the interfaces are consistently higher than the cement pastes.
- Fracture toughness reduces with increasing w/c ratio at all ages; for example, 28 d values are on average 40% to 50% lower for w/c = 0.55 in comparison with w/c = 0.30.
- Fracture toughness of pastes reach an approximate plateau level after about 7 days for lower w/c (0.3 and 0.4) pastes. For higher w/c (0.5 and 0.55) pastes, this plateau is reached at a later age of about 28 days.
- In contrast to pastes, fracture toughness values of interfaces show the trend of a continuous increase over the 91 day maximum test age period. This is shown more clearly in Figure 3.
- For the lower w/c ratio interface specimens, andesite appears to have a slight advantage over the other two rock types - the differences are, however, generally small. In contrast, for the higher w/c ratio interface specimens, andesite has no such advantage, and all three rock types appear to behave similarly.

Discussion

The higher fracture toughness of the interface specimens in comparison with the pastes is consistent with previous results. It raises the question of the interpretation of the interfacial region as being the "weak link" in concrete. The present tests have to do with an ideal situation in which pure paste is cast against a naturally fractured and therefore rough rock surface. In this case, bleeding effects and accumulation of water in the interfacial region would be negligible or absent, particularly for the lower w/c ratio pastes. This has been noted by Diamond et al[6], who, using SEM, found no obvious evidence of a transition zone between rock surface and bulk paste in samples prepared as in the present tests.

This method of characterising the interfacial zone using fracture testing must therefore be considered to represent the intrinsic or potential mechanical properties of such a zone. Clearly, interfacial zones in real concretes will differ due to effects mentioned above, and could be expected to show lower values (for the purposes of modelling, say) than are measured in the present tests. However, these results are of great interest in indicating that, potentially, paste/aggregate interfaces may not necessarily be intrinsically weaker or less tough than the paste phase. The present values also reflect the effect of rock surface roughness which is extremely

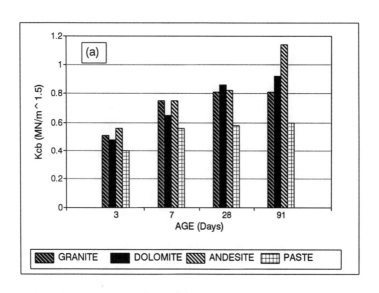

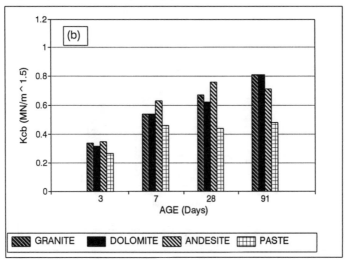

FIGURE 2 - Fracture toughness K_{CB} for specimens of w/c = (a) 0.3 (b) 0.4

important in real concretes, especially those with crushed aggregates. The questionof a "weak link" interfacial zone must also be related to the stress concentrations set up by hard aggregates in a softer matrix, causing cracks to propagate in this zone.

The relatively lower interface properties at early ages (3 days and less) could be considered to have implications for incipient crack formation at aggregate interfaces in immature concretes,

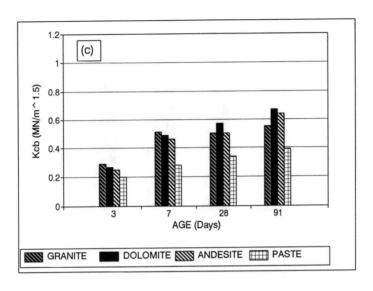

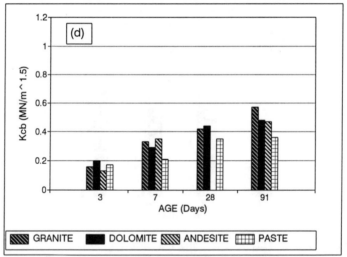

FIGURE 2 - Fracture toughness K_{CB} for specimens of w/c = (c) 0.5 (d) 0.55

due for example to early thermal and shrinkage strains.

Work of Fracture R_{CB}

These results are given in Figures 4 (a) - (d). In comparison with K_{CB} values, a somewhat different picture emerges.

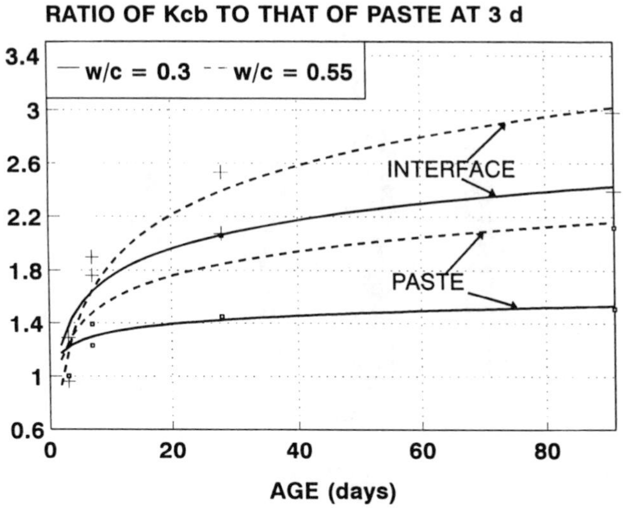

RATIO OF Kcb TO THAT OF PASTE AT 3 d

FIGURE 3 - Ratio of K_{CB} of pastes and interfaces to K_{CB} of paste at 3 d

- The work of fracture of pastes does not show the same clear trend of lower values in comparison with interfaces. The trend would seem to indicate that, as w/c ratio increases, paste values generally become equal to or greater than interface values.
- R_{CB} values reduce with increasing w/c at all ages; 28 d values are on average 30%-40% lower for w/c = 0.55 in comparison with w/c = 0.30.
- Work of fracture values do not approach plateau values with increasing age as is the case with K_{CB}. At greater ages, reductions in fracture energy are possible reflecting greater brittleness of both pastes and interfaces.
- The high variability of results for R_{CB}, particularly at higher w/c ratios, precludes firm conclusions being drawn on the influence of rock type. Taken overall, granite and andesite interfaces appear to exhibit higher values of fracture energy.

Discussion

Work of fracture values can be considered to characterise the post cracking behaviour of a composite, in contrast with fracture toughness which describes the resistance to crack initiation. The results discussed above indicate that, in the "ideal" system modelled here, post-peak crack propagation would not necessarily be favoured energetically in the interface zone. This is borne out by observation of the failure modes where cracks may propagate along the interface itself, or deviate into the paste phase. Clearly, in real systems where water films form around aggregate particles during mixing, the interfacial zone may better represent a region for low energy crack propagation.

The results also provide some insight into the relative roles of interfaces in normal strength

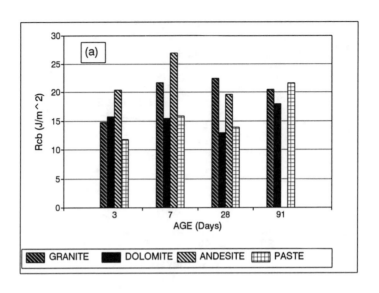

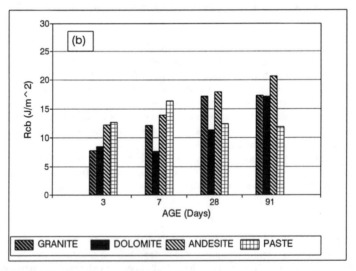

FIGURE 4 - Work of fracture R$_{CB}$ for specimens of w/c = (a) 0.3 (b) 0.4

as compared with high strength concretes. Interfacial zones in HSC are required to exhibit mechanical properties comparable if not superior to the paste[7], and this is borne out in the present tests. The results also suggest that improvements could be made in the properties of normal strength concretes by modifying (i.e. improving) the properties of the interfacial zone.

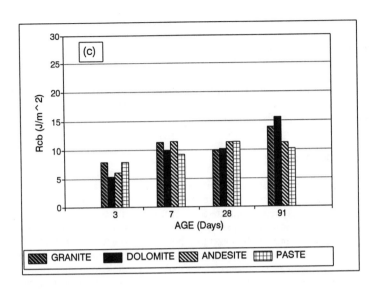

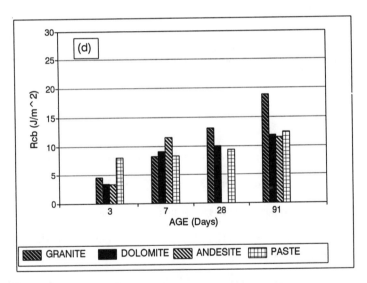

FIGURE 4 - Work of fracture R$_{CB}$ for specimens of w/c = (c) 0.5 (d) 0.55

Comparison with previous 28 d values

Comparison of 28 d, w/c = 0.3 data in Figures 2 and 4 with data in Table 1 shows that :

- For rock, the present specimens exhibit substantially higher K$_{CB}$ values, while R$_{CB}$ values are very similar.

- For paste, present and previous values are comparable.
- For paste/rock interfaces, K_{CB} values for dolomite and granite are substantially higher than before, while R_{CB} for dolomite has also improved.

The differences noted above are to be expected in view of the natural variabilities of rock, and the fact that different areas of a quarry will be worked at different times. The implications for cementitious composites using crushed rock aggregates is that the interface region may represent the most variable phase within the composite.

CONCLUDING REMARKS

The present study, in which paste/rock interfaces have been tested at w/c varying from 0.3 to 0.55, at ages from 3 d to 91 d, has shown, broadly, similar trends in comparison with previous work. In particular, the study has demonstrated the variability inherent in rock fracture values, which may be reflected in values for paste/rock interfaces. It has also shown that interface toughness is not necessarily intrinsically lower than pure paste, which raises questions concerning the "weak link" interface interpretation of concretes. The results seem to indicate that, at least for use of crushed rock aggregates in concrete, it should be possible to modify the properties of the interface within a broad range of values so as to suit the particular engineering requirements of the composite.

Acknowledgements

This study was undertaken with financial assistance from the South African Cement Industry and the Foundation for Research Development.

REFERENCES

1. M.G. Alexander and D.E. Davis. Civ. Eng. in South Africa, **34** (5), 161-170 (1992).

2. M.G. Alexander. Cem. and Conc. Res., **13** (3), 567-575 (1993).

3. M.G. Alexander and S. Mindess. Accepted for publ., Cem. and Conc. Res., (1994).

4. M.G. Alexander, S. Mindess and L. Qu, in Proc. RILEM Int. Conf. on Interfaces in Cementitious Composites, Toulouse, France, (E & FN Spon, London, 1992) 129-138.

5. M.G. Alexander, S. Mindess, S. Diamond and L. Qu. Accepted for publ., Mats. and Struc., (1994).

6. S. Diamond, S. Mindess, L. Qu and M.G. Alexander, in Proc. RILEM Int. Conf. on Interfaces in Cementitious Composites, Toulouse, France, (E & FN Spon, London, 1992) 13-22.

7. B.J. Addis and M.G. Alexander, Cem. and Conc. Res., **24** (5), 975-986 (1994).

FRACTURE MECHANICAL PROPERTIES OF ROCKS AND MORTAR/ROCK INTERFACES

MANOUCHEHR HASSANZADEH
Division of Building Materials, Lund Institute of Technology,
Box 118, S221 00 Lund, Sweden

ABSTRACT

This study has determined the fracture mechanical properties of 9 types of rock, namely fine-, medium- and coarse-grained granites, gneiss, quartzite, diabase, gabbro, and fine- and coarse-grained limestones. Test results show among other things that quartzite has the highest compressive strength and fracture energy, while diabase has the highest splitting tensile strength and modulus of elasticity. Furthermore, the strength and fracture energy of the interfacial zone between the rocks and 6 different mortars have been determined. The results showed that, in this investigation, the mortar/rock interfaces are in most cases weaker than both mortars and rocks.

INTRODUCTION

The introduction of superplasticizers in concrete technology has made it possible to produce concrete with low water/binder ratios. Consequently, it has been possible to produce concrete which has much higher compressive and tensile strength than normal concrete. High strength is not the only advantage of high-performance concrete: the increased impermeability of the concrete, due to low water/binder ratio, has positive effects on its durability. The disadvantage of high-performance concrete in comparison with normal-strength concrete is its increased brittleness [2, 3]. The increase in brittleness is caused by the fact that the ability of the material to store elastic energy has increased due to higher strength, while its ability to consume the stored energy when it is released, during the fracture process, has not increased in the same order.

The ratio between the energy which is consumed by the material during fracture and the maximum elastic strain energy which can be stored per unit volume of the material under uniaxial tensile stress condition is proportional to $l_{ch} = EG_F/f_t^2$, E = modulus of elasticity, G_F = fracture energy and f_t = tensile strength. l_{ch} which has the dimension length is, according to the "Fictitious Crack Model" [1, 4, 6], a material property and is called characteristic length. The lower the l_{ch}, the higher the brittleness of the material. In normal-strength concrete G_F and f_t are primarily governed by the mechanical properties of the cement paste and the shape and the amount of the aggregates, not by the mechanical properties of the aggregates. However, as the cement paste becomes stronger the influence of the mechanical properties of the aggregates on the mechanical properties of concrete increases. Since approximately 75% of the volume of concrete is occupied by aggregates, among which 50% are coarse aggregates, it is important to determine to what extent the mechanical properties of concrete are influenced by the mechanical properties of the aggregates. Furthermore, it is important to determine to what extent it is possible to design concrete composition with the desired mechanical properties by choosing aggregates of different origin. A major investigation is going on at the Division of Building Materials, Lund, in order to understand the role of the aggregates in governing the mechanical properties of concrete. This paper deals with a part of the investigation.

MECHANICAL PROPERTIES OF ROCKS

Rock types

This investigation included 9 types of rock, namely fine-, medium- and coarse-grained granites (GR_f, GR_m and GR_c), gneiss (GN), quartzite (QZ), diabase (DB), gabbro (GB), and fine- and coarse-grained limestones (LS_f and LS_c). The rocks were drilled from different fields in southern parts of Sweden. A petrological examination has been performed for the rocks, which is summarized in Tables I and II.

Table I. The constituents of the rocks. Values are in % by weight for limestones and % per volume for the remaining rocks.

Rock type ⇒	GR_f	GR_m	GR_c	GN	QZ	DB	GB	LS_f	LS_c
Quartz	30.7	26.8	21.7	32.8	95.3	0.0	0.0	-	-
K-feldspar	35.3	34.2	43.8	36.5	1.3	0.0	0.0	-	-
Plagioclase	29.7	30.7	29.8	22.7	3.2	61.0	68.0	-	-
Mica	4.3	5.7	4.5	3.6	0.0	1.8	10.7	-	-
Hornblende									
Opaque	0.0	1.1	0.0	1.2	0.0	2.2	4.8	-	-
Apatite	0.0	0.0	0.2	2.4	0.0	0.0	0.0	-	-
Olivine	0.0	0.0	0.0	0.0	0.0	10.2	11.7	-	-
Pyroxene	0.0	0.0	0.0	0.0	0.0	24.8	4.8	-	-
$CaCO_3$	-	-	-	-	-	-	-	87.1	97.8
Balance	0.0	1.5	0.0	0.8	0.2	0.0	0.0	12.9	2.2

Table II. Density (kg/m³) and SiO_2-content (% by weight) of the rocks.

Rock type ⇒	GR_f	GR_m	GR_c	GN	QZ	DB	GB	LS_f	LS_c
Density (kg/m³)	2640	2640	2620	2640	2620	3030	3020	2680	2660
SiO_2 (% kg/kg)	75.4	74.8	71.9	71.8	98.3	43.7	40.0	-	-

For this investigation, 130 - 500 mm long rock cores with a diameter of 60 mm were drilled. In total, 7 m of cores were drilled for each rock type. The cores were used to make two types of samples which differed in length, namely short samples of 120 mm and long samples of 320 mm. As far as cracks detectable by magnifier (300 % magnification) are concerned the surfaces of the samples were crack free. All samples were cut by diamond saw to the appropriate length and stored in water shortly after drilling.

The mechanical properties which have been determined are the static and dynamic modulus of elasticity E_s and E_d, fracture energy G_F, splitting tensile strength f_{st} and compressive strength f_c. The long samples were used to determine the fracture energy of the rocks. After fracture energy tests, one half of the 320 mm specimens was used to determine the dynamic modulus. The short samples were used to determine the static modulus, splitting tensile strength and compressive strength of the rocks.

The static modulus was determined in compression. The displacements were measured by means of three LVDT which were placed in such a way that they formed an equilateral triangle surrounding the cross-section of the specimen. The displacements were measured over a 70 mm gauge length. The use of three LVDT makes it possible to detect the rotation of the cross-section which may occur due to undesired eccentricity of the applied load. The specimens were loaded in load-control at the rate of 10 kN/s. The maximum applied load was 90 kN. The modulus is calculated as a secant modulus within the interval 10 - 80 kN. It should be noted that the rocks behaved linear-elastically within the limits of the applied load. The dynamic modulus was determined by means of measuring the fundamental transverse frequencies of the rock cylinders.

The fracture energy was determined by means of a stable three-point bend test on notched cylindrical rods (distance between supports = 300 mm, notch depth = 30 mm). Fracture energy is defined as the work performed by the external load including the work performed by the weight of the specimen divided by the crack surface, see [7]. The notch was introduced by means of a diamond saw. The tests were performed in displacement-control. In order to obtain stability, the "Crack Mouth Opening Displacement" (CMOD) was chosen as controlling quantity. The CMOD was measured by a clip gauge which was attached 26 mm from the tip of the notch, i.e. 4 mm from the lower edge of the specimen. The tests were performed with two different CMOD rates, namely 10^{-3} and 10^{-4} mm/s. Three tests were performed for each rock type and CMOD rate. However, since no significant rate effects were observed in these tests, the results will be presented as a mean value of 6 tests, i.e. no attention will be paid to rate effects.

The rate of loading in the compression and splitting tests was 10 and 1 kN/s respectively.

The net flexural strength (f_{net}) of the rocks, i.e. ultimate stress at the tip of the notch, was calculated according to the linear elastic strength theory, by means of the maximum load obtained in the fracture energy test. The characteristic length (l_{ch}) of the rocks was calculated by equating the E and f_t to E_s and f_{st}.

Test results and discussions

Results are presented in Table III. Values for f_c, f_{st}, f_{net}, E_s, E_d, and G_F are mean values of 3, 3, 6, 3, 3 and 6 tests, respectively.

The results show that the quartzite and diabase have the highest compressive and splitting tensile strength while coarse-grained limestone has the lowest. Furthermore, neglecting the limestones, the coarse-grained granite has the lowest compressive and splitting strength among the rocks included in this investigation.

Compressive and splitting strength of quartzite is higher than granite rocks, while its net flexural strength is not. Although the main reason is of course their different geological

Table III. Fracture mechanical properties of rocks.

Rocks ⇓	f_c MPa	f_{st} MPa	f_{net} MPa	E_d GPa	E_s GPa	G_F N/m	l_{ch} m
GR_f	233.7	14.3	15.6	79.3	59.1	124.4	0.036
GR_m	160.0	11.1	11.8	79.9	63.3	134.9	0.071
GR_c	153.4	9.6	13.5	73.5	61.2	159.0	0.106
GN	183.3	11.6	12.0	72.6	59.4	115.6	0.052
QZ	332.0	15.0	13.9	76.0	60.2	162.8	0.044
DB	311.9	15.9	17.2	123.3	104.6	128.7	0.054
GB	256.7	14.3	16.9	104.1	93.7	133.6	0.062
LS_f	160.4	7.2	10.6	67.0	57.8	50.6	0.057
LS_c	51.9	4.6	6.5	64.5	52.3	74.3	0.203

background, the results may indicate that the shear resistance of the quartzite is higher than that of the granites. In compression and splitting tests, low shear strength may lead to partial shear failure, which can have substantial influences on the compressive and splitting tensile strength.

In normal-strength concrete which has weak cement paste, the mechanical properties of the coarse aggregates have no significant influence on the fracture behaviour of concrete. However, as the strength of the cement paste increases, the influence of the mechanical properties of the coarse aggregates on the fracture behaviour of the concrete becomes more significant. For instance in compression tests, since both splitting tensile cracking and inclined shear cracking are the governing failure mechanisms, the use of aggregates which have high tensile and shear strength is essential in order to obtain concrete compositions which have high compressive strength. It has been shown [5] that using granite aggregates instead of quartzite reduces the compressive strength of the concrete. The reduction which has been observed is approximately 20 MPa for some concrete compositions with W/B ratio below 0.37.

Furthermore, the results show that diabase and gabbro have the highest net flexural strength and modulus of elasticity, while the coarse-grained limestone has the lowest. Quartzite and coarse-grained granite have the highest fracture energy, while the fine- and coarse-grained limestone have the lowest fracture energy. The most brittle rock in this investigation is the fine-grained granite, while the most ductile rock is the coarse-grained limestone.

PROPERTIES OF THE MORTAR/ROCK INTERFACES

Preparation of specimens and test methods

In this investigation the tensile strength and fracture energy of the mortar/rock interface have been determined. The tensile strength has been determined by means of splitting tests on cylinders consisting of half mortar and half rock, see Fig. 1. The half rock specimens were obtained by means of splitting tests on the short specimens. The loading rate was 1 kN/s.

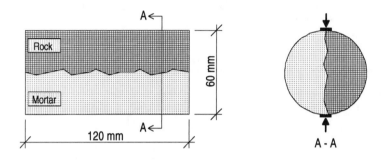

Fig. 1. Determination of bond-strength by means of splitting test, test setup.

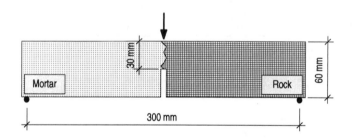

Fig. 2. Determination of fracture energy of bond, test setup.

The fracture energy has been determined in the same way as for the rocks. However in this case half of the specimens consisted of mortar which was cast against the rock, see Fig. 2. The bending tests were performed in CMOD control. The rate of CMOD was 10^{-4} mm/s. The net flexural strength of the mortar/rock interface is calculated in the same way as in the case of the rocks.

Six different mortar compositions have been used. The constituents of the mortars and their proportions are shown in Table IV. The cement which was used in this investigation was a moderate heat type of standard Portland cement. The silica fume (Si) was of the micro pozzolan type. The sand (Sa) was washed sea bottom sand with fractions within the interval 0.5-3 mm, of which 78 and 7 percent passed through sieves of 2 mm and 0.5 mm respectively. The superplasticizer (Sp) was a melamine-based solution with 33% by weight dry content.

The specimens for the splitting tests were cast in the position which is shown in Fig. 3a. The casting was performed in the following order: one of the halves of a split cylinder was placed in a cylindrical mould and the remaining part of the mould was filled by mortar. The specimens for fracture energy tests were cast in the position which is shown in Fig. 3b. A cylindrical tube was placed on top of the rock specimen with the fractured plane in upright position. The tube was filled with mortar and vibrated by the vibro-tube. Both types of specimens were unmoulded a day after casting and were cured for 12 weeks in lime-saturated water.

Table IV. The composition of the mortars.

Mortar→	I	II	III	IV	V	VI
C (kg/m^3)	455.90	600.20	638.60	568.40	604.40	624.20
Si (kg/m^3)	0.00	0.00	0.00	28.40	30.20	31.20
Si/C	0.00	0.00	0.00	0.05	0.05	0.05
W (kg/m^3)	250.80	240.10	191.60	238.70	190.40	163.90
W/(C+Si)	0.55	0.40	0.30	0.40	0.30	0.25
Sa (kg/m^3)	1595.70	1500.60	1596.40	1491.90	1586.60	1638.60
Sa/(C+Si)	3.50	2.50	2.50	2.50	2.50	2.50
Sp (kg/m^3)	0.00	0.00	14.40	0.00	14.30	24.60
Sp/(C+Si) (%)	0.00	0.00	2.25	0	2.25	3.75

C = Cement, Si = Silica fume, W = Water, Sa = Sand, Sp = superplasticizer

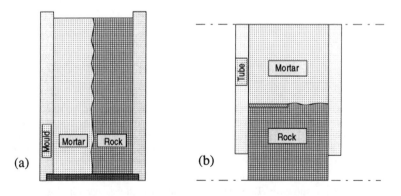

(a)   (b)

Fig. 3. Casting of specimens for splitting test (a), and bending test (b) on bond.

Test results and discussions

Results of the splitting tests are shown in Fig. 4. Every point on the figures is the mean value of three tests. The solid lines in the figures show the mean value of the results, taking all the results together. The dashed lines show the mean value of the results of the mortars with W/B ratios 0.4 and 0.3, taking all rock types together. Due to scatter of the results it is difficult to draw any reliable conclusions. As can be observed, although in some cases the results indicate that the lower the W/B ratio, the higher the bond-strength, the results do not show conclusively that the bond-strength is significantly influenced by the variation of the W/B ratio when the W/B ratio is lower than 0.4. Furthermore, it cannot be observed that replacement of 5% of the cement by silica fume has any significant influence on the bond-strength; compare the dashed lines.

Results of the fracture energy tests of the bond are shown in Fig. 5. The same conclusions as above may be drawn for the fracture energy tests. Furthermore, the results indicate that the grain size of the rock has some effect on the fracture energy of the bond. Compare, for instance, the fracture energy of the fine-, medium- and coarse-grained granite rocks with each other. Also compare the fracture energy of the coarse-grained limestone with the other rock types. The fracture surface of the coarse-grained limestone usually showed much higher roughness than the other rocks.

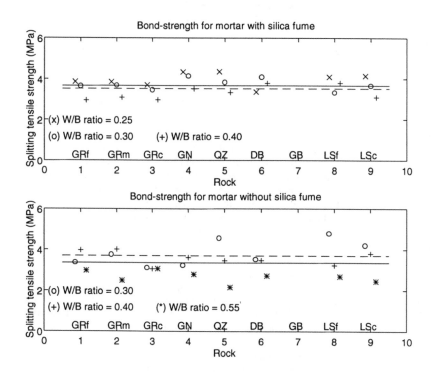

Fig. 4. Determination of bond-strength by means of splitting test, test results.

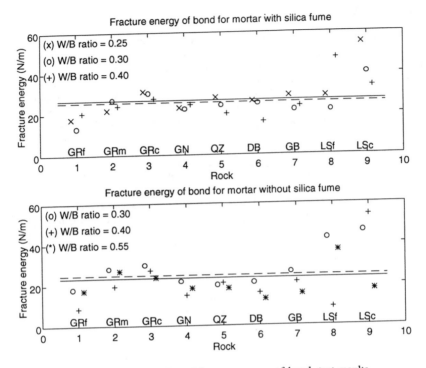

Fig. 5. Determination of fracture energy of bond, test results.

Results of the net flexural strength of the bond are shown in Fig. 6. As can be observed the bond-strength in the case of quartzite and fine-grained limestone is somewhat stronger than the bond-strength in the case of other rocks. As for the rest the same conclusions as in the case of splitting tests can be drawn for the flexural strength.

It should be noted that in all cases the fracture occurred in the interfaces, except for a few tests which involved the coarse-grained limestone. In these cases the crack passed through the rock. It is interesting to note that the results of the splitting tests and the bending tests, with the exception of a few cases, show the same tendencies. The exception is that the quartzite and fine-grained limestone show somewhat higher bond-strength in bending tests. The reason may be differences in the testing methods which may have caused different modes of failure. For instance in the splitting tests, some parts of the interface may have been exposed to both tensile and shear stresses, while this type of stress condition is not likely to occur in bending tests. Furthermore, the roughness of the rock (crack) surface has different influences in the splitting tests than the bending tests.

The mica content of the rocks may also have influenced the test results. As can be observed the rocks which do not contain mica (QZ, LS_f and LS_c) show somewhat higher bond-strength than the rocks which contain mica. However, since the mica content on the fracture surface is unknown this cannot be shown conclusively.

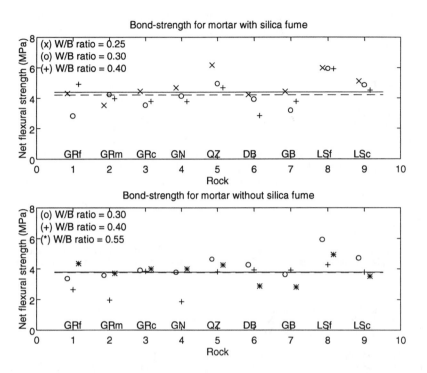

Fig. 6. Determination of bond-strength by means of bending test, test results.

CONCLUSIONS

a) Among the rocks tested in this investigation, diabase and quartzite have the highest
    compressive and splitting tensile strength, while the coarse-grained limestone has the
    lowest. Furthermore, diabase and gabbro have the highest modulus of elasticity, while
    the coarse-grained limestone has the lowest.

b) Among the rocks tested in this investigation, quartzite and coarse-grained granite have
    the highest fracture energy, while the fine-grained limestone has the lowest. Further-
    more, the most brittle rock in this investigation is the fine-grained granite, while the
    most ductile rock is coarse-grained limestone.

c) In this investigation, the mortar/rock interfaces were in all cases, with the exception of
    a few cases involving coarse-grained granite, weaker than both mortars and rocks.

d) Although in some cases the results indicate that the lower the W/B ratio, the higher the
    bond-strength, the results do not show conclusively that the bond-strength is signi-

ficantly influenced by the variation of the W/B ratio when the W/B ratio is lower than 0.4. Furthermore, it cannot be observed that replacement of 5% of the cement by silica fume has any significant influence on the bond-strength.

ACKNOWLEDGEMENTS

The author would like to acknowledge the financial support received from the Swedish Consortium on High-Performance Concrete. The Consortium consists of Cementa, Elkem Materials, Euroc Beton, NCC Bygg, SKANSKA, Strängbetong, The Swedish Council for Building Research (BFR) and the Swedish National Board for Industrial and Technical Development (NUTEK). Furthermore, the author would like to thank Dr Ulf Söderlund at the Geological Institute, Lund University, who carried out the petrological examination.

REFERENCES

1. P. J. Gustafsson, Fracture Mechanics Studies of Non-yielding Materials Like Concrete, PhD thesis, Division of Building Materials, Lund Institute of Technology, Lund, Sweden (1985).

2. A. Haghpassand, Fracture Mechanical Parameters of Normal- and High-Strength Concrete (in Swedish), Report No. TVBM-5024, Division of Building Materials, Lund Institute of Technology, Lund, Sweden (1992).

3. M. Hassanzadeh and A. Haghpassand, in Utilization of High Strength Concrete, edited by I. Holand and E. Sellevold (Norwegian Concrete Association, P.O. Box 2312 Solli, N-0201, Oslo, 1992) pp. 1092-1099.

4. A. Hillerborg, Application of Fracture Mechanics to Concrete, Report No. TVBM-3030, Division of Building Materials, Lund Institute of Technology, Lund, Sweden (1988).

5. B. Persson, (private communication), Division of Building Materials, Lund Institute of Technology, Lund, Sweden (1993).

6. P.E. Petersson, Crack Growth and Development of Fracture Zones in Plain Concrete and Similar Materials, PhD thesis, Division of Building Materials, Lund Institute of Technology, Lund, Sweden (1981).

7. RILEM Draft Recommendation, 50-FMC Committee Fracture Mechanics of Concrete, Materials and Structures, 18 (106), 285-290, (1985).

CRACK STABILITY IN THE FRACTURE OF CEMENTITIOUS MATERIALS

S. Tandon[*], K.T. Faber[*] and Z.P. Bazant[**],
[*]Dept. of Materials Science and Engineering,
[**]Dept. of Civil Engineering,
Robert R. McCormick School of Engineering and Applied Science,
Northwestern University, Evanston, Il 60208.

ABSTRACT

The aim of the present study is to investigate the stability of crack propagation in cementitious materials. Tests were conducted on bend specimens in three–point and four–point loading conditions. Three–point bend specimens showed stable crack growth for mortar, normal strength and high strength concrete specimens. Alternatively, four–point bend specimens showed catastrophic failure for mortar and quasi–catastrophic failure for normal strength and high strength concrete specimens. Results will be discussed in relation to brittleness number model and specific microstructural features including the interfacial transition zone between the cement paste and the aggregate and the attendant toughening mechanisms.

1. INTRODUCTION

The measurement of fracture toughness and resistance-curve (or R–curve) behavior in brittle and quasi-brittle materials often requires the stable growth of a single crack through the specimen. The assurance of crack stability is of particular interest in the study of cementitious materials, such as cement paste, mortar and concrete, where process zone shielding has been documented to give rise to R–curve behavior [1-3]. Specifically, the process zone phenomena in these materials include crack bridging by unhydrated cement grains in cement paste [4], crack bridging by aggregates [5] and microcracking [6] in concrete.

A number of issues influencing crack stability during fracture tests have been investigated already. For example, theoretical models have been developed to predict the stability of shrinkage cracks in concrete and reinforced concrete [7]. Geometric factors, such as notch depth in notched-beam specimens [8] and specimen size [9,10] have been explored both experimentally and analytically. In geometrically similar specimens, large samples tend to fail catastrophically while small specimens show stable crack growth [9,10]. Long starter notches improve the stability of crack extension in notched–beam tests [8]. One interesting well–established property is that crack extension during three–point bend tests is inherently more stable than four–point bending [11]. What is not well understood are the microstructural changes, particularly in process zone phenomena, that differ under the two loading conditions. The intent of this work is to examine, firstly, crack stability in three– and four–point bend testing of mortar, normal strength and high strength concrete, and, secondly and more importantly, the differences in the microstructural aspects of fracture which accompany the change in loading.

2. EXPERIMENTAL DETAILS

The materials used for this study– mortar, normal strength concrete (NSC) and high strength concrete (HSC)–were designed and mixed in the laboratory. The mix ratio (by weight) of the normal strength concrete, was cement: sand: gravel: water = 1: 2: 2: 0.6. The mix ratio of the high strength concrete, by weight, was cement: sand: gravel: water: silica fume = 1: 2: 2: 0.3: 0.3. A

Mat. Res. Soc. Symp. Proc. Vol. 370 © 1995 Materials Research Society

water reducing agent[*] (88.5 ml) was added to one batch (~0.3 ft.^{3}) of the high strength concrete mix. Silica fume[†] was used as a mineral admixture to strengthen the interface between the aggregate particles and the matrix. The mix ratio (by weight) of the mortar, was cement: sand: water = 1: 2: 0.6. Type I portland cement, P (pea–) type gravel and ASTM # 2 sand were used. The maximum aggregate size in the mixes of normal and high strength concrete was 9.5 mm.

Single-edge notched beams were cast, with beam depth, d = 76 mm. For all the beam specimens, the length–to–depth ratio was equal to 8/3, the ratio of notch length, a_o, to depth was 1/3, and specimen thickness was 38 mm. All the specimens were compacted by rodding and vibration. The specimens were left in the molds during the first 24 h and were removed and cured in water until the time of testing. The notches were cut with a diamond band saw and were 1.8 mm wide. Companion cylinders of 76 mm diameter and 152 mm length were cast. These cylinders were capped with a sulphur compound and cured under water with the notched specimens. The cylinders were tested in compression after 28 days of curing. Normal strength concrete cylinders failed at an average maximum compressive strength of 46.4 MPa, with a standard deviation of 5.4 percent. High strength concrete cylinders had an average compressive strength of 73.2 MPa, with a standard deviation of 3.3 percent.

Bend specimens were loaded on a servo–electric mechanical testing machine[‡] at a constant crosshead displacement rate of 2 μm/min. During each test, the load, crosshead displacement and crack mouth opening displacement (CMOD) were monitored at 1 Hz with a data acquisition system.[**] The crack opening displacement was measured using a clip gage.[††] For the three–point bend tests, the outer span, l_o, was 188 mm whereas for the four–point bend tests, the outer span was 188 mm and the inner span, l_i, was 94 mm. Four to ten specimens were tested for each loading condition (three– or four–point bend test), for each material.

A camera was attached to a traveling microscope mounted on the base of the testing machine. At the end of the test, the motion of the crosshead was arrested and crack pictures were taken. The specimen was unloaded and its surfaces were examined under an optical microscope for evidence of any side cracking or any other difference in the crack profiles on the front (surface facing the camera) and back surfaces.

3. RESULTS

Figure 1(a) shows the load versus CMOD plot of a normal strength concrete specimen tested in three–point bending. Crack growth in this specimen is stable in contrast to normal strength concrete specimen tested in four–point bending (Figure 1(b)). In this experiment a sudden load drop occurs in the softening branch of the load–CMOD curve and the crack opens by a few hundred microns followed by a stable crack growth with a slowly decreasing load carrying capacity. To explore the effect of the inner span on the stability of the test, the inner span was reduced to 15 mm (the allowable minimum inner span roller distance on the fixture). The load–CMOD plot of this test is shown in Figure 1(c). Stability of the crack growth is improved over the four–point bend test with the wide inner span.

---

[*] W.R.Grace, Daracern-100, Cambridge, MA
[†] W.R.Grace, WRDA-19, Cambridge, MA
[‡] MTS 808, Materials Testing Systems, Minneapolis, MN
[**] Macintosh SE computer with an Analog Connection SE data acquisition card
[††] MTS, COD GAGE 632.03E-30, Minneapolis, MN

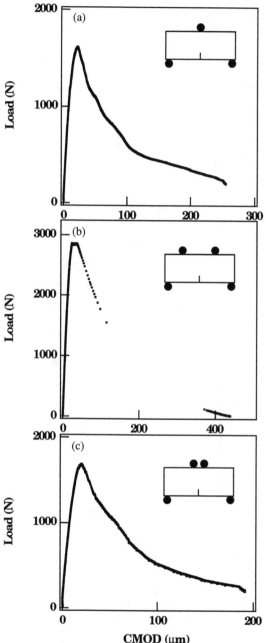

Figure 1: Load versus CMOD of normal strength concrete specimens tested in (a) three–point bending (b) four–point bending and (c) four–point bending with reduced inner span.

Similar results are demonstrated in high strength concrete (Figure 2), however, in four–point bending the extent of instability is larger than for normal strength concrete with crack opening of nearly 500 μm prior to crack arrest. In mortar, crack arrest never occurs in four–point bending, as seen in Figure 3(b). In contrast, three–point bending results in stable crack extension in mortar (Figure 3(a)).

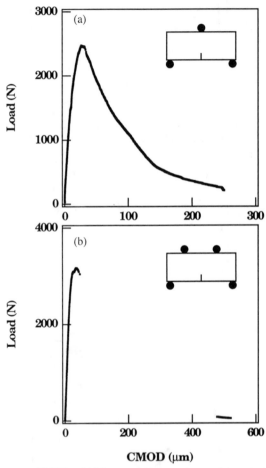

Figure 2: Load versus CMOD of high strength concrete specimens tested in (a) three–point bending and (b) four–point bending.

Figure 4 shows a fully developed crack profile in normal strength concrete recorded at the end of the four–point bend test prior to unloading the specimen. Different sites marked A to E show various microfracture toughening mechanisms active in the material. All these sites are shown in the insert at a higher magnification. Point A shows offset crack being bridged by the untorn ligaments in the materials. Grain localized bridging is evident at point B. Point C shows an example of the debris wedging, likely a deflection and localized branching of the main crack

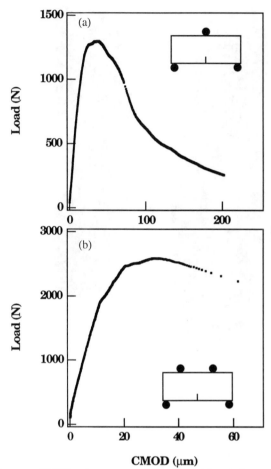

Figure 3: Load versus CMOD of mortar specimens tested in (a) three–point bending and (b) four–point bending.

are illustrated in area D. Furthermore, these cementitious materials have a "damage zone" of the discontinuous cracks at the end of the main crack. Area E shows the "damage zone" observed in the specimen. Similar toughening mechanisms are active in the specimens tested in three–point bending.

Optical micrographs were taken of the crack profiles on the front and back surfaces of the tested specimens. As shown in the Figure 5(a), on the back surface of the normal strength con-crete specimen tested in the four–point bending, a crack parallel to the main crack is seen at a distance of 4-5 mm from the main crack and extends over a length of approximately 20 mm. This second crack is not seen on the front surface so it does not extend through the whole thickness of the specimen. A similar big mass of the concrete between the two cracks was seen in the four–point bend specimen tested with reduced distance (15 mm) between the upper rollers (Figure

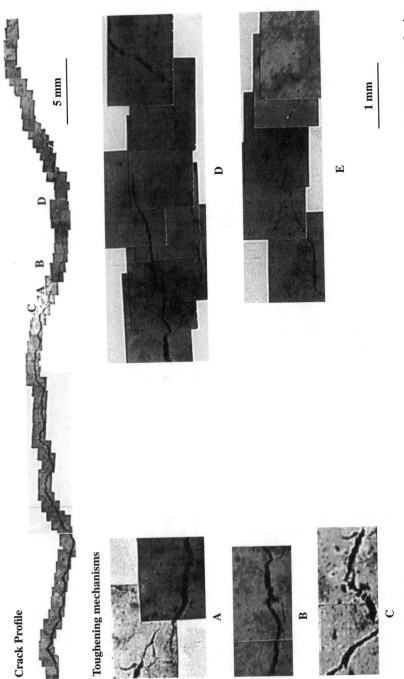

Figure 4: Crack profile of a normal strength concrete specimen recorded at the end of the four point bend test. Inserts A-E show toughening mechanisms active in the material at a higher magnification. **A**–Untorm ligaments, **B**–Grain bridging, **C**–Debris wedging, **D**–Deflection and localized branching and **E**–Damage zone.

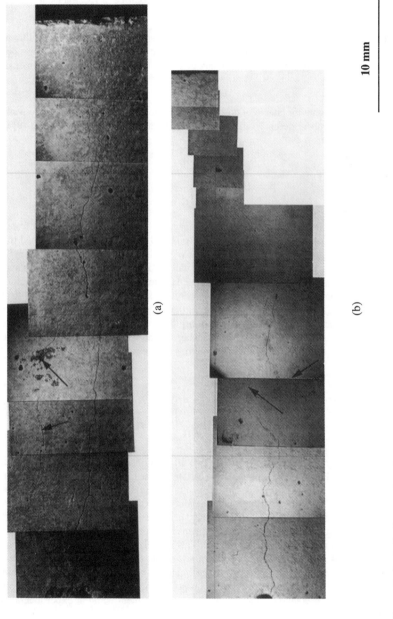

10 mm

(a)

(b)

Figure 5: Optical micrographs of the back surface of the normal strength concrete specimen tested in (a) four–point bending and (b) four–point bending with reduced inner span. Arrows point to secondary cracks on the surface.

5(b)). No such parallel cracks were seen on either surface of the specimens tested in three–point bending. At higher magnification, several disconnected cracks were found in the specimens loaded in four–point bending. No such cracks were seen in three–point bend specimens at the same magnification.

Analysis of the pictures taken prior to unloading high strength concrete show similar microfracture processes as seen in the normal strength concrete specimens. The four–point bend specimen shows a highly tortuous crack. Secondary cracking is seen along with the main crack.

4. DISCUSSION

To explain the fracture behavior (stable or unstable) of the concrete specimens of different sizes and shapes, many researchers defined various brittleness numbers. Bazant and Pfeiffer [9] defined a brittleness number, β as:

$$\beta = B^2 g_f(\alpha_o) \frac{f_t^2 d}{G_f E_c} \tag{1}$$

where B is a constant which gives the plastic load capacity of the specimen calculated from f_t, $g_f(\alpha_o)$ is a nondimensional energy release rate equal to $[k(\alpha_o)]^2$ and $k(\alpha_o)$ can be obtained from the fracture mechanics handbooks [12]; α_o is the relative notch length, G_f is the fracture energy, E_c is the elastic modulus, f_t is a strength parameter usually taken as direct tensile strength and d is cross–section dimension of the specimen. The brittleness number, β, serves as an indicator of the type of fracture behavior. A specimen with a high brittleness number fails suddenly, with audible noise, and crack growth is unstable. A lower brittleness number is indicative of a material which demonstrates stable crack growth. We use the brittleness number to compare the observed fracture behavior of three– and four–point bend specimens of the same type of concrete. The ratio of two brittleness numbers for these different geometries can be calculated as:

$$\frac{\beta_{3pt}}{\beta_{4pt}} = \left(\frac{B_{3pt}}{B_{4pt}}\right)^2 \left(\frac{k(\alpha_o)_{3pt}}{k(\alpha_o)_{4pt}}\right)^2 \left(\frac{(G_f)_{4pt}}{(G_f)_{3pt}}\right) \tag{2}$$

where the ratios of fracture energy of concrete in two different loading conditions can be calculated from experimental load–CMOD plots shown earlier and the ratio of $k(\alpha_o)$ values is a geometry and load dependent quantity. Table 1 lists the results of the ratio of two brittleness numbers. If the material behaves more brittle in four–point bending than the three–point bending, its brittleness number will be higher and the ratio of the two brittleness numbers will be lower than 1. Our calculated values suggest that for mortar specimens the four–point bend test is significantly less stable with respect to the three–point bend test. This is in agreement with our observation of catastrophic failure of mortar four–point bend test and stable crack growth in the three–point bend test. For normal and high strength concrete specimens the four–point bend test is less stable than the three–point bend test and stability of the four–point bend test improves if the upper roller distance is decreased. These analytical results are in good agreement with our experimental observations and similar to observations by Asghari and Barr [11].

Table 1: Ratio of brittleness numbers for three– and four–point bend tests

	Mortar	Normal Strength Concrete	High Strength Concrete
$B_{3\,pt}/B_{4\,pt,\,94}$*	0.50	0.50	0.50
$k\,(\alpha_o)_{3\,pt}/k\,(\alpha_o)_{4\,pt,\,94}$*	0.94	1.05	1.28
$(G_f)_{3\,pt}/(G_f)_{4\,pt}$*	0.74	0.42	0.93
$\beta_{3\,pt}/\beta_{4\,pt,\,94}$*	0.30	0.66	0.44
$\beta_{3\,pt}/\beta_{4\,pt,\,15}$†	–	0.85	–

* For the four–point bend test with inner span as 94 mm

† For the four–point bend test with inner span as 15 mm (B ratio = 0.92, k(α_o) ratio = 0.98, G_f ratio = 0.96)

In the four–point bend geometry, a larger volume of the specimen is under uniform tensile stress compared to the three–point bend geometry (Figure 6). Microstructurally, this is manifested in greater damage or observed secondary cracking in the four–point bend specimens, and consequently toughening, as denoted by $(G_f)_{3pt}/(G_f)_{4pt}$ in Table 1. The increase in toughness, however, does not translate into a relative decrease in brittleness with respect to three–point bending. Instead, we examine the energetics of the fracture process. Since the stored elastic energy in the four–point bend specimen is greater than in the three–point bend specimen, the elastic energy released during fracture of the four–point bend specimen is higher. If this released elastic energy is greater than that required to grow the crack and produce the attendant damage through the specimen, the crack will grow in an unstable manner and the specimen will fail catastrophically as observed in the case of the mortar specimens. If material absorbs the released energy by debonding the particles at the aggregate/matrix interface (NSC) or by fracturing the particles (HSC), the crack could grow unstably but later be arrested due to the presence of inclusions. When the upper roller distance in the four–point bend test is reduced, crack stability improves, by limiting the size of the damage zone (and smaller magnitude of the stored elastic energy), as seen in Figure 1(c) for normal strength concrete specimen.

5. SUMMARY

1. The three–point bend specimens of the size studied here are found to exhibit stable crack growth in all three materials– mortar, normal strength concrete and high strength concrete.

2. The four–point bend specimens exhibit unstable crack growth. For mortar four–point bend specimens, fracture is catastrophic, whereas for normal and high strength concrete specimens fracture is quasi–catastrophic.

3. Reducing the distance between the upper rollers improves the stability of the four–point bend test.

4. Experimental observations of the load–CMOD relation agree very well with the results predicted by the brittleness number concept.

5. In four–point bend specimens a wider zone is subjected to high tensile stress, which causes more cracking in the specimen and makes crack growth unstable.

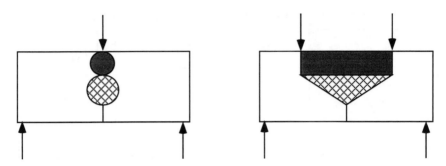

Figure 6: Regions of maximum stress in three– and four–point bend specimens.(dark: compression, cross–hatched: tension)

REFERENCES

(1) B.L. Karihaloo, A. Carpinteri, and M. Elices, Advn. Cem. Bas. Mat. **1**(1), 92-105 (1993).

(2) S.P. Shah and C. Ouyang, J. Am. Ceram. Soc. **74**(11), 2727-2738 (1991).

(3) S. Mindess, in Toughening Mechanisms in Quasi-Brittle Materials Edited by S.P. Shah, (Kluwer Academic Publishers, 1991) p. 271-286.

(4) L. Struble, P. Stutzman, and E.R. Fuller Jr., "Microstructural Aspects of the Fracture of Hardened Cement Paste," J. Am. Ceram. Soc., **72**[12] 2295-2299 (1989).

(5) J.G.M. Van Mier, Toughening Mechanisms in Quasi-Brittle Materials Edited by S.P. Shah, (Kluwer Academic Publishers, 1991) p. 329-335.

(6) H. Horii, in Fracture Processes in Concrete, Rock and Ceramics Edited by J.G.M. Van Mier, J.G. Rots, and A. Bakkar, (E. & F. N., 1991) p. 95-110.

(7) Z.P. Bazant and L. Cedolin, Stability of Structures: Elastic, Inelastic, Fracture and Damage Theories (Oxford University Press, 1991) p. 984.

(8) A. Carpinteri, J. Struct. Eng., **108**(4), 833-848 (1982).

(9) Z.P. Bazant and P.A. Pfeiffer, ACI Mat. J., 1987, 463-480.

(10) Z.P. Bazant and R. Gettu, ACI Mat. J., **89**(5), 456-468 (1992).

(11) A. Asghari and B. Barr, Fracture Processes in Concrete, Rock and Ceramics Edited by J.G.M. Van Mier, J.G. Rots, and A. Bakkar, (E. & F. N., 1991) p. 515-522.

(12) H. Tada, P.C. Paris, and G.R. Irwin, The Stress Analysis of Cracks Handbook 1973.

Acknowledgment

This work was supported by the Center for Advanced Cement-Based Materials at Northwestern University.

INTERFACE TRANSITION ZONE AND THE ELASTIC MODULUS
OF CEMENT-BASED COMPOSITES

PETER I. SIMEONOV* AND S.H. AHMAD
Department of Civil Engineering, NC State University, Raleigh, NC 27695
* Permanent address: Central Laboratory of Physico-Chemical Mechanics,
Bulgarian Academy of Sciences, Sofia 1113, Bulgaria

ABSTRACT

The influence of the Interface Transition Zone (ITZ) on the elastic modulus of concrete is demonstrated as a divergence of the experimental data from the general trend of the theoretical Hashin-Shtrikman bounds. This divergence is well related to the W/C of the composite. With reduction of W/C the influence of ITZ decreases and for values close to 0.4 and lower it is insignificant.

The formation of the ITZ is characterized by a transfer of water from the matrix to the surface of the aggregates. As a result of this a highly porous ITZ is formed while the matrix remains with a reduced porosity. This process can also be described as a transfer of material properties. For some compositions the balance of this transfer can approach zero. The imbalance in this process is more pronounced at higher W/C.

The effect of Interface Transition Zone can be successfully simulated by the help of recently derived Hashin's variational bounds for two-phase composites with imperfect interfaces.

INTRODUCTION

The properties and the volume fraction of the phases - the matrix cement paste and the aggregates are the main factors which determine the deformation behavior of the cement-based composites. Unfortunately the proposed two-phase composite models fail to adequately predict the elastic deformation of concrete. One of the possible reasons for that could be the influence of the interface transition zone (ITZ) which is not taken into account in these models.

A powerful approach in the analysis of composite materials is the bounding of the effective elastic properties between upper and lower bounds. For statistically isotropic and homogeneous material, applying the variational method, Hashin and Shtrikman [1] derived improved bounds (HS), which are considered the best (the closest) bounds for estimating the effective modulus of elasticity of a two-phase composite [2]. Although for high moduli ratio they are set quite apart they still can serve as a good test for a two-phase composite elastic behavior.

A number of experimental studies confirm the validity of HS bounds [2] for different composite materials. First Mandel and Dantu [3] made an experimental verification of HS bounds for cement composites and found that their results lie below the lower HS bound. Low results in such investigations can be due to non

Mat. Res. Soc. Symp. Proc. Vol. 370 © 1995 Materials Research Society

compensated porosity or anisotropy in the specimens [4] but can also be due to the effect of the ITZ.

Monteiro [5] suggested that HS bounds can be used to asses the significance of the ITZ influence on the E-modulus of concrete - if the experimental results "are outside these bounds that will mean the influence of the interface layer is significant and will show the need of three-phase models". Nilsen and Monteiro [6] analyzed Hirsch's data by calculating HS bounds and concluded that concrete cannot be considered a two phase material and the ITZ should be included as a third phase. In support of this were also their own experimental results on mortars [7].

Cohen Goldman and Chen [8] used an alternative approach varying the specific surface at a constant volume of the aggregates. Their experiments on dynamic modulus of elasticity clearly demonstrated the significance of ITZ and its weakening effect in cement mortars as well as the role of silica fume for ITZ modification.

In this paper the the effect of ITZ is demonstrated as a divergence of the experimental data from the general trend of HS bounds. This divergence is well related to W/C and can be successfully simulated by the recently derived variational bounds for two-phase composites with imperfect interfaces.

ANALYSIS OF EXPERIMENTAL DATA

Numerous papers can be found in the literature on experimental investigations with cement concretes and mortars which contain data for the modulus of elasticity of the aggregate (Ea) and the matrix (Em) (paste or mortar) and the modulus of elasticity of the composite (concrete or mortar) (Ec) for different volume fractions of the aggregate (Va).

In this paper, such specific data from Hirsch [9], Mandel and Dantu [3], Counto [10], Anson and Newman [11] and Sock, Hannant and Williams [12] are considered. A more detailed information on these data sets and the corresponding values for the lower and upper HS bounds computed for each set of data can be found in [13].

Calculation of HS Bounds

HS bounds are calculated for every set of experimental data according the following equations:

$$K_m = E_m/(3(1-2n_m)) \tag{1}$$
$$G_m = E_m/(2(1+n_m)) \tag{2}$$
$$K_a = E_a/(3(1-2n_a)) \tag{3}$$
$$G_a = E_a/(2(1+n_a)) \tag{4}$$
$$K_{c(-)} = K_m + V_a/(1/(K_a-K_m)+3V_m/(3K_m+4G_m)) \tag{5}$$
$$K_{c(+)} = K_a + V_m/(1/(K_m-K_a)+3V_a/(3K_a+4G_a)) \tag{6}$$
$$G_{c(-)} = G_m + V_a/(1/(G_a-G_m)+6V_m(K_m+2G_m)/(5G_m(3K_m+4G_m))) \tag{7}$$
$$G_{c(+)} = G_a + V_m/(1/(G_m-G_a)+6V_a(K_a+2G_a)/(5G_a(3K_a+4G_a))) \tag{8}$$
$$Ec_{(-)} = 9K_{c(-)}G_{c(-)}/(3K_{c(-)}+G_{c(-)}) \tag{9}$$
$$Ec_{(+)} = 9K_{c(+)}G_{c(+)}/(3K_{c(+)}+G_{c(+)}) \tag{10}$$

where:

E - Young's modulus;
K - bulk modulus;
G - shear modulus;
n - Poisson's ratio;
V - relative volume of the phase;
m - matrix;
a - aggregate;
c - composite;
(-) - value for lower bound;
(+) - value for upper bound.

For calculation of HS bounds the bulk and the shear moduli of the matrix and the aggregates must be known but usually they are not provided by the researchers as part of the experimental data. However they can be evaluated by equations (1-4) using the data for the Young's moduli and knowing the Poisson's ratios of the phases.

Poisson's ratio for cement paste is close to 0.25 [3,11,5] and is not influenced by W/C [11]. For mortars the average value of 0.2 is assumed based on the data of Anson and Newman [11]. The average value of 0.15 is assumed for the coarse rock aggregates. The values of 0.25 for glass, 0.30 for steel and 0.45 for lead are assumed as used by Mandel and Dantu [3]. Calculations show that HS bounds are not very sensitive to inaccuraces in the Poisson's ratio. Changes in the range 0.1-0.2 for one of the phases influence the results for HS bounds usually less than 1.0%.

From HS bounds for K and G the bounds for Young's modulus (E) can be calculated by (9) and (10). This procedure will give the correct bounds since E is an increasing function both of K and G [14].

Interpretation of experimental data with HS bounds

To check if mortars and concretes behave as two-phase composites a comparison of experimental data to HS bounds is analyzed.

Fig.1 shows representative sets of data from different authors for cement-based composites with various W/C, various type of matrix and aggregates in a broad range of Va (0.2-0.8). Generally speaking the experimental data follow the main trend of the theoretical HS bounds for a two-phase composite. At the same time most of the points lie very close and some of them below the lower HS bound.

For the purpose of a generalized assessment and comparison the data is normalized to the lower HS bound (LHS). Analysis shows a nearly normal distribution of data to the LHS bound and no significant trend of divergence for the data as a whole with increase of Va. However if the individual data sets are analyzed an interesting phenomenon is observed. With increase of Va most of the data sets tend to diverge from LHS but at different angles.

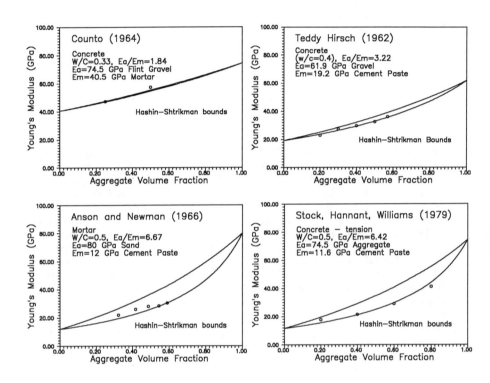

FIG.1.
Experimental data sets with corresponding HS bounds.

The Influence of ITZ on the Elastic Modulus of Concrete

Fig.2 shows the data sets of **Fig.1**, normalized to LHS bound. For every set of data the divergence from LHS is characterized by the slope of the linear regression line. This value gives the rate of relative change of the elastic modulus of the composite per unit volume change of the aggregate, or this can be considered a characteristic for the influence of ITZ.

It can be seen from **Fig.2** that the slopes for the different sets of data are not the same. An explanation for this can be found considering some of the main factors influencing the formation and the properties of the ITZ.

400

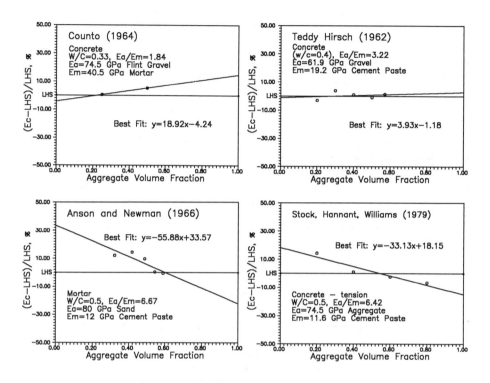

FIG.2.
Linear regression of data sets, normalized to the lower HS bound.
The slope of the regression line gives the effect of the ITZ formation.

W/C is the factor governing the structure and the properties of the cement paste in the composite (mortar, concrete) and it is also the main factor influencing the properties and the thickness of ITZ. In order to evaluate the influence of this factor on the effect of ITZ on Ec, 23 data sets from five authors are analyzed and the results are shown on **Fig.3**. The data for concretes and for mortars are presented separately since the specific surfaces of the aggregates differ significantly. Most of the coarse aggregates used, have relatively close values of specific surface - 3-6 cm^2/cm^3 and a correct comparison between them can be made. For the sand, the specific surface values are around 100 cm^2/cm^3 and the ITZ is expected to be in thinner layers.

Fig.3. shows what is the influence of W/C on the rate of change of the normalized values of Ec in % per unit Va, or in other words, how the effect of ITZ on Ec depends on W/C. It is clearly demonstrated that there is a notable effect of ITZ on the elastic modulus of concrete and that this effect is well related to the W/C. It can be seen that at high W/C the effect of ITZ is negative and with the reduction of W/C this effect is reduced and even turns into positive at low W/C. The linear regression of data for concrete (**Fig.3a**) and for mortar (**Fig.3b**) shows that the effect of ITZ approaches zero at a value 0.41 for concrete and 0.37 for mortar.

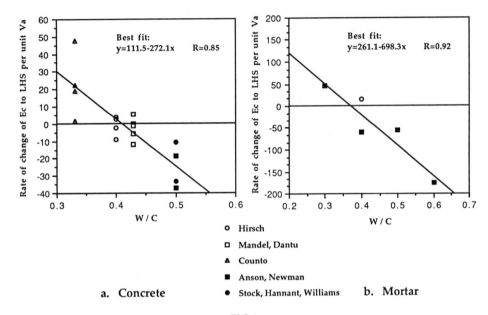

a. **Concrete**

○ Hirsch
□ Mandel, Dantu
▲ Counto
■ Anson, Newman
● Stock, Hannant, Williams

b. **Mortar**

FIG.3.
The influence of W/C on the effect of ITZ
on the modulus of elasticity of cement based composites.

Simulation of ITZ effect by Hashin's bounds for two-phase composites with imperfect interfaces and Discussion of the results

Recently, Hashin [15], using the variational method, has derived bounds for the elastic properties of two-phase composites with imperfect interfaces (HB). He assumes that the imperfect interface can be described and modeled as a compliant thin interfacial layer or interphase, and its parameters can be evaluated by interphase thickness and properties. For all finite values of the interphase parameters (imperfect interface) Hashin's bounds are lower than the Paul bounds while for infinite values (perfect interface) they reduce to Paul bounds.

In contrast to the case with perfect interface conditions (like Paul bounds, Hashin-Strikman bounds) Hashin's bounds depend on interface geometry. In the case where the inclusions are with spherical shape and known size distribution they can be represented by equal inclusions with equivalent radius where the the bounds assume the following form:

$$K_{(+)} = K_m V_m + K_a V_a / (1 + 3K_a / D_n A) \tag{11}$$
$$G_{(+)} = G_m V_m + G_a V_a / (1 + 5G_a / (2D_n + 3D_s + D_t)A) \tag{12}$$
$$K_{(-)} = (V_m / K_m + V_a / K_a + 3V_a / D_n A)^{-1} \tag{13}$$
$$G_{(-)} = (V_m / G_m + V_a / G_a + 2V_a (2/D_n + 3/D_s + 3/D_t)/5A)^{-1} \tag{14}$$

where:
D_n, D_s and D_t - interface parameters,
A - equivalent radius of the spherical inclusions.

Hashin [16] has shown that for an isotropic thin compliant interphase the interface parameters are given by:

$$D_n = (K_i + 4G_i/3)/T \tag{15}$$
$$D_s = D_t = G_i/T \tag{16}$$

where:
i - indicates interphase material,
T - interphase thickness.

Hashin's bounds for two-phase composites with imperfect interfaces like Paul bounds are not close enough (especially for high moduli ratios of the two phases) to be used for prediction of composite properties. Nevertheless they can be used to simulate and analyze the influence of ITZ on the elastic properties of cement-based composites. For the purpose of the current analysis the median of the calculated bounds is used as a more clear presentation of the main trend in the relation Ec - Va.

ITZ is simulated as a thin isotropic interphase with equivalent elastic properties. The influence of ITZ thickness and modulus of elasticity on the elastic properties of the composite is analyzed.

In the examples (**Figs.4 a-d**) cement mortars with W/C = 0.4, Em=17GPa and Ea=80GPa are considered based on the data of Anson and Newman [11]. The equivalent radius of the sand which is with normal size distribution is assumed to be 0.315 mm with corresponding specific surface 95 cm^2/cm^3 and the Poisson's ratio of ITZ is assumed to be the same as that of the matrix cement paste (0.25).

On **Fig.4a** in dashed lines are shown the upper and the lower HS bounds for this composite. For comparison Hashin's upper and lower bounds are computed for the case of perfect bond composite T=0μ (Paul bounds) and the case of ITZ with thickness T=20μ at the assumption that its equivalent modulus of elasticity is 50% of Em. It is evident that the introduction of ITZ layer causes a significant divergence of the bounded values from the trend Ec-Va for a perfect bond composite.

The influence of ITZ thickness (T) is demonstrated on **Fig.4b**. The median of Hashin-Sthtrikman bounds (in dashed lines), is the best estimate of the expected trend in the relation Ec-Va for a two-phase composite. For T=0 (perfect bond), the median of HB lay close to the median of HSB practically following the same trend. With the introduction of a 5μ layer of ITZ the main trend goes lower with the increase of Va. The bigger the ITZ thickness the bigger is the divergence of the HB median from the expected trend for a perfect bond composite. The thickness of ITZ is a function of W/C as well as of aggregate size. For higher W/C at a constant size distribution of the aggregates ITZ will be thicker and the divergence will be bigger.

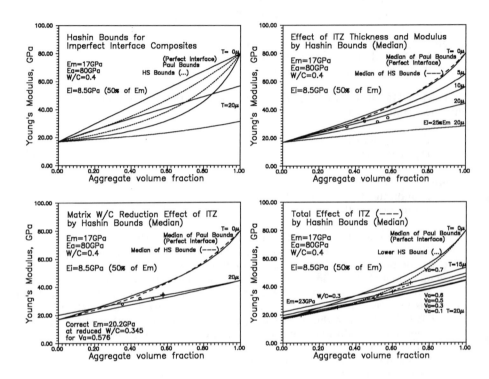

FIG.4.
Simulation of ITZ effect on modulus of elasticity of cement-based composites
with Hashin's bounds for composites with imperfect interfaces

Fig.4b shows also how a change in the equivalent elastic modulus of ITZ (Ei) will affect Ec. The HB median is calculated for ITZ with thickness 20μ and modulus of elasticity 50% and 25% of Em. The weaker ITZ the bigger the divergence from the trend for a perfect bond composite. Ei in cement-based composites is function of W/C. The higher the W/C of the composite the lower the Ei and the bigger the expected divergence.

The properties of ITZ are well related to the properties of the matrix. At the same time the formation of new interfaces also can influence in some extent the properties of the matrix or their properties are interrelated.

With the addition of the aggregates in the matrix a new surface appears. The free water from the matrix cement paste tend to move towards this new surface forming zones with high W/C, high porosity and lower modulus of elasticity. These zones have a layered structure and a gradient of properties in direction normal to the surface of the aggregates. That is why ITZ can be described only as a phase with some average or equivalent properties. At the same time the formation of a ITZ reduces the water content and the porosity in the matrix cement paste and increases its modulus of elasticity. This effect, is notable especially for high volume fractions of aggregates with relatively high specific surface and in some cases it can compensate to a significant extent the effect of ITZ.

Considering this, a more realistic trend for the modulus of elasticity of the composite will be obtained after the appropriate correction of Em in HB. An example is shown on **Fig.4c** where the correction of Em is calculated for aggregate volume fraction 70% at the following assumptions: average thickness of ITZ is 20μ, $Ei=8.5GPa$ is 50% of Em (which corresponds to matrix with $W/C=0.6$). The calculated volume of ITZ is 14% of the total volume or 47% of the volume of the matrix. W/C in the matrix is reduced to 0.284 which corresponds to $Em=24GPa$.

The reduction of W/C in the matrix is a function of the volume and the equivalent W/C of ITZ, or for a constant specific surface it is a function of Va. At the same time the volume and the porosity of ITZ depend on W/C of the matrix. That is why the two-phase modeling or bounding of the relation Ec-Va with fixed values for Em is not successful. A more realistic approach must take into account also the transfer of properties (elasticity) from the volume of cement paste transformed into a weaker ITZ to the decreasing volume of the stiffening matrix. In other words for cement composites Em, T and Ei must be considered functions of Va. The increase of Va and the volume of ITZ leads to increase of Em and Ei and reduction of T. This effect is shown on **Fig.4d** as elevation of HB median with the increase of Va. The dashed line gives the resultant trend for Ec revealing the influence of ITZ.

The formation of ITZ can be described as a rearrangement of the structure, transfer or redistribution of properties in the matrix in which process the formation of zones with higher porosity around the aggregates is accompanied by a reduction of porosity in the matrix cement paste. In some cases (low W/C) these interrelated processes can be in balance and macroscopically the composite can behave elastically similar to a two-phase composite with perfect interface or what is lost in the formation of ITZ is gained back in the improved matrix. For higher W/C the weakening effect of ITZ prevails which can result in a significant reduction of Ec.

In the light of this analysis the following interpretation of **Fig.3** is suggested: at higher W/C ($W/C>0.4$) the ITZ is with very low Ei and in a thicker layer and respectively in bigger relative volume to a unit interface surface. With the increase of Va it causes reduction of Ec that is bigger than the increase of Em caused by the reduction of the real W/C in the bulk cement paste. The overall result is a notable divergence from LHS. At W/C close to 0.4 the effect of ITZ is compensated by the increase in Em due to the decrease of the real W/C. As a result there is no notable effect on Ec and with the increase of Va it tends to follow LHS. At lower W/C ($W/C<0.35$) the ITZ is thinner and the values of Ei and Em are close. With the increase of Va the influence of ITZ is either negligible or positive which together with the increased Em as a result of the reduction of the real W/C causes a notable relative increase of Ec to LHS.

Besides W/C there are some other factors which can have a significant role in formation of ITZ and its properties. Chemical composition, fineness of grinding, amount, chemical and physical nature of additives and admixtures can significantly change the ability of cement paste to release and transfer the free water to the surface of aggregates. A good example is the influence of silica fume, which with its high specific surface immobilize the free water in the matrix cement paste and in this way strongly reduces the thickness and the volume of ITZ, which is a possible explanation of the experimental results of Cohen Goldman and Chen [8]. In high strength and high performance concretes as a result of the use of HRWR (W/C<0.3) and application of silica fume the thickness of ITZ is very small. In this case its properties can be significantly influenced by the aggregate type and finally affect the elastic properties of the composite.

CONCLUSIONS

There is a significant effect of ITZ on the elastic properties of concrete. It is demonstrated as a divergence of the experimental data from the general trend of Hashin-Shtrikman bounds with the increase of aggregate volume and respectively the ITZ volume. This effect is well related to W/C.

The properties of ITZ and the matrix in cement-based materials are interrelated. The formation of ITZ is a process of transfer of water leading to redistribution of the structure and the properties of the matrix. As a result the effect of ITZ can be partially or in some cases completely compensated by the changes in the matrix.

Hashin-Shtrikman bounds prove to be an effective tool for analysis of the elastic behavior of composite materials enabling the detection of such tiny influences as that of the ITZ in concrete.

Hashin's bounds for two-phase composites with imperfect interfaces can be successfully applied for simulation and modeling of the effect of ITZ on the modulus of elasticity of cement-based composites.

REFERENCES

1. Z.Hashin and J.Shtrikman, J.Mech.Phys.Solids, 11, 127 (1963)
2. Z.Hashin, J.Appl.Mech., 50, 481 (1983)
3. J.Mandel and P.Dantu, Ann. Ponts Chaussees, 133, 115 (1963)
4. J.P.Watt and R.J.O'Connell, Phys.Earth Planet.Inter., 21, 359 (1980)
5. P.G.M.Monteiro, Cem.Con.Res., 21, 947 (1991)
6. A.U.Nilsen and P.J.M.Monteiro, Cem.Con.Res., 23, 147 (1993)
7. A.U.Nilsen and P.J.M.Monteiro, Cem.Con.Res., 24, 194 (1994)
8. M.D.Cohen, A.Goldman and W.-F.Chen, Cem.Con.Res. 24, 95, (1994)
9. T.J.Hirsch, ACI Journal, 59, 427 (1962)
10. U.J.Counto, Mag.Con.Res., 16, 129 (1964)
11. M.Anson and K.Newman, Mag.Con.Res., 18, 115 (1966)
12. A.F.Stock, D.J.Hannant and R.I.T.Williams, Mag.Con.Res., 31, 225 (1979)
13. P.I.Simeonov, S.H.Ahmad, Cem.Con.Res. (in press) (1995)
14. R.M.Zimmerman, Mech.Res.Commun., 19 (6), 563, (1992)
15. Z.Hashin, J.Mech.Phys.Solids, 40, 767, (1992)
16. Z.Hashin, J.Mech.Phys.Solids, 39, 745, (1991)

A Method for Estimating the Dynamic Moduli of Cement Paste-Aggregate Interfacial Zones in Mortar

MENASHI D. COHEN, TURNG-FANG F. LEE, AND ARIEL GOLDMAN
Purdue University
School of Civil Engineering
West Lafayette, IN 47907

ABSTRACT

The objective of this paper is to propose a method to estimate the average values of the dynamic modulus of elasticity and the dynamic shear modulus of cement paste-aggregate interfacial zones in mortar by applying the Logarithmic Mixture Rule (LMR). Both portland cement mortars (PC mortars) and portland cement mortars with silica fume (SF mortars) are investigated and compared. The influence of silica fume on the dynamic moduli of interfacial zone is also examined. Results indicate that for the specific ingredients and mix design used, the dynamic modulus of elasticity of interfacial zone falls between 0.4 and 2.0 ($\times 10^6$ psi) for PC mortar and 1.2 to 2.2 ($\times 10^6$ psi) for SF mortar. These values are lower than the values obtained for PC mortar (4.2 $\times 10^6$ psi), PC paste (2.7 $\times 10^6$ psi), SF mortar (4.4 $\times 10^6$ psi), and SF paste (2.5 $\times 10^6$ psi).

INTRODUCTION

Concrete is a porous multiphase composite material with aggregate, bulk cement paste and interfacial zones of cement paste-aggregate as its important constituents. Normally, the interfacial zone is the weakest link among the three main components of concrete. Therefore, it may exhibit a significant influence on the mechanical properties of concrete.

The interfacial zone is composed of mainly calcium hydroxide crystals, C-S-H gel, ettringite, pores and microcracks. These natural pores and microcracks may affect the values of dynamic moduli of the interfacial zone under unstressed state. After applying tensile stresses, the microcracks formed at the paste-aggregate interfacial zone are the cause of prepeak-nonlinear behavior. Besides, the transition from linear to nonlinear response of a cementitious material in tension is primarily governed by the extent of available interfacial microcracks [1]. At stressed state, the static moduli of interfacial zone may be more suitable to describe the physical properties of interfacial zone. In general, the higher the dynamic moduli, the higher the corresponding static moduli will be obtained.

From a mechanical point of view, the extent of microcracks in the paste-aggregate interfacial zone is governed by the degree of strain mismatch between aggregate and bulk cement paste under the applied loading. For a stress equilibrium condition, the strain difference between aggregate and paste along interfacial zone can be related to the discrepancy of the moduli of elasticity and the shear moduli as well among aggregate, interfacial zones, and bulk cement paste.

In making a high-strength concrete, one important strategy is to focus on how to increase the stiffness of interfacial zones of cement paste-aggregate to a comparative level as bulk paste and aggregate. This may be achieved by adding a mineral admixture such as silica fume, fly ash, or

Mat. Res. Soc. Symp. Proc. Vol. 370 © 1995 Materials Research Society

waste furnace slag to the concrete mixture. The remaining question is how much the moduli of the interfacial zones are increased. In practice, it is hard to find an instrument to measure with accuracy these moduli values due to the small thickness of interfacial zones (10 - 50 µm). So, a technique is evolved to estimate the values of these moduli.

From a microscopic point of view, the local value of dynamic modulus may vary depending on the local porosity of interfacial zones which may be quite large near the aggregate surface. However, the technique developed in this paper is to assess the average values of the dynamic modulus of elasticity and the dynamic shear modulus of interfacial zones in mortar. These moduli are the indices to indicate how compact the interfacial zones are, after mortars being mixed. They can be used to estimate the moduli consistency among aggregate, interfacial zones, and bulk cement paste.

EXPERIMENTAL

There are several factors that may affect the dynamic moduli of interfacial zones. These include W/C, curing age (related to degree of hydration and porosity), aggregate content, and the specific surface area of aggregate, the mixture of silica fume in the mix design, etc. Among the factors, for simplicity, W/C of 0.52 was fixed. The sand in the mortar specimens is high purity natural quartzite from Baraboo, Wisconsin. The specific surface area of sand is calculated by using the well-known equation: $surface\ area = 6/(\rho \cdot d_{mean})$, where ρ is the density of sand (R. D. = 2.65 for quartzite) and d_{mean} is the mean diameter of sand grains. The volume content of sand was kept constant at 37%. Five different size fractions of sand were used for PC mortar specimens (#4-#8, #8-#16, #16-#30, #30-#50, and #50-#100 ASTM sieve no.), each specimen containing one size fraction. The calculated specific surface areas for these size fractions of sand are 6.4, 12.8, 25.4, 50.3, and 100.6 cm^2/g respectively. The calculated average thicknesses of paste shell surrounding one sand particle are 984, 490, 246, 125, and 62 µm respectively. Besides, three size fractions were used for SF mortar specimens (#4-#8, #16-#30, and #50-#100) as indicated in Table 1 of reference [6].

Two groups of mortar specimens were prepared. One was the PC mortar specimens group acting as the control group, the other was the SF mortar (with 10% of silica fume substituted for portland cement by mass basis) specimens group. Similar to mortar, there were also two groups of paste specimens prepared for testing. All of the specimens were cured in saturated lime-water at 23° C.

METHODOLOGY

The first step in estimating the average moduli values of the interfacial zones is to determine the average thickness of interfacial zone. The bulk paste and the interfacial zone are heterogeneous. Both contain pores, microcracks, and solid phase. These components are subject to change with degree of hydration, water content, and environmental conditions such as temperature, humidity, etc.

The difference between bulk paste and interfacial zone is variable and sometimes the boundary is quite vague, so it is not easy to measure the thickness of interfacial zone accurately. By using scanning electron microscopy (SEM) for investigating lightweight aggregate and paste, a 60 µm thickness of interfacial zone was reported by Khokhrin [2]. Also, measuring by SEM,

Bentur and Cohen [3] suggested the thickness of interfacial zone to be about 20 to 50 μm. Winslow et al. [4] proposed a value in the range of 15 to 20 μm as measured by mercury intrusion method. These values are consistent with the 10 - 50 μm range as indicated by Mehta and Monteiro [5]. Based on the above values, the calculation of the average dynamic moduli of paste-aggregate interfaces will consider values of 10, 30, or 50 μm as interfacial zone thickness.

The basic idea of this investigation is to apply LMR model for mortar specimen and then to calculate the average dynamic moduli of interfacial zone. So, the second step is to expand the LMR for modulus of elasticity to three phases. Thus:

$$\log E_m = V_s \log E_s + V_p \log E_p + V_i \log E_i \qquad (1)$$

where E_m is dynamic modulus of elasticity of mortar (determined experimentally).
  E_s is dynamic modulus of elasticity of sand (determined experimentally).
  E_p is dynamic modulus of elasticity of bulk paste (determined experimentally).
  E_i is dynamic modulus of elasticity of interfacial zone (calculated according to equ. (1)).
  V_s is volume fraction of sand (controlled by mixture proportioning).
  V_p is volume fraction of bulk paste (controlled by mixture proportioning and by assumption of V_i).
  V_i is volume fraction of interfacial zone (assumed value).

Rearranging equation (1) yields:

$$E_i = 10^{\frac{\log E_m - V_s \log E_s - V_p \log E_p}{V_i}} \qquad (2)$$

The dynamic moduli of elasticity in the above equations (1) and (2) can also be substituted by the dynamic shear moduli (G_m, G_s, G_p, and G_i), respectively, in order to calculate the dynamic shear modulus of interfacial zone.

The specific surface area was calculated. Then, it is quite straight forward to calculate the volume of interfacial zones according to the assumed interfacial zone thickness 10, 30, or 50 μm. The mix design of the mortar is known. So, the volume content of sand can be determined. Since the total volume of mortar can be measured, by subtracting the volumes of sand and interfacial zones from it, we get the volume of bulk paste. So, V_s, V_p, and V_i can be calculated. By applying resonant frequency method, the dynamic modulus of elasticity and shear modulus of mortar (E_m and G_m) can be determined. Also the dynamic modulus of elasticity and shear modulus of paste (E_p and G_p) can be determined with resonant frequency method by testing the separately prepared paste specimens of the same W/C ratio and silica fume content as in the mortar. The dynamic modulus of elasticity and shear modulus of sand (E_s and G_s) were determined by ultrasonic pulse velocity method by measuring the longitudinal and shear velocities that transmit in the rock with hydrostatic pressure of 100 atm. The reason for applying high hydrostatic pressure on rock in measuring the pulse velocities is to eliminate the effect of pore spaces in the rock that should be much less in the sand. The dynamic modulus of elasticity of the sand used is 14×10^6 psi, and the dynamic shear modulus of the sand is 6.5×10^6 psi. Thus, the remaining unknowns are the dynamic modulus of elasticity and shear modulus of interfacial zone (E_i and G_i) which can be computed through equation (2).

SENSITIVITY OF MODEL

The error sensitivity of E_i to the specific surface area of sand S for a fixed volume content of sand is estimated to be:

$$\frac{dE_i}{dS} = \left(\frac{V_s \log E_s + V_p \log E_p - \log E_m}{V_i^2} \frac{dV_i}{dS} + \frac{0.4343}{E_m V_i} \frac{dE_m}{dS} - \frac{\log E_p}{V_i} \frac{dV_p}{dS} \right) \ln 10 \times 10^{\frac{\log E_m - V_s \log E_s - V_p \log E_p}{V_i}} \quad (3)$$

This error sensitivity depends on mix design and the constituent material properties. For one of the mix designs, the error sensitivity of E_i to the specific surface area of sand is tabulated in Table 1;

Table 1. The error sensitivities of E_i to different specific surface areas of sand S

Specific surface area of sand S (cm²/g)	6.4	12.8	25.4	50.3	100.6
Error sensitivity $\left\| \frac{dE_i}{dS} \right\| \left(\frac{10^6 \, psi}{cm^2 \, / \, g} \right)$	4.4×10^{-1}	1.7×10^{-2}	1×10^{-2}	6×10^{-3}	1.8×10^{-3}

From the data of the Table 1, it can be concluded that the higher the specific surface area, the smaller the absolute value of error sensitivity, thus the more accurate the calculated results. This is the reason why the largest specific surface area of sand (= 100.6 cm²/g) in the mix design was chosen to calculate the dynamic moduli.

It must be pointed out that an increase of the specific surface area would lead to the increase in the volume of interfacial zone. However, the increase of the specific surface area also reduces the dynamic modulus of elasticity for PC mortar but it does not lead to reduction in SF mortar as can be obtained from Figure 2. of reference [6]. In this Figure, the dynamic moduli of elasticity of mortar specimens vary depending on three factors: the specific surface area of sand, the curing age, and whether there is silica fume added to the specimens or not. So, the dynamic moduli of elasticity of interfacial zones will also vary depending on these three factors as can be inferred from equation (1).

RESULTS

1. Figure 1 presents the relationship between the dynamic moduli of elasticity and curing age (1, 3, 7, 14, 28, and 56 days) obtained experimentally for mortar and paste. The calculated values of moduli of elasticity for the interfacial zones assuming 10, 30, or 50 μm thickness are plotted as well.

2. Figure 2 shows the relationship between the dynamic shear moduli and curing age for the same conditions as indicated in results 1.

CONCLUSIONS

1. Figure 1: According to the assumed thicknesses of interfacial zones; 10 to 50 μm and the fixed W/C ratio (= 0.52), the dynamic modulus of elasticity of interfacial zone should fall between 0.4 and 2.0 ($\times 10^6$ psi) for PC mortar and 1.2 to 2.2 ($\times 10^6$ psi) for SF mortar (for 10% replacement by weight of cement with silica fume). The data and calculations suggest that silica fume actually

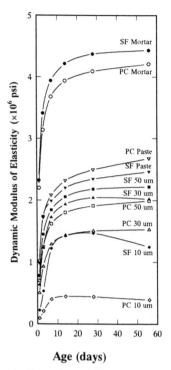

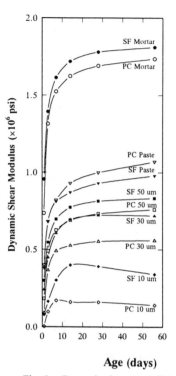

Fig. 1.- Dynamic moduli of elasticity ($\times 10^6$ psi) versus age (days) for mortar, paste, and interfacial zones with assumed thickness of 10, 30, or 50 μm with/without silica fume. The dynamic modulus of elasticity of sand is 14×10^6 psi. W/C = 0.52.

Fig. 2. - Dynamic shear moduli ($\times 10^6$ psi) versus age (days) for mortar, paste, and interfacial zones with assumed thickness of 10, 30, or 50 μm with/without silica fume. The dynamic shear modulus of sand is 6.5×10^6 psi. W/C = 0.52.

increases the stiffness of interfacial zone. If the thickness of the interfacial zone is assumed to be 10 μm, the calculated value of the dynamic modulus of elasticity of the interfacial zone with silica fume mixture is about 3 times larger by comparing to that of pure portland cement. If the interfacial zone is assumed to be 50 μm thick, silica fume increases the dynamic modulus value of the interfacial zone by about 12% compared to that of pure portland cement.

2. The dynamic modulus of elasticity of PC paste is lower than that of SF paste with 10% substitution of silica fume. For mortar, the trend is the opposite, Figure 1. This suggests that the benefit derived in increasing the stiffness from the addition of silica fume may be limited to the interfacial zones.

3. For any one of the assumed thicknesses of the interfacial zone, the calculated dynamic modulus of elasticity of interfacial zone is lower than that of bulk paste or mortar, Figure 1. Physically there exists one unique model with the specific volume fractions for interfacial zones and bulk cement paste. Mathematically three assumed thicknesses of interfacial zones were chosen for

411

calculations. If the assumed thickness of interfacial zone is larger (i.e., 50 μm), the volume fraction of bulk cement paste will be smaller. The assumed volume of interfacial zone will cover the regions which are more condensed and should be included in bulk cement paste. In consequence, the calculated dynamic modulus of elasticity of interfacial zone is approaching that of bulk cement paste.

4. In reference to Figure 2, the dynamic shear modulus shows the same trends as the dynamic modulus of elasticity and the same conclusions (1, 2, and 3) apply. The dynamic shear modulus of interfacial zone falls between 0.1 and 0.7 ($\times 10^6$ psi) for PC mortar and 0.3 to 0.8 ($\times 10^6$ psi) for SF mortar. The data suggests, therefore, that silica fume increases the rigidity of the interfacial zone. If the thickness of the interfacial zone is assumed to be 10 μm, silica fume will increase the rigidity of interfacial zone to be about 2.4 times by comparing to that of pure portland cement. If the interfacial zone is assumed to be 50 μm thick, silica fume increases its rigidity by about 9% compared to that of pure portland cement as well.

5. Similar to the dynamic modulus of elasticity, the dynamic shear modulus of PC mortar is lower than that of SF mortar with 10% substitution of silica fume. For paste, the trend is the opposite, Figure 2. This also suggests that silica fume's improvement of rigidity may be limited to the interfacial zones.

6. In Figure 1 and 2, there are several curves show that the values of moduli are decreasing with curing age after 14 or 28 days. This may be due to the deviation of experimental data or the error propagation of mathematical calculations. It is hard to see this phenomenon with physical sense.

ACKNOWLEDGMENTS

This research was supported in part by grants from the National Science Foundation (MSS-9202123, Program Director: Dr. K. P. Chong), INDOT, the Purdue Research Foundation, and the NSF-ACBM center. Thanks are due to Professor Nikolas I. Christensen, Department of Geosciences, Purdue University, for supplying the experimental data of the dynamic modulus of elasticity and shear modulus of Baraboo, Wisconsin, quartzite rock.

REFERENCES

1. B. L. Karihaloo, A Carpinteri, and M. Elices, *Fracture Mechanics of Cement Mortar and Plain Concrete*, Advanced Cement Based Materials, Vol. 1, No. 2, pp. 92-105, Dec. 1993.

2. N. K. Khokhrin, *The Durability of Lightweight Concrete Structural Members*, Kuibyshev, U.S.S.R. 1973.

3. A. Bentur and M. D. Cohen, *Effect of condensed Silica Fume on the Microstructure of the Interfacial Zone in Portland Cement Mortars*, Journal of the American Ceramic Society, Vol. 70, No. 10, pp. 738-743, 1987.

4. D. N. Winslow, M. D. Cohen, D. P. Bentz, K. A. Snyder, and E. J. Garboczi, *Percolation of Interfacial Zone Pores in Cement Mortar and Concrete*, Cement and Concrete Research, Vol. 24, pp. 25-37, 1994.

5. P. K. Mehta and P. J. M. Monteiro, *Concrete - Structure, Properties, and Materials*, 2nd ed., (Prentice Hall, New Jersey, 1993), p. 18.

6. M. D. Cohen, A. Goldman and W.-F. Chen, *The Role of Silica Fume in Mortar: Transition Zone versus Bulk Paste Modification*, Cement and Concrete Research, Vol. 24, pp. 95-98, 1994.

EFFECT OF THE TRANSITION ZONE ON
THE BULK MODULUS OF CONCRETE

MELANIE P. LUTZ* and PAULO J. M. MONTEIRO**
*Dept. of Materials Science & Mineral Engineering, Univ. of California, Berkeley, CA 94720
**Dept. of Civil Engineering, Univ. of California, Berkeley, CA 94720

ABSTRACT

In concrete, non-uniformities in the hydration process, due to the the "wall effect" produced by the aggregate (inclusion) particles, lead to an interfacial transition zone (ITZ) that is characterized by an increase in porosity near the inclusions. This increase in porosity may in turn be expected to cause a local decrease in the elastic moduli. We have modeled the effect of the ITZ by assuming that the elastic moduli vary smoothly in the vicinity of the inclusions, according to a power law. The exponent in the power law is chosen based on the estimated thickness of the ITZ. For this model, a closed-form expression can be found for the overall effective bulk modulus. The predicted bulk modulus of the concrete depends on known properties such as the elastic moduli of the bulk cement paste and the inclusions, the volume fraction of the inclusions, as well as on the elastic moduli at the interface. By comparing the model predictions to measured data, we can obtain estimates of the elastic moduli at the interface. Application of this inverse procedure to a set of data from the literature on mortar containing sand inclusions leads to the conclusion that the modulus at the interface is about 15-50% lower than in bulk cement paste.

INTRODUCTION

In concrete, non-uniformities in the hydration process lead to a "transition zone" that is characterized by an increase in porosity near the inclusions, along with other microstructural differences [1,2]. Although little quantitative analysis has been done to study the elastic moduli in this zone, it seems clear that one effect of the locally-increased porosity will be to cause the elastic moduli of the cement paste to decrease near the inclusions, as compared to the moduli of bulk cement paste. If this were indeed the case, one implication would be that it may not be possible to accurately model the elastic behavior of concrete under the traditional assumption that it consists of inclusion particles imbedded in a homogeneous matrix phase. This implication is consistent with the recent findings of Nilsen and Monteiro [3], who analyzed the experimental data of Hirsch [4] and showed that the measured elastic moduli could not be rationalized under the assumption that the concrete consisted of two homogeneous phases (aggregate and cement paste), since the measured moduli violated the Hashin-Shtrikman bounds [5]. They proposed that concrete could be more accurately modeled as a three-phase material in which each aggregate particle is surrounded by an interfacial transition zone, which in turn is imbedded in a matrix consisting of "bulk" cement paste. The work of Scrivener and Gartner [6], in which computerized image analysis was used to quantify the porosity gradient in the transition zone, shows that there is no clear demarcation between the transition zone and the unaltered cement paste. Hence, it is likely that the mechanical properties of this transition zone are not uniform, but vary gradually as a function of distance from the inclusion.

In this paper we model the elastic properties of concrete by assuming that the aggregate particles are spherical, and are surrounded by a matrix in which the elastic moduli are described by a constant term plus a term that varies with radius according to a power law. The thickness of the interfacial transition zone in the model is controlled by the exponent in the power law, and therefore can be varied from occupying a very thin layer around each inclusion to occupying most of the region between inclusions. The idealization that the

Mat. Res. Soc. Symp. Proc. Vol. 370 © 1995 Materials Research Society

inclusions are spherical is justified by the fact that slight variations from sphericity are of minor importance, particularly if the inclusions are stiffer than the matrix [7]. We then utilize the closed-form analytical solution for the effective bulk modulus of this type of material that has recently been found by Lutz and Zimmerman [8]. The predicted effective bulk modulus depends on "known" parameters such as the volume fraction of sand and the elastic moduli of the sand and the bulk cement paste, as well as on the elastic moduli at the interface with the inclusions. If the model is used to "invert" measured data, the elastic moduli at the interface can be inferred. This in effect yields a quantitative and non-destructive means of estimating the properties of the interfacial transition zone.

MODEL OF THE TRANSITION ZONE AS A RADIALLY-INHOMOGENEOUS REGION

We propose a conceptual model of concrete as shown in Fig. 1. The aggregate particles are assumed to be spherical, with radius a. Outside of each inclusion, the elastic moduli vary smoothly with radius, and approach those of the bulk cement paste as r increases. The precise manner of variation of the moduli will not generally be known, and the assumption of a smooth variation is of course an approximation. The main requirements of the equations used to model the variation in moduli are that they decay away from the interface, and asymptotically level off to some constant values that represents the bulk cement paste. Furthermore, for the model to be sufficiently general and flexible, the values of the elastic moduli at the interface, as well as the thickness of the interphase zone, should be adjustable parameters. With these conditions in mind, and guided in part by the findings of Theocaris [9], we will assume that the two moduli K and G vary according to

$$K(r) = K_{cp} + (K_{if} - K_{cp})(r/a)^{-\beta},$$ (1)

$$G(r) = G_{cp} + (G_{if} - G_{cp})(r/a)^{-\beta},$$ (2)

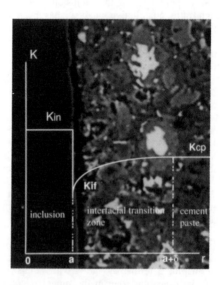

Fig. 1. Schematic diagram of the elastic moduli variation described by eqs. (1,2), superimposed over a photograph of an interfacial transition zone.

where a is the radius of the inclusion, the subscript cp refers to the bulk cement paste, and the subscript if refers to the interface with the inclusion. The parameter β controls the rate at which the moduli decay away from the inclusion; larger values of β correspond to interphase zones that are more localized. In order for the analytical solution found by Lutz and Zimmerman [8] to be applicable, β must be an integer. As the moduli variations will never be known precisely, this restriction poses no practical limitation on the applicability of the model. Finally, note that in solving the elasticity equations, it is convenient to use the moduli K and G instead of the commonly-used engineering parameters E and ν. However, these parameters are related to each other through the identities

$$E = \frac{9KG}{3K+G}, \qquad \nu = \frac{3K-2G}{6K+2G}. \tag{3}$$

Lutz and Zimmerman [8] found a closed-form solution for the displacements, stresses and strains around an inclusion such as shown in Fig. 1 that was subjected to far-field hydrostatic loading. They then used this solution to derive the following expression for the effective bulk modulus of a material that contained a volume fraction c of inclusions:

$$\frac{K_{eff}}{K_{cp}} = \frac{1+(4G_{cp}/3K_{cp})fc}{1-fc}, \tag{4}$$

where

$$f = \frac{3(K_{in}-K_{if})\sum\limits_{n=0}^{\infty}\Gamma_{n\beta} + [K_{if}+(4/3)G_{if}]\sum\limits_{n=0}^{\infty}n\,\beta\Gamma_{n\beta}}{3(K_{in}-K_{if})\sum\limits_{n=0}^{\infty}\Gamma_{n\beta+3} + [K_{if}+(4/3)G_{if}]\sum\limits_{n=0}^{\infty}(n\,\beta+3)\Gamma_{n\beta+3}}, \tag{5}$$

and the subscript in refers to the inclusions. The coefficients Γ are found from the following recursion relation:

$$\Gamma_{n+\beta} = \frac{-\{(K_{if}-K_{cp})[n^2+(\beta-3)n+\beta]+(4/3)(G_{if}-G_{cp})[n^2+(\beta-3)n-2\beta]\}}{(n+\beta)(n+\beta-3)[K_{cp}+(4/3)G_{cp}]}\Gamma_n, \tag{6}$$

with $\Gamma_0 = \Gamma_3 = 1$. From Γ_0 and Γ_3, the recursion relation (6) will generate $\{\Gamma_\beta, \Gamma_{2\beta}, ...\}$ and $\{\Gamma_{\beta+3}, \Gamma_{2\beta+3}, ...\}$, which are the only Γ coefficients that are needed in eq. (5). In the limiting case where the interphase zone is homogeneous, all the coefficients except Γ_0 and Γ_3 will vanish, and $K_{if} \to K_{cp}$, etc., in which case $f \to 3(K_{in}-K_{cp})/(3K_{in}+4G_{cp})$, and eq. (4) reduces precisely to the result that has been found previously for the effective bulk modulus of a material composed of spherical inclusions in an *homogeneous* matrix [7,10]; this expression also coincides with the theoretical Hashin-Shtrikman lower bound [5].

APPLICATION OF MODEL TO THE DATA OF ZIMMERMAN ET AL.

Zimmerman et al. [11] performed measurements of compressional wavespeeds V_p and shear wavespeeds V_s on a suite of mortar specimens containing varying concentrations of sand inclusions. The bulk modulus can be found from the two wavespeeds using the relationship [11]

$$K_{eff} = \rho_{eff}\left(V_p^2 - \frac{4}{3}V_s^2\right), \tag{7}$$

where the effective density ρ_{eff} is related to the densities of the cement paste and sand inclusions by

$$\rho_{eff} = (1-c)\rho_{cp} + c\rho_{in}. \tag{8}$$

Note that we are assuming that density of the cement paste is uniform; more sophisticated models might attempt to account for the variation of density in the ITZ.

The measured bulk moduli of the eleven specimens are shown in Fig. 2, as a function of the sand concentration. If the matrix of each specimen was homogeneous, the effective bulk moduli of the mortar would be expected to lie near the Hashin-Shtrikman lower bound. This is because the effective modulus of a composite is mainly controlled by the matrix phase, not the inclusions, and in the case of mortar the bulk modulus of the matrix phase is lower than that of the inclusion phase. All of the experimental values lie below the Hashin-Shtrikman bounds, which correspond to the curve that is labeled $D = 0$ (D is defined below). In order to evaluate the bounds, and to evaluate the moduli predicted by the variable-moduli interphase model, values are needed for the elastic moduli of both the cement paste and the sand inclusions. The elastic moduli of the bulk cement paste were measured by Zimmerman et al. [11] to be $K_{cp} = 20.8$ GPa and $G_{cp} = 11.3$ GPa; these values correspond in Fig. 2 to the datum point at zero sand concentration. The elastic moduli of the sand inclusions are $K_{in} = 44.0$ GPa and $G_{in} = 37.0$ GPa [7].

Table 1. Elastic moduli and density of mortar specimens tested by Zimmerman et al. [11].

Material	K (GPa)	G (GPa)	E (GPa)	ν	ρ (kg/m³)
Sand inclusions	44.0	37.0	86.7	0.17	2700
Cement Paste	20.8	11.3	28.7	0.27	2120
Interface	20.8(1-D)	11.3(1-D)	28.7(1-D)	0.27	2120

In order to attempt to rationalize the fact that the data fall below the lower bounds for a two-component material, we assume that the moduli in the matrix can be described by eqs. (1,2). As we would like to limit the number of adjustable parameters in our model, while focusing on the effect that the lowered modulus in the ITZ has on the macroscopic elastic behavior, we will assume that K and G vary at the same relative rate near the inclusions, which implies (see eq. (3)) that the Poisson's ratio is uniform throughout the ITZ. As the variation in moduli is assumed to be due to the presence of microporosity in the ITZ, the assumption that K and G vary at the same rate is in accordance with micromechanical models of the effect of porosity on elastic moduli [12]. As the modulus K_{if} is expected to lie between 0 and K_{cp}, we will parametrize the interface modulus by defining $K_{if} = (1-D)K_{cp}$, where D is a *local* damage parameter. Under these assumptions, eqs. (1-3) can be used to express the elastic moduli in the matrix as

$$K(r) = K_{cp} + (K_{if} - K_{cp})(r/a)^{-\beta} = K_{cp}[1 - D(r/a)^{-\beta}], \tag{9}$$

$$G(r) = G_{cp} + (G_{if} - G_{cp})(r/a)^{-\beta} = G_{cp}[1 - D(r/a)^{-\beta}]. \tag{10}$$

The final parameter needed to implement the model is β. Although in the present model the ITZ does not have a clearly defined thickness, we will assume that the "outer boundary" of the transition zone, as determined by visual observation, corresponds to the location at which the moduli perturbation has decayed to 1% of its value at the interface. In other words, the outer edge of the ITZ will be located at some distance $r = a + \delta$, defined by

$$\frac{K(r = a + \delta) - K_{cp}}{K_{if} - K_{cp}} = 0.01, \tag{11}$$

where a is the inclusion radius and δ is the "thickness" of the ITZ (see Fig. 1). Eqs. (9) and (11) can be combined to yield an estimate of β:

$$[(a + \delta)/a]^{-\beta} = 0.01 \quad \Rightarrow \quad \beta = \frac{\ln(100)}{\ln(1 + \delta/a)}. \tag{12}$$

The sand grains used in the mortar specimens tested by Zimmerman et al. [11] had radii of about 50 µm. A typical value for the ITZ thickness δ, based on the porosity gradients measured by Scrivener [13], is about 40 µm. Using these values in eq. (12) yields $\beta = 7.83$. As β must be an integer in eqs. (5,6), we will therefore take $\beta = 8$. Note that as the logarithm is a slowly-varying function, the estimate of β given by eq. (12) is not very sensitive to our arbitrary use of a 1% moduli perturbation as the defining condition for the ITZ thickness.

Estimation of the effective bulk modulus is now accomplished as follows. We first choose a value of the local damage parameter D, which is equivalent to specifying that the moduli at the interface are lower by a factor of $(1 - D)$ than those in the bulk cement paste. Using the elastic moduli parameters discussed above, which are summarized in Table 1 for convenience, we then compute the Γ coefficients from eq. (6), after which we find K_{eff} from eqs. (4,5). The computed values of K_{eff} are plotted in Fig. 2, as functions of inclusion concentration, for various values of D. If desired, the calculation procedure outlined above could be iterated for each measured value of K_{eff}, until a value of D was found that led to an exact match be-

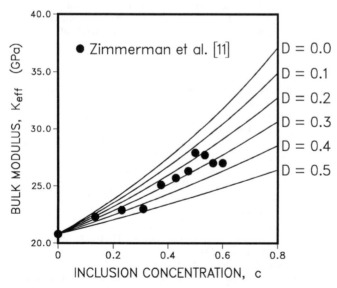

Fig. 2. Bulk modulus of mortar containing a volume fraction c of sand inclusions, each surrounded by a radially-inhomogeneous transition zone. The calculation procedure is described in the text; the parameters are listed in Table 1. By comparing the data to the theoretical curves, we can estimate the elastic modulus at the interface.

tween the model predictions and the measured modulus. Given the assumptions and idealizations inherent in the model, however, this degree of numerical precision does not seem warranted. Nevertheless, the curves in Fig. 2 are sufficiently closely-spaced that the best-fitting values of D can be estimated to two significant figures by visual interpolation. These best-fitting values of D all fall in the range of 0.15-0.5. If one value of D were chosen to fit all ten measured values, $D = 0.3$ would seem to be the best choice.

SUMMARY AND DISCUSSION

A conceptual model of concrete has been proposed in which the aggregate particles are modeled as spheres surrounded by a radially-inhomogeneous matrix. The elastic modulus in the matrix is represented by a constant term plus a term that decays with radius according to a power law. The exponent in the power law is determined by the thickness of the interfacial transition zone (ITZ), which can be estimated by visual observation. The effective bulk modulus is then found using the exact solution recently derived by Lutz and Zimmerman [8]. The only open parameter in the model is the bulk modulus of the cement paste at the interface with the inclusions, K_{if}. By matching the model predictions to experimental data [11] on mortar specimens containing varying amounts of sand inclusions, we were able to estimate the bulk moduli at the interface. The estimated values of K_{if} were 15-50% lower than those of bulk cement paste. The model assumes that the interfacial transition zones around each inclusion are similar; in reality, the damage parameter D represents an average value taken over all the inclusions. Nevertheless, despite this and other idealizations, the model offers a micromechanically-based explanation for the fact that the measured elastic moduli fall below the Hashin-Shtrikman bounds for a two-component material.

ACKNOWLEDGEMENTS

The work of M. Lutz was supported by California Legislative Grant Account No. 442402-19900 and by National Science Foundation Graduate Engineering Education for Women and Minorities Fellowship No. EID-9018414. The work of P. Monteiro was supported by NSF-8957183.

REFERENCES

1. S. Mindess, in: Materials Science of Concrete, edited by J. P. Skalny (American Ceramic Society, Westerville, OH, 1989), pp. 163-180.
2. P. K. Mehta and P. J. M. Monteiro, Concrete: Structure, Properties, and Methods, 2nd ed. (Prentice-Hall, Englewood Cliffs, NJ, 1993).
3. A. U. Nilsen and P. J. M. Monteiro, Cement Concr. Res. 23, 147-151 (1993).
4. T. J. Hirsch, J. Amer. Concrete Inst. 59, 427-451 (1962).
5. Z. Hashin and S. Shtrikman, J. Franklin Inst. 271, 336-341 (1961).
6. K. L. Scrivener and E. M. Gartner, in Bonding in Cementitious Composites, edited by S. Mindess and S. P. Shah, (Mater. Res. Soc. Proc. 114, Pittsburgh, PA, 1988) pp. 77-85.
7. G. T. Kuster and M. N. Toksöz, Geophysics 39, 587-606 (1974).
8. M. P. Lutz and R. W. Zimmerman, submitted for publication.
9. P. S. Theocaris, in: Composite Interfaces, edited by H. Ishida and J. L. Koenig (North-Holland, Amsterdam, 1986), pp. 329-345.
10. T. Mori and K. Tanaka, Acta Metall. 21, 571-574 (1973).
11. R. W. Zimmerman, M. S. King, and P. J. M. Monteiro, Cement Concr. Res. 16, 239-245 (1986).
12. R. W. Zimmerman, Mech. Maters. 12, 17-24 (1991).
13. K. L. Scrivener, in: Materials Science of Concrete, edited by J. P. Skalny (American Ceramic Society, Westerville, OH, 1989), pp. 127-161.

THE INTERFACIAL TRANSITION ZONE: "DIRECT" EVIDENCE ON COMPRESSIVE RESPONSE

DAVID DARWIN
Department of Civil Engineering, University of Kansas, Lawrence, KS 66045

ABSTRACT

There is little question that the strength of the interfacial transition zone (ITZ) between cement paste and aggregate affects the compressive strength of concrete. The key question, rather, is to what degree? It is difficult to directly measure the response of the overall composite to changes in interfacial properties, since it is difficult to isolate interfacial strength as the only variable.

Research on the effects of interfacial strength on the compressive response of concrete that comes the closest to providing direct evidence is summarized. The studies, dating to the 1950's, include both experimental and analytical efforts aimed at isolating the effects of the ITZ, as well as experimental efforts that are considered to provide strong indirect evidence. The research shows that the ITZ plays a measurable role in the response of concrete to compressive stress, but that its role is overshadowed by the properties of the cement paste and aggregate constituents of concrete and the heterogeneous nature of the composite.

INTRODUCTION

It is widely recognized that significant differences in structure exist between bulk cement paste and cement paste located in proximity to aggregate particles[1-9]. Paste near an aggregate surface exhibits a smaller fraction of unhydrated cement particles (out to about 40 μm) and greater porosity (especially out to about 10 μm) than cement paste in regions located farther from the aggregate[9]. This region is referred to as the interfacial transition zone (ITZ) and has an estimated thickness of 15 to 50 μm, depending on the method of estimation[3, 9, 10].

Early studies using fractured surfaces of mortar and concrete indicated that the zone consists of duplex layers, dominated by calcium hydroxide crystals with c-axes normal to the aggregate surface[11]. More detailed studies using flat polished surfaces indicate that the material in the interfacial zone is principally calcium silicate hydrate[8], with an average calcium hydroxide (CH) content only slightly higher than surrounding paste. The predominance of CH in earlier observations is likely due to the use of fractured surfaces, which are not representative of the true full cross-section. CH will appear more often on fractured surfaces due to the ease of cleavage along its basal plane.[8]

The observed differences in structure, combined with tests of interfacial strength[12, 13], has led to the conclusion that the interfacial transition zone is the "weak link" in the strength of concrete. It is often[6, 7, 14-17] but not universally[18-20] assumed that the ITZ plays a dominant role in the compressive as well as tensile strength of concrete. This concern has led to a continuing interest in the interfacial transition zone.

This paper summarizes research aimed at directly establishing the role of the interfacial transition zone in controlling the compressive strength of concrete. Experience has shown that it is difficult to measure changes in strength due to specific changes in interfacial properties. However, such measurements have been attempted in a number of studies, with various levels of success.

The studies used to evaluate the contribution of interfacial strength to the compressive strength of concrete can be placed in three categories:

1. Experimental studies in which the strength of the interface between aggregate and paste is artificially modified (interfacial properties may or may not be measured) and the effect on compressive strength of concrete is determined.

2. Finite element studies in which the properties of cement paste or mortar, aggregate, and the

419

ITZ are modeled. Of particular interest are those studies in which the properties of the ITZ are varied and the effects on the analytical response are determined.

3. Experimental studies in which the paste constituent is modified (usually by replacement of cement by silica fume) in a way that will result in changes in the ITZ. The resulting changes in the properties of the concrete are measured.

The following evaluation of these studies is aimed at determining the degree to which changes in the interfacial region affect the compressive strength of concrete.

EXPERIMENTAL STUDIES MODIFYING INTERFACIAL BOND STRENGTH

There are two aspects of interfacial strength: 1) the strength of the paste in the interfacial transition zone itself and 2) the bond strength between cement paste and aggregate. The experimental studies discussed in this section have involved attempts to modify the latter without changing the properties of the paste within the ITZ.

Studies of the effect of paste-aggregate bond strength on the compressive strength of concrete date at least to the 1950's[21, 22]. Studies by Dantinne[22], Shah and Chandra[23], and Nepper-Christen-sen and Nielson[24] used relatively thick, soft coatings on coarse aggregate particles to reduce bond strength. Dantinne[22] used a thick bituminous coating; Nepper-Christensen and Nielson[23] used a 13 μm coating of a soft plastic; and Shah and Chandra[23] used a 66 μm coating of silicon rubber. Darwin and Slate[25, 26] analyzed the effect of these coatings and found that they isolated the aggregate from the surrounding mortar, resulting in behavior more representative of a material with a large number of voids in the concrete matrix than a material containing an aggregate with reduced interfacial strength. Thus, the apparently large impact of changes in interfacial properties observed in those studies[22-24] must be disregarded.

Aware of problems in the earlier studies, Darwin and Slate[25, 26] and Perry and Gillott[27, 28] designed studies to determine the effect of paste-aggregate strength on compressive strength that would not isolate the coarse aggregate from the surrounding mortar. Darwin and Slate coated natural coarse aggregate particles with a 10 μm coating of polystyrene and demonstrated that the thickness and stiffness of the coating were such that the aggregate particles were not isolated from the surrounding mortar matrix. They ran separate tests to determine the tensile and shear bond strength of coated aggregate using sawed pieces of sandstone and limestone. Their tests demonstrated that they were able to reduce the tensile strength to about a third and the shear strength to about a fifth of that obtained at a typical mortar-aggregate boundary[12, 13]. They evaluated the stiffness and strength of concrete mixes and measured microcracking in concretes with and without coated coarse aggregate. Concrete strengths ranged from 21 to 31 MPa. In the load tests, Darwin and Slate observed no change in the initial stiffness, but about a 10 percent reduction in compressive strength for concrete containing coated aggregate compared to concrete with uncoated aggregate (Fig. 1). Using both microscopic and x-ray analyses, they observed no appreciable difference in interfacial microcracking in the two concretes at compressive strains of 0.0010, 0.0018, and 0.0022. They did, however, observe a small, yet significant increase in mortar microcracking in the concrete containing the coated coarse aggregate.

Perry and Gillott[27, 28] cast a series of concrete mixes containing glass marbles and artificially prepared quartzite coarse aggregate particles with different degrees of surface roughness. The glass marbles had either smooth or roughened surfaces. The roughened surfaces were obtained by placing the particles in ball mill containing either 80 or 1000 mesh silicon carbide grit. The quartzite particles were manufactured from solid rock and had a smooth barrel shape. The particles had polished surfaces or surfaces roughened in a manner similar to that used for the glass spheres. The maximum roughness for both aggregates corresponded to a "center line average" value of 4 to 5 μm. Concrete strengths ranged from 25 to 57 MPa. For each aggregate type, the rougher the coarse aggregate surface, the higher the compressive strength. The differences in strength, however, were low, matching the differences observed by Darwin and Slate[25, 26]. The smooth-surfaced glass spheres produced a compressive strength equal to 91.6 percent of that obtained with the 80 mesh roughened glass spheres, and the polished quartzite particles produced a concrete with

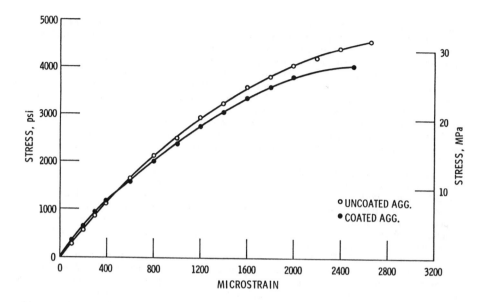

Fig. 1 Stress-strain curves for concrete comparing the effects of uncoated and coated coarse aggregate[25,26]

a strength equal to of 92 percent of that obtained with the roughened quartzite particles.

In another study, involving polymer-impregnated concrete, Carino[29] measured the effects of polymer impregnation on the bond between mortar and aggregate. Measuring interfacial tensile strength using modified tension briquets and interfacial shear strength using inclined aggregate slabs in mortar prisms, he observed that polymer impregnation does not increase interfacial bond strength, but does increase the tensile strength of mortar. Carino attributed the increase in the compressive strength in concrete obtained from polymer impregnation to the increased mortar strength, downgrading the importance of interfacial bond.

In another attempt to increase the compressive strength of concrete by increasing interfacial bond, Popovics[15] used a series of surface treatments on coarse aggregate. His procedures resulted in increases in the compressive strength of some mixes, but in a reduction in compressive strength in the majority of cases. Popovics pointed out that his lack of success was probably due to a failure at the interface at some distance from the actual contact region between paste and aggregate. The result was that the surface treatments did not adequately increase the actual interfacial strength, which was in fact governed by the surrounding cement paste, raising the point that to improve bond strength, the strength of the full interfacial transition zone must be increased, not just the strength at the contact surface between paste (or mortar) and aggregate.

Overall, the four studies described above demonstrate that 1) significant decreases and moderate increases in interfacial strength at the paste-coarse aggregate boundary, compared to values obtained in normal strength concrete, have a measurable but relatively low effect on compressive strength, and 2) increases in compressive strength can be obtained by increasing the strength of the mortar constituent (including the ITZ) alone.

FINITE ELEMENT STUDIES

Finite element models have been used to gain an improved understanding of the response of concrete to compressive load[30-35]. These models include separate representations for the mortar, aggregate, and interfacial zones. Three of these studies, by Buyukozturk[30], Maher and Darwin[31, 32], and Stankowski, Runesson, Sture, and Willam[35], provide special insight into the relative importance of the interfacial zone on the compressive response of concrete.

Buyukozturk[30] and Maher and Darwin[31, 32] represented concrete as consisting of one or more cylindrical aggregate inclusions in a mortar matrix. In Buyukozturk's model, nonlinear response was limited to cracking at the interface and through the mortar, while Maher and Darwin's model also included a nonlinear compressive response for the mortar. The models have an advantage over experimental concretes, since the effect of each material parameter on compressive strength can be measured independently. In addition, changes in interfacial strength can be strictly enforced. For example, perfect bond can be achieved, without concern for failure elsewhere within the ITZ.

The models were used to study microcracking and the stress-strain response of a physical model of concrete tested by Buyukozturk[30] that contained nine aggregate inclusions. As shown in Fig. 2 [36], Buyukozturk's finite element model did not adequately represent the nonlinear behavior of the physical model, while Maher and Darwin's model closely replicated the full nonlinear response, demonstrating that bond and mortar microcracking alone are not enough to represent material response. In both models, the presence of the aggregate inclusions resulted in compressive strengths below that of the mortar constituent alone.

Maher and Darwin[31, 32] went on further with a simpler model containing a single aggregate inclusion to evaluate the effect of paste-aggregate bond strength on the strength of concrete. As shown in Fig. 3, they observed that an increase in bond strength from normal values to perfect bond (no failure at the interface) resulted in only a 4 percent increase in the compressive strength of the model. A decrease to zero interfacial strength resulted in a decrease in compressive strength of just 11 percent.

In a more recent study, Stankowski et al.[35] represented concrete as a series of polygonal aggregate particles in a mortar matrix (Fig. 4). Using this realistic representation of concrete, they obtained only a 7 percent increase in compressive strength for a model with perfect interfacial bond

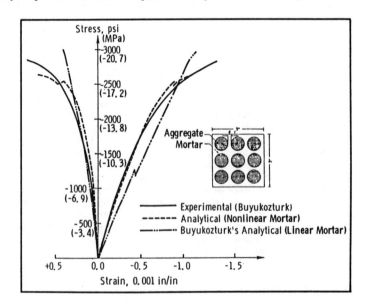

Fig. 2 Stress-strain curves for concrete model [36]

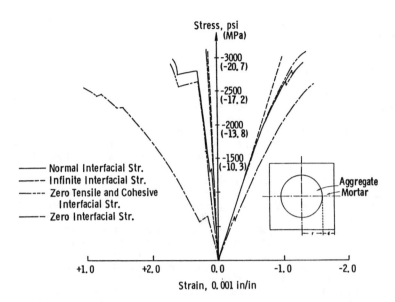

Fig. 3 Stress-strain curves for finite element models of concrete with different values of mortar-aggregate bond strength[31, 32]

Fig. 4 Finite element model of mortar and aggregate[35]

strength compared to one with typical values of tensile and shear strength at the interface. Not surprisingly, the same change in interfacial strength produced a 29 percent increase in tensile strength.

The key observations from the analytical studies measuring the effects of direct changes in interfacial properties match the experimental observations indicating that the bond strength between the interfacial region and aggregate plays a less than dominant role in the compressive strength of concrete.

EXPERIMENTAL STUDIES MODIFYING PASTE CONSTITUENT

The use of silica fume as a cement replacement has been shown to result in significant increases in the strength of concrete[7, 14, 16, 17, 19, 20]. There has been, however, a continuing controversy as to the causes of this increase in strength. The key questions involve the effect of silica fume on the bond strength between paste and aggregate, on the density and strength of the interfacial zone, and on the properties of the cement paste constituent as a whole. Much of this research has been reviewed elsewhere[19, 20], and it is clear that silica fume does result in a denser interfacial transition zone[4, 6, 7].

A popular approach in studies to evaluate the effect of silica fume on the compressive strength of concrete has been to cast cement paste and mortar or concrete specimens with similar silica fume replacements of cement and then to compare the relative changes in compressive strength obtained for paste and the other material. Some studies have demonstrated no increase in the strength of cement paste with the addition of silica fume, but significant increases in the strength of concrete[7, 16, 17]. While concrete is usually weaker than cement paste with the same water-cement ratio[7, 17, 37], the concretes containing silica fume in these studies[7, 16, 17] have exhibited compressive strengths higher than both the corresponding cement paste and the concretes without silica fume. These observations have led the investigators to conclude that the increase in compressive strength with the addition of silica fume has been due to an increase in the strength of the ITZ.

In contrast, other studies have demonstrated increases in the compressive strength of cement paste, mortar and concrete with the partial replacement of cement by silica fume[19, 20], leading to the conclusion that increases in the strength of the cement paste constituent are the primary reason for increases in the strength of concrete and mortar.

A key drawback in comparisons of cement pastes, mortars and concretes containing silica fume is that the amount of superplasticizer (used to provide workability) that works well in one material will not work well in another. As demonstrated by Cong, Gong, Darwin, and McCabe[20], the dosage of superplasticizer that provides adequate workability and results in the highest strength for concrete provides too much fluidity for the corresponding cement paste, resulting in segregation and bleeding in the plastic material and a significant decrease in strength.

More importantly, as pointed out earlier in this paper, the structure of bulk cement paste is significantly different from the cement paste within the interfacial transition zone[1-9]. Thus, a direct comparison between the response of pastes and the response of concretes to the addition of silica fume will not provide useful information on the role of the interfacial transition zone in the compressive strength of concrete.

A more useful comparison of the effects of changes in the ITZ can be obtained by comparing the strengths of mortar and concrete cast with similar silica fume replacements of cement. In this case, the effects of differences in the ITZ at the mortar-coarse aggregate boundary can be determined without the complication of excess bleeding that may occur with bulk cement paste. Such a comparison was carried out by Cong et al.[20] who studied silica fume replacement in cement paste, mortar, and concrete. One method of comparison that they used to establish the role of the ITZ was to compare the ratio of concrete strength to mortar strength (f'_c/f'_m) as a function of mortar strength. Thus, any changes brought about in the compressive strength of mortar due to silica fume replacement of cement would be considered separately from changes in the interfacial transition zone between the mortar and the coarse aggregate.

As illustrated in Fig. 5, Cong et al.[20] observed that the ratio f'_c/f'_m is virtually independent of the degree of silica fume replacement and the water-cementitious material ratio. The figure shows the results for concretes and mortars with water-cementitious material ratios of 0.33 and 0.39 tested at ages of 3, 7, and 28 days. The mixes involved contained 1) no admixtures, 2) superplasticizer alone, or 3) superplasticizer and a 15 percent cement replacement by silica fume. All of the mortars had identical volumes of fine aggregate and all of the concretes had identical vol-

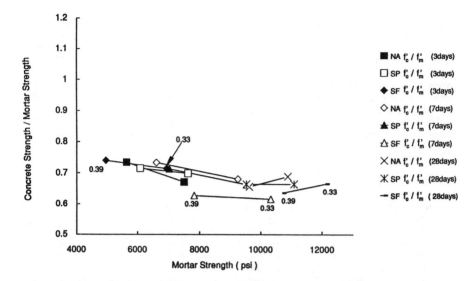

Fig. 5 Ratios of concrete strength to mortar strength as a function of mortar strength (NA = no admixtures; SP = superplasticizer; SF = superplasticizer and silica fume)[20]

umes of fine and coarse aggregate. The presence of coarse aggregate results in a drop in strength of 26 to 39 percent. The results plot in a narrow range and, if the results for the materials containing silica fume (SF) tested at 7 days are removed, the results could be characterized as falling in a *very* narrow range. The principal factor controlling the ratio f'_c/f'_m is mortar strength, independent of age, water-cementitious material ratio or admixture. The results in Fig. 5 lead to the conclusion that, for the materials used in the study, the increases in the density of the ITZ around coarse aggregate particles obtained through the use of silica fume had no measurable impact on the strength of concrete, again pointing out the small impact of changes in interfacial properties on the strength of the composite material.

FINAL COMMENTS

A few final points can be obtained from the early microcracking studies in which the tensile and shear bond strengths between cement and aggregate were measured[12, 13]. The tensile strength between paste and aggregate decreased at a slower rate than the compressive strength of cement paste with an increase in water-cement ratio. The tensile and shear strengths between mortar and coarse aggregate were largely insensitive to water-cement ratio for water-cement ratios ranging between 0.36 and 0.75. This lack of sensitivity in bond strength to changes in water-cement ratio provides strong support for the matrix, rather than the interface, as the principal controlling factor in the strength of concrete.

SUMMARY

The research summarized in this paper emphasizes studies designed to directly measure the effects of the properties of the interfacial transition zone on the compressive response of concrete. The research involves experimental studies in which the interfacial strength between aggregate and paste is artificially modified, finite element studies in which the properties of the constituents and the ITZ are modeled, and experimental studies in which the paste constituent is modified by partial replacement of cement with silica fume. Both the experimental studies in which the strength of the interface is artificially modified and the finite element studies demonstrate that changes in interfacial strength provide a measurable but relatively small change in compressive strength. Most of the studies in which the ITZ is modified by changing the paste constituent do not provide clear information on the role of the ITZ in the compressive strength of concrete. One study, however, demonstrates that it is the strength of the mortar constituent, rather than the density of the ITZ, that dominates the strength of concrete and the heterogeneous nature of the composite.

Overall, it may be concluded that the ITZ plays a measurable role in the response of concrete to compressive stress, but that the role is overshadowed by the properties of the cement paste and aggregate constituents of concrete and the heterogeneous nature of the composite.

REFERENCES

1. J. Farran, *Rev. Matér Constr.* (490-491) 155; (492) 191 (1956).
2. R. Javelas, J. C. Maso, J. P. Ollivier, and B. Thenoz, *Cem. Concr. Res.* **5**, 285 (1975).
3. S. Diamond, S. Mindess, and J. Lovell in *Liasons Pâtes de Ciment Matériaux Associés*, (Proc. RILEM Colloq.), (Laboratorie de Génie Civil, Toulouse, France, 1982) p. C42.
4. M. Regourd in *Very High Strength Cement-Based Materials*, edited by J. F. Young (Mater. Res. Soc. Proc. **42**, Pittsburgh, PA, 1985) pp. 3-17.
5. K. L. Scrivener and P. L. Pratt in *8th Intl. Congress Chem. Cement* (Rio de Janeiro, 1986) p. 466.
6. A. Bentur and M. D. Cohen, *J. Amer. Cer. Soc.* **70** (10), 738 (1987).
7. A. Bentur, A. Goldman, and M. D. Cohen in *Bonding in Cementitious Composites,* edited by S. Mindess and S. P. Shah (Mater. Res. Soc. Proc. **114**, Pittsburgh, PA, 1988) pp. 97-104.
8. K. L. Scrivener and E. M. Gartner in *Bonding in Cementitious Composites,* edited by S. Mindess and S. P. Shah (Mater. Res. Soc. Proc. **114**, Pittsburgh, PA, 1988) pp. 77-85.
9. K. L. Scrivener, A. K. Crumbie, and P. L. Pratt in *Bonding in Cementitious Composites,* edited by S. Mindess and S. P. Shah (Mater. Res. Soc. Proc. **114**, Pittsburgh, PA, 1988) pp. 87-88.
10. D. N. Winslow, M. D. Cohen, D. P. Bentz, K. A. Snyder, and E. J. Garboczi, *Cem. Concr. Res.* **24** (1), 25 (1994).
11. B. D. Barnes, S. Diamond, and W. L. Dolch, *Cem. Concr. Res.* **8** (2), 233 (1978).
12. T. C. Hsu and F. O. Slate, *J. Amer. Concr. Inst.* **60** (4), 465 (1963).
13. M. A. Taylor and B. B. Broms, *J. Amer. Concr. Inst.* **61**, (8), 939 (1964).
14. C.-Y. Huang and R. F. Feldman, *Cem. Concr. Res.* **15**, (2), 285 (1985).
15. S. Popovics, *Mat. and Struct.*, RILEM, **20** (115), 32 (1987).
16. A. M. Rosenberg and J. M. Gaidis, *Concr. Intl.* **11** (4), 31 (1989).
17. A. Goldman and A. Bentur, *ACI Matls. J.* **86** (5), 440 (1989).
18. S. Mindess in *Bonding in Cementitious Composites,* edited by S. Mindess and S. P. Shah (Mater. Res. Soc. Proc. **114**, Pittsburgh, PA, 1988) pp. 3-10.
19. D. Darwin, Z. Shen, and S. Harsh in *Bonding in Cementitious Composites,* edited by S. Mindess and S. P. Shah (Mater. Res. Soc. Proc. **114**, Pittsburgh, PA, 1988) pp. 105-110.
20. X. Cong, S. Gong, D. Darwin, and S. L. McCabe, *ACI Matls. J.* **89** (4), 375 (1992).
21. J. Farran, *Publ. Technique* No. 78 (Centre d'Etudes et de Recherches de l'Industrie des Liants Hydrauliques, Paris, 1956).
22. R. Dantinne, *Bull.* No. 8 (Centre d'Etudes, et d'Essasis Scientifiques du Genie Civil, Liége, France, 1956).
23. S. P. Shah and S. Chandra, *J. Amer. Concr. Inst.*, **65** (9), 770 (1968).
24. P. Nepper-Christensen and T. P. H. Nielson, *J. Amer. Concr. Inst.*, **66** (1), 69 (1969).
25. D. Darwin, M.S. Thesis, Cornell University 1967.
26. D. Darwin and F. O. Slate, *J. of Matls.*, ASTM **5** (1), 86 (1970).
27. C. Perry and J. E. Gillott, *Res. Report* No. CE 77-4 (University of Calgary, Alberta, Canada

1977).
28. C. Perry and J. E. Gillott, *Cem. Concr. Res.* **7** (5), 553 (1977).
29. N. J. Carino, *Cem. Concr. Res.* **7** (4), 439 (1977).
30. O. Buyukozturk, Ph.D. Thesis, Cornell University 1970.
31. A. Maher and D. Darwin, *CRINC Report* SL-76-02 (University of Kansas Center for Res. 1976).
32. A. Maher and D. Darwin in *Proc.* First Intl. Conf. on Math. Modeling (St. Louis, 1977) **III**, p. 1705.
33. E. Yamaguchi and W.-F. Chen, *Report*, CE-STR-89-16 (Purdue University 1989).
34. W.-F. Chen and E. Yamaguchi in *Micromechanics of Failure of Quasi-Brittle Materials*, edited by S. P. Shah, S. E. Swartz, and M. L. Wang (Elsevier Applied Science, London and New York, 1990) p. 265.
35. T. Stankowski, K. Runnesson, S. Sture, and K. J. Willam in *Micromechanics of Failure of Quasi-Brittle Materials*, edited by S. P. Shah, S. E. Swartz, and M. L. Wang (Elsevier Applied Science, London and New York, 1990) p. 285.
36. D. Darwin, presented at First Intl. Conf. on Math. Modeling, St. Louis, MO 1977 (unpublished).
37. J. L. Martin, D. Darwin, and R. E. Terry, *SM Report* No. 31 (University of Kansas Center for Res. 1991).

MODELLING THE D.C. ELECTRICAL CONDUCTIVITY OF MORTAR

E.J. GARBOCZI[*], L.M. SCHWARTZ[**], and D.P. BENTZ[*]
[*]National Institute of Standards and Technology, Building Materials Division, 226/B350, Gaithersburg, MD 20899
[**]Schlumberger-Doll Research, Old Quarry Road, Ridgefield CT 06877-4108.

ABSTRACT

The interfacial zone separating cement paste and aggregate in mortar and concrete is believed to influence many of the properties of these composites. This paper presents a theoretical framework for quantitatively understanding the influence of the interfacial zone on the overall electrical conductivity of mortar, based on realistic random aggregate geometries. These same ideas may also be used to approximately predict the fluid permeability of mortar.

INTRODUCTION

The D.C. electrical conductivity of mortar and concrete is an important measure of ionic diffusivity [1], via the Nernst-Einstein relation [2]. Diffusivity is of interest in connection with a range of issues related to durability, such as sulfate attack and chloride ion-induced corrosion of steel reinforcing bars [3]. Much recent work has been done on understanding how the microstructure of cement paste determines its electrical conductivity [4-9]. However, relatively little work has been done on how the conductivity of concrete depends on quantities like the number and arrangement of aggregate particles, and on the cement paste:aggregate interfacial zone [6,10-12].

In this paper, we are concerned with the (approximately 10 μm-1000 μm) length scale that adequately describes a typical mortar [13,14]. Within this framework, mortar (and concrete) can be viewed as a three-phase composite [15-17]: bulk cement paste, aggregate, and interfacial zone cement paste [see Figure 1], where all three phases can be thought of as uniform continuum materials. In such a three-phase composite model, the volume fraction assigned to the interfacial zone phase depends on what thickness is taken to define the boundary between the interfacial zone and the bulk cement paste. For values of the interfacial zone thickness around 20 μm, the interfacial zone cement paste occupies 20-30 % of the total cement paste volume, and therefore 10-15% of the total mortar volume [14]. Since the interfacial zone cement paste occupies a significant volume fraction, the different microstructure [18,19] and therefore different physical properties of this phase will certainly have an influence on the overall behavior of the mortar/concrete composite [20,21], especially since recent modelling and mercury injection experimental work have shown that the interfacial zone cement paste phase forms a continuous percolating channel [14,22].

RANDOM THREE-PHASE MODEL FOR MORTAR

More details of the following model are given in Ref. [23]. For the purposes of electrical conduction, the aggregate grains are simply inert obstacles to the flow of current. The basic

Mat. Res. Soc. Symp. Proc. Vol. 370 © 1995 Materials Research Society

model is then defined by two parameters: 1) the structure of the interfacial layer, and 2) the electrical contrast between this layer and the bulk cement paste. In this paper we replace the variable conductivity (because of varying porosity) interfacial zone region with a shell of fixed width and constant conductivity, σ_s, for the sake of simplicity. All the other cement paste outside the shell of thickness h is given the bulk cement paste value of conductivity, σ_p. Several authors have carried out experiments with planar or cylindrical aggregate shapes. A range of values for σ_s/σ_p have been found, from about 10 [11], assuming h = 20 μm, to 12-15 [12], assuming h = 100μm. This latter value for h seems much too large, since SEM investigations of the interfacial zone generally find that the porosity of the interfacial zone decreases to the bulk cement paste value by a distance of 30-50 μm from the aggregate grain surface [18,19].

Since there is no definitive value established experimentally for the value of σ_s / σ_p, we have chosen to allow this parameter to vary in the following computations, and studied the dependence of the composite conductivity, for a given sand content, on the value of σ_s / σ_p. However, based on the mercury intrusion and modelling results of Ref. [14], we have chosen h = 20 μm as the best value for the width of the interfacial zone. We emphasize that extracting the value of σ_s / σ_p from experiments will require an assumption for the value of h.

To obtain a model of a mortar with a realistic random sand grain arrangement, we used the system studied in Ref. [14] and shown in Fig. 1. From the experimently measured sand grain

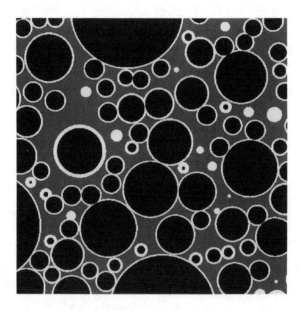

Figure 1: The structure of mortar, as represented by the random aggregate model. There are four sizes of sand grains (black), with diameters given in the text, and bulk cement paste matrix (gray). The thickness of the interfacial zone region (white) is 20 μm. The total volume fraction of sand is 54%.

size distribution used there, we chose four representative diameters: 250, 500, 750, and 1500 μm [23], with fractional volumes of 0.1895, 0.2233, 0.2317, and 0.3555 respectively of the total sand volume. For the maximum sand grain volume fraction studied, 54%, a total of 6500 particles were randomly placed and interfacial zones, of thickness 20 μm, were added to each grain.

Figure 2 shows the connectivity of the interfacial zone phase as a function of sand volume fraction. The y-axis, labelled "Fraction Connected", is just the fraction of the total interfacial zone volume, at a given sand volume fraction, that is part of a connected path across the sample. The interfacial zone phase first becomes partially connected at a sand volume fraction of about 36%, and essentially all of the interfacial zones are connected to each other by a sand volume fraction of 51%.

The conductivity of the random sand grain mortar model was computed using a random walk algorithm, used extensively in studies of disordered porous media [24] and composite materials [23,25-27].

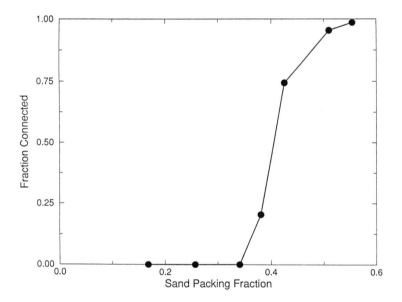

Figure 2: Percolation curve for random sand grain model, for a 20μm thick interfacial zone. The x-axis is the sand volume fraction, and the y-axis is the fraction of the interfacial zone phase that is contained in its percolating cluster.

ELECTRICAL CONDUCTIVITY RESULTS

Figure 3 shows the overall composite conductivity as a function of the ratio of the interfacial zone cement paste conductivity to the bulk cement paste conductivity, for a mortar with a sand volume fraction of 54%. The conductivity at $\sigma_s/\sigma_p = 1$ is that which would be obtained if the interfacial zone cement paste had the same porosity and therefore the same conductivity as the bulk cement paste. The presence of the insulating sand grains in this case reduces the overall normalized conductivity from 1 to 0.35. This is consistent with a 3/2 power law found in suspensions of spheres and many porous rocks, where the normalized conductivity goes like the 3/2 power of the conductive phase volume fraction. In this case, $(0.46)^{3/2} = 0.31$ [28,29].

The composite conductivity can be thought of as the result of a competition between the insulating sand grains, which tend to lower the overall conductivity, and the interfacial zone cement paste shells, which tend to raise the overall conductivity, when $\sigma_s > \sigma_p$. We note that in Fig. 3, the overall shape of the curve is concave down. The curve could at most be straight. This would be the case if the two phases, interfacial zone and bulk cement paste, were exactly in parallel. Since the microstructure is such that the two cement paste phases are not exactly in parallel, then the curve must be sub-linear, or concave down. As $\sigma_s/\sigma_p \to \infty$ the curve will asymptotically go to a straight line, as the interfacial zone phase will eventually dominate the conductivity.

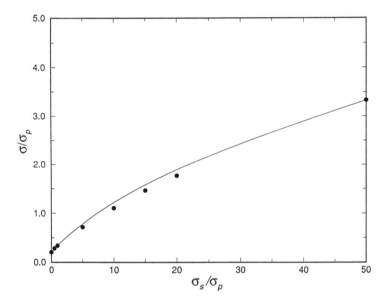

Figure 3: Composite conductivity for the random mortar model is plotted vs. the interfacial zone conductivity. [Both are normalized by bulk paste conductivity.] The solid dots are the numerical data, the solid line is the Padé approximant explained in Ref. [23].

To achieve an overall conductivity that is equal to the bulk cement paste conductivity, the value of σ_s/σ_p must be equal to approximately 8, as can be seen in Fig. 3. At $\sigma_s/\sigma_p = 20$, the overall conductivity is about 1.8. The solid line in Fig. 3 is a Padé approximant [23]. The fit to the data points is not perfect. However, the asymptotic slope of the Padé approximant for large values of σ_s/σ_p should be accurate, so that this curve can be safely extrapolated to predict the effect on σ/σ_p of much higher values of σ_s/σ_p. This becomes important below when we consider the effect of the interfacial zone on the fluid permeability of mortar.

A second important data set, more easily obtained by experimental techniques, will be obtained by measuring the conductivity of a mortar as a function of sand volume fraction, for a fixed value of σ_s/σ_p. One important aspect of such a data set is the dilute limit, where the sand volume fraction is low. The composite conductivity in this regime contains important information about the conductivity and size of the interfacial zone. This is the case because exact analytical calculations can be made of the influence on the overall conductivity of a few sand grains surrounded by an interfacial zone shell placed in a matrix [23]. The conductivity for a low volume fraction of sand (less than 10%) is an explicit function of h/b, where b is the sand grain radius, and of the parameter σ_s/σ_p [23]. For a given value of b and σ_s/σ_p, the conductivity takes the form

$$\frac{\sigma}{\sigma_p} = 1 + m(\frac{h}{b}, \frac{\sigma_s}{\sigma_p}) \; c \; , \quad c = total \; sand \; volume \; fraction \qquad (1)$$

When there is a distribution of sand sizes, m is replaced by $<m>$, the average of m over the different values of b, for a given value of h and σ_s/σ_p.

From the results of Ref. [23], we find that at $\sigma_s/\sigma_p \approx 8.26$, for the sand size distribution used in this paper, $<m> = 0$ [23]. At this value, adding a few sand grains would, to leading order in the sand volume fraction, have no effect on the overall conductivity, keeping its value at the bulk cement paste value. In Fig. 3, it was shown that a value of $\sigma_s/\sigma_p \approx 8$ was required to make the composite conductivity equal to the bulk cement paste conductivity at a sand volume fraction of 55%. This implies that there is information about the shape of the conductivity vs. sand volume fraction curve for this particular sand size distribution and interfacial zone width obtainable without further computation. For $\sigma_s/\sigma_p \leq 8.26$, such a curve must start out with negative slope, and σ/σ_p will always lie under the bulk cement paste conductivity. For $\sigma_s/\sigma_p > 8.26$, the curve will start out with positive slope and always remain above the bulk cement paste conductivity. Using different sand size distributions will change this cutoff value but the overall picture should be similar, even for the large aggregate size distributions typical of concrete.

Figure 4 shows the results of computations of the overall mortar conductivity for different values of σ_s/σ_p (1,5,20) as a function of sand volume fraction. The predictions of the previous paragraph are shown to hold true, as 1 and 5 are smaller than 8.26, and 20 is larger than 8.26. The computed numerical values of $<m>$ also agree within a few percent with the predicted values [23].

COMPARISON WITH EXPERIMENTAL EVIDENCE

Recently chloride diffusivity measurements have been made for cement pastes and mortars with several different sand volume fractions [30]. The ratio of the overall diffusivity to the cement paste diffusivity is, via the Nernst-Einstein relation, the same as the ratio σ/σ_p [2]. It was found that this ratio was on the order of 1-2, for both 0.4 and 0.5 w/c mortars, at about 50% sand content. Using Fig. 3, this result implies, assuming that the sand particle size distribution was similar to that used in our random mortar model, that σ_s/σ_p was roughly between

10 and 20, in agreement with the experimental results discussed earlier on flat aggregate interfaces [11-12].

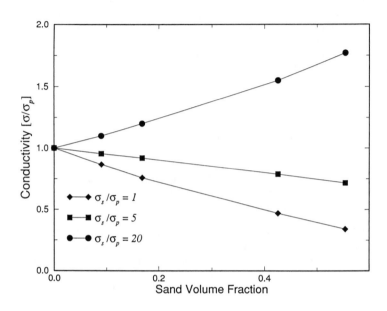

Figure 4: Composite conductivities (calculated by random walk simulations) for the random mortar model are shown as a function of sand concentration for several values of the interfacial zone conductivity. [Normalization is as in Fig. 3.]

Another paper that attempted to measure the conductivity of mortar as a function of sand volume fraction found completely different results [10]. We do not feel these results are trustworthy, as only a single frequency A.C. measurement was used, so that the true D.C. conductivity may not have been measured [5]. Also, the measured conductivities were significantly lower that the simple 3/2 power law [28,29] would predict, assuming that the interfacial zone conductivity was the same as that of the bulk cement paste matrix. In addition, the theoretical equation used in Ref. [10] does not give the correct dilute sand limit, as given in eq. (1) [23], so that even if the measurements were correct, any deductions made regarding the interfacial zone conductivity would be in error.

It is interesting to try to extend the modelling work to include fluid permeability. This can be done *approximately* via an electrical analogy between Ohm's law and Darcy's law [23,31]. The parameter K_s/K_p, the ratio of the permeability of the interfacial zone and the permeability of the bulk cement paste matrix, then takes the place of the conductivity parameter, σ_s/σ_p, and the equation to be solved is still Laplace's equation. One way to estimate the quantity K_s/K_p is to make use of the Katz-Thompson equation, which predicts the permeability of a porous medium in terms of its conductivity and a critical pore radius characteristic of the largest connected pores in the material, defined by a mercury intrusion experiment [1,32]. Neglecting constants of proportionality, the Katz-Thompson equation is

$$k \sim \frac{\sigma}{\sigma_o} d^2 \qquad (2)$$

where σ/σ_o is the conductivity of the porous material relative to the conductivity σ_o of the conductive pore fluid it contains, and d is the critical pore diameter. If we assume that the value of d for interfacial zone cement paste is about 10 times as large as that for the bulk cement paste, in rough agreement with the available mercury intrusion evidence [14], and take the interfacial zone conductivity to be about 10 times larger than that of the bulk cement paste, in rough agreement with experiments on flat aggregates, then we would expect that the ratio K_s/K_p would be about 1000.

The largest value of K_s/K_p or σ_s/σ_p computed in Fig. 3 was only 50, but we can use the fitted Padé approximant, which should be more accurate in the large K_s/K_p limit, to estimate the overall permeability to be about 35 times that of the bulk cement paste. Data in Ref. [30] indicates that the permeabilities of mortars with about 50% sand are about 20-60 times higher than the cement paste matrix, again in reasonable agreement with the model prediction.

FUTURE WORK

An important goal of future work, based on the techniques and results of this paper, will be to develop a general equation for the conductivity/diffusivity of concrete. The inputs to such an equation would be: 1) the chemistry, water:cement ratio, and degree of hydration of the cement, in order to determine the bulk matrix conductivity, 2) interfacial zone characteristics, 3) aggregate size distribution and volume fraction, and 4) computations like those in this paper. With such an equation, the conductivity/diffusivity of a concrete could be determined with good accuracy at the mix design stage, based on fundamental physical parameters.

ACKNOWLEDGMENTS

We are grateful to Daniel Hong for useful conversations concerning the random walk algorithm in two-conducting-phase materials. Part of this work has been supported by the National Science Foundation, via the Science and Technology Center for Advanced Cement-Based Materials. LMS thanks NIST for partial support of his sabbatical leave from Schlumberger.

REFERENCES

1) E.J. Garboczi, Cem. and Conc. Res. 20, 591 (1990).
2) A. Atkinson and A.K. Nickerson, J. Mater. Sci. 19, 3068-3078 (1984).
3) H.F.W. Taylor, Cement Chemistry (Academic Press, San Diego, 1990).

4) E.J. Garboczi and D.P. Bentz, J. Mater. Sci. 27, 2083-2092 (1992).

5) R.T. Coverdale, B.J. Christensen, T.O. Mason, H.M. Jennings, D.P. Bentz, and E.J. Garboczi, "Interpretation of the impedance spectroscopy of cement paste via computer modelling Part I: Bulk conductivity and offset resistance," J. of Materials Science (1994), in press.

6) H.W. Whittington, J. McCarter, and M.C. Forde, Mag. of Conc. Res. 33, 48-60 (1981).

7) R.T. Coverdale, B.J. Christensen, T.O. Mason, H.M. Jennings, and E.J. Garboczi, "Interpretation of the impedance spectroscopy of cement paste via computer modelling Part II: Dielectric response," J. of Materials Science 29, 4984-4992 (1994).

8) B.J. Christensen, T.O. Mason, H.M. Jennings, D.P. Bentz, and E.J. Garboczi, MRS Soc. Symo. Proc. Vol. 245 (1992), pp. 259-264.

9) R.A. Olson, B.J. Christensen, R.T. Coverdale, S.J. Ford, H.M. Jennings, T.O. Mason, and E.J. Garboczi, "Interpretation of the impedance spectroscopy of cement paste via computer modelling Part III: Microstructural analysis of frozen cement paste," submitted to J. Mater. Sci.

10) X. Ping and T. Ming-shu, Il Cimento 85, 33 (1988).

11) X. Ping, J.J. Beaudoin, and R. Brousseau, Cem. and Conc. Res. 21, 515-522 (1991).

12) D. Bretton, J.-P. Ollivier, and G. Ballivy, in Interfaces in Cementitious Materials (E. and F.N. Spon, London, 1993), pp. 269-278.

13) S. Mindless and J.F. Young, Concrete (Prentice-Hall, Englewood Cliffs, NJ, 1981).

14) D.N. Winslow, M. Cohen, D.P. Bentz, K.A. Snyder, and E.J. Garboczi, Cem. and Conc. Res. 24, 25-37 (1994).

15) A. Goldman and A. Bentur, Cem. and Conc. Res. 23, 962-972 (1993).

16) M.D. Cohen, A. Goldman, and W.-F. Chen, Cem. and Conc. Res. 24, 95-98 (1994).

17) A.U. Nilsen and P.J.M. Monteiro, Cem. and Conc. Res. 23, 147-151 (1993).

18) K.L. Scrivener, in Materials Science of Concrete I (American Ceramic Society, Westerville, Ohio, 1989).

19) B.D. Barnes, S. Diamond, and W.L. Dolch, J. Amer. Ceram. Soc. 62, 21-24 (1979).

20) J.F. Young, in ACI SP-108 Permeability of Concrete, ed. David Whiting and Arthur Hall (American Concrete Institute, Detroit, 1988) pp.1-18.

21) Y.F. Houst, H. Sadouki, and F.H. Wittmann, in Interfaces in Cementitious Composites, ed. J. Maso (E & F.N. Spon, London, 1993), pp. 279-288.

22) D.P. Bentz, E. Schlangen, and E.J. Garboczi, in Materials Science of Concrete IV (Amer. Ceram. Soc., Westerville, Ohio, 1994).

23) E.J. Garboczi, L.M. Schwartz, and D.P. Bentz, "Modelling the influence of the interfacial zone on the D.C. conductivity of concrete," submitted to Journal of Advanced Cement-Based Materials.

24) L.M. Schwartz and J.R. Banavar, Phys. Rev. B 39, 11965 (1989).

25) I.C. Kim and S. Torquato, Phys. Rev. A 43, 3198-3201 (1991).

26) D.C. Hong, H.E. Stanley, A. Conoglio, and A. Bunde, Phys. Rev. B 33, 4564 (1986).

27) L.M. Schwartz, E.J. Garboczi, and D.P. Bentz, "Interfacial transport in porous media: Application to D.C. electrical conductsivity of mortars," submitted to Phys. Rev. E.

28) R.E. De La Rue and C.W. Tobias, J. of the Electrochemical Society 106, 827-833 (1959).

29) P.Z. Wong, Physics Today 41, 24-35 (1988).

30) P. Halamickova, R.J. Detwiler, D.P. Bentz, and E.J. Garboczi, "Water permeability and chloride ion diffusion in portland cement mortars: Relationship to sand content and critical pore diameter," submitted to Cem. and Conc. Res.

31) N. Martys, D.P. Bentz, and E.J. Garboczi, Physics of Fluids A 6, 1434-1439 (1994).

32) A.J. Katz and A.H. Thompson, J. of Geophys. Res. 92, 599 (1987).

INTERFACIAL ZONE PERCOLATION IN CONCRETE: EFFECTS OF INTERFACIAL ZONE THICKNESS AND AGGREGATE SHAPE

D.P. BENTZ*, J.T.G. HWANG+, C. HAGWOOD*, E.J. GARBOCZI*, K.A. SNYDER*, N. BUENFELD**, AND K.L. SCRIVENER**
* Building and Fire Research Laboratory, Building 226, Room B-350, National Institute of Standards and Technology, Gaithersburg, MD 20899 USA
+ Cornell University, Ithaca, NY 14853 USA
** Imperial College of Science and Technology, London, SW7 2BP ENGLAND

ABSTRACT

Previously, a hard core/soft shell computer model was developed to simulate the overlap and percolation of the interfacial transition zones surrounding each aggregate in a mortar or concrete. The aggregate particles were modelled as spheres with a size distribution representative of a real mortar or concrete specimen. Here, the model has been extended to investigate the effects of aggregate shape on interfacial transition zone percolation, by modelling the aggregates as hard ellipsoids, which gives a dynamic range of shapes from plates to spheres, to fibers. For high performance concretes, the interfacial transition zone thickness will generally be reduced, which will also affect their percolation properties. This paper presents results from a study of the effects of interfacial transition zone thickness and aggregate shape on these percolation characteristics.

INTRODUCTION

In recent years, the importance of the cement paste-aggregate interfacial transition zone (ITZ) in influencing concrete properties has been recognized, as witnessed by this symposium's topic. It has been shown that, even when air voids are not considered, concrete must still be considered as a three-phase material (paste, aggregate, and ITZ) to properly investigate elastic properties [1, 2], transport coefficients [3], and drying shrinkage [4]. Consideration of the ITZ is important because the microstructure of the ITZ in ordinary concrete is different than that of the bulk paste. It contains more and larger pores and less anhydrous cement and calcium silicate hydrate gel [5, 6]. Therefore, its response to mechanical and thermal loadings and chemical ingress will also differ from that of bulk paste.

The magnitude of the influence of these microstructural differences on the composite's properties depends on two parameters: the relative difference in properties between ITZ and bulk paste, and the volume fraction and connectivity (percolation) of the ITZ regions [3]. This paper addresses only the latter of these topics, studying the influence of ITZ thickness and aggregate particle shape on the percolation of ITZ in a mortar or concrete. Experimental evidence of this phenomenon of interfacial transition zone percolation has been provided by mercury intrusion porosimetry studies [7, 8, 9] and by experimental comparisons of the transport properties of mortars with those of their constituent cement pastes [10].

COMPUTER MODEL

The computer model employed in this study is an extension of that employed in the study described in [8] in which particle shapes were limited to spheres. Based on recent analytical work [11], a new computer program was written to extend the model to triaxial ellipsoids, with semi-axes of lengths a, b, and c and any arbitrary orientation. Therefore, one may vary particle shapes from spheres (air voids) to cylindrical ellipsoids (fibers) to triaxial ellipsoids (aggregates). Given any size distribution of particles and scale factors a, b, and c, the program places a requested number of particles in a three-dimensional cubic volume, typically 10 mm on a side, such that no particles overlap. The program proceeds from the largest to the smallest particles. Interfacial transition zones are modelled as "soft shells" of a constant thickness t surrounding each "hard core" aggregate particle [8]. The ITZ may overlap one another and also may intersect the aggregate particles. After all particles have been placed, the program uses a burning algorithm [12] to determine the connectivity of the ITZ regions and volumetric point sampling to determine the volumes occupied by aggregates, bulk cement paste, percolated ITZ regions, and disconnected ITZ regions. By varying the number of particles placed in the box, the percolation behavior can be mapped out as a function of aggregate content.

The computer program was validated by determining the excluded volume of triaxial ellipsoids of different aspect ratios [13]. The excluded volume for an ellipsoid is the volume surrounding the ellipsoid where the center of an arbitrarily oriented second identical ellipsoid cannot be placed without some portion of the second ellipsoid overlapping a portion of the first one. For a variety of aspect ratios, the excluded volumes determined using the computer program, by sampling 500,000 random configurations, were within 0.5% of the exact analytical result [14].

In this study, three interfacial transition zone thicknesses (10, 22.5, and 50 μm) and five combinations of aspect ratios (1:1:1, 1.25:1:0.8, 1.5:1:0.667, 2:1:0.5, 1.59:0.794:0.794) were studied. The aspect ratios were chosen to maintain a constant particle volume (a*b*c=1) while varying the aggregate surface area. Surface areas were calculated using formulas presented in [15], which can also be found more recently in [16]. In a given execution of the simulation, the aspect ratio was held constant, but it could be easily varied to follow some distribution of particle shapes. The aggregate particle size distribution (PSD) for the mortars corresponded to that of a typical sand used in concrete, as measured experimentally in [17]. The particle diameters ranged between 75 and 4750 μm. For the 1000 mm^3 cubic volumes used in this study, between seven and twenty thousand aggregate particles were required to represent an actual mortar. Fig. 1 shows two-dimensional slices from three-dimensional microstructures for systems with 16,000 particles (aggregate volume fraction=43%), an ITZ thickness of 50 μm, and the two aspect ratios indicated.

RESULTS AND DISCUSSION

The ITZ thickness, t, of 22.5μm was chosen as an estimate of a typical thickness to be encountered in a mortar or concrete [8]. Fig. 2 shows a plot of the fraction of "ITZ paste" percolated vs. the aggregate content for the five aggregate particle aspect ratios investigated in this study. One clearly notices that as the aggregates become more elongated (higher aspect ratios), significantly lower aggregate contents are needed to form a percolated ITZ path across the microstructure. This result is in agreement with percolation studies of totally overlapping particles [13, 18]. Interestingly, when one plots these results against

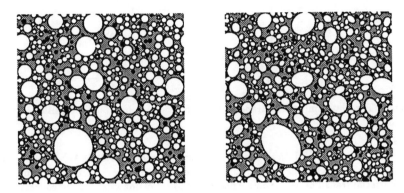

Figure 1: Two-dimensional slices from three-dimensional microstructures for aspect ratios of 1:1:1 (left) and 1.25:1:0.8 (right) (white: aggregate, grey: bulk cement paste, black: ITZ cement paste).

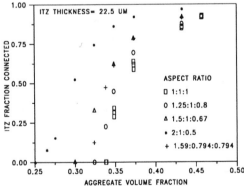

Figure 2: Percolated interfacial transition zone fraction vs. aggregate content. Three data sets are shown for the 1:1:1 aspect ratio system to provide some indication of variability.

the ITZ volume fraction (or the aggregate surface area), the results for the different aspect ratios appear to converge to a single curve as shown in Fig. 3. This suggests that for a given aggregate PSD and ITZ thickness, it is the volume fraction of ITZ paste, and not the aggregate content, which uniquely determines the percolation behavior of these systems.

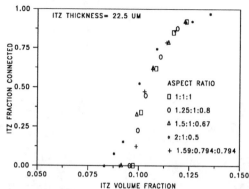

Figure 3: Percolated interfacial transition zone fraction vs. interfacial transition zone volume fraction.

Results for the three ITZ thicknesses are summarized in Table I. As would be expected, decreasing t increases the aggregate content necessary to achieve percolation, but decreases the interfacial transition zone volume needed. While plotting ITZ fraction connected against ITZ volume fraction for different aspect ratios gives a universal curve for all three ITZ thicknesses studied, plotting against aggregate surface area (Fig. 4) generates universal curves only for the two thicker interfacial transition zones. For the 10 μm ITZ thickness (dotted lines in Fig. 4), the plot against aggregate surface area exhibits a greater degree of dispersion which contrasts with the results obtained for t values of 22.5 and 50 μm. Since to a first order, ITZ volume is proportional to aggregate surface area, this suggests that the volume fraction occupied by regions composed of two or more **overlapping** ITZ may be greater for the systems with the smallest ITZ thickness. For example, if the average width of overlapping interfacial transition zone regions remains relatively constant as t varies, the fraction of total ITZ volume contained in overlaps will increase with decreasing ITZ thickness (e.g., 2 μm out of 10 μm vs. 2 μm out of 50 μm).

TABLE I: Volume Fractions at Percolation vs. Interfacial Transition Zone Thickness

ITZ Thickness (μm)	Agg. Vol. Frac. Range(%)[a]	ITZ Vol. Frac.(%)[a]
10	44-50	6
22.5	30-37	10
50	16-24	16

[a]Values for fraction connected = 50%

A third scenario which can be addressed using this simulation model is the effect of aggregate PSD on ITZ percolation. Using spherical particles and a t value of 22.5 μm, Fig. 5 shows the effects of removing the smallest particles (< 150 μm in diameter) from the original aggregate PSD. Physically, this would be equivalent to sieving the aggregate and discarding

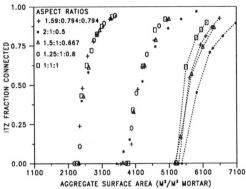

Figure 4: Percolated interfacial transition zone fraction vs. aggregate surface area for $t=50$, 22.5, and 10 (dotted lines) μm from left to right.

the portion which passed through a sieve with an opening size of 150 μm. In coarsening the aggregate PSD, one finds that a greater volume of aggregate, but a smaller volume of ITZ paste (indicated by the points connected by dotted lines in Fig. 5) is needed to percolate the system. In this case, one can imagine maintaining percolation by replacing an already percolated cluster of small aggregate particles by a single larger aggregate particle, which would indeed increase aggregate volume while somewhat decreasing ITZ volume.

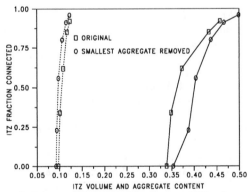

Figure 5: Percolated interfacial transition zone fraction vs. ITZ volume (dotted lines) and aggregate volume fraction (solid lines) for $t=22.5$ μm showing effect of removing finest aggregate particles.

CONCLUSIONS

A computer experiment has been performed to study the effects of interfacial transition zone thickness and aggregate shape on the percolation of ITZ in mortars. Both parameters can have significant effects on the percolation characteristics of these systems. For a given aggregate particle size distribution and ITZ thickness, the curve of ITZ fraction connected vs. ITZ volume fraction was found to be relatively independent of particle shape, while elongated aggregate shapes resulted in lower aggregate contents being required for ITZ percolation than spherical ones. Future experimental studies will concentrate on a systematic assessment of the effects of ITZ connectivity on transport coefficients, such as ionic diffusivity, permeability, and sorptivity, both for mortars made with typical sands and with spherical aggregates.

REFERENCES

[1] Nilsen, A.U., and Monteiro, P.J.M., *Cem. and Conc. Res.* **23**, 147 (1993).

[2] Cohen, M.D., Goldman, A., and Chen, W.F., *Cem. and Conc. Res.* **24**, 95 (1994).

[3] Garboczi, E.J., Schwartz, L.M., and Bentz, D.P., "Modelling the Influence of the Interfacial Zone on the Conductivity and Diffusivity of Mortar," submitted to *J. of Advanced Cement-Based Mat.*

[4] Neubauer, C.M., Jennings, H.M., and Garboczi, E.J., "Modelling the Effect of Interfacial Zone Microstructure and Properties on the Elastic Drying Shrinkage of Mortar," submitted to *J. of Advanced Cement-Based Mat.*

[5] Scrivener, K.L., Bentur, A., and Pratt, P.L., *Advances in Cem. Res.* **1** (4), 230 (1988).

[6] Bentz, D.P., Stutzman, P.E., and Garboczi, E.J., *Cem. and Conc. Res.* **22** (5), 891 (1992).

[7] Feldman, R.F., *Cem. and Conc. Res.* **16** (1), 31 (1986).

[8] Winslow, D.N., Cohen, M.D., Bentz, D.P., Snyder, K.A., and Garboczi, E.J., *Cem. and Conc. Res.* **24** (1), 25 (1994).

[9] Bourdette, B., Ringot, E., and Olliver, J.P., "Modelling of the Transition Zone Porosity," accepted by *Cem. and Conc. Res.*

[10] Halamickova, P., Detwiler, R.J., Bentz, D.P., and Garboczi, E.J., "Water Permeability and Chloride Ion Diffusion in Portland Cement Mortars: Relationship to Sand Content and Critical Pore Diameter," submitted to *Cem. and Conc. Res.*

[11] Hwang, J.T., and Hagwood, C., "Necessary and Sufficient Conditions for Two Ellipses to Have an Intersection," to be submitted.

[12] Stauffer, D., Introduction to Percolation Theory, (Taylor and Francis, London, 1985).

[13] Garboczi, E.J., Snyder, K.A., Douglas, J.F., and Thorpe, M.F., "Continuum Percolation in Three Dimensions: Ellipsoids of Revolution," submitted to *Phys. Rev. E.*

[14] Rallison, J.M., and Hardin, S.E., *J. of Coll. and Int. Sci.* **103**, 284 (1985).

[15] Moran, P.A.P., in Statistics and Probability: Essays in Honor of C.R. Rao, edited by G. Kallianpur et al.(North-Holland, Amsterdam, 1982) pp. 511-518.

[16] Maas, L.R.M., *J. of Comp. and Appl. Mathematics* **51**, 237 (1994).

[17] Ludirdja, D., PhD thesis, University of Illinois, 1993.

[18] Garboczi, E.J., Thorpe, M.F., DeVries, M.S., and Day, A.R., *Phys. Rev. A* **43**, 6473 (1991).

MOISTURE PERMEABILITY OF MATURE CONCRETE
AND CEMENT PASTE

GÖRAN HEDENBLAD
Lund Institute of Technology, Div. of Building Materials, Box 118, S-22100 Lund, Sweden

ABSTRACT

Moisture permeabilities (δ_v) of mature concrete and cement paste with water-cement ratio (w_0/C) 0.4, 0.5 and 0.6 have been determined as functions of the relative humidity (RH). These δ_v can be used to calculate δ_v of the aggregates including the interfacial transition zone (ITZ). δ_v of granite has been preliminarily determined as a function of RH.

EXPERIMENTS

The tested materials comprise concrete, cement mortar and cement paste. This paper only report some results from concrete and cement paste. The compositions of the concretes are shown in Table I.

Table I: Composition of concrete with different w_0/C.

w_0/C	cement, C kg/m^3	water, w_0 kg/m^3	sand/gravel kg/m^3	crushed stone 8-18 mm, kg/m^3
0.4	418	167	990	810
0.5	368	184	990	810
0.6	328	197	990	810

For cement paste 3 different w_0/C were tested, namely 0.4, 0.5 and 0.6. There could be some segregation in the specimens. The experiments are described in detail in [1].

THEORY

The density of the moisture flow rate, under isothermal conditions, is given by

$$g = \delta_v \frac{\partial v}{\partial x} \qquad (1)$$

g = the density of moisture flow rate (kg/(m^{2}s))
v = the humidity by volume (in the the pores of the material) (kg/m^3)
δ_v = the moisture permeability with regard to humidity by volume (m^2/s)

δ_v for many materials is not a constant, it varies with the moisture content in the material

Mat. Res. Soc. Symp. Proc. Vol. 370 © 1995 Materials Research Society

and probably with the temperature. In [1] it is shown that for concrete, cement mortar and cement paste δ_v also probably depends on the size of the specimen or the gradient in the specimen. It may also depend on whether the material is under desorption or absorption.

RESULTS

When a material contains different kinds of salts, e.g. potassimum hydroxide (KOH) the isothermal equlibrium is affected. It is shown in [1] that the maximum RH, in products made of cement that contains alkali, is lower than 100 %. The maximum RH depends on the type of cement and on w_0/C. The measured results can be used to calculate the moisture permeability of the interfacial transition zone as a function of RH.

Results for concrete

In Table II the moisture permeabilities (δ_v) are shown as functions of the relative humidity for specimens with a height of about 100 mm. The calculated maximum RH for different w_0/C are also given in Table II and in Table III.

Table II: Moisture permeability for mature concrete with different w_0/C.

RH (%)	$\delta_v \cdot 10^6$ (m²/s)		
	$w_0/C = 0.4$	$w_0/C = 0.5$	$w_0/C = 0.6$
33 - 65	0.13	0.14	0.17
80	0.28	0.33	0.38
86	0.39	0.59	0.73
90	≈0.53	1.0	1.5
93	≈0.6	1.7	3.2
95	≈0.7	2.8	7.5
96	≈0.7	4.2	8.5
97	-	9	14
97.6	-	-	22

Moisture permeability for cement paste

The moisture permeabilities for cement pastes with different w_0/C are given in Table III. For higher RH the moisture distributions in the specimens are estimated (and then the moisture permeability). The estimated RH-distribution curve gradually and contiuously follows the measured RH -distribution curve. More details are given in [1].

Moisture permeability for granite

Preliminary results for granite show that δ_v is about $0.006 \cdot 10^{-6}$ m²/s between 33 to 94 % RH, between 94 to 97.5 it is about $0.05 \cdot 10^{-6}$ and between 97.5 and 100 % RH it is about $1 \cdot 10^{-6}$ m²/s. The crushed stone, used for the concrete, consits of quartzite and one estimation is that the moisture permeabilities for the two materials are of the same magnitude, see also [2].

Table III: Moisture permeability for mature cement paste with different w_0/C.

RH (%)	$\delta_v \cdot 10^9$ (m²/s)		
	$w_0/C = 0.4$	$w_0/C = 0.5$	$w_0/C = 0.6$
33 - 65	0.3	0.65	1.0
80	0.5	1.5	1.8
86	0.6	2.1	2.8
90	0.65	2.6	4.3
93	0.7	3.0	6.4
95	0.72	3.4	8.5
96	0.75	3.6	10
97	-	3.9	13.5
97.6	-	-	16.8

A COMPOSITE MODEL OF MOISTURE TRANSPORT IN CONCRETE

The model presented below is a <u>very</u> simplified model but maybe it can give a qualitative understanding of moisture transport in concrete. Hillerborg [3] has proposed a simple composite model for materials, which is given by

$$\delta_v^n = V_1 \cdot \delta_{v1}^n + (1-V_1) \cdot \delta_{v2}^n \tag{2}$$

δ_v = the moisture permeability of the composite   (m²/s)
δ_{v1} and δ_{v2} = the moisture permeability of component 1 and component 2   (m²/s)
V_1 = the volume content of component 1 $0 < V_1 < 1$ (1)
n = a constant that depends on the ratio δ_{v1}/δ_{v2}, where δ_{v1} is for the particle phase and δ_{v2} is for the continous phase in the material.

In concrete the cement paste is the continuous phase and the aggregate is the particle phase. When a material consists of more than two phases you can first calculate δ_v for two phases. Then you can take this δ_v and combine with the third phase and so on. When n = 1 the model is the parallel model and with n = -1 it is the series model. Normally n lies between -0.5 and +0.5. Hillerborg proposes n-values and they are shown in Figure 1 as a function of log $(\delta_{v1}/\delta_{v2})$.

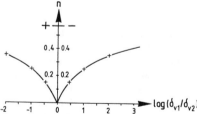

Figure 1: The values of n as a function of log $(\delta_{v1} / \delta_{v2})$.

δ_v is measured for concrete and cement paste with w_0/C 0.6 as function of RH. With eq. (2) it is possible to calculate δ_v of the "aggregate". δ_v of the aggregate is not only the aggregate but it also includes the moisture permeability of ITZ. The volume fraction is 0.68

(V_1). The calculations of the moisture permeability of the "aggregate" are shown in Table IV.

Table IV. Calculation of the moisture permeability of the "aggregate". (w_0/C=0.6)

RH (%)	paste $\delta_v \cdot 10^6$ (m²/s)	concrete $\delta_v \cdot 10^6$ (m²/s)	n (1)	aggegate $\delta_v \cdot 10^6$ (m²/s)	aggregate δ_v / δ_{33} (1)
33	1.0	0.17	0.29	0.046	1
80	1.8	0.38	0.26	0.13	2.8
86	2.8	0.73	0.24	0.32	7
90	4.3	1.5	0.20	0.84	18
93	6.4	3.2	0.13	2.3	50
95	8.5	7.5	0.06	7	152
96	10	8.5	0.04	8	172
97	13.5	14	-0.03	14	309
97.6	16.8	22.5	-0.09	26	563

When the calculated δ_v of the aggregate is compared with the measured δ_v of granite it is seen that the calculated values is at least 10 times bigger.

A model for the moisture transport in concrete

The solid aggregate is considerd to be impermeable and all moisture transport in connection with the aggregate takes place in the ITZ. One aggregate is shown in Figure 2.

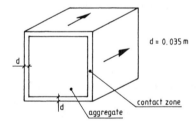

Figure 2: A cubical aggregate with its interfacial transition zone.

All the aggregates are assumed cubic. For w_0/C 0.5 and 0.4 Scrivener [4] has shown the porosity and the thickness of the ITZ . The average porosity is 17 % for w_0/C 0.5 and 12 % for w_0/C 0.4. All the different sizes of aggregates are supposed to have the same thickness of the ITZ (d = 0.035 mm.). The moisture flow in ITZ is considerd to be unidimensional and takes place on 4 sides of the 6 sides of the aggregate. There are pores in the ITZ and only 1/3 of them are considerd to be in the flow direction. The other 2/3 are considerd to be in the perpendicular diections. The porosity in the flow direction for w_0/C 0.6 is calculated as

$$P = \frac{(sieve\ size+2d)^3 - (sieve\ size)^3}{(sieve\ size+2d)^3} \cdot 4/6 \cdot 0.17 \cdot 1/3 \qquad (3)$$

The moisture permeability of the aggregate is calculated with the parallel model

$$\delta_v = \Delta P \cdot \delta_v(ITZ) \qquad (4)$$

The moisture permeability in ITZ at low RH is assumed to be the same as in air ($25 \cdot 10^{-6}$ m²/s). From 80 % RH the calculated δ_v at 33 % RH is increased with the factor given in Table IV column 6. This is done because at higher RHs there are also other transport mechanisms than pure diffusion. The calculated properties of the "aggregate" at 33% RH for w_0/C 0.6 are shown in Table V.

Table V: Calculated properties of the "aggregate" at 33 % RH.

sieve size mm.	% material pass. each sieve, M	ΔM %	Weight kg/m³	Porosity in the flow direct. (P)	ΔP (-)	$\delta_v \cdot 10^6$ of the aggregate m²/s	volume fraction of the aggregate	Fraction number
		4						
0.125	4			0.0278				
		11	109		0.0238	0.595	0.0411	1
0.25	15			0.0198				
		17	168		0.0161	0.403	0.0634	2
0.5	32			0.0123				
		20	198		0.00962	0.241	0.0747	3
1	52			0.00694				
		20	198		0.00532	0.133	0.0747	4
2	72			0.00370				
		15	149		0.00281	0.0703	0.0562	5
4	87			0.00192				
		13	129		0.00145	0.0363	0.0487	6
8	100			0.00098				
8	0			0.00098				
		50	405		0.00081	0.0204	0.1528	7
12	≈50			0.00065				
		50	405		0.00055	0.0136	0.1528	8
18	100			0.00044				

δ_v of the concrete is calculated with eq.(2). First the resulting δ_v of the cement paste and fraction number 1 is calculated. With this δ_v and δ_v of fraction 2 the resulting δ_v is calculated and so on. The calculations are shown in Table VI. The model used is much simplified, but it give a qualitative understanding. At low RH, below about 80 %, all fractions of aggregate decrease the moisture permeability. When RH is over about 80 % the smallest aggregates (the ITZ) are more permeable than the cement paste. The higher RH is, the more permeable the aggregates are. In Figure 3 the measured δ_v of cement paste and concrete with w_0/C 0.6 and 0.4 are shown. The "theoretical" δ_v according to Table VI is also shown. The "theoretical" δ_v is not purely theoretical since the quotients in column 6 in Table IV are used as input data in the calculations of δ_v. For concrete with w_0/C 0.4 δ_v is calculated from δ_v of the cement paste and δ_v of the "aggregate" from w_0/C 0.6, but corrected for the porosity in ITZ.

Table VI: "Theoretical" moisture permeability of concrete with $w_0/C = 0.6$.

Fraction number	V_1 in eq. (2)	$\delta_v \cdot 10^6$ (m²/s)				
		RH=80%	RH=86%	RH=90%	RH=93%	RH=95%
C-paste		1.83	2.77	4.3	6.4	8.5
1	0.109	1.81	2.90	4.74	7.4	10.4
2	0.144	1.69	2.88	5.05	8.5	13.0
3	0.145	1.49	2.68	4.95	8.9	14.9
4	0.127	1.28	2.37	4.54	8.6	15.5
5	0.087	1.11	2.11	4.11	8.0	15.0
6	0.070	0.98	1.87	3.70	7.3	14.1
7	0.180	0.68	1.32	2.66	5.4	11.1
8	0.153	0.49	0.98	2.0	4.2	8.9

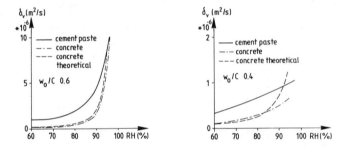

Figure 3: Moisture permeabilities of concrete and cement paste as functions of RH. Measured and theoretical curves. $w_0/C = 0.6$ and 0.4.

CONCLUSIONS

The reported moisture permeabilies for concrete, cement paste and granite can be used to calculate the moisture transport properties of the interfacial transition zone. A very simplified model for the moisture permeability of a composite material is presented.

REFERENCES

1. G. Hedenblad, Moisture Permeability of Mature Concrete, Cement Mortar and Cement Paste, Ph.D. Thesis, Lund Institute of Technology, Lund, Sweden (1993).

2. T.C. Powers, L.E. Copeland, J.C. Hayes and H.M. Mann, Permeability of Portland Cement Paste, J. Am. Con. Inst. Vol. 26, 285 (1954), PCA Bull. No. 53 (1955).

3. A. Hillerborg, Compendium of Building Materials FK (in Swedish), Lund Institute of Technology, Division of Building Materials, Lund, Sweden (1986).

4. K.L. Scrivener, The Microstructure of Concrete, Materials Science of Concrete, Editated by J.P. Skalny, The Am. Ceramic Society (1989).

TRANSFER PROPERTIES OF INTERFACIAL TRANSITION ZONES AND MORTARS

BOURDETTE B.*, J.P. OLLIVIER ** AND E. REVERTEGAT*
* CEA / DESD Saclay 91 191 Gif sur Yvette, France
** LMDC, INSA-UPS Génie civil Complexe Scientifique de Rangueil. 31 077 Toulouse, France

ABSTRACT

It seems probable that the higher porosity of the Interfacial Transition Zone (ITZ) affects concrete durability. An experimental program has thus been developed to provide more information in this area and to evaluate the influence of the transition zone on the transfer properties. A model based on the mercury intrusion porosity results shows that the ITZ is about 1.5 to 2.5 times more porous than OPC cement paste and in this zone the tritium diffusion coefficient is a factor of three higher.

The tritium diffusion coefficient has been measured for mortars containing different sand concentrations. We have shown that the binder phase diffusivity decreases with the sand concentration down to a minimum, then increases when the ITZ are interconnected. Under these conditions, the sand volume concentration is above 50% and for higher concentrations, the diffusivities of the binder phase and pure cement paste are similar. The effects of ITZ on the porosity compensate for the increasing tortuosity due to the interposition of particle aggregates.

Measurement of the calcium leaching rates from mortar with high sand concentrations and pure cement paste with the same W/C ratio confirms these results: the leaching kinetics of the binder phase in mortar are very close to those of the pure cement paste.

INTRODUCTION

To produce durable concretes, we look for processes for yielding low porosity materials. The microstructure of concrete is very heterogeneous and much more complex than that of a pure cement paste. In particular, the aggregates are surrounded by a transition region, a higher porosity zone. When the concrete is put in contact with a weakly ionized solution that is frequently renewed continuous leaching occurs. That is, there is dissolution of hydrates and liquid phase transfer of elements from the concrete to the aggressive solution by diffusion. The instantaneous chemical reactions coupled with a diffusion process result in degradation kinetics with dependence on the square root of the time [1]. We report here on a study of the effects of the transition zones on concrete leaching.

EXPERIMENTAL PROGRAM

Materials

The cement used in preparing the mixes is an ordinary Portland cement (OPC) with the chemical composition given in Table 1.

To show the effects of the Interfacial Transition Zone(ITZ) on leaching, two mixes were selected: a pure cement paste and a mortar in which the ITZ are interconnected. The ITZ

Mat. Res. Soc. Symp. Proc. Vol. 370 © 1995 Materials Research Society

connections in a mortar depend on their thickness and their content and grade. Using a model [2,3], the fraction of connected ITZ may be calculated when the sand volume fraction is above the percolation threshold. With the sand used in our study and assuming a 30-μm thickness for the transition zone, the percolation threshold corresponds to a sand volume fraction of 50%, and 97% of the transition zones are connected for the selected sand volume fraction of 52%.

Oxide	CaO	SiO$_2$	Al$_2$O$_3$	Fe$_2$O$_3$	SO$_3$
Content(%)	62.9	20.6	5.8	3.6	3.1

Table 1 : Cement Chemical Composition

The effects of the sand volume fraction on the diffusion coefficient were studied to analyze the leaching results. Three other mixtures were prepared with 20, 40 and 62% sand.

The cement pastes and mortars were prepared using the same W/C ratio (0.4) to demonstrate clearly the effects of the ITZ microstructure on diffusion and the deterioration process.

The materials were cast in cylindrical molds (pure cement paste: diameter 70 mm and height 100 mm; mortar: diameter 110 mm and height 220 mm), then removed from the molds and cured in lime-water at 20°C for three months. A single grade - 0.16/3.15 mm - of inert clean quartz sand was used.

Experimental procedure for leaching studies

We studied the amounts of elements leached (calcium, silicon and sulfur) as a function of the square root of the time for the pure paste and mortar. The form and dimensions of the leached specimens were such as to respect the following conditions:

- a core of sound material must remain during the length of the experiment (around 8 months) ;
- the leaching must be unidirectional over a constant diffusing surface area.

The mortar cylinder (70 x 100 mm) was produced by coring the 110 x 220 mm specimens so that the transition zones were in direct contact with the aggressive solution. As described, the pure cement paste was formed in a 70 x 100 mm mold. The experimental arrangement (Figure 1) makes it possible to follow the leaching kinetics of a pure cement paste or mortar cylinder in a constant composition aqueous solution. Each cylinder was suspended and immersed in 2-liter containers with CO$_2$-free, deionized water at pH 8.5.

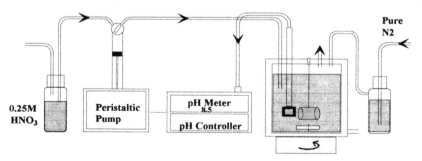

Figure 1 : Experimental arrangement for determining leaching kinetics

EXPERIMENTAL RESULTS AND DISCUSSION

The concentrations of leached calcium, silicon and sulfur are studied as a function of $t^{0.5}$ for the pure paste and mortar. The curves showing a plot of the amounts leached versus the square root of the time give a straight line. Obtaining this time dependence for these amounts indicates that leaching of the elements in the cylindrical specimens obeys Fick's laws for a unidirectional diffusion process in a semi-infinite medium.

The leaching rate of an element i in a given material corresponds to the slope of the straight line obtained in the plot of the amount leached against the square root of the time. This amount leached is that of the element i that has diffused to the sample surface. We have calculated the slopes of the straight lines for leaching of calcium, silicon and sulfur for the mortar and the pure paste. However, the leaching capacity of the two materials can be compared only if the rates are calculated using surfaces that have a true diffusing character. This is why the leaching rates of the elements in the mortar were recalculated using the mortar surface area less that of the non-porous aggregates. These then correspond to leaching rates of the different elements for the cement paste in the mortar.

We have theoretically determined the rate of each element leached for the cement paste in the mortar from experimentally obtained leaching results for the mortar. If it is assumed that the ratio of the diffusion surface areas is identical to the ratio of the diffusion volumes, the leaching rates for each element in the cement paste are calculated as follows :

$$J_{PM} = \frac{S_M}{S_{PM}} J_M = \frac{V_M}{V_{PM}} J_M$$

J_M and J_{PM} are, respectively, the leaching rates of the mortar and the cement paste in the mortar $(mol.dm^{-2}.d^{-0.5})$;

V_M and V_{PM} are, respectively, the volume of the mortar and the volume of the cement paste contained in the volume V_M (dm^3) ;

S_M and S_{PM} are, respectively, the surface area of the mortar and the surface area of the cement paste contained in a volume V_M (dm^3).

The leaching rates of each element for the pure paste, for the mortar and the cement paste in the mortar (i.e., pure paste with transition zones) are given in Table 2.

	Pure paste $(mol.dm^{-2}.d^{-1/2})$	Mortar $(mol.dm^{-2}.d^{-1/2})$	Cement paste in the mortar $(mol.dm^{-2}.d^{-1/2})$
Calcium	$1.14.10^{-2}$	$6.08.10^{-3}$	$1.27.10^{-2}$
Silicon	$2.32.10^{-4}$	$1.18.10^{-4}$	$2.46.10^{-4}$
Sulfur	$1.75.10^{-4}$	$3.31.10^{-5}$	$6.89.10^{-5}$

Table 2 : Calcium, silicon and sulfur leaching rates for the pure paste, mortar and pure paste in the mortar (Portland cement, W/C ratio 0.4)

It can be seen that less calcium, silicon and sulfur are leached from a mortar than from a pure paste This can be easily understood, since for an identical volume of material, the mortar initially contains much less paste. On the other hand, when we calculate the leaching rates for each element for the paste in the mortar, we observe that there is very little difference between these rates and those obtained with the pure paste. There is a difference between the two results for sulfur which can perhaps be due to measurement uncertainties since the concentrations of sulfur measured in the leaching solution are of the order of mg.dm^{-3}.

From the characteristics of the cement paste in the mortar, we would have thought that the mortar would be leached much more rapidly than a pure paste since it contains transition zones which:

- are **very porous regions** : from the calculation model for the transition zone porosity using results from mercury porosimetry [2], we have shown that the transition zones for this mortar are around 2.5 times more porous than the surrounding cement matrix. Thus, they give the cement paste in the mortar a higher porosity (33%) than that of the pure paste (27%) having the same W/C ratio as the mortar. In addition, the transition zones are highly interconnected and there are therefore possibilities for transfer by higher porosity pathways inside the material;

- have a **3-times higher diffusivity** than that of the porous network in the cement matrix. We determined the diffusion coefficient for tritium through the transition zone using the same method as that developed by Breton et al. [4]. This experiment consisted of determining the diffusion coefficient through plates of pure paste and mixed granite-cement paste plates. These mixed plates were prepared from granite plates with holes varying in number and diameter [1] which were filled with cement paste. The area of the interfaces increases with the number of holes making it possible to vary the contribution of the ITZs to diffusion through the plates. We can then derive the diffusion capacity of the transition zone.

The presence of aggregates, which constitute obstacles to the migrating species, is shown by an increase in the tortuosity of the mortar microstructure with respect to that of a pure paste. As the particles which diffuse cannot cross the aggregates with zero porosity, they go around them which increases their diffusion path length. The leaching results show that one of the roles played by the transition zones in degrading the mortar consists in suppressing the increase in the tortuosity of the material texture. This is because the leaching rates of the elements in the mortar cement paste are slightly higher than those of the pure paste which has a lower tortuosity.

To verify this hypothesis, we experimentally determined the diffusion coefficient of mortar cement paste containing a variable volume percentage of sand. We compared these results with the diffusion coefficient of the pure paste, which would be the diffusion coefficient of the mortar paste if there were no aggregates.

Experimentally, we prepared mortar cylinders (70x100 mm) containing different sand concentrations, which we sectioned in 4-mm thick plates, after 3 months curing in lime water.

If it is assumed that the ratio of the diffusion surface areas is the same as the diffusion volumes ratio, the coefficient of the mortar paste is calculated as follows:

$$De_{PMs} = \frac{V_M}{V_{PM}} De_{Ms} \iff D_{PMs} = \frac{V_M}{V_{PM}} \cdot \frac{1}{\phi_{PMs}} De_{Ms} = \frac{De_{Ms}}{\phi_{Ms}}$$

De_{PMs} and De_{Ms} are, respectively, the effective diffusion coefficients in the sound mortar cement paste and in the sound mortar ;
D_{PMs} and D_{Ms} are, respectively, the diffusion coefficients in the sound mortar cement paste and in the sound mortar ;
ϕ_{PMs} and ϕ_{Ms} are, respectively, the porosity in the sound mortar cement paste and in the sound mortar ;
V_M and V_{PM}, are respectively, the volume of mortar and the volume of cement paste contained in the volume V_M (dm^3).

The diffusion coefficient for the mortar cement paste D_{PMs} is calculated from the effective diffusion coefficient in the mortar De_{Ms} and its porosity ϕ_{Ms} as these are directly available experimental results. The variations of the experimental and theoretical diffusion coefficients of the mortar cement paste for different mortars as a function of their volume fractions of sand (20, 40, 52 and 60%) are shown in Figure 2. The theoretical diffusion coefficient of the mortar cement paste corresponds to the diffusion coefficient of the pure sound paste ($D_{PPs} = 8.74.10^{-5}$ dm^2.d^{-1}).

It can be seen that the experimental diffusion constant values are less than the theoretical values for mortars containing 20% and 40% sand. From Ringot's model [2], there is no percolation of the transition zones in these mortars. The addition of aggregates results in an increase in the tortuosity which is not compensated for by the presence of a network of interconnected transition zones with higher diffusivity. Due to this, we observe a decrease in the diffusion coefficient.

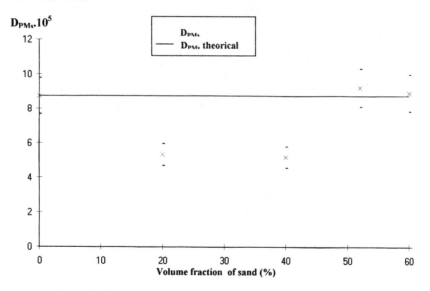

Figure 2 : Variation of the mortar cement paste diffusion coefficient with the volume fraction of sand

On the other hand, for the mortar with 52% sand, the experimental diffusion coefficient value is slightly higher than the theoretical value. This mortar, which is above the percolation threshold of the transition zones (threshold estimated at around 50% sand), contains a network of interconnected transition zones which increases the diffusivity of the elements through the mortar. Let us recall that the diffusion coefficient of a diffusing species i in a medium k is a function of the tortuosity as given by:

$$D_{i,k} = f \, D_{vi} \qquad f \leq 1$$

$D_{i,k}$ and D_{vi} are, respectively, the diffusion coefficient of the species i in the medium and in pure water;
f is a restrictive factor which takes into account the tortuosity of the medium in which the species diffuses.

We deduce that when the mortar contains a sand concentration above the percolation threshold of the transition zones, the tortuosity increase due to the presence of the aggregates is suppressed by interconnected transition zones : assuming a 30μm thickness, 97% of the transition zones are connected for the sand volume fraction of 52%.

For the mortar with 60% sand the diffusion coefficient is lower than that with 52% sand since even if all the transition zones are interconnected in the first case, the porous volume in which the particles can migrate becomes small.

This phenomenon was observed by Houst et al [5], who noted a progressive decrease in oxygen and carbon dioxide diffusion coefficients with the amount of sand contained on the mortar, then a sudden increase in the diffusion coefficients for mortars containing a certain volume percentage of sand. They, as we, attribute the diffusion coefficient increase to the formation of a network of interconnected transition zones. It can be noted that the diffusion measurements for oxygen reported by these authors show that the aggregate concentration has a much greater effect than in our diffusion measurements with tritiated water. These deviations can be explained by the differences in the experimental conditions. The oxygen measurements require a certain extent of desiccation which can result in microcracks. It is known that these develop preferentially at interfaces between the cement and the aggregates. In contrast, our mortar samples were always saturated.

CONCLUSIONS

The linear variations of the amounts leached for each element in the pure paste and mortar with the square root of the time shows that leaching kinetics are governed by the diffusion equations for a semi-infinite medium with unidirectional diffusion.
When the transition zones in the mortar belong to a percolation cluster, the result is that the mortar cement paste has slightly higher leaching kinetics than that of the pure paste for the same W/C ratio. On the other hand, when the sand concentration in the mortar is below the percolation threshold, the aggregates act as a barrier by increasing the tortuosity of the material, thus slowing down the leaching kinetics.

References

[1] B. BOURDETTE, « Durabilité du mortier : prise en compte des auréoles de transition dans la caractérisation et la modélisation des processus physiques et chimiques d'altération », Thèse de l'INSA de Toulouse, 1994

[2] B. BOURDETTE, E. RINGOT, J.P. OLLIVIER, « Modelling of the transition zone porosity », to published in Cement and Concrete Reasearch

[3] D.P. BENTZ, E.J. GARBOCZI and P.E. STUTZMAN, « Computer modelling of interfacial zone in concrete », Interfaces in Cementitious Composites, RILEM Proceedings 18, Edited by J.C. Maso, E & FN Spon, Toulouse, p. 107-116, 1992

[4] D. BRETON., J.P. OLLIVIER, G. BALLIVY, « Diffusivité des ions chlore dans la zone de transition entre pâte de ciment et roche granitique », Interface in Cementitious composites, RILEM Proceedings 18, Edited by J.C. MASO, E & FN SPON, pp. 279-288, 1992

[5] Y. F. HOUST, F. H. SADOUKI, F. H. WITTMANN, « Influence of aggregate concentration on the diffusion of CO_2 and O_2 », Interface in Cementitious Composites, RILEM Proceedings 18, Edited by Maso J.C., E & FN SPON, pp. 279-288, 1992

EFFECT OF THE ITZ ON THE LEACHING OF CALCIUM HYDROXIDE FROM MORTAR

Christophe CARDE[*] and Raoul FRANÇOIS[**]
[*]CEA, CEN Saclay, 91191 Gif-sur-Yvette CEDEX, FRANCE
[**]LMDC INSA-UPS Génie Civil, complexe scientifique de Rangueil, 31077 Toulouse CEDEX, France

ABSTRACT

This paper deals with the effect of the ITZ on the leaching process of cement based materials. In order to characterize this effect, we have performed experiments on mortar samples and pure cement paste samples. Two aggressive waters were used, the first one with pH=4.5 and the second one with pH=8.5. These two aggressive waters were modelled by a solution of ammonium nitrate. Compression tests were conducted on micro-cylinder samples (10, 12 and 14 mm of diameter) because of the slow kinetics of degradation due to the leaching. In the case of a sample made with cement paste only, the leaching process leads to a uniform degradation of the samples. The thickness of this degraded zone is the same whatever the size of the sample. The decrease in compression strength is thus completely explained by this degraded zone. In the case of a sample made with mortar, the leaching process leads also to a degradation zone whose thickness is the same whatever the size of the sample. But, the decrease in mechanical properties and the increase in irreversibility couldn't be explained for the different sizes of samples by the same thickness of leaching zone. This is due to the fact that the chemical attack progresses through the interfacial zones because they have a higher porosity than the bulk paste. The relative volume of interfacial zones linked together with the aggressive environment is more important in the case of the smaller sample.

INTRODUCTION

The research program in progress tries to characterize the deterioration of the mechanical properties of the concrete surrounding radioactive wastes, due to the water flow during storage. The chemical attack of this small amount of ionized water is essentially a leaching of the calcium hydroxide and a progressive decalcification of the CSH. The slow kinetics of these chemical reactions leads us to increase the agressivity of the environment, first in decreasing the pH to 4.5 (by admixture of HNO_3) then secondly by using a solution of NH_4NO_3. The chemical attack due to NH_4NO_3 is also a leaching but quicker than those obtained with the water. Then, the use of ammonium nitrate allows us to model both the aggression of pure water and acid water, but also to study the durability of reinforced concrete used by the factories which produce nitrate fertilizer.
In this paper, the results obtained on mortar samples and cement paste samples exposed to an aggressive solution of NH_4NO_3 are presented. These results show that the ITZ seems to play a significant role in the deterioration process due to the leaching.

EXPERIMENTAL PROGRAM

Materials

The mortar and the cement paste (same E/C ratio = 0.5) were both made with the same OPC cement. The mortar composition is given in Table 1. The chemical composition of the cement is given in Table 2.

gravel (silica)	0.6/1.2 mm	70% - 881 kg/m^3
gravel (silica)	1.2/2.5 mm	30% - 377 kg/m^3
Normal Portland Cement		629 kg/m^3
water		315 kg/m^3

Table 1 : Mortar composition

Chemical composition of cement						
SiO$_2$	Al$_2$O$_3$	Fe$_2$O$_3$	CaO	MgO	SO$_3$	Na$_2$O
20.2	4.9	3	63.4	0.67	3.2	0.25

Table 2 : Chemical composition

The mixtures were poured into cylindrical moulds (11x22 cm). The demoulding was done after 24 hours, then the samples were cured for 27 days immersed in water saturated in lime at 20°C (± 1°C).

Samples

Because of the slow kinetics of leaching, we have been obliged to work with samples of small sizes. The samples used are cylinders whose diameters are 10,12 and 14 mm with a ratio h/ϕ = 2 (h height of the sample). After curing, the samples are extracted from the test pieces by means of a diamond tipped core lubricated with water. For each sample dimension, two series of samples have been made, the first one which has been immersed in the aggressive solution (treated series), the other one which has been kept in an endogenous environment (control series).

Leaching process

The aggressive environment used is a solution of ammonium nitrate NH_4NO_3 (concentration =50%.). This process induces mainly a total leaching of the lime and a progressive decalcification of CSH. The leaching process leads to an uniform degradation of the samples. The peripheral degraded zone delimited by the dissolution front of calcium hydroxide, is more porous than the bulk sample and also less resistant.

The chemical attack of the ammonium nitrate leads to the development of a soluble calcium nitrate, a not very soluble nitro-aluminate of calcium and an ammoniac emanating NH_3[1]. Because of this emanation, the solutions of ammonium nitrate work like diluted acids. The degradation is governed by a diffusion mechanism and can be described by the Fick's law relating the degraded thickness to the square root of time $\sqrt{t}$. Figure 1 shows the linear variation of the degraded thickness, measured out with a microscope, and the loss of mass of the sample in relation to the square root of immersion time in the aggressive solution.

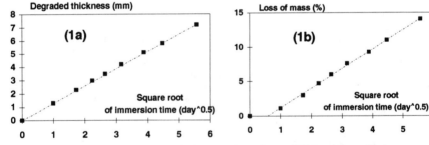

Fig. 1 : Variation of degraded thickness (1a) and the loss of mass (1b) in relation with the square root of immersion time in the solution of NH4NO3 (50%)

The variation of the thickness of this degraded zone is the same for the samples of mortars and the samples of cement.

Mechanical tests

Both the treated samples and the control samples have been subjected to a compressive load in order to measure their compressive resistances. The device used is a Hounsfield press with a maximal capacity of 50kN. The displacement speed is controlled during the load. The force applied on the sample and the longitudinal displacements are measured during the test. The displacement measured is the average of three displacement recorded during the test by means of three transducers fixed on the clamping plate with an angular location of 120° (fig. 2), so the bending effects cannot disturb the measure of the displacement.

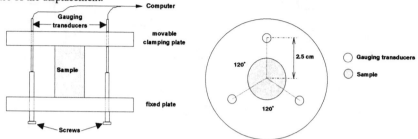

Fig. 2 : Experimental device for the compressive tests

The average longitudinal strain εl_{avr} is calculated using the three measured displacements.

$$\varepsilon l_{avr} = \frac{d_1 + d_2 + d_3}{3} \cdot \frac{1}{h} \quad (\mu m/m)$$

where　　　d1, d2, d3 are the displacements recorded by means of the transducers 1, 2 and 3 (μm)
　　　　　　h is the height of the sample (m)

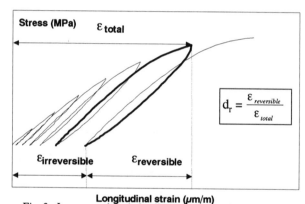

$$d_r = \frac{\varepsilon_{reversible}}{\varepsilon_{total}}$$

Fig. 3 : Incremental cycle loading to measure the degree of reversibility

The compressive strength is evaluated by dividing the maximal load F_{ult} by the area of the sample S

$$\sigma_c = \frac{F_{ult}}{S} \quad (MPa)$$

The simultaneous recording of the force and the average displacement allows the plotting of the curve $\sigma = f(\varepsilon)$. Two loading methods are used. The first one is a simple standard loading, the second one is a incremental cyclic loading which allows us to measure the degree of reversibility d_r of the materials [2]. This parameter characterized the irreversibilities due to the loading. The calculation of this parameter at the end of each cycle allows us to study the increase of the irreversibilities in relation to the % of the stress rupture σ_c. (fig. 3).

EXPERIMENTAL RESULTS

The samples were degraded for 4 days in the aggressive 50% concentration solution of NH_4NO_3. So as to homogenize the aggressive solution, it is subjected to continuous shaking. After treatment, the part of the base area which is degraded and would modify the displacement and the measured irreversibilities is eliminated from the samples.
The thickness of the degraded zone is 2.8 mm.

The values of the compressive strength (average of 8 specimens) measured in the treated group and the control group are given in Table 3.

	$\phi = 10$ mm			$\phi = 12$ mm			$\phi = 14$ mm		
	control	treated	$\Delta\sigma$	control	treated	$\Delta\sigma$	control	treated	$\Delta\sigma$
Cement paste	49.2	17.9	31.3	52.3	27.0	25.3	53.0	30.0	23.0
Mortar	48.2	12.8	35.4	53.5	25.6	27.9	49.2	27.2	22.0

Table 3 : compressive strength in MPa

These results show that the decrease of stress $\Delta\sigma$ is more important for the mortar than for the cement paste whereas the thickness of the degraded zone is the same. This phenomenon is more pronounced when the size of the sample decreases. If we suppose that the chemical attack leads to the development of two separated zones, a peripheral one which is degraded and the bulk which is sound, we can model (in a simplified way) the distribution of stresses in the sample. Figures 4a and 4b show respectively the model when the degraded zone is assumed to be unresisting and when it is assumed there is a gradient of stresses in the degraded zone.

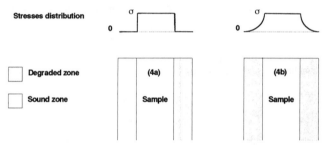

Fig. 4 : Simplified models of the distribution of stresses in a degraded sample

The use of one or other model allows us to conclude in the same way the comparison between the cement and the mortar samples. So, to compare the behavior between both types of samples, the first model will be used.

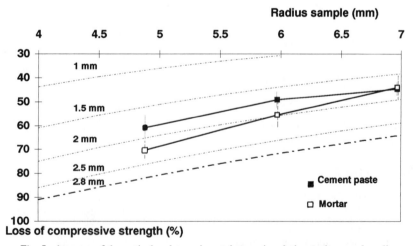

Fig. 5 : decrease of theoretical and experimental stress in relation to the sample radius

Figure 5 shows the decrease of the theoretical strength calculated for different thicknesses of the degraded zone, with the model used and the decrease of experimental strengths evaluated in % in relation to the resistance measured on the control group. The compressive strength of the control group is assumed to be a constant.

These results show: first, the model used overestimates the decrease of compressive strength. The experimental values are all located above the theoretical curve corresponding to the real thickness degraded (2.8 mm); nevertheless, the model seems to be a good representation of the variation in strength in relation to the size of the sample. Second, the loss of compressive strength measured on the cement paste samples can be explained with the same thickness of the degraded zone whatever the size of the sample. On the other hand, in the case of the samples made with mortar, the decrease in the mechanical properties cannot be explained by the same thickness of the leaching zone; the decrease is more pronounced when the size of the sample decreases. The theoretical thicknesses which explain the loss of strength are shown on Table 4.

	$\phi = 10$ mm	$\phi = 12$ mm	$\phi = 14$ mm
Cement paste	1.85	1.75	1.80
Mortar	2.25	2.00	1.75

Table 4 : Theoretical degraded thickness (mm)

Why is the behavior of the cement paste and the mortar different? A cement paste is a one phase material whereas the mortar is a three phase material. In mortar, there is aggregate, a bulk paste and an interfacial transition zone (ITZ) between the aggregate and the paste. It is well known that the ITZ is the weak point of the mortar and the differences observed between the experimental results obtained on mortar and on paste could be explained by the presence of the ITZ. In this way, Bourdette [3] has found that the diffusivity of the ITZ is 4.3 times more important than the bulk paste for some samples which have been cured for 3 months, Breton [4] has revealed a ratio between 12 and 15 after a cure of 10 days. So, we can venture the hypothesis that in the case of the mortar, the chemical attack doesn't lead only to a peripheral degradation but also progresses through the interfacial zones, especially if the different ITZ are linked together. Furthermore, this phenomenon is more pronounced when the size of the sample decreases, i.e. the ratio Φ/D_{max} decrease (where Dmax is the maximal diameter of the aggregates).

This hypothesis seems to be confirmed both by the different behavior near the rupture and the variation of the degree of reversibility between the mortar and the cement paste.

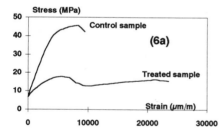 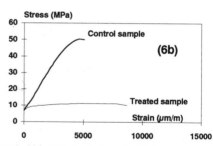

Fig. 6 : Curves $\sigma = f(\varepsilon)$ for a treated sample (6a) and a treated mortar (6b)

In the case of the cement paste, Figure 6a show two peaks: the first one corresponds to the delamination of the leaching zone, the second one corresponds to the rupture of the sound bulk paste. In the case of the mortar, the behavior is different, the rupture is more ductile without any pronounced peak (fig. 6b), which seems to be due to a damage located in the whole mortar.

Figures 7 and 8 show the development of the degree of reversibility for the cement paste and the mortar. In the case of the cement paste, the irreversibilities are weak and almost the same as the control group. On the other hand, in the case of the mortar, the irreversibilities are more important when the size of the sample decreases. These findings confirm the previous hypothesis of further damage to the ITZ beyond the peripheral degraded zone. This damage is more important in the case of a small sample because the relative volume of ITZ linked together is higher.

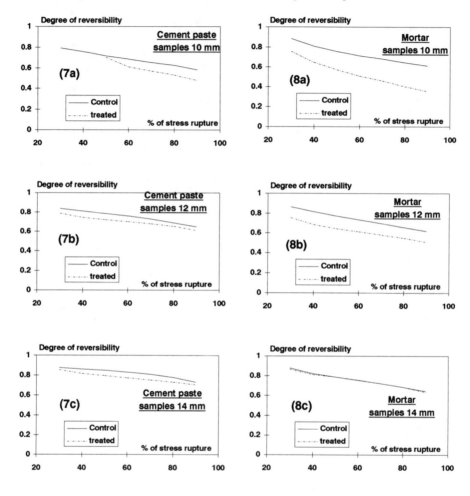

Fig. 7 : Variation of the degree of reversibility for the paste samples

Fig. 8 : Variation of the degree of reversibility for mortar samples

CONCLUSION

The deterioration of the cement paste and the mortar exposed to the action of the ammonium nitrate was manifested by a peripheral zone of less resistance. The thickness of this zone is apparently the same for both materials.

Nevertheless, the mechanical tests performed on both materials have shown a qualitative difference of behavior near the rupture. In the case of the cement paste, the rupture is preceded by a delamination of the leaching zone whereas in that of the mortar it is more ductile.

Quatitatively, the loss of resistance, in the case of the paste, can be explained by the same thickness of the degraded zone whatever the size of the sample. In the case of the mortar, the degraded thickness which can explained by the loss of resistance is more important when the size of the sample is small. This is probably due to to the fact that the deterioration of the ITZ progresses beyond the visible front of leaching.This hypothesis can also be explained by the variation of the degree of reversibility.

REFERENCES

1 . LEA F.M., Magazine of Concrete Research, Vol. 17, n 52, Sept. 1965, p 115-116.

2 . MASO J.C., Le Béton Hydraulique, Paris, ed. by Presses de l'Ecole Nationale des Ponts et Chaussées, Paris, 1982, p273-293.

3 . BOURDETTE B., PhD thesis, INSA Toulouse, 1994.

4 . BRETON D., OLLIVIER J.P., BALLIVY G., Interfaces in Cementitious Composites, Proceedings of the RILEM International Conference, ed. by J.C. Maso RILEM Proceedings 18, 1992, pp. 269-278.

EFFECT OF THE ITZ DAMAGE ON DURABILITY OF REINFORCED CONCRETE IN CHLORIDE ENVIRONMENT

Raoul FRANÇOIS and Ginette ARLIGUIE
LMDC INSA-UPS Génie Civil, complexe scientifique de Rangueil, 31077 Toulouse CEDEX, France

ABSTRACT

This paper deals with the effect of the ITZ on the service life of reinforced concrete. In the case of reinforced concrete structures, the penetration of chlorides does not depend only on concrete transfer properties but also on the loading applied, on the state of strains and on the exposure to the aggressive environment.
In order to take into account these different parameters, we have performed experiments on reinforced concrete elements, over a long period. The samples used have to be of an adequate size (3 meters long) and stored in a salt fog in a loading state so as to be representative of the actual operating conditions of the reinforced concrete structures.
The bending of the beams leads to the development of cracks which are neither preceded nor accompanied by microcracks, but the cement paste-aggregate interfaces are damaged in the tensile areas.
The service loading of reinforced concrete has two consequences : firstly, a cracking with widths ranging between 0.05 mm and 0.5 mm according to the intensity of the mechanical strength applied. Secondly, a damage of the ITZ in the tensile areas causing an increase of chloride penetration directly proportional to the intensity of the stress applied to the beam.
The model of the development of corrosion, worked out in relation with time and based on our results, emphasizes the influence of the paste-aggregate interface damage on the duration of the service life.

INTRODUCTION

The deterioration of reinforced concrete is essentially due to corrosion of the reinforcement. In the development of this phenomenon, the cracks which for the most part are clearly visible, have been quickly implicated. With this mind, the rule books have emphasized the importance of the crack widths as a durability criterion. These rules have lead to an increase in the quantity of steel reinforcement (50 kg/m^3 in 1950 to 160 kg/m^3 in 1990) in concrete in order to control this cracking. The overcost due to these measures has lead to a development of research on this topic. The most recent developments show that the cracks, as long as their width doesn't exceed 0.5 mm, are not the essential factor in the corrosion process. The concrete cover quality and the cover width seem to play the most significant role. Nevertheless, there is a relation between the quality of concrete and the cracking because a crack is only the visible part of the concrete damage. So in the area of the concrete structure where the cracks have occurred, the quality of concrete has decreased due to the damage.
The study developed at LMDC has been conceived with the aim of taking into account the real state of concrete: both the damage and the cracking. The model of the development of corrosion, elaborated in relation to time and based on our results, emphasizes the influence of the paste-aggregate interface damage to the duration of the service life.

EXPERIMENTAL PROGRAM

The experimental program has been already described in previous works but for the purposed of making the present work more comprehensible we will give a summary of its essential details. The reinforced concrete samples used are beams three meters in length (15x28 cm cross-section) kept in a confined salt fog for seven years. The type A beams and the type B beams have a different

Mat. Res. Soc. Symp. Proc. Vol. 370 © 1995 Materials Research Society

reinforcement (Fig. 1). Concrete composition and its compressive strength at 28 days are given in Table 1.

The types A and B correspond to a 4 cm maximum and 1 cm minimum concrete cover respectively, in accordance with French regulations at the time of manufacturing (1983) [1]. The type A beams respect the international requirements (E.C.2) for storage in an aggressive environment. The type B beams do not respect the international requirements for storage in an aggressive environment, because one of the aims of this study is to improve actual requirements to know if these are justified. These beams are superimposed upon one other and separated by metallic wedges. Their construction is in compliance with current standards.

These reinforced concrete elements are exposed, in a loading state, to salt fog.

For each specimen, the concrete composition is:

rolled gravel (silica + limestone)	5/15 mm	1220 kg/m^3
sand	0/5 mm	820 kg/m^3
Normal Portland Cement: HP (high perform.)		400 kg/m^3
water		200 kg/m^3

Table 1: Concrete composition

Water content was adjusted to obtain 7 cm in slump-test for every concrete mixture. The average cylinder specimen compression stress was measured at 45 MPa at 28 days.

Chemical composition of cement

	SiO$_2$	Al$_2$O$_3$	Fe$_2$O$_3$	CaO	MgO	SO$_3$	Na$_2$O
% weight	21.4	6	2.3	63	1.4	3	0.5

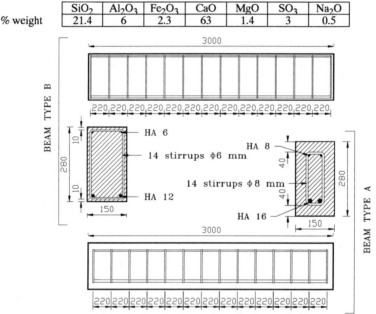

Fig. 1 : Lay-out of reinforced concrete beams

PENETRATION OF CL⁻ IONS INTO THE CONCRETE

Either exogeneous or endogeneous, the cl⁻ ions inside concrete are of two types: bound and free.

The bound cl⁻ ions are fixed either by adsorption or by chemical reactions with some cement components. In the latter case, what results is monochloroaluminate of calcium $C_3A,CaCl_2,10H_2O$.

The free cl⁻ ions can migrate into the concrete and reach the reinforcement and are responsible for the corrosion of the reinforcement.

Thus, the method used to measure the chlorides in concrete must allow the distinction between bound cl⁻ ions, harmless to the steel, and free cl⁻ ions. The selective separation can be achieved by using differences between chlorides solubilities. The free cl⁻ ions can be extracted by cold water: this is not the case, under certain conditions, for the bound cl⁻ ions.

Consequently, the samples must be taken out from the beams using a dry process so as to avoid any leaching. For this, the sampling is done by regular drilling at a depth of 10 mm. The powder obtained from this process is carefully collected and stored in a sealed container before being measured.

The choice of the localization of the various samplings has been dictated by the wish to demonstrate the influence of different parameters, such as the intensity of the mechanical stress applied and the loading state of the concrete.

Effect of strain state

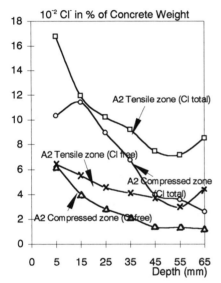
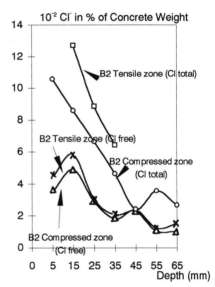

Fig. 2b: Penetration of cl⁻ ions at the middle of the beam B2 according to the stress in the sampling zone

Fig. 2a: Penetration of cl⁻ ions at the middle of the beam A2 according to the stress in the sampling zone

Figure 2 shows the chloride content for some sample taken at a distance of 125 cm from the end of the beam, i.e. almost in the middle of the beam. In each case, the sample has been extracted from 6 cm under the upper surface of the beam (corresponding to the compressed zone) and 6 cm above the lower surface of the beam (corresponding to the tensile zone).

We notice that the tensile areas have a higher chloride content than the compressed areas.

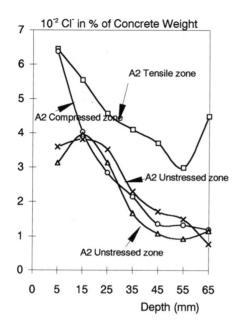

Figure 3 shows the variation in free chloride content according to samples extracted from the tensile zone, the compressed zone and from the end of the beam (unstressed zone). We notice that the content of free chloride is much higher in the tensile zone than in the other zones.

These results are in accordance with those obtained by R. François and al [2]. They are due to damage to the aggregate-paste interface subsequent to the loading which leads to an increase in porosity in the aggregate-paste transition area. As a result, the penetration of aggressive agents is more important in the tensile area than in the compressed area.

Fig. 3: Penetration of cl⁻ ions in various stressed zones of the beam A2

Effect of intensity of the applied mechanical strain

Two loading levels are used. According to French standards, the first one (M_1 = 13.5 kN.m) corresponds to the maximal loading vis à vis mechanical strength for type B beam and to the maximal loading vis à vis durability for type A beam. The second one (M_2 = 21.2 kN.m) corresponds to the maximal loading vis à vis mechanical resistance for A beam and to 80% of rupture for type B beam.

The tensile stress of the reinforcement is about 176 MPa for the type A1 beams and 274 MPa for the A2 beams; the tensile stress of the reinforcement is about 210 MPa for the type B1 beams and 325 MPa for the B2 beams.

For a same state of strain (tension), and a same type of beam (A or B), Figure 4 compares the increase in the chloride content between the two levels of strain applied.

The type A and the type B beams which are loaded to the first level are called A1 and B1, the other which are loaded to the second level are called A2 and B2.

In tensile area, the penetration of cl⁻ ions increases with the stress intensity:

       Chloride contents in A2 > contents in A1
       Chloride contents in B2 > contents in B1

This could be due to the fact that in the tensile area, increasing the strain intensity increases the damage of the aggregate-paste interface. As a result, this damage facilitates the penetration of the chlorides.

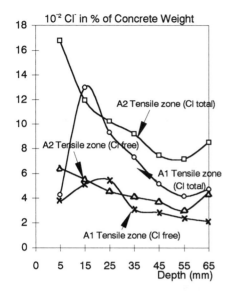

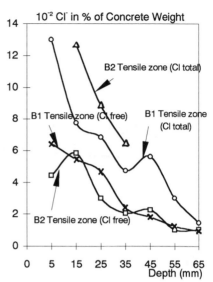

Fig. 4a: Penetration of cl⁻ ions at the middle of the beams A according to the loading level

Fig. 4b: Penetration of cl⁻ ions at the middle of the beams B according to the loading level

EFFECT OF THE ITZ ON THE CORROSION PROCESS

The ITZ has an effect on the corrosion process because in real life conditions the reinforced concrete is loaded, and this loading leads to a damage of the paste-aggregate interface [3].

In this case, the corrosion process of cracked reinforced concrete has been divided into four periods (fig. 5) as presented in a previous work [4]. These different periods can be summarized thus:

Incubation period--during this period, the aggressive agents penetrate through the primary cracks to reach quickly the reinforcement.

Initiation period--At the bottom of the crack, the percentage of cl⁻ ions has reached the depassivation threshold of the steel. The corrosion progresses along the reinforcement in the decoherence areas (mechanical debonding or decoherence due to bleeding).

Induction period--The rust products seal the decoherence areas and the bottom of the cracks. However the quantity of this expansive product is not sufficient to create a secondary cracking. The sealing of cracks slows down strongly the development of corrosion.

Propagation period--All concrete cover is contaminated by chlorides, so corrosion restarts. The development of rust products induces a secondary cracking.

When the reinforced concrete is not cracked, the corrosion process can be divided into two periods [5] which are shown in dotted lines on fig. 5.

With the Tuutti model (in the case of uncracked concrete), the duration of the service life is in relation to the duration of the incubation period. With our model (in the case of cracked concrete) , the duration of the service-life is in relation to the duration of the induction period because both the incubation period and the initiation period are very short (less than one year). As shown on fig. 5, the duration of the induction period is reduced in comparison with uncracked concrete by the damage of the ITZ in the tensile areas of the reinforced concrete. This is due to the fact that the damage of ITZ leads to an increase of chloride penetration. And this increase is directly proportional to the intensity of the stress applied to the beam.

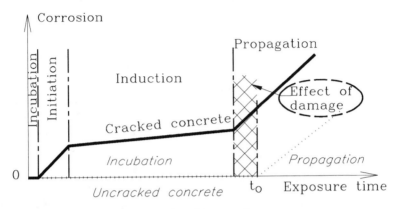

Fig. 5: Modeling of the corrosion process

CONCLUSION

The cracking is the visible sign of the mechanical degradation of the reinforced concrete, and is accompanied by damage to the concrete localized at the interface paste-aggregate.
This study, which has been conducted using some reinforced concrete elements of large sizes and stored under loading conditions has allowed us to build a network of cracks corresponding to the one found in real structures. Then, the penetration of the chloride has been studied taking into account the real state of concrete (cracking and damage due to loading) and the way it changes with time. This is not possible when using samples of small sizes as used in most research programs. This study has shown that the cracking is not the essential factor in the corrosion of the reinforced concrete.
On the other hand, the damage of the ITZ of concrete in the tensile areas leads to an increase of chloride penetration which could reduce the duration of the induction period and then of the service-life duration.
The model presented displays this phenomenon qualitatively, but complementary studies are necessary to quantify the real effect of the ITZ damage.
It seems obvious that the load applied to a reinforced concrete beam plays a significant role in the penetration of aggressive agents and then in the corrosion of its reinforcement. The model of the development of corrosion, worked out in relation with time and based on our results, emphasizes the influence of primary cracking on the duration of the incubation period and the influence of the paste-aggregate interface damage on the duration of the induction period. Any environmental conditions to which the concrete is exposed will also have a decisive influence.

REFERENCES

1 . French Regulations for Reinforced Concrete Structures - BAEL 83
2 . R. François, J-C. Maso - Effect of damage in reinforced concrete on carbonation or chloride penetration, Cement and Concrete Research, Vol. 18, pp. 961-970, 1988.
3 . R. François, J-C. Maso - Etude de la microfissuration de poutres en béton armé sollicitées en flexion simple, Annales de l'I.T.B.T.P. n°464 - mai 1988.
4 . R. François, G. Arliguie - Durability of loaded reinforced concrete in chloride environment, Third CANMET/ACI International Conference on Durability of Concrete. SP145-30, Nice 05/1994.
5 . K. Tuutti - Corrosion of steel in concrete, CBI reseach report 4, 1982.

THE CORROSION OF STEEL REINFORCEMENT IN CARBONATED CONCRETE UNDER DIFFERENT HUMIDITY REGIMES

ANASTASIA G. CONSTANTINOU AND KAREN L. SCRIVENER
Department of Materials, Imperial College of Science, Technology and Medicine,
London, SW7 2BP, England

ABSTRACT

In this investigation, four concretes (two mix designs and two curing regimes), containing reinforcing bars at covers 11 mm and 20 mm were carbonated fully in 100% CO_2 and 65% relative humidity (RH). They were then placed in two different humidity environments (65% RH and 90% RH). The progress of the corrosion of the samples was followed electrochemically (using the linear polarisation technique) and microscopically (using a scanning electron microscope).

Electrochemical monitoring of the corrosion rate showed that corrosion started very soon after placement in the humidity regimes and generally increased with time. The water/cement ratio, the concrete cover of the reinforcement and the curing of the concrete had a significant effect the length of time taken to carbonate the specimens. However, the effect of these variables on the corrosion rate was negligible. In addition, the corrosion rate showed no dependance on the humidity regimes.

Microstructural examination of the samples revealed the existence of corrosion products at the steel/concrete interface after 6 months in the humidity regimes. There was more corrosion product in the samples at 90% RH. The samples with the low w/c showed more extensive cracking in the interfacial region.

INTRODUCTION

With regard to reinforcement corrosion, carbonation of concrete is not generally considered to be as damaging as chloride contamination. However, although the effects of carbonation may take longer to develop, once the carbonation front has reached the steel reinforcement corrosion will proceed if the humidity is high enough. The corrosion of the steel reinforcement results in the cracking and eventual spalling of the cover due to the internal stresses created as a result of the increase in volume associated with the transformation of steel to rust [1].

In carbonated concrete, the corrosion process is thought to be generalised and homogeneous producing, over the long term, a reduction in the cross-sectional area of the steel and a significant amount of oxides which may crack the cover or diffuse through the pores to the surface of the concrete. However, the relationship between the cracking of the concrete cover, the spatial distribution of corrosion and the deposition of the corrosion products are not well understood.

Good quality concrete will carbonate more slowly and will have a lower permeability once carbonated. However, once corrosion is initiated, the effect on the corrosion rate is not important. The major factors affecting the quality of the concrete are the w/c ratio, the degree of curing and the reinforcement cover.

In this investigation, concrete samples were prepared with two different w/c ratios and cured under two different regimes. After carbonation they were placed in two different humidity

Mat. Res. Soc. Symp. Proc. Vol. 370 © 1995 Materials Research Society

followed electrochemically (using the linear polarisation technique) and microscopically (using a scanning electron microscope).

EXPERIMENTAL

The samples were cast with w/c ratios of 0.56 (cement content 315 kg/m^3) and 0.65 (cement content 275 kg/m^3) in cylindrical moulds 100 mm diameter by 200 mm in length, with steel reinforcement bars (8 mm diameter) at cover depths of 11 mm and 20 mm. The samples were demoulded after one day, then one set was air cured while the other was cured in damp sacks for a further 7 days. The samples were then carbonated in 100% CO_2 and 65% RH at 20 °C. The progress of carbonation was checked regularly by spraying phenolphthalein indicator on parallel control samples cast without reinforcement.

After full carbonation, the samples were placed under two humidity regimes: 65% RH and 90% RH. Electrochemical monitoring of the samples for corrosion began one month after placement in the humidity regimes and continued at monthly intervals thereafter.

The samples were monitored electrochemically using the linear polarisation resistance technique. The steel reinforcement bars are used as working electrodes, and an external counter electrode as a reference electrode. The corrosion potential (E_{corr}) is first measured with respect to a stable reference electrode (calomel). A small potential step, $\Delta E = -15$ mV, is then applied about the E_{corr} value via a counter electrode (stainless steel mesh) and the resulting current, ΔI, is measured. The polarisation resistance, R_p, is given by:

$$R_p = \frac{\Delta E}{\Delta I}$$

The overall corrosion current, i_{corr}, flowing between anodic and cathodic regions of the reinforcing steel is given by:

$$I_{corr} = \frac{B}{R_p} \qquad \text{where B is the Stern-Geary constant, for active corrosion} = 26 \text{ mV.}$$

The corrosion rate is therefore:

$$i_{corr} = \frac{I_{corr}}{A} \qquad \text{where A = the corroding area, 32.7 cm}^2.$$

Six months after being placed in the humidity regimes, one sample from each batch was removed from the tank for microscopical examination. Sections of the reinforced samples were cut using a Buhler Isomet Plus precision saw at a speed of 3800 rpm. and a load of 380 grams with a high concentration diamond wafering blade. The sections were then resin impregnated, lapped with 9 μm alumina powder and polished down, using diamond paste, to ¼ μm grit size and carbon coated. A JEOL-35CF scanning electron microscope (in the back scattered mode) was used to examine the samples.

RESULTS AND DISCUSSION

Electrochemical Measurements

In this paper only the i_{corr} measurements are presented which indicate the corrosion rate of the samples. The E_{corr} measurements indicate the thermodynamical probability of corrosion, but are highly susceptible to small changes in the surrounding environment (e.g. humidity and temperature changes) [2]. On the Figures 2 - 5, two horizontal lines separate the passive regime from the zone of active corrosion i.e. below 0.1 $\mu A/cm^2$ the corrosion rate will not be concidered dangerous to the structure.

In order to assess the reproducibility of the electrochemical measurements, readings were made on six nominally identical specimens with w/c 0.65. The results are shown in Figure 1. There is also a variability within values of identical samples and this is shown on figure 1.

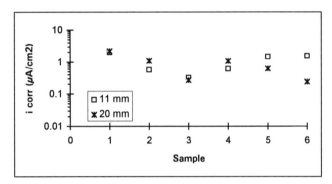

Figure 1
Variability
of
electrochemical
measurements

Figures 2 - 5 show the electrochemical measurements for the samples, all of which appear to be actively corroding regardless of preparation or exposure conditions. When the variability of the measurements is considered, there do not appear to be any significant differences between the corrosion rates.

For the different concretes this is perhaps not surprising. Increasing the length of curing and decreasing the w/c ratio will lower the permeability of the concrete and it will take longer for the concrete to carbonate to the depth of the reinforcement [3]. This was in fact observed in the preliminary treatment of the samples: times for carbonation under these accelerated conditions are shown in Table I:

Table 1: Carbonation times for the samples examined:

w/c ratio	Curing (days)	Carbonation Time (mnths)
0.65	1	3
0.65	8	6
0.56	1	6
0.56	8	9

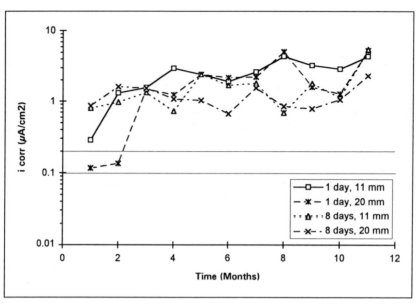

Figure 2. Electrochemical measurements for concretes the w/c=0.65 exposed at 90%RH

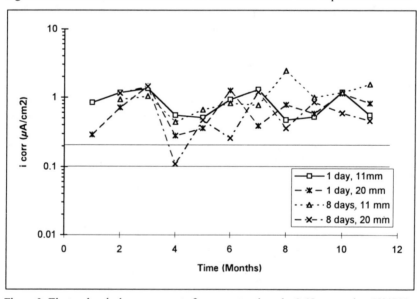

Figure 3. Electrochemical measurements for concretes the w/c=0.65 exposed at 65%RH

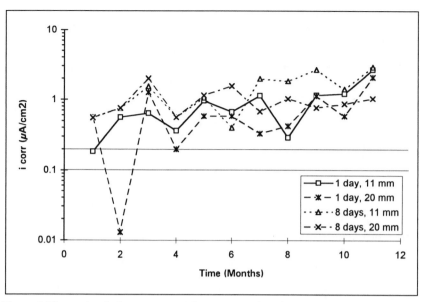

Figure 4. Electrochemical measurements for concretes the w/c=0.56 exposed at 90%RH

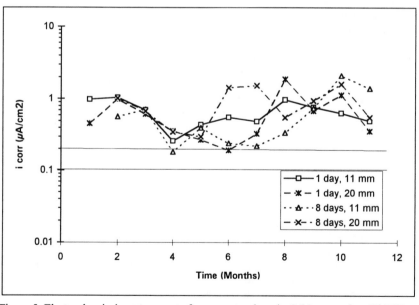

Figure 5. Electrochemical measurements for concretes the w/c=0.56 exposed at 65%RH

However, once the concrete is fully carbonated the rate of corrosion will only be limited by the availability of moisture or of oxygen, which are unlikely to be affected over the range of w/c ratios and curing conditions examined here.

The results are perhaps more unexpected with regard to the different relative humidities. Again there appear to be no significant difference between the two humidities examined. It might have been predicted that the steel in the concretes exposed at 65% RH would corrode less rapidly then that in concrete exposed at 90% RH. The relative humidity would become a significant factor if it falls under a certain threshold value, becoming so low that the concrete dries out and the conductivity of the concrete drops. On the other hand, if the pores of the sample become saturated with water (i.e. the sample is completely immersed in water), the ingress of oxygen will be limited and the corrosion process will be unable to progress.

Microstructures

Figure 6 shows the carbonated microstructure of the sample with w/c = 0.56 (the white part is the steel reinforcement). This microstructure displays the expected features of a carbonated concrete. It has a very dense microstructure due to the deposition of the carbonation products in the pores, and dark relics can be seen from parts of the anhydrous grains depleted in calcium [4]. Figure 7 shows the microstructure of the sample with w/c = 0.56 after exposure to 65% RH for 6 months. The sample shows some corrosion products forming at the steel/concrete interface and moving into the paste particularly around the paste/aggregate interface. , Figure 8 shows the microstructure of the sample with w/c = 0.56 after exposure to 90% RH for six months. More extensive corrosion has occurred in this section and cracking can be observed at the interface between steel/concrete as well as further into the concrete paste. Figure 9 shows the microstructure of the sample with a w/c of 0.65 in 90% RH. Again, corrosion products can be observed, extending into the paste as well as some cracking at the steel/concrete interface.

Quantitative studies of the interface between the steel and concrete by the authors have shown that the microstructure is highly variable from section to section, therefore no firm conclusions can be drawn regarding the relative degrees of corrosion in the different samples. It is noticeable, however, that the lower w/c, leads to a higher amount of cracking in the interfacial zone. This could be due to the fact that in the w/c=0.65, more porosity is available to be filled by the corrosion products and hence more corrosion will be needed before any serious cracking can occur.

CONCLUSIONS

1. The water/cement ratio, the concrete cover of the reinforcement as well as the curing regime are only significant when the samples are carbonating but have no effect on the rate at which the steel is corroding.
2. The humidity conditions in which the samples are placed after full carbonation did not affect the rate at which the samples corroded.
3. After six months in the humidity regimes, both batches of samples (w/c = 0.56 and 0.65) showed corrosion when examined microscopically. All the samples, irrespective of their humidity conditions, showed similar amounts of corrosion products developing in the region adjacent to the steel interface and moving into the paste.

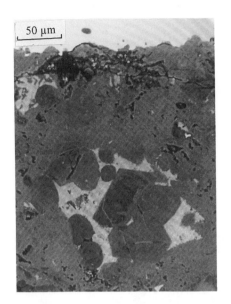

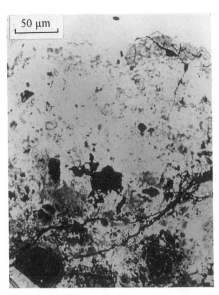

Figure 6: Microstructure of the sample with w/c = 0.56 after carbonation, before corrosion

Figure 8: Microstructure of the sample with w/c = 0.56 after exposure to 90% RH for six months

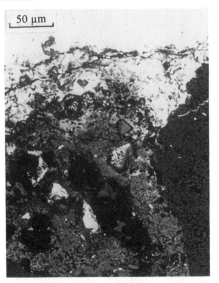

Figure 7: Microstructure of the sample with w/c = 0.56 after exposure to 65% RH for 6 months

Figure 9: Microstructure of the sample with a w/c of 0.65 in 90% RH for six months

4. The samples with w/c = 0.56 showed more cracking in the paste adjacent to the reinforcement because the concrete was more dense.

ACKNOWLEDGMENTS

The authors would like to thank Dr. M.A.Sanjuan for helpful comments.

REFERENCES

1. A. Rosenberg, C.M. Hansson and C. Andrade, "Materials science of concrete-Vol.1". (Ed. J.S. Skalny), 285-313, 1989. The American Ceramic Society.
2. J. Rodriguez, L.M. Ortega, A.M. Garcia, L. Johansson and K. Peterson in "Concrete across Borders", Denmark, June 1994.
3. N.I. Fattuhi, Materials and constructions, Vol.19, No.110,
4. A.G. Constantinou, M.Sc. Thesis, University of London, 1991.

FIBRE-CEMENT PASTE TRANSITION ZONE

VAHAN AGOPYAN* AND HOLMER SAVASTANO JÚNIOR**
*Escola Politécnica, University of São Paulo, 05508-900 São Paulo, SP, Brazil.
**Faculdade de Zootecnia e Engenharia de Alimentos, University of São Paulo, 13630-000 Pirassununga, SP, Brazil.

ABSTRACT

The characteristics of fibres and paste of ordinary Portland cement transition zone are analysed and correlated to the mechanical properties of the produced composites. The water-cement ratio of the matrix varies from 0.30 to 0.46 and the age of the specimens varies from 7 to 180 days. Composites of vegetable fibres (coir, sisal and malva) are compared with those of chrysotile asbestos and polypropylene fibres. The analysis is made by backscattered electron image (BSEI) and energy dispersive spectroscopy (EDS). Mechanical tests evaluate the composite tensile strength and ductility.

Mainly for vegetable fibre composites the transition zone is porous, cracked and rich in calcium hydroxide macrocrystals. These results are directly associated with the fibre-matrix bonding and with the composite mechanical performance. Further studies considering the impact performance of the composites compare the porosity of the transition zone with the toughness of the composites.

INTRODUCTION

Studies of hydrophobic fibres [1], aggregates [2] and steel re-bar [3] in cement matrix showed that the transition zone of these materials is approximately 40 micrometers thick and that it has moderate portlandite (calcium hydroxide crystals) accumulation. Observations were made with scanning electron microscopy (SEM) and with X-ray microanalysis and with these techniques the evolution of porosity and of hydrated products on the interfacial surface was registered for a period of time.

However with water absorbent fibres, such as the vegetable fibres, the transition zones in cementitious matrices have different porosity and concentration of the hydrated products [4].

This research analyses the basic characteristics of the portland cement paste and vegetable fibres of malva (Urena **lobata** Linn.), sisal (Agave **sisalana** Perrine) and coir (Cocos **nucifera** Linn.) composites and compare them with those composites of fibres of chrysotile asbestos and of polypropylene.

The vegetable fibres present some singular characteristics: (i) high water absorption (near 100% or even more); (ii) impurities, such as powder and husk; and (iii) heterogeneous mechanical properties [5]. These singularities affect the composites mechanical properties as well as their microstructure.

The importance of vegetable fibres is due to its great availability in Brazil and in other developing countries. Building components reinforced with vegetable fibres can be technically feasible, with cost reduction as high as 15% [6].

EXPERIMENTAL WORK

The matrix consisted of portland cement paste only so that there is no interference of aggregates in the transition zone between fibres and matrix. The water-cement ratios are 0.30, 0.38 and 0.46 and the study ages of the specimens are 7, 28, 90 and 180 days.

The selected fibres for this study are: chrysotile asbestos type 4Z (QUEBEC SCREEN TEST classification), polypropylene (filament), coir, malva and sisal. All the fibres do not have any type of treatment.

The water absorption of the vegetable fibres is higher than 100%, with more than 60% of absorption occuring in the first 15 minutes of immersion, as can be seen in Table I. This characteristic affects directly the behaviour of the fibres immersed in the fresh matrix. The durability problem of these vegetable fibres is discussed elsewhere [7].

Fibre lengths vary from 15 to 30 mm, with the exception of asbestos fibres which are 5 mm long. The fibres are randomly dispersed in the cement paste matrix. The fibre volume for the composites with asbestos and polypropylene fibres is 1% and for the vegetable fibres composites is 4%; these volumes are large enough to change the properties of the matrices.

The composite was prepared in an ordinary pan mixer and the specimens were casted on a vibrating table without compaction.

Table I. Water absorption by vegetable fibres at room temperature (ASTM C127/88).

Fibre/time	Water absortion (%)				
	5 min	15 min	1 h	8 h	24 h
malva	133.6	160.3	186.4	156.6	156.6
sisal	89.3	88.4	95.4	96.8	92.2
coir	43.2	52.9	58.3	72.2	80.4

RESULTS AND COMMENTS

Microstructural analysis

The fibre-matrix transition zone is analysed through scanning electron microscopy with backscattered electron image (BSEI). The energy dispersive spectroscopy (EDS) analysis is also applied to identify the major chemical elements.

The BSEI is recommended for specimens with a smooth and polished surface. The study of this type of surface is better than that one of fractured surface. The latter has the inconvenience of crossing, mainly, the composite elements of less strength. The BSEI still has the advantage of identifying the different surface regions analysed, by way of contrasting the atomic number: the higher the atomic number of the chemical element, the clearer the image and vice-versa [8].

The specimens were prepared following the recommendations of Kjellsen [9]; this preparation consists basically of: a) stoppage of cement hydration with acetone; b) impregnation of the specimens with a low viscosity resin in a vacuum chamber; c) polishing the analysed surface with silicon carbide and later with a diamond paste; d) recovering the surface with a carbon layer of a few micrometers, as the composite is not an electrical conductor.

For the malva composite, at the age of 7 days and with a water-cement ratio of 0.38, the transition zone is highly porous and has portlandite macrocrystals in the thickness of about 50 micrometers (Figure 1). This zone is clearly fissured probably due to strength lower than the matrix itself. Some microcracks cross even the portlandite crystals and calcium hydroxide plates can also be noticed in a perpendicular direction to the analysed surface.

The fibres were seen to be almost completely debonded from the matrix. This phenomenon is probably due to the fact that there is a large amount of water in the transition zone because of high water absorption of the fibre. When the composite begins to dry, the fibres shrink and separate them from the matrix thus impairing their action as reinforcement.

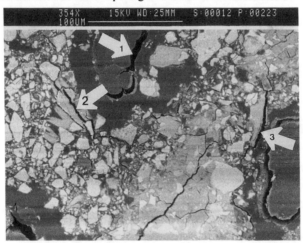

Fig. 1. BSEI of composite of malva fibre, water-cement ratio of 0.38, age of 7 days. Arrow 1: fibre debonding; arrow 2: microcracks; arrow 3: portlandite macrocrystal.

The sisal fibre is also debonded from the matrix and the transition zone is highly porous, even at the age of 180 days. Figure 2 shows a section of a sisal fibre (dark area at left) with a transition zone of about 100 micrometers in thickness.

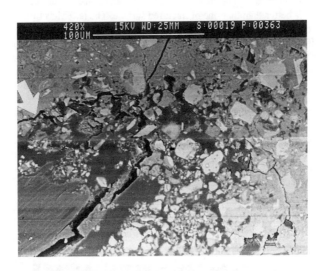

Fig. 2. BSEI of composite of sisal fibre, water-cement ratio of 0.38, age of 180 days. Arrow: high porosity and microcracked transition zone.

Figure 3 presents coir fibre composite with 0.30 water-cement ratio. The high cracking rate of the matrix close to the fibres has not shown reduction with the decrease of water-cement ratio, therefore the matrix cracking seems to be mainly a function of the fibre absorption rate.

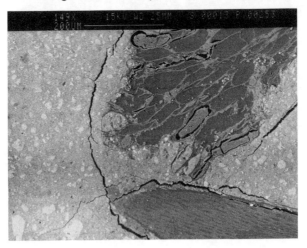

Fig. 3. BSEI of composite of coir fibres, water-cement ratio of 0.30, age of 28 days. High cracking rate close to the fibre.

For asbestos fibres, a 10 micrometers portlandite layer can be seen in Figure 4 (the light gray shaded region), with no increase in porosity of the interface. This is different than Akers and Garret's results [10] probably because their composites were prepared with pressure.

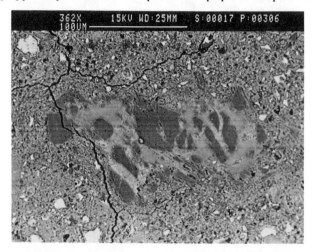

Fig. 4. BSEI of composite of asbestos fibre, water-cement ratio of 0.46, age of 28 days.

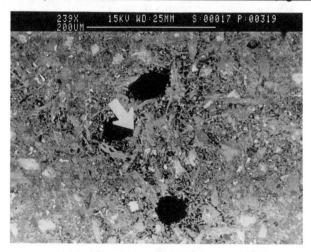

Fig. 5. BSEI of composite of polypropylene fibre, water-cement ratio of 0.38, age of 90 days. The arrow indicates transition zone with orientaded portlandite crystals and limited increase of porosity.

The efficiency of polypropylene fibres clearly presented by the mechanical properties (Figure 6) can also be confirmed by microstructural analysis as shown on Figure 5: there is no fibre debond and the porosity increase at transition zone is very limited when compared with that of the vegetable fibres composites.

<u>Bending test</u>

The adopted method is the four point bending test, following the RILEM recommendation [11] for cement composites. The specimens were small beams 300 mm long with cross section of 150 mm x 15 mm, tested over a span of 270 mm. Figure 6 presents the specific energy according to the age. Specific energy is total energy absorbed by the fracturated surface. As it was expected, the short asbestos fibres induce a small energy absorption of their composites; the best results are obtained with the polypropylene fibres. The vegetable fibre composites present a satisfactory performance when compared with asbestos ones and non-reinforced matrix.

The effect of age of the composites has been observed in those ones containing vegetable fibres, which present a reduction of the ductility with the increase of age because of deterioration of the fibres in alkaline medium.

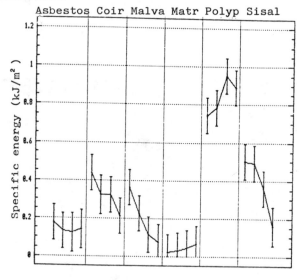

Fig. 6. Bending test: effect of the age on specific energy with water-cement ratio of 0.38; for each fibre, the age sequence is 7, 28, 90 and 180 days.

<u>Direct tensile test</u>

The ASTM C 190 test method for mortars is applied for the composites. In Figure 7, as was expected the tensile strength is reduced with the increase of water-cement ratio. The best results are achieved with asbestos fibres, probably due to the dense transition zone (Figure 4).

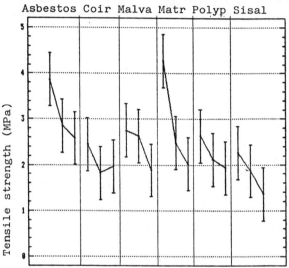

Fig. 7. Direct tensile test: effect of water-cement ratio on the strength at the age of 28 days; for each fibre, the w/c sequence is 0.30, 0.38 and 0.46.

4 CONCLUSIONS

The transition zone for vegetable fibre reinforced composites is different from the matrix itself. The following characteristics are pointed out:
- Thickness of up to 100 micrometers, which can be identified by the intensive cracking and also by the increase of porosity and portlandite concentration. These aspects contribute to reduce the durability of the vegetable fibre-cement composite. During the mixing of the composites, the porous vegetable fibres absorb a great amount of water, making the transition zone more pronounced than when asbestos or polypropylene fibres are used.
- Effect of the water-cement ratio, as the higher values induce higher accumulation of portlandite;
- Effect of the age of hydration, as the porosity tends to be higher when the hydration begins.

- Excessive porosity and debonding of the vegetable fibres, which contribute to the reduction of static mechanical performance of these composites.

The relatively high specific energy of vegetable fibre composites confirms the transition zone influence on the material ductility. These results confirm the high toughness values of composites reinforced with vegetable fibres at impact experiments [12], by the fibres vibration and interfacial microfissures in the high porous transition zone. Nevertheless further studies are required to measure the relation between the porosity of transition zone and the composite toughness.

5 REFERENCES

1. A. Bentur, S. Diamond and S. Mindess, Cem. Concr. Res. **15**, 331 (1985).
2. P. M. Monteiro, PhD thesis, University of California, Berkeley, 1985.
3. S. Mindess, I. Odler, and J. Skalny, in Proc. 8th Int. Congr. on the Chemistry of Cement, (Abla, Rio de Janeiro, 1986) **1**, pp.151-157.
4. H. Savastano Jr. and V. Agopyan, in Fibre Reinforced Cement and Concretes, edited by R. N. Swamy, (RILEM Proc. 17, E&FN Spon, London, 1992), pp. 1110-1119.
5. H. Savastano Jr., in Vegetable Plants and their Fibres as Building Materials, edited by H. S. Sobral, (RILEM Proc. 7, Chapman and Hall, London, 1990), pp. 150-157.
6. V. Agopyan and V. M. John, Building Research and Information, **20**, 233 (1992).
7. V. Agopyan, in Natural Fibre-Reinforced Cement and Concrete, edited by R. N. Swamy (Blackie, Glasgow, 1988) pp. 208-242.
8. P. J. Goodhew and F. J. Humphreys, Electron microscopy and analysis, 2nd. ed. (Taylor & Francis, London, 1988).
9. K. O. Kjellsen, R. J. Detwiler and O. E. Gjorv, Cem. Concr. Res., **21**, 388 (1991).
10. S. A. S. Akers and G. G. Garret, J. Mater. Sci., **18**, 2200 (1983).
11. RILEM Technical Committee 49 TFR, Structures and Materials, **17**, 441 (1984).
12. H. S. Ramaswamy, B. M. Ahuja, and S. Krishnamoorthy, Int. J. Cem. Comp. and Lightweight Concr., **5**, 3 (1983).

ACKNOWLEDGMENTS

The authors would like to thank Professor Paulo S. C. Pereira da Silva of the Metallurgical and Materials Engineering Department of EPUSP, for his help with the microstructural analysis. They are also in debt with the help of Dr. Maria A. Cincotto and Mr. Pedro C. Bilesky.

THE MICROSTRUCTURE OF WOOD FIBER REINFORCED CEMENTITIOUS COMPOSITES

X. LIN, M.R. SILSBEE, D.M. ROY, The Pennsylvania State University, Intercollege Materials Research Laboratory, University Park, PA 16802.

ABSTRACT

Wood fiber reinforcing of cement matrices is an economic and an efficient approach to producing high performance cementitious composites. In this study, wood fiber reinforced cementitious composites (WFRCs) were made by using both conventional and novel processing styles. Wood fibers exhibited a considerable ability to improve the flexural strength and the toughness of WFRC when an adequate content of the fibers was used. The morphologies of various type of wood fibers and fracture surface of WFRC were examined by scanning electron microscope (SEM) and an environmental scanning electron microscope (ESEM). The microstructures of wood fiber and WFRC were correlated with their mechanical properties. Results indicate a significant interfacial bonding between the cement matrix and the wood fibers.

INTRODUCTION

Plain cement, a widely used traditional construction material, is stiff, durable, has adequate compressive strength, but is weak in flexural and impact strengths due to its brittleness and low strain capacity, which have greatly limited its application in may aspects. Using fibers to strengthen brittle cement matrices is an old concept. Ancient Egyptians used straw to reinforce adobe bricks. Now a wide range of fiber types has been employed in reinforcing cement matrices. The type of fiber materials have included steel, polymer, ceramic and natural products which are well documented by Bentur et al. [1].

Wood fiber, one type of natural fiber was proposed as a reinforcement in cement-based building materials almost 100 years ago, while cellulose has been used as an economic substitution for asbestos for nearly 50 years [2]. The development and production of WFRCs have attracted increasing attention in the last 10-20 years [3]. It was reported that wood fiber reinforced cement and concrete panels have served as both non-structural and structural, and as both internal and external building materials in Finland, Indonesia, Malaysia, New Zealand and Australia [4,5]. The forms of wood used in cement have been wood chips, flakes, excelsior-thin ribbon-like strands and fibers. In Europe, South America, and Asia, cement has been used as a bonding agent for wood excelsior boards, particleboard, fiberboard, and wood-cement composite products for nearly 50 years. In the United States, manufacturing cement-bonded wood composite products has been limited to excelsior board made from a mixture of southern pine and portland cement [6]. WFRC can be used as wall sections, masonry blocks and precast units etc.

In this study, wood fibers were employed in both ordinary portland cement and macro-defect-free (MDF) cement systems using paddle mixer and high shear mixer, respectively. This study was focused on the microstructure of various wood fibers and WFRC samples in order to understand the mechanism of WFRC's performance in the aspects of mechanical, physical and chemical properties.

Mat. Res. Soc. Symp. Proc. Vol. 370 ⊚ 1995 Materials Research Society

EXPERIMENTAL

Materials

Wood fibers used in this study were bleached hardwood pulp (Wickliff), unbleached recycled kraft paper and recycled newspaper. The water soluble polymers and chemical additives used to pretreat the fibers were: PVAc (polyvinyl acetate)[1] , PEG (polyethylene glycol with molecular weight of 1000)[2] and Acryl 60 (Acrylic polymer in aqueous emulsion)[3] and Silane (alkylakoxysilane in water)[4]. ASTM type III portland cement[5] was employed in this study: The composition of cement is given elsewhere [6]. Type I ordinary portland cement (OPC-I)[6] and Polyacrylamide (PAM) with high molecular weight of greater than 10,000,000[7] were used for making wood fiber reinforced macro-defect-free cement (WFRC-MDF).

Preparation of fibers and WFRC sample

The details of preparation of wood fibers and of WFRC samples were described in our previous paper [7]. When chemical treatment of the fibers was desired the fibers were impregnated in the polymer or other aqueous solution for 1/2 hour, followed by hand squeezing, filtering, air drying and milling. The volume percent of polymers (or other chemical additive) in the aqueous solution used were none, 1% PVAc, 10% PEG, 10% Silane and 10% Acryl 60 respectively.

Table I gives the basic formulations employed in wood fiber reinforced traditional cement system. Samples were molded in the size required for testing (e.g. 25.4 by 25.4 by 127 mm bars for flexural strength testing), then demolded after one day of curing, and continuously cured (at 38 °C,100% relative humidity, ambient pressure, and the desired temperature) until testing.

The typical formulation for wood fiber reinforced macro-defect-free (MDF) cement system used in this study is shown in Table II. The sample was prepared using a high shear mixer. The PAM was dissolved in water, then mixed

Table I, Basic formulations employed in traditional cement system.

Sample	Weight % of components			
	cement	de-ionized water	fibers	superplasticizer
Control*	75.99	22.80	0.00	1.21
WFRC+	67.38	26.95	4.24	1.43

*W/C = 0.30, + W/C = 0.40.
Superplasticizer- Borem 100 HMP, supplied by Boremco Specialty Chemicals

with the cement, fibers and glycerin by Brabender high shear mixer at ambient condition, at 40-60 RPM for about 4 minutes, followed by extrusion at about 10-30 RPM. The extruded sheet was cut into bars of about 3 by 5 by 35 mm. Samples were cured for 1 day in a sealed plastic bag and 6 days in an open bag, following by drying at 80°C for 4 hours.

---

[1] from Borden Chemical
[2] supplied by Aldrich Chemical Co., Inc
[3] from Thoro System Products, Inc
[4] supplied by Hydrozo Incorporated, Inc
[5] from Lehigh Portland, Cement Co
[6] from Keystone
[7] from Polysciences, Inc

Table II, Basic formulations employed in MDF cement system.

Sample	OPC-I	PAM	De-ionized water	wood fibers	Glycerin
	Weight % of components				
Control	78.5	8.0	13.0		0.5
WFRC-MDF	74.2	8.0	13.0	4.3	0.5

Property Measurements

The microstructure of wood fibers and the fracture surface of WFRCs have been examined using a ISI DS 130 scanning electron microscope (SEM) and an Electroscan ES-2 environmental scanning electron microscope (ESEM).

For wood fiber reinforced MDF cement samples, the flexural strength was tested using an Instron Testing Machine at a deflection rate of 0.1 mm/min.

RESULTS AND DISCUSSION

Morphologies of wood fibers

Wood is composed of 40-44 wt. % of high molecular weight cellulose, 15-35 wt. % of lower molecular weight hemicellulose, 18-25 wt. % of thermoplastic lignin and 5-30 wt. % of extractives. Extractives include oils, fatty acids, terpenes and terpenoids, phenols, tannins, gums, waxes, resins and starches, sugars and salts, which are water soluble and alkaline soluble [5]. Wood fibers are obtained through a pulping process in which lignins are removed.

Wood fiber has a hollow center structure, and it can be a collapsed flat ribbon shape or be an open hollow cylindrical shape. Figures 1-3 show the microstructure of hardwood fibers, softwood fibers and newspaper fibers respectively. The images show how flexible the wood fibers are, that they can be bent and collapsed to various degree. They are conformable which makes them behave like a sponge. Newspaper fibers are mixtures of softwood and hardwood fibers, in which hardwood fibers comprise a larger fraction. Kraft paper is composed of softwood fibers. The softwood fibers had an average diameter of 30 µm and average length of 3.0 mm [5] while the hardwood fibers were shorter and smaller than the softwood fibers. The shape and the morphology of the fibers not only be varies with different sources but also can be influenced by the mechanical processing to which they have been subjected. Hardwood fibers with relatively thicker cell wall are less easy to be collapsed than softwood fibers. Newspaper fibers have been delaminated to a certain extent since they have been subjected to more processing. The typical effects resulting from the mechanical beating of fibers can be illustrated by comparison of the microstructure of a fiber before and after mechanical beating as shown in figure 4 and 5 [8]. Delaminated fibers and curly fibers offer more opportunity for mechanical bonding in the composite [9,10].

Performance of wood fibers in cement matrix

Figure 6 shows wood fibers distributed in cement matrix. Wood fibers are distributed randomly in three dimensions acting as bridges to restrict further prolongation of a crack or dislocate a flaw-track at a higher level of deformation.

The mechanical performance of WFRC can be investigated by comparing its load-deflection behavior with that of an unreinforced cement. Based upon the three- point-bending flexural strength test, the typical load versus deflection curves corresponding to the fiber content

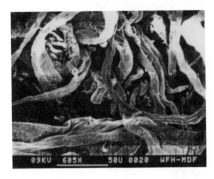

Figure 1, SEM microstructure of hardwood fibers.

Figure 2, ESEM microstructure of softwood fibers.

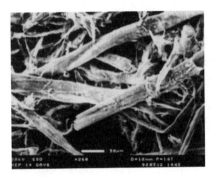

Figure 3, ESEM microstructure of a newspaper fibers.

Figure 4, SEM microstructure of a wood fiber before mechanical beating.

applied to cement matrix are illustrated in figure 7, in which fiber content is increased from 0.0 wt.% at left to 11.3 wt.% at right.

As shown in figure 7, for a plain cement matrix, a small flaw can be easily developed into a completely failed crack, so it's typical load-deflection curve stops at the first crack indicating that it has no ability to hold load with further deformation. For the WFRC, its-load deflection curve is extended after the first crack. Two trends can be clearly noticed on these curves: 1) as the fiber content increases, the flexural strength increases to an optimum level then decreases, and 2) the load-deflection curves expand progressively with increased fiber content. The deformation ability or toughness of the composite increases with fiber content. Wood fibers have considerable ability to increase the flexural strength of toughness of the cement composite. The effect of fiber content on the mechanical properties of WFRC was discussed elsewhere [11].

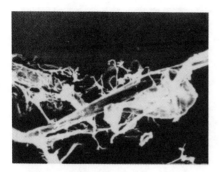

Figure 5, Microstructure of a wood fiber after mechanical beating.

Figure 6, Microstructure of a 7 days aged WFRC shows distribution of wood fibers in cement matrix.

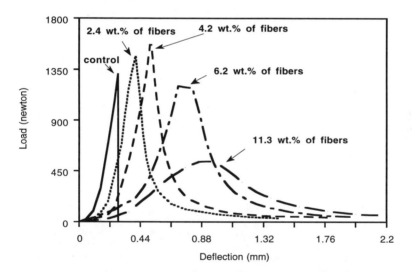

Figure 7, Effect of fiber concentration on the load-deflection curve. WFRC with 1% PVAc treated kraft fibers were cured at 38°C and high relative humidity for 7 days. Three-point-bending on 25.4x25.4x127 mm bar with span length of 76.2 mm.

The failure behavior of wood fibers and the interfacial bond between wood fiber and cement

Figures 6, 8 and 9 show the microstructures of the WFRC samples cured for 7 days. A large fraction of the fibers pulling out of the fracture surface can be observed in the 7-day-old samples. While figure 10-11 show that one-year-old samples have a large proportion of broken fibers on the fracture surface, three types of fiber failure can be further observed 1) failure at the

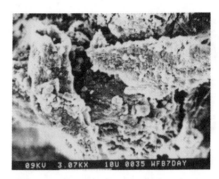

Figure 8, wood fibers pulling out from the cement matrix. Seven days aged kraft fiber reinforced type III portland cement.

Figure 9, the cement hydration products formed on the surface of wood fibers. Seven days aged kraft paper fiber reinforced type III portland cement .

cross-section of the fiber lumen, 2) delaminating of the fiber cell wall, and 3) part of the fiber cell wall cracks. Our previous study indicates that WFRC had higher toughness at earlier curing period and the toughness of WFRC decreases as curing time increases [12]. Which is due to higher proportion of fiber pull out during failure since the interfacial bond between fiber and cement is weak when cement matrix is still friable. As the cement matrix gets stronger with increased curing time, the bond between fiber and cement gets stronger, hence fiber fractures, instead of pulls out, under loading. Fracturing of fiber under load limits the strain capacity of the composite and results in a decreased toughness of WFRC.

Figures 9 and 11 show typical cement hydration products growing on the surface of wood fibers in the type III cement composite. The hollow structure of the wood fiber allows it to absorb a certain amount of water during mixing with the cement paste. In return, the diffusion of water from the interior of the fibers to the surface of the fibers could lead to an increase in cement hydration.

Figure 10, The fracture surface of one year cured WFRC sample indicate wood fiber fracturing instead of pulling out. (Kraft paper fiber reinforced type III portland cement).

Figure 11, Delaminated and cracked wood fiber in the fracture surface of one-year aged WFRC (Kraft paper fiber reinforced type III portland cement).

SEM and ESEM images show that hydrate products covered nearly the whole surface of the fibers. The phenomena suggest that the weak zone is not at the interface of the fiber and the cement matrix, but at the matrix near the interface and at the cell wall of fibers where the fracture occurs. There is a significant interfacial bond between the wood fiber and the cement matrix. This bond gets stronger as the curing time increases. It also makes the composite brittle resulting a decreased toughness of WFRC as increased curing time [12].

Reinforcement of MDF cement with wood fibers

Macro-defect-free (MDF) cement technique is an attractive approach for synthesizing high-performance polymer-cement composites by viscous plastic processing. Macro-defect-free (MDF), refers to the absence of large pores (>200 mm) [13-15]. Various types of wood fibers: hardwood pulp, recycled newspaper and recycled kraft paper were used in macro-defect-free (MDF) cement system to examine their efficiency as reinforcement. The treatments with each type of wood fiber were: untreated, 10 vol.% silane-aqueous solution and 10 vol.% A60-aqueous solution. The formulation is based on Table II, in which 4.2 wt.% of cement was replaced by wood fibers for WFRC-MDF.

SEM pictures show the microstructure of the fracture surface of the wood fiber reinforced MDF cement. Figure 12 shows the wood fiber dispersed in the MDF cement matrix. High shear mixing offers an approach to achieve the efficient dispersal of fiber in the MDF paste. Fibers acting as bridging in the matrix are shown in figure 13, in which a fiber is blocking a crack path in the matrix. The delaminated fibers in figure 14 indicates the weaker adhesion in the fiber wall corresponding to a stronger adhesion along the interfacial zone of the fiber and the matrix. Figure 15 shows the polymer matrix dispersing on the fracture surface of a sample, and it suggests that the polymer matrix is a weak zone.

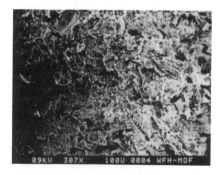

Figure 12, Uniformly dispersed hardwood fibers in OPC-PAM matrix.

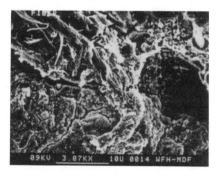

Figure 13, A hardwood fiber blocks a crack path in the OPC-PAM matrix.

The flexural strengths of samples are illustrated in figure 16. The results indicate that all the wood fibers reinforced MDF samples have higher flexural strengths, up to 97 MPa. compared to the control MDF cement sample at 75 MPa. The newspaper fibers reinforced samples exhibited the highest strength. Newspaper fiber reinforced macro-defect-free cement samples have the highest flexural strength compared with the other type of wood fibers reinforced samples. Recycled newspaper fibers, generally a combination of a larger portion of hardwood and small portion of softwood fibers, have been subjected to many mechanical and

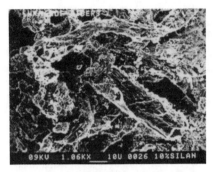

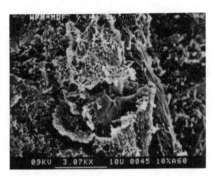

Figure 14, Delaminated 10% silane treated hardwood fiber in OPC-PAM matrix.

Figure 15, A dispersed polymer membrane surrounding a 10% A 60 treated hardwood fiber in the MDF cement matrix.

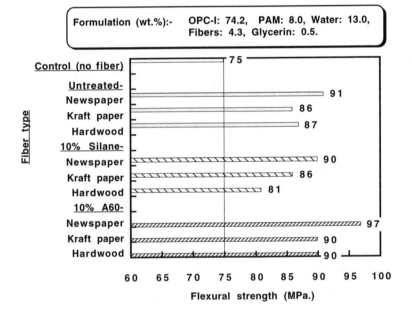

Figure 16, Flexural strength of various wood fiber reinforced MDF cements.

chemical treatments. From which a significant quantity of fibers are de-laminated forming micofibrils structure (see figure 3), that make them more conformable and more easily to be entangled forming a mechanical bond. In addition, delaminated fibers and softwood fibers have a higher surface area providing more opportunities for bonding to the cement matrix. Chemical treatments may enhance the durability of the fibers as a result of coating.

SUMMARY

Wood fibers with hollow structures can be bent and collapsed to various degrees. The flexible and conformable nature of wood fibers makes them easy to mix with cement paste in both conventional and novel processing styles. WFRC is an economic and efficient approach to improving the flexural strengths and the toughness of cementitious composites. In the fracture surface of WFRC samples, a large percentage of the wood fiber surface was covered by cement hydration products, which indicates a significant interfacial bonding between the cement matrix and the wood fibers. The interfacial bonding between cement matrix and wood fibers increased with age constraining fiber-pull out, resulting in embrittlement of the composite.

Newspaper fiber reinforced macro-defect-free cement samples have the highest flexural strength compared with the other type of wood fiber reinforced samples. Recycled newspaper fibers, coming from the most economic source, generally contain a larger portion of hardwood and small portion of softwood fibers, and have been subjected to many mechanical and chemical treatments. From which a significant quantity of fibers are de-laminated forming microfibrils that make them more conformable and more easily to be entangled forming a mechanical bond. In addition, delaminated fibers have a higher surface area providing more opportunities for bonding to the cement matrix.

ACKNOWLEDGMENTS

The authors would like to acknowledge the financial support of United States Department of Agriculture by grant # 91-37103-6540. The authors would also like to acknowledge the contributions of Professor P. Blankenhorn, K. Kessler and Maria DiCola of Penn State's Forest Products Lab to this work. Thanks also go to Darlene Wolfe-Confer and Maria Klimkiewicz for their technical assistance.

REFERENCES

1. A. Bentur and S. Mindess, *Fiber Reinforced Cementitious Composite*, (Elsevier Science Publishers, New York, 1990) p.9.
2. R. N. Swamy, *Natural Fiber Reinforced Cement and Concrete*, (Blackie and Son, London, 1988), p. v-vi.
3. Y. Chen, C. K. Park, M. R. Silsbee, and D. M. Roy in *Advanced Cementitious Systems: Mechanisms and Properties,* edited by F.P.Glasser, G.J. McCarthy, J.F.Young, T.O.Mason and P.L.Pratt (Mater. Res. Soc. Proc. 245, Pittsburgh, PA, 1992) pp. 229-234.
4. A. Sarja, in *Natural Fiber Reinforced Cement and Concrete*, edited by R. N. Swamy (Elackie and Son, London 1988), pp.63-91; Concrete International, 11(7), 45-49(1989).
5. R. S. P. Coutts, in *Natural Fiber Reinforced Cement and Concrete*, edited by R. N. Swamy (Elackie and Son, London 1988), pp.1-62; J. Mater. Sci. Lett. 6, 955 (1987).
6. A. W. C. Lee and P. H. Short, Forest Prod. J. 39(10), 68(1989).
7. X. Lin, M.R. Silsbee, K. Kessler, P.R. Blankenhorn, and D.M. Roy, Cement and Concrete Research 24(8),1558 (1994).
8. K. Kessler (private communication).
9. A. J. Michell and G. Freischmidt, J. Mat. Sci. 25, 5225 (1990).
10. R. S. P. Coutts and P. Kightly, J. Mat. Sci. 19, 3355 (1984).
11. X. Lin, M.R. Silsbee and D.M. Roy, presented at 96th Annual Meeting of American Ceramic Society, Indianapolis, 1994.

12. X. Lin, M.R. Silsbee, K. Kessler, P.R. Blankenhorn, and D.M. Roy, "Wood Fiber Reinforced Cementitious Composites and Chemical Pretreatments of Wood Fibers" - to be published.
13. I. Titchell, J. Mat. Sci. 26, 1199 (1991).
14. M. R. Silsbee, D. M. Roy and J. H. Adair, in *Specialty Cements with Advanced Properties,* edited by B. E. Sheetz et al. (Mater. Res. Soc. Proc. 179, Pittsburgh, PA 1991) pp. 129-144.
15. D. M. Roy, Science, 235, 651 (1987).

FIBER-MATRIX INTERFACIAL AREA REDUCTION IN FIBER REINFORCED CEMENTS AND CONCRETES AT MODERATE TO HIGH FIBER VOLUME FRACTIONS

GEBRAN N. KARAM
Arlington, MA 02174
formerly Department of Civil and Environmental Engineering, Massachusetts Institute of Technology, Cambridge, MA 02139

ABSTRACT

The area and properties of the fiber-matrix interface in fiber reinforced cements and concretes determines the amount of stress transferred forth and back between the cement paste and the reinforcement and hence controls the mechanical properties of the composite. Fiber-fiber interaction and overlap of fibers with fibers, voids and aggregates can dramatically decrease the efficiency of the reinforcement by reducing this interfacial area. A simple model to account for this reduction is proposed and ways to integrate it in the models describing the mechanical properties of short fiber reinforced concretes are presented. The parameters of the model are evaluated from previously published data sets and its predictions are found to compare well with experimental observations; for example, it is able to predict the non-linear variation of bending and tensile strength with increasing fiber volume fraction as well as the existence of an optimal fiber content.

INTRODUCTION AND BACKGROUND

Short randomly distributed fibers are added to brittle matrices to increase their toughness, tensile and bending strengths, to control the shrinkage and thermal cracking, and to improve their post cracking behavior. Due to their potential importance and economic impact as construction materials, a wide variety of fiber reinforced systems has been investigated over the past years. Researchers have studied experimentally the properties of cements and concretes reinforced with steel, asbestos, mica, wollastonite, glass, carbon, polypropylene, polyethylene, acrylic, nylon, wood and other natural cellulosic fibres.

A review of the literature has identified two common observations : the decrease in density or the increase in void content of the composite in direct proportion to the increase in fiber volume fraction; and the nonlinear increase in the bending and tensile strength of the composite with increasing fiber volume fractions. A maximum strength is reached at an optimal fiber volume beyond which, the strength of the composite decreases with increases in the fiber content. Table I lists some of the sources that have reported these observations for the strength of fiber cement composites or their toughness with a description of the composite type and the observed optimal fiber volume. Additional evidence of these observations has been reported throughout the tremendous fiber cement composites literature. Table I is only intended as an indicative list.

Mat. Res. Soc. Symp. Proc. Vol. 370 © 1995 Materials Research Society

Table I: Optimal fibre volume for different fibre cement composites

Reference	Fibre-Matrix Composite	Optimal Fibre Volume (%)
[1]	Abaca fiber reinforced cement (air-cured)	8-9%
[2]	Asbestos reinforced mortar	15%
[3]	Banana fiber reinforced cement	8-14%
[4]	Carbon fiber reinforced cement and mortar	4-5%
[5]	Carbon fiber reinforced cement	3-5%
[6]	Coconut fiber reinforced mortar	3%
[7]	Coir fiber reinforced cement	14-15%
[8]	Eucalyptus wood fiber reinforced cement	8-10%
[9]	Flax fiber reinforced cement and mortar	7.5-9.5%
[10]	Glass fiber reinforced cement	6-8%
	Glass fiber reinforced concrete	2-3%
[11]	Mica flake reinforced cement and mortar	3-4%
[12]	Pinus radiata wood pulp reinforced cement	9%
[13]	Polypropylene fiber reinforced concrete	0.3-0.5%
[14]	Sisal pulp reinforced cement	7-8%
[15]	Steel fiber slurry infiltrated concrete	2-4%
[16]	Wollastonite micro-fiber reinforced cement	10-14%
[17]	Wood fiber reinforced cement	6-8%
[18]	Wood fiber reinforced cement (autoclaved)	7-8%
[19]	Wood kraft pulp fiber mortar (air-cured)	8%
[20]	Wastepaper fiber reinforced cement	10-12%

The available models (e.g. the rule of mixtures) which describe the composite strength properties as a function of those of the matrix, the fibers, and the fiber-matrix interface, fail to predict the nonlinear behavior at moderate to high fiber volume fractions (> 2%). The experimental observations reported above and the shortcoming in the models can be attributed to two main reasons [21]: first the weakening of the matrix strength with the increase in porosity associated with added fiber volumes; and second, the decrease in fiber reinforcement efficiency due to the reduction of the fiber-matrix interfacial area caused by fiber-fiber, fiber-void, and fiber-aggregate interaction and overlap. Few attempts have been made by researchers to account for these two phenomena. Beaudoin [11] and Andonian et al. [17] have proposed to reduce the matrix strength contribution to account for increasing porosity with increasing fiber volume fraction. However they have not accounted for the reduced efficiency of the fiber reinforcement. Akihama et al. [4] and Toutanji et al. [22] on the other hand, have neglected the effect of the added voids on the matrix properties, and have proposed two different empirical exponential decay functions to model the decrease in the fiber reinforcement efficiency with increasing fiber volume content.

In the following paragraphs, the fiber-matrix interfacial area reduction is investigated, and a model proposed in a recent study by Karam [21] is briefly described. Previously published data sets on fiber reinforced cement composites are then used to test the model. Finally the predictive capabilities and potential applications of the model are discussed.

FIBER-MATRIX INTERFACIAL AREA REDUCTION

The interfacial bond between the matrix and the fibers has been identified as a major factor in determining the strength of the composite. The interfacial condition is responsible for transmitting the stresses from the matrix to the fibers and back. The external loads are usually applied to the matrix, while the contribution to the composite strength comes from both matrix and fibers. Thus, the interfacial surface plays a critical role in the composite behavior. In the preparation of fiber cement composites, the fibers and the aggregates are mixed with the cement slurry by mechanical agitation. It is generally assumed that this process will result in a random and more or less uniform distribution of the fibers inside the mass of the composite, and that the fibers will be completely coated by the matrix. This is not generally true, as fibres tend to clump together entrapping water filled spaces which then turn into voids upon curing. Fibers will also overlap with each other and with voids, and will interact with aggregates. They will have less than total interfacial contact with the matrix. Increasing the fiber volume in the composite increases clumping and overlap. The fiber surfaces that are not in direct contact with the matrix do not transfer stresses between the matrix and the fiber. Those fibers with partial or no contact with the matrix are partially or totally inefficient in reinforcing it.

Figures 1, 2 and 3 show SEM micrographs of plane sections through two commercial wood fiber reinforced cement products with about 13% fiber volume fraction [23]. Figure 1 shows the general microstructure with fibers and voids. Figure 2 shows a fiber clump with the entrapped voids and fiber-fiber and fiber-void overlap. Figure 3 presents a close up showing fiber-fiber overlap and partial fiber-matrix contact.

A new efficiency factor for the strength contribution of the fibers based on interfacial surface reduction was proposed by Karam [24] for unidirectional fiber composites and later [21] generalized to short fiber reinforced composites. It is summarized in the following. Consider an ideal composite made only of a homogeneous matrix and fibers, and discretized into a regular array of cubes. Furthermore, consider the fibers to be inclusions of a regular symmetrical geometry such as a cube of the same size into which the matrix was discretized. The fibers are assumed to be randomly distributed in the composite, and can fully occupy one location as shown in Figure 4. Then, the probability of finding a fiber at a given location is equal to V_f, the fiber volume fraction in the composite, and the probability of finding two contiguous fibers is $V_f * V_f = V_f^2$. The probability of one side of a fiber being overlaped by a neighbor is equal to V_f^2 which is the probability of two fibers being contiguous. Since each side of the ideal cubic fiber has the same probability of being covered by a neighbor, it is found [21] that, on the average, a fiber will have V_f^2 of its surface covered by neighboring fibers. Assuming that the contribution of a fiber to the strength of the composite is proportional to its surface in physical contact with the matrix, the effective surface area of the fibers becomes $A_{effective} = A_{initial} - A_{overlapped} = A_{initial}(1-V_f^2)$. This is equivalent to saying that V_f^2 is the proportion of inefficient fibers, or that the fiber volume in the composite, V_f, has a reinforcing effect equal to an effective volume given by:

$$V_{f\ effective} = V_f(1-V_f^2) \tag{1}$$

In reality the fibers are not cubic but elongated, and they not only overlap with each other but they also interact with the voids and aggregates present in the matrix. Most of the interaction of the fibers with voids is with the additionally entrained or entrapped voids, the volume fraction of which was observed to be proportional to the fiber volume fraction [1-20]. The major effect of the aggregates is to reduce the available volume inside the composite where the fibers can be distributed. The concentration of the fibers in the physically accessible space results in a higher fiber-fiber and fiber-void interaction probability. To account for all these effects, it is proposed to define the amount of inefficient fibers as $n^2 V_f^2$, proportional to V_f^2, where n (>1) is a fiber efficiency factor relating the real composite to the ideal one. The effective fibre volume becomes:

Figure 1. Micrograph showing the microstructure of a commercial wood fiber cement.

Figure 2. Micrograph showing a fiber clump in a commercial wood fiber cement.

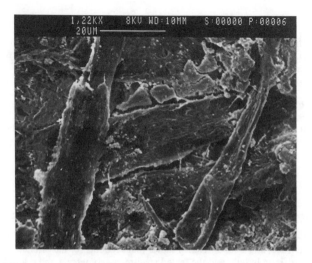

Figure 3. Micrograph showing fiber-fiber overlap in a commercial wood fiber cement.

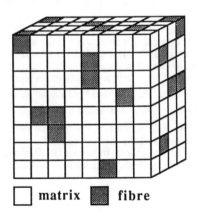

☐ matrix ▨ fibre

Figure 4. Ideal composite model.

$$V_{f \text{ effective}} = V_f(1-n^2V_f^2) \qquad (2)$$

The fiber efficiency factor n can be determined from microstructural measurements on micrographs of the fiber reinforced composite using methods described elsewhere [21, 24].

MODEL FOR TENSILE AND BENDING STRENGTH

A modified rule of mixtures that accounts for both the fiber efficiency reduction with higher fiber volumes and the matrix strength loss with increased voids was proposed [21]:

$$\sigma_c = \sigma_{mo}\exp(-4\Delta V_o)(1-V_f) + \sigma_f V_f(1-n^2V_f^2) \qquad (3)$$

where σ_c is the strength of the composite, and σ_{mo} is the strength of the matrix at zero fiber content. $\sigma_f = \phi\tau(l/d)$ is the fiber strength contribution with τ being the strength of the fiber matrix interface, and ϕ an efficiency factor that accounts for the orientation of the short fibers, their finite length, and the tested specimen geometry. l and d are the length and diameter of the fibers assumed to have a circular cross section. ΔV_o is the increase in void volume fraction assumed to be directly proportional to the fiber volume. It is proposed to model this increase as $\Delta V_o = \alpha V_f$ where α is determined from experimental data. The strength of the matrix is assumed to decrease exponentially with ΔV_o as shown in Equation 3.

Equation 3 contains four constants: σ_{mo}; α; σ_f; and n all of which have to be determined prior to the application of the model. σ_{mo}, the strength of the unreinforced matrix, is the most readily available and easily measured parameter. α can be obtained by linear regression of the composite void volume fractions versus the fiber volume fractions incorporated in the matrix. For this purpose it is necessary to measure the porosity or the density of the composite at different fiber volumes. About a third of the references surveyed in Table I did not report any density or porosity for their fiber reinforced cement composites. The rest reported the density but only a few references reported measures of both porosity and density. $\sigma_f = \phi\tau(l/d)$ can be estimated from the values of its factors, but it is difficult to estimate ϕ, and experimentally demanding to determine the fiber-matrix interface strength τ from single fibre pull out tests [2, 25]. n can be estimated from independent measurements on plane surface micrographs of the composites [21]. However it was not common so far for researchers in the field of fiber reinforced cement composites to analyze the microstructure of their composites. This information is missing in all the surveyed literature [1-20, 22].

Model Application

To test the model presented by Equation 3, data on the bending strength of wood fiber reinforced cements was collected from two different studies [8, 17]. In addition to the variation of the bending strength with fiber volume fraction, these sources reported the

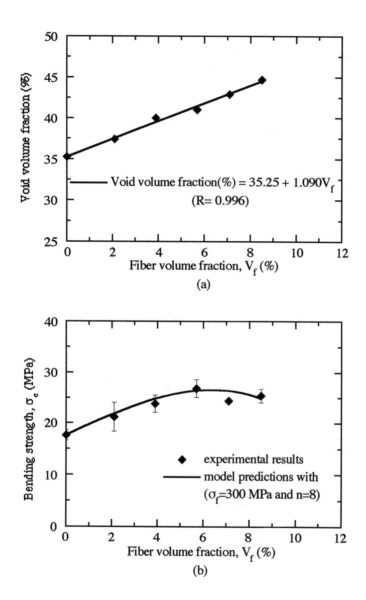

Figure 5. (a) Void volume fraction vs fiber volume fraction, and (b) Bending strength vs fiber volume fraction, for data on wood fiber cement from Andonian et al. [17]

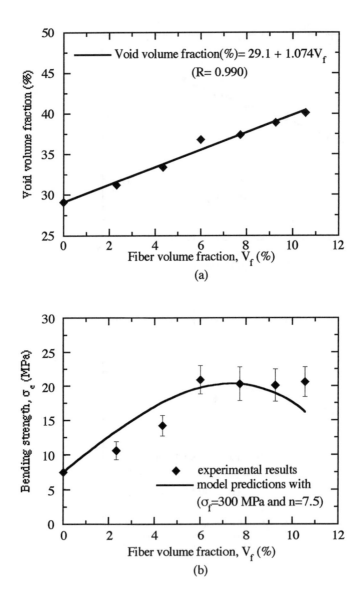

Figure 6. (a) Void volume fraction vs fiber volume fraction, and (b) Bending strength vs fiber volume fraction, for data on wood fiber cement from Coutts [8].

change in porosity with fiber additions. The value of α can be readily obtained from the slope of the regression analysis of void volume versus fiber volume. No information on the microstructure was available to estimate the fiber efficiency factor, n. It is assumed that n will take values between 7 and 8 similar to what Karam [21] has reported for other wood fiber reinforced cements. The only remaining unknown parameter is the fiber strength contribution σ_f. When all the other parameters known, σ_f can be obtained from Equation 3 by using the experimentally measured composite bending strength at a fiber volume fraction in the middle of the range investigated. Figures 5a and 6a show the linear variation of the void volume with the fiber volume. Figures 5b and 6b show the prediction of the model for the bending strength of the two data sets for estimated values of σ_f and n. Predictions and experimental results compare well.

DISCUSSION AND CONCLUDING REMARKS

The proposed modified rule of mixtures incorporates the fiber efficiency decrease due to fiber-matrix interfacial area reduction, and the matrix loss of strength due to entrained voids. It also captures the nonlinear variation of the bending strength of wood fiber reinforced cements with fiber volume fraction. It can be applied to other fiber reinforced cement systems if the necessary information is collected. Density and porosity measures are needed to estimate the matrix strength loss with fiber additions. The choice of a matrix strength loss model can affect the predictions, but in general the contribution of the matrix is limited and this effect is minimal. The information on the microstructure of the composite is important since the fiber efficiency factor n controls the nonlinearity of the model. The fiber-matrix interfacial area reduction is a major cause for the fiber reinforcement efficiency decrease. More work is needed to shed additional light on this topic. Finally the model proposed can be used for the optimization of commercial fiber cement products. For a given fiber-matrix system, the fiber content at which maximum strength is reached, i.e., the optimal fiber content, can be found by solving for the fiber fraction yielding $\partial \sigma_c / \partial V_f = 0$. Compaction and casting pressure reduce both the voids in the matrix and those entrapped between the fibers, which results in stronger composites [7, 26]. The four parameters (σ_{mo}, α, σ_f and n) are affected differently by different industrial processes. Integrating these effects into a model that predicts composite strength allows for the properties optimization and cost minimization.

REFERENCES

1. R.S.P. Coutts and P.G. Warden, Int. J. of Cem. Comp. and Lightweight Concr. **9** (2), 69-73 (1987).
2. Y.W. Mai, J. Mat. Sci. **14**, 2091-2101 (1979).
3. W.H. Zhu, B.C. Tobias, R.S.P. Coutts and G. Langfors, Cem. and Concr. Comp. **16**, 3-8 (1994).
4. S. Akihama, S. Tatsuo and T. Banno, Int. J. of Cem. Comp. and Lightweight Concr. **8** (1), 21-33 (1986).
5. S.B. Park, B.I. Lee and Y.S. Lim, Cem. and Concr. Res. **21**, 589-600 (1991).
6. P. Paramasivam, G.K. Nathan and N.C. Das Gupta, Int. J. of Cem. Comp. and Lightweight Concr. **6** (1), 19-27 (1984).
7. L.K. Agarwal, Cem. and Concr. Comp. **14**, 63-69 (1992).
8. R.S.P. Coutts, J. Mat. Sci. Let. **6** (8), 955-957 (1987).
9. R.S.P. Coutts, Int. J. of Cem. Comp. and Lightweight Concr. **5** (4), 257-262 (1983).

10. D.J. Hannant, <u>Fibre Cements and Fibre Concretes</u>, (John Wiley and Sons, New York, 1978).
11. J.J. Beaudoin, Cem. and Concr. Res. **13**, 153-160 (1983).
12. R.S.P. Coutts, R.S.P. and P.G. Warden, J. Mat. Sci. Let. **4**, 117-119 (1985).
13. Z. Bayasi and J. Zeng, ACI Mat. J. **90** (6), 605-610 (1993).
14. R.S.P. Coutts and P.G. Warden, Cem. and Concr. Comp. **14**, 17-21 (1992).
15. P. Balaguru and J. Kendzulak , <u>American Concrete Institute SP105-ACI</u>, 247-268 (1984).
16. N.M.P. Low and J.J. Beaudoin, Cem. and Conr. Res. **22**, 981-989 (1992).
17. R. Andonian, Y.W. Mai and B. Cotterell, Int. J. of Cem. Comp. and Lightweight Concr., **1** (3), 151-158 (1979).
18. R.S.P. Coutts, Composites **15** (2), 139-143 (1984).
19. R.S.P. Coutts, Composites **18** (4), 325-328 (1987).
20. R.S.P. Coutts, Int. J. of Cem. Comp. and Lightweight Concr. **11** (3), 143-147 (1989).
21. G.N. Karam, J. of Comp. Tech. and Res. **16** (2), 154-160 (1994).
22. H.A. Toutanji, T. El-Korchi, R.N. Katz and G.L. Leatherman, Cem. and Concr. Res. **23** , 618-626 (1993).
23. G.N. Karam, SMCE thesis, Massachusetts Institute of Technology, 1990.
24. G.N. Karam, Composites **22** (2), 81-84 (1991).
25. Y. Wang, V.C. Li and S. Backer, J. Mat. Sci. **26** , 6565-6575 (1991).
26. R.S.P. Coutts and P.G. Warden, Cem. and Concr. Comp. **12**, 151-156 (1992).

INTERFACIAL DEBONDING AND SLIDING IN BRITTLE CEMENT-MATRIX COMPOSITES FROM STEEL FIBER PULLOUT TESTS

M. JAMAL SHANNAG*, WILL HANSEN* AND RUNE BRINCKER**
*University of Michigan, Department of Civil Engineering, 2330 G.G. Brown, Ann Arbor, MI-48109
**Aalborg University, Department of Building Technology and Structural Engineering, Sohngaardsholmsvej 57, Aalborg 9000, Denmark

ABSTRACT

A specially designed single fiber pullout apparatus was used to provide simultaneous results on total fiber displacement versus load in addition to monitoring the fiber displacement at the free end. In this apparatus the fiber was going through the entire specimen, which made it possible to determine the point of complete debonding. To control the embedment length a plastic tube was inserted around the fiber. The described fiber pullout test method coupled with an appropriate analysis provides a quantitative determination of interfacial properties which are relevant to toughening of brittle materials through fiber-reinforcement. The technique was used on a high strength cement-based matrix called the Densified Small Particle system (DSP), and on an ordinary strength matrix. Other parameters investigated included fiber embedment length and fiber volume fraction in the cement matrix. The results indicate that: (1) the dense DSP matrix has significantly improved interfacial properties as compared to the ordinary strength matrix; (2) the major energy of pullout in both systems is due to sliding; and (3) both the debonding energy and sliding energy increase with fiber embedment length. These results are important in the understanding of the role of steel fibers in improving the tensile properties of high performance fiber reinforced composites.

INTRODUCTION

In steel fiber reinforced cementitious composites, the properties of the interface between the fiber and the matrix material strongly affect the strength and toughness of the composite. It is well known that fibers primarily contribute to the post cracking response of the composite by bridging the crack and providing resistance to crack opening. The resistance to crack propagation provided by the fibers depends on the mechanical properties of the matrix, fibers, and the fiber - matrix interface, as well as the fiber length, orientation, volume content, and spacing [1-3].

When a crack moves through matrix-containing fibers, energies are absorbed in a number of different ways. Energy absorbing mechanisms include matrix cracking, fiber-matrix interfacial debonding, frictional sliding, fiber fracture, fiber elongation across the cracked surface, load redistribution in the cracked matrix, and finally fiber pullout [4].

A fiber pullout test is usually designed to obtain the bond strength and energies associated with fiber debonding and sliding of the fiber-matrix interface. In this test, a fiber is cast into a cement-based matrix and loaded in tension until the fiber debonds and is withdrawn.

The purpose of this paper is to provide a new method for analyzing interfacial properties from single fiber pull-out curves from which the point of complete debonding is accurately determined. An ordinary strength mortar and high strength mortar (DSP) are investigated. The analysis allows determination of the frictional shear strength, the debonding energy, and the sliding energy up to point of complete debonding.

EXPERIMENTAL PROGRAM

The experimental program was carried out with a view to study the effect of the matrix strength and fiber embedment length on the interfacial bond. Two different matrices were used. DSP cement based matrix with a compressive strength of 150 MPa, was proportioned by weight as cement to sand to water ratio (c:s:w). The mix designs are: 1:1.2:0.18. The ordinary mortar matrix with a compressive strength of 40 MPa was proportioned (c:s:w) as 1:1.2:0.45. Hardened stainless steel piano wire 0.19 mm in diameter with tensile strength of 2,990 MPa was used as the pullout fiber. Fiber embedment lengths of 6, 12, 18 mm were used [5].

The pullout specimens had a prismatic shape of 25x23x19 mm. The specimens were carefully prepared by mixing in a Hobart type laboratory mixer. Prior to casting the steel wire was cleaned with acetone and encapsulated with a plastic tube of 2 mm in diameter which provide a 6 mm crack length. Varying embedment lengths were obtained by inserting a plastic tube from the other end. The wire was then inserted through the middle of the specimen mold and kept in straight position without any significant pretensioning.

The pullout specimens were vibrated and stored at room temperature for 24 hours prior to demolding. They were then placed in hot water at 80° C for additional 24 hours, and stored in a sealed container to avoid evaporation of water until time of testing.

Test procedure

The pullout apparatus used in the testing program is illustrated in Fig. 1. The specimen was held in place by means of a steel plate. The top end of the wire was clamped in a slotted block which allowed free movement of the specimen until load was applied to the wire. The bottom end of the wire was held by a specially designed grip. The grip attachment was connected to a 200 N load cell. The load was applied manually at a constant rate. The displacement of the top end of the fiber was monitored by means of two linear variable differential transducers (LVDT) attached 3 mm above the top of the specimen. This length, known as the free fiber length was kept constant in all tests. The displacement of the bottom end of the fiber was monitored by another LVDT.

Preliminary tests were conducted to ensure that no slip occurred between the specimen and the grips even at the maximum pullout loads. The value of the force, as well as the movement at both ends of the fiber, were recorded by a data acquisition system, and stored in computer file. The pullout load versus top and bottom displacement was then developed and plotted, using such data files.

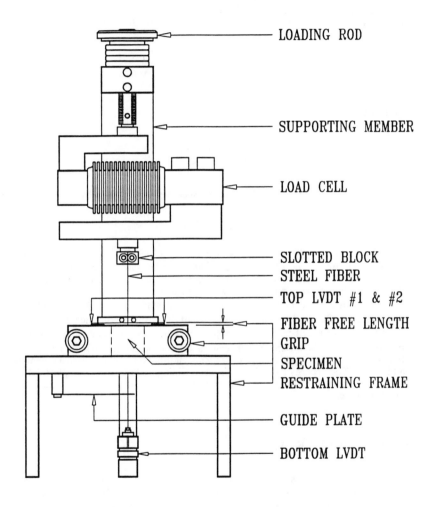

Fig. 1: Schematic of the single fiber pull-out test

NEW MODEL FOR ANALYZING PULL-OUT CURVES

Figure 2 is a schematic illustration of a single fiber pull-through test in which the force F is measured versus total displacement u for a fiber embedment length L. a is the interfacial crack length, and b is the free fiber length.

For the case of pure friction the following equilibrium equation (5) can be obtained for the relationship between total displacement, u, and total force, F:

$$u = \frac{1}{2}\frac{F}{E_f A_f}\frac{F}{q_f} + \frac{F}{E_f A_f}\left(b + \frac{1}{2}\frac{F}{E_f A_f}\frac{F}{q_f}\right) \tag{1}$$

The matrix elastic contribution u_e [5] is given by:

$$u_e = \frac{F(1+v)}{\pi l E}log\left(\frac{R}{r}\right) \tag{2}$$

where E_f , A_f , are the fiber stiffness, area respectively, q_f is given by:

$$q_f = 2\pi r\tau_f \tag{3}$$

where τ_f is the frictional bond strength, and r is the fiber radius. R is the radius of the bulk material around the fiber obtained from shear lag theory. E, v are the Young's modulus and the Poisson's ratio of the bulk material around the fiber respectively.

Thus, the pullout curve is a horizontally lying parabola symmetrical placed around the u-axis cut off at the constant level $F = q_f l$ reached when $a = l$. Once the location of complete debonding is obtained for a given set of data from the displacement versus force of the free end of the fiber τ_f is estimated. Then the curve for pure friction is estimated from equation 1. The difference between the total response from which the elastic fiber stretch has been subtracted is due to debonding contribution. This is shown schematically in Figure 2. Total energy from debonding and sliding up to point of complete debonding are obtained respectively as the area between the two curves, and the area under the frictional curve.

RESULTS AND DISCUSSION

Figures 3 and 4 show results of the frictional shear strength versus embedment length, for the DSP and the ordinary mortar system, respectively. An average value of about 4.4 MPa was obtained for the DSP system and an average value of about 1.9 MPa was obtained for the ordinary strength mortar. The DSP system is expected to have a stronger interface than the ordinary strength mortar system due to the much denser micro structure.

Due to the nature of pullout tests more scatter is expected in data than for bulk properties. The difficulties in obtaining results for short embedment lengths is visible.

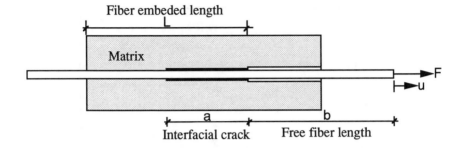

a. Schematic of fiber pullout

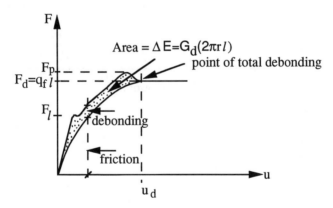

b. Combined debonding and friction

Fig. 2: Proposed model for estimating the debonding energy and
frictional bond stress of the interface

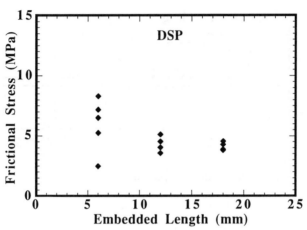

Fig. 3: Frictional shear stress versus fiber embedded length for DSP system

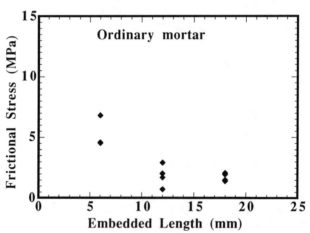

Fig. 4: Frictional shear stress versus fiber embedment length for ordinary mortar system

Typical total pullout curves are shown for up to 1mm displacement in Figure 5 for the DSP matrix and in Figure 6 for the ordinary strength matrix for three different embedment lengths. An exploded version is shown of these two figures up to the point of complete debonding in Figures 7 and 8. As expected the peak load and displacement at complete debonding increase with embedment length for both systems, but the DSP system has 2-3 times the pullout load as regular strength mortar. This system seems to indicate a more rapid decay in load capacity following the peak load before reaching a near constant value for large displacement. The DSP system however remains more stable. This may be due to load induced(frictional) decay in the weaker mortar. For large displacement (about 0.5 to 1mm) both systems seem to reach a plateau of near constant frictional load. This is expected since the fiber embedment length remains constant in a pull-through test.

Analysis of energies due to pull-out show that both systems are frictionally controlled (Figures 9 and 10) in the range where debonding and friction occurs simultaneously. As expected the energies of debonding and friction increase with embedment length for the two systems. The reason for estimating the frictional energy up to a total displacement determined by complete debonding is because this deformation state corresponds closely to end of multiple cracking and peak load in high performance FRC's.

The ratios of frictional to debonding energy are nearly constant and the same for both systems indicating same nature of bonding to steel in both systems except for varying degrees of porous micro structure.

CONCLUSIONS

1. The model for analyzing for pure friction provides a simple direct way of estimating the debonding and frictional energies, and frictional strength of the fiber-matrix interface from the experimental pullout curves.

2. Analysis of the pullout curves for both the high strength mortar (DSP) and ordinary strength mortar show that the frictional energy is dominant over debonding energy up to complete debonding. Both the frictional and debonding energies increase with embedment length.

ACKNOWLEDGMENTS

The authors gratefully acknowledge the financial support of the National Science Foundation Center for Advanced Cement-Based Materials (ACBM) and the support from Aalborg University, Denmark. Thanks are due to Mrs. Henning Andersen and Nick Flint, Department of Building Technology and Structural Engineering, at Aalborg University for help in designing the test setup and in carrying out the pullout tests.

REFERENCES

1. Stang, H., Li, Z. & Shah, S.P., "Pull-out problem: stress versus fracture mechanical approach." ASCE J. Eng. Mech., 116(1990) pp. 2136-2150.
2. Mandel, J., Wei, S. & Said, S., "Studies of the properties of the fiber-matrix interface in steel fiber reinforced mortar." ACI Materials J., 84(1987) pp. 101-109.
3. Naaman, A. E., Namur, G., Alwan, J. M. & Najm, H., "Fiber Pullout and Bond Slip. I: Analytical Study; II: Experimental Validation," ASCE J. of Structural Engineering, 117 (1991) pp. 2769-2800.

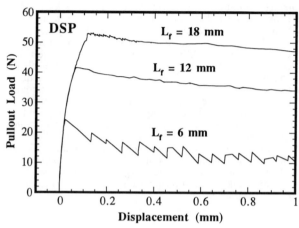

Fig. 5: Typical pullout curves for DSP system, L_f = 6, 12, and 18 mm, fiber dia.= 0.19 mm

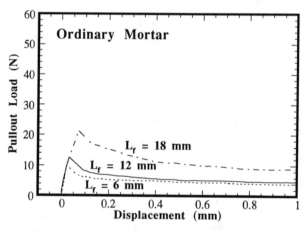

Fig. 6: Typical pullout curves for ordinary mortar, L_f = 6, 12, and 18 mm, fiber dia. = 0.19 mm

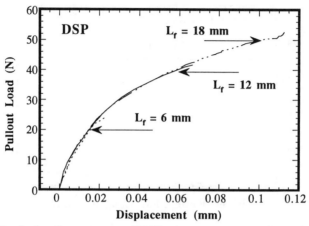

Fig. 7: Typical pullout curves of DSP system up to complete debonding for 6, 12, and 18 mm embedment lengths, fiber diameter 0.19 mm

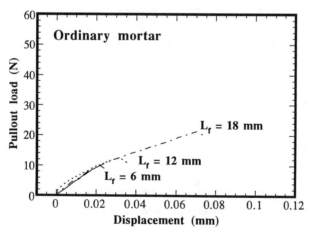

Fig. 8: Typical pullout curves of ordinary mortar system up to complete debonding for 6, 12, and 18 mm embedment lengths, fiber dia. 0.19 mm

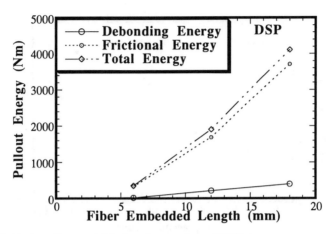

Fig. 9: Debonding and frictional energies of DSP system up to complete debonding point

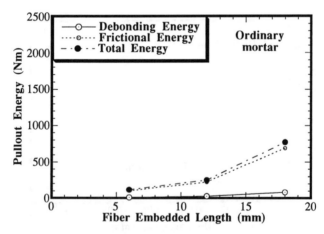

Fig. 10: Debonding and frictional energies of ordinary mortar system up to complete debonding point

4. Kim, J.K., Zhou, L.M., Mai, Y.W.,"Interfacial debonding and fiber pullout stresses, Part III." Journal of materials science 28(1993) pp. 3923-3930.
5. Shannag, M., "Tensile behavior of fiber reinforced cement based DSP", Ph.D. Thesis, Department of Civil and Environmental Engineering, University of Michigan, Ann Arbor, 1994, 220 pp.

DAMAGE EVOLUTION OF FIBER/MORTAR INTERFACE DURING FIBER PULLOUT

YIPING GENG AND CHRISTOPHER K.Y. LEUNG
Department of Civil and Environmental Engineering, Massachusetts Institute of Technology, Cambridge, MA 02139, USA

ABSTRACT

This study aims at connecting the change in interfacial properties of FRC during fiber pullout to the microstructural features of the interface. The interfaces between steel, polypropylene and nylon fibers and mortar are investigated by SEM and EDXA at four stages during fiber pullout (a) before debonding; (b) right after debonding; (c) pullout 1 mm and (d) pullout 10 mm. Since fiber pullout is found to be sensitive to lateral compression, microscopic studies are carried out on fibers pulled out with and without lateral compression.

For steel FRC, significant damage occurs at the interfacial transition zone (ITZ) of the mortar. Regardless of the presence or absence of lateral compression, it is found that the 'abrasion and grinding effect' is the major mechanism leading to the rapid decline of the pullout force and interfacial friction. Based on SEM and EDXA results, an interfacial damage evolution process is proposed.

For polymeric FRC's, the damage occurs mainly on the polypropylene and nylon fibers and the ITZ of the mortar experiences much less damage. The surface of the polypropylene fiber is shaved and abraded during the pullout process. The surface of the nylon fiber is peeled and shaved which contributes to the increase of the pullout force and interfacial friction.

INTRODUCTION

The performance of fiber reinforced concrete (FRC) is strongly affected by the fiber debonding/pullout behavior. For different fibers, the pullout curves may exhibit totally different trends. For example, the interfacial friction, which is very sensitive to interfacial microstructures, may either decrease rapidly (steel FRC [1,2]) or increase slightly (polymeric FRC [3]) after total debonding. In order to understand the variation during fiber debonding and pullout, it is necessary to investigate microstructural features at the interface, especially the damage evolution of the microstructure during fiber pullout.

Steel Fiber-Concrete Interface

It has been found that the microstructure near the cement paste, mortar or concrete interface is considerably different from the bulk matrix [4]. An SEM study by Bentur, et al. [5] concluded that the interfacial zone around steel fibers can be characterized by several layers (Fig. 1). The steel fiber surface is in contact with a thin duplex film layer

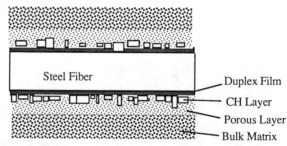

Figure 1. Microstructure of the steel fiber/cement interface [5]

(1 μm) consisting of a sub-layer of CH next to the steel and a C-S-H sub-layer. More than half of the area outside this duplex film is occupied by a thicker layer of dense CH crystals which is

about 10 to 20 µm in thickness. The remaining area scattered among the dense CH crystals is a porous zone that contains some C-S-H gel and possibly some ettringite particles (AFt: calcium aluminum trisulfate). Around the thick CH layer is a distinct porous layer of C-S-H gel, and only beyond that porous layer is the bulk cement paste microstructure observed. The region surrounding the fiber thus contains a very porous and weak layer parallel to the fiber at a distance of at least 10 µm away from the fiber surface.

Although the microstructure of the fiber-mortar interface has been reasonably understood, the microstructural change of interface during the fiber pullout process, such as interfacial debonding and frictional wear are known in considerably less detail. To the authors' knowledge, only Pinchin and Tabor [6] attempted to explain the decrease in interfacial friction of steel fiber-mortar during the pullout process based on the surface compaction of hydrated cement paste observed by Soroka and Sereda [7].

Pinchin and Tabor attribute the significant friction decrease during a small amount of steel fiber pullout to densification or compaction, but not to wear, on the mortar surface. They argued that the compaction in their pullout test occurs on a very fine scale in the order of 0.1-0.3 µm near the embedded steel fiber and is difficult to detect. Since their conclusions are based on the final stage of the pullout test, i.e., at total fiber pullout, it does not reflect the whole pullout process.

Polymeric Fiber-Concrete Interface

Polymeric fibers have high yield strength, but unlike steel fiber, they have lower elastic moduli and transverse strengths than cementitious materials. Therefore, the interfacial damage mechanism of polymeric fiber-mortar is different from that of steel fiber-mortar. Polymeric fibers also have less corrosion in harsh environments, but high creep effect. Based on these properties, polymeric fibers have been used to reinforce FRC in the early stage when the matrix is weak, and of low modulus, or to resist impact load and corrosion.

Polypropylene fiber was the first polymeric fiber applied to concrete in forms of monofilament or fibrillated film. Baggott and Gandhi [8] studied continuous monofilament polypropylene fiber (340 µm) reinforced cement beam under tensile load. They observed defects of up to 10 µm on the polypropylene fiber interface. One typical damage observed was the chiseling out of a long shaving of fiber by matrix particles (Fig. 2).

The application and study of nylon FRC so far are not as extensive as polypropylene FRC although nylon fiber exhibits good toughness and durability. Wang, et al. [3] investigated the nylon-cement interface and observed peeling and fibrillation at the fiber surface (Fig. 3). They concluded that the increased interfacial friction during fiber pullout is due to the increase in surface abrasion.

In this study, the interfacial microstructures and the damage evolution of steel, nylon and polypropylene FRC's are investigated microscopically with scanning electron microscopy (SEM) and energy dispersive X-ray analysis (EDXA). Different interfacial damage mechanisms are observed by comparing the interfacial microstructural evolution of the three types of FRC.

Fig. 2. Shaving on Polypropylene Fiber [8]

Fig. 3. Peeling & Fibrillation on Nylon Fiber [3]

EXPERIMENTAL PROCEDURE

Specimen Preparation

The specimen is a mortar block with an embedded fiber (Fig. 4). The fibers used include steel wire (low carbon and cold drawn steel), nylon monofilament (Nylon 66) and polypropylene monofilament (relaxed low draw). The steel fiber is 0.5 mm in diameter. The nylon and polypropylene monofilaments (0.5 mm in diameter) are 2143 deniers and 1549 deniers, respectively. Mortar is made from Type III Portland cement and mortar sand. The

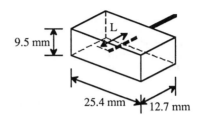

Figure 4. Single Fiber Pullout Specimen (L = 10 mm)

water/cement/sand ratio is 0.5:1:2. Tests are performed with saturated surface dry (SSD) specimens at the age of 7 days. The material properties are listed in Table 1.

Table 1. Coefficients of Cement Mortar and Fibers

	Diameter (mm)	Young's Modulus (GPa)	Tensile Strength (MPa)	Other Strength (MPa)
Cement Mortar	---	22.3	3.6 (splitting)	40 (Compres.)
Steel	0.5	200	1242	1040 (yield)
Nylon	0.5 (equivalent)	6 [9]	451	---
Polypropylene	0.5 (equivalent)	4 [9]	391-462	---

Testing Procedure

A 2-D loading device is used to perform the fiber pullout test [2]. Fiber is pulled out under displacement control with a loading rate of 0.159 mm/min. Two sets of specimens are tested with zero lateral compression and a constant lateral compressive stress (σ_c = 20 MPa). At different stages of pullout (Fig. 5), interfacial examinations are carried out:
- (a)   before debonding (sliding distance = 0), no pulling is required
- (b)   after debonding (sliding distance ~ 0.1 mm), sliding stops immediately after the peak load is reached
- (c)   pullout 1 mm (sliding distance = 1 mm)
- (d)   pullout 10 mm (sliding distance = 10 mm)

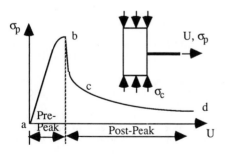

Figure 5. Four Loading Stages

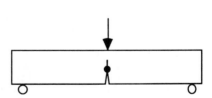

Figure 6. Split the Specimen after Loading

The tested specimens are split by three point bending (Fig. 6) to expose fiber-mortar interfaces. In order to identify the microstructural features of the fiber-mortar interfaces at each stage of pullout, the fiber and mortar groove surfaces are gold-coated for SEM analysis. SEM and EDXA are carried out near the fiber embedded end and the mortar groove exit since these locations generally experience the most interfacial interaction.

EXPERIMENTAL RESULTS AND DISCUSSION

Figure 7 shows the typical pullout curves (fiber stress vs. displacement at the pulling end of the fiber) for steel, nylon and polypropylene fibers. Compared with the tensile strengths of the fibers (Table. 1), the peak loads are much lower. Therefore, fibers can be considered within their elastic ranges except that damage may occur at the fiber surface. The post-peak pullout behavior show different trends for different fibers:

(a) steel fiber pullout -- decrease rapidly
(b) nylon fiber pullout -- increase slowly
(c) polypropylene fiber pullout -- decrease slowly

With lateral compression, the peak pullout load increases. However, for the steel fiber, the post-peak load drop is also more rapid.

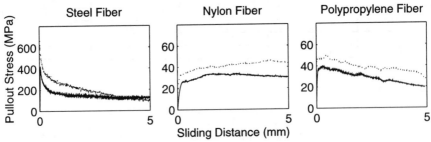

Figure 7. Typical Pullout Curves (Solid: $\sigma_c = 0$; Dotted: $\sigma_c = 20$ MPa)

Steel Fiber-Mortar Interface

Figure 8 shows the mortar surface when pullout is under zero lateral compression. Each picture in the figures corresponds to one of the four pullout stages.

At stage (a), the interface is separated by tensile debonding. Most of the mortar surface follows the topology of the steel surface. Small pieces of cement mortar (CH crystals and C-S-H fibrous blocks) are spalled from the matrix, similar to the observation of Pinchin and Tabor [6] and Bentur, et al. [5]. The result indicates that interfacial fracture usually occurs along the steel-mortar interface, but occasionally occurs within the interfacial transition zone where the tensile strength of the mortar is less than the chemical bond at the interface.

At stage (b), shear debonding can also result in some mortar adhered to the steel surface since shear debonding is the same as tensile debonding at the atomic scale. A small amount of sliding grinds the adhered pieces to small particles of size up to 2 μm.

The irregular spalling pieces are ground to much smaller particles under further pullout at stages (c) and (d) and becomes more regular in their shapes. The decrease of particle size is more rapid in the early sliding stage, which agrees well with the rapid decrease of the post-peak pullout curves. At the final stage (d), the surface is smoothed with very fine particles filling the surface cracks and only the scratch lines can be observed.

The same observation is repeated for the case with lateral compression (Fig. 9). By comparison of stages (b), (c) and (d) for the cases with and without compression in Fig. 8, one

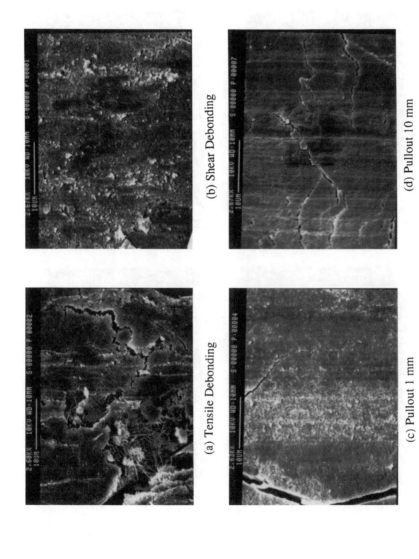

(a) Tensile Debonding

(b) Shear Debonding

(c) Pullout 1 mm

(d) Pullout 10 mm

Figure 8. Mortar Surface at Steel-Mortar Interface under no Compression ($\sigma_c = 0$)

523

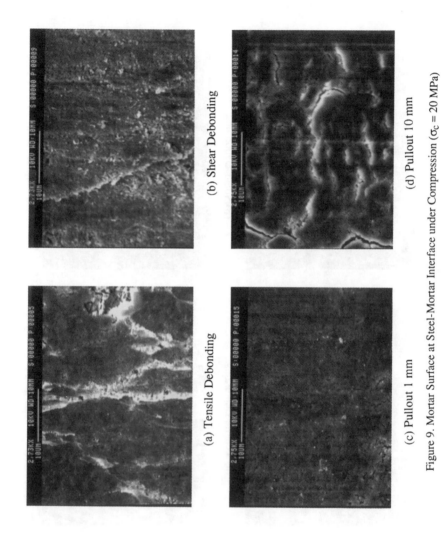

(a) Tensile Debonding

(b) Shear Debonding

(c) Pullout 1 mm

(d) Pullout 10 mm

Figure 9. Mortar Surface at Steel-Mortar Interface under Compression (σ_c = 20 MPa)

can observe that the particle size decreases more rapidly when compression is applied. The lateral compression accelerates the abrasion process due to severe grinding.

In order to quantify the surface damage, energy dispersive X-ray analysis (EDXA) is carried out. From the ratio of the area under the Ca and Si peaks on the X-ray spectrum, one can deduce the change in CH/C-S-H on the matrix surface during the fiber debonding/pullout process. Figure 10 shows the Ca/Si ratio at the four pullout stages for mortar surface. The Ca/Si ratio at the steel-mortar interface increases in the first three stages and decreases at the last stage. Based on the SEM observations in Figs. 8 and 9, there are two possible contributions to the increase and the decrease:

(a) debonding occurs preferentially in the weak area at the interface or in the matrix, where the brittle CH phase is rich. In the early pullout stages, the brittle CH phase may be crushed when it interacts with the steel surface and is smeared over the matrix surface, leading to an increase in the Ca/Si ratio.

(b) the decrease of the Ca/Si ratio can be attributed to abrasion. On further fiber sliding, the crushed CH phase is ground into much finer particles and partly swept into the pores and cracks on the mortar surface, hence exposing the underlying C-S-H phase. This process can reduce the Calcium count and increase the Silicon count.

One can see in Fig. 10 that debonding and an inevitable small amount of sliding at stage (b) increases the Ca/Si ratio. Further sliding at stage (c) is a process of crushing and smearing the CH phase, which increases the Ca/Si ratio. The long sliding at stage (d) is a process of sweeping and cleaning the CH phase, which decreases the Ca/Si ratio.

The results of EDXA also verify the abrasion effect. In Fig. 10, the Ca/Si ratio remains the same for the cases both with and without compression at stage (a). Since the fiber is not pulled at stage (a) and no abrasion exists at the interface, one can conclude that surface compaction will not alter the Ca/Si ratio. Nevertheless, the Ca/Si ratio during fiber sliding varies significantly for cases both with and without compression, which confirms that the surface microstructures changes significantly during fiber sliding. This microstructural evolution is caused by the surface abrasion (grinding effect).

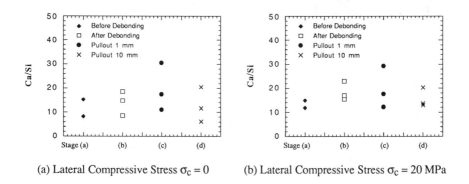

(a) Lateral Compressive Stress $\sigma_c = 0$ (b) Lateral Compressive Stress $\sigma_c = 20$ MPa

Figure 10. Ca/Si Ratio at Matrix Surface of Steel-Mortar Interface

Based on microscopic observation and X-ray analysis, the pullout process can be characterized by a simple model in Fig. 11. Brittle CH layer is first abraded followed by abrasion of C-S-H gel. This process results in a change of Ca/Si ratio on the surface due to the grinding and sweeping of CH phase into pores and cracks.

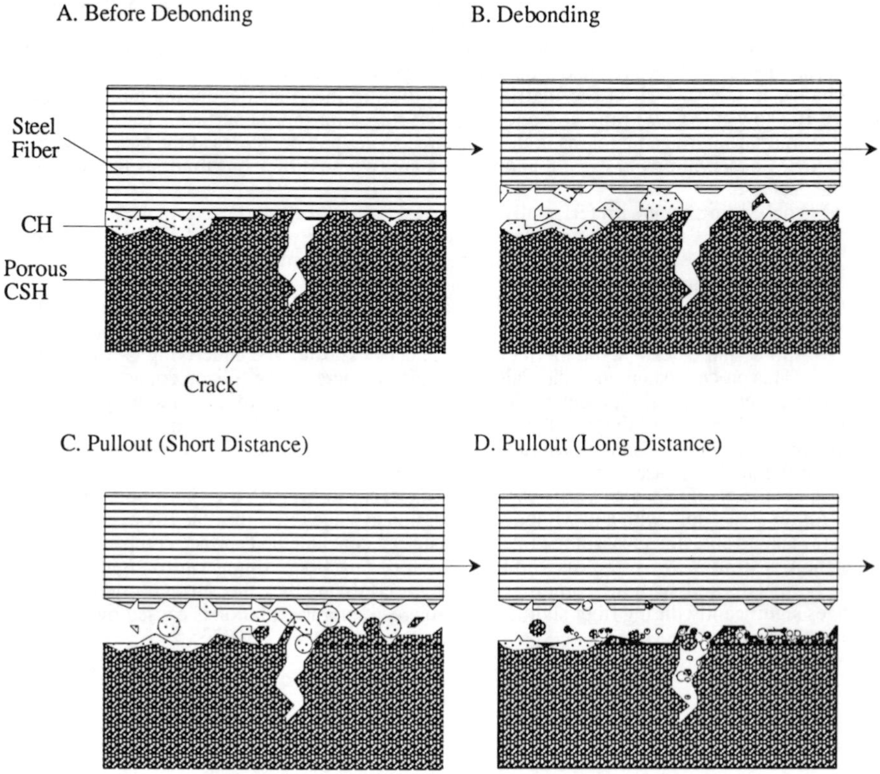

A. Before Debonding   B. Debonding

Steel Fiber

CH

Porous CSH

Crack

C. Pullout (Short Distance)   D. Pullout (Long Distance)

Figure 11. Damage Evolution at Steel Fiber-Mortar Interface

Polymeric Fiber-Mortar Interface

During mixing of nylon FRC, cement particles can penetrate into the rough surface of the nylon fiber to form a hydrogen bond. The hardened mortar can either tear away some of the nylon surface or spall from the matrix and adhere to the fiber. The low asperity polypropylene fiber has less peeling because of the weaker van der Waal's bonds at the interface. After total debonding, the stiffer mortar surface causes surface peeling in both nylon fiber (Fig. 12(a)) and polypropylene fiber (Fig. 13(a)). For both nylon and polypropylene fiber FRC's, the debonding strength is much lower than the interfacial friction and the friction play a more important role.

When the fiber is pulled under compression over a long distance (10 mm), severe peeling and shaving are observed on the nylon fiber surface (Fig. 12(b)) and the black holes are observed on the polypropylene surface (Fig. 13(b)). These black holes are similar to those observed on a thin polymer film failed in tension. The polypropylene fiber surface is subjected to shear load by the mortar matrix (surface stretch microscopically) during pullout. Under the low pulling rate and the long pullout distance, the fiber surface undergoes significant molecular alignment and creeping which results in a pseudo volumetric increase. In order to maintain a constant volume, small holes are generated by the Poisson's effect, i.e., the surface ligaments get thinner and break into small holes.

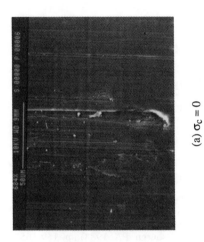

(a) $\sigma_c = 0$

(b) $\sigma_c = 20$ MPa

Figure 12. Nylon Surface

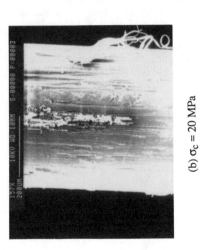

(a) $\sigma_c = 0$

(b) $\sigma_c = 20$ MPa

Figure 13. Polypropylene Surface

Since nylon fiber is hydrophilic, water may penetrate into the nylon surface and causes fiber swelling. The increase of the fiber radius due to peeling and swelling causes a significantly higher interfacial compression and increases the post-debonding load over sliding distance. On the other hand, the polypropylene fiber is hydrophobic. The surface is plowed with long whiskers and the long scratch lines. The scratch lines are similar to those on steel fiber surface caused by surface hardening. The lower surface asperity of the polypropylene fiber results in the post-debonding decrease and randomness of pullout load.

CONCLUSIONS

The microstructural features of the steel, nylon and polypropylene fiber-mortar interfaces during fiber debonding and pullout are studied. The damage at the steel fiber-mortar interface includes mortar abrasion. The damage at the nylon and polypropylene fiber-mortar interfaces is fiber surface peeling and the mortar surface experiences very little damage. By applying lateral compression to the mortar during fiber pullout, the abrasion effect becomes more severe for steel fibers. In both nylon and polypropylene, peeling becomes more significant and black holes can occur on the polypropylene surface.

REFERENCE

1. Naaman, A.E. and Shah, S.P. in ASCE J. of Structural Division, 102, ST8, pp.1537-1548 (1976)
2. C.K.Y. Leung, and Y. Geng, to appear in ACI Special Publication on Interfaces in Cementitious Materials, Edited by O. Buyukozturk and M. Wecheratana (1994)
3. Y. Wang, V.C. Li, and S. Backer in Bonding in Cementitious Composites, edited by S. Mindess and S.P. Shah, (Mater. Res. Soc. Proc. 114, Pittsburgh, PA, 1988) pp. 159-165
4. S. Mindess in Materials Science of Concrete I, Edited by Jan P. Skalny (The American Ceramic Society, Inc., Westerville, OH, 1989) pp. 163-180
5. A. Bentur, S. Diamond, and S. Mindess, J. of Mater. Sci., 20, 3610 (1985)
6. D.J. Pinchin, and D. Tabor, J. of Mater. Sci., 13, 1261 (1978)
7. I. Soroka, and P.J. Sereda, Proceedings of the 5th International Symposium on the Chemistry of Cement, (3, Tokyo, 1968) pp. 67-73
8. R. Baggott, and D. Gandhi, J. of Mater. Sci., 16, 65 (1981)
9. P.N. Balaguru, and S.P. Shah, Fiber Reinforced Cement Composites, McGraw-Hill, Inc, New York (1992) 108-109

BOND PROPERTIES OF MICRO-FIBERS IN CEMENTITIOUS MATRIX

Amnon Katz[1] and Victor C. Li
Advanced Civil Engineering Materials Research Laboratory
Department of Civil and Environmental Engineering
University of Michigan, MI 48105-2125, U.S.A.

ABSTRACT

Using a new technique for pullout test of microfibers, the interfacial bond properties of two carbon fibers having a diameter of 10 μm and 46 μm were tested and compared with those of high modulus polyethylene and steel fibers having a diameter of 42 μm and 18 μm, respectively. The fibers were embedded in cement matrices of different water to binder ratio and silica-fume content.

By studying the complete pullout load-pullout displacement curve and by using Environmental Scanning Electron Microscope (ESEM), it was found that the bond mechanism is mainly of friction for the fine and smooth carbon and polyethylene fibers. For the large carbon fiber and steel fiber, mechanical anchorage is the main bond mechanism.

INTRODUCTION

Interfacial fiber-matrix properties serve as an important parameter with the use of fibers in composite materials. Most composite models are based on the pullout behavior of a single fiber, assuming a constant frictional bond between the fiber and the matrix, independent of slip. P-u (pullout load vs. displacement) relationship was developed by Li [1], assuming constant friction (Fig. 1).

The pre-peak P-u relationship typically shows a non-linear behavior due to the debonding process and fiber stretching with increasing load, while the post-peak relationship shows a linear decrease as the embedded length of the fiber becomes shorter when the fiber is pulled out (line a in Fig. 1).

However, studies done by Wang et al. [2] on Nylon fibers and Naaman et al. [3] on steel fibers showed slip-hardening or slip-weakening behavior, respectively, as the fiber-matrix interaction is changed due to damage of the interface, associated with the relative stiffness of fiber and matrix. For stiff matrix and relatively weak fiber it might be slip-hardening behavior (line c in Fig. 1), due to peeling of the fiber by the stiff matrix (Wang et al. [2]). For a stiff fiber and a relatively weak matrix it might be slip-weakening behavior (line b in Fig. 1), due to rapid destruction of the matrix around the fiber as it is pulled out (Naaman et al. [3]).

---

[1]Present address: National Building Research Institute, Technion - Israel Institute of Technology, Haifa 32000, Israel.

Mat. Res. Soc. Symp. Proc. Vol. 370 © 1995 Materials Research Society

In this work, the bond properties of micro-fibers were studied by using the direct pullout technique developed by Katz and Li [4] and by using an Environmental Scanning Electron Microscope (ESEM) for different matrix compositions and micro-fiber types, in order to determine the pullout behavior of this kind of fibers for different matrices.

EXPERIMENTAL

Four micro-fiber types were tested in the experimental program: Fiber A - high modulus polyethylene fiber ('Spectra 900') having a diameter of d=42 μm and modulus of elasticity of E_f=120 GPa; Fiber B - steel fiber, d=18μm and E_f=210 GPa; Fiber C - carbon fiber, d=10 μm and E_f=240 GPa; and Fiber D - carbon fiber, d=46 μm and E_f=175 GPa. Two water to binder (cementitious materials) ratios were tested: low (w/b=0.35) and high (w/b=0.50). The effect of silica-fume, which improves the density of the matrix and possibly also the interface, was also studied in the experimental program. The composition of the different mixes is listed in Table 1. All fiber types were tested for all the matrices in Table 1 at the age of 14 days.

Test procedure

Direct pullout testing of a fine brittle fiber, such as carbon fiber, from a cementitious matrix, is difficult to carry out as fiber orientation is changed due to local bending while casting or the fiber breaks during the pullout test due to local bending of the fiber close to the matrix or the gripper of the pullout device.

A new testing procedure, developed by Katz and Li [4], enables the preparation of specimens with high accuracy in fiber orientation and better control of the matrix around the fiber. Other test technique currently known (Wang et al. [5] and Larson et al. [6]), were found to be inadequate for obtaining the complete pullout behavior of micro-fibers, as they determine only the critical embedment length of the tested fiber. Thus a special pullout test technique had to be developed.

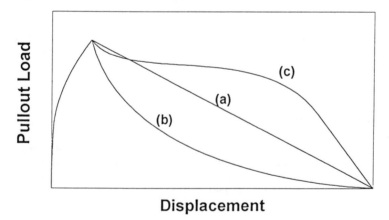

Fig. 1. Typical P-u curve based on friction (a), together with the effect of slip-weakening (b) and slip-hardening (c).

	Mix 1 0.35 LD	Mix 2 0.35 HD	Mix 3 0.5 LD	Mix 4 0.5 HD
Portland cement type III	200 gr	180 gr	200 gr	180 gr
Water	70 gr	46 gr	100 gr	76 gr
Silica-fume (slurry, 51% solids)	0 gr	41 gr	0 gr	41 gr
Superplasticizer	0 gr	4 gr	0 gr	2 gr
Water/binder	0.35	0.35	0.50	0.50
Silica-fume/binder	0	0.10	0	0.10

By using this procedure, a specimen with 5-15 fibers is prepared, as can be seen in Fig. 2a. The specimen is then sawn along the dashed line to provide a small specimen of a desired matrix thickness, L (Fig. 2b). Using a sensitive load-cell a pullout test is then performed, yielding the complete P-u curve of the fiber.

Bond strength was calculated according to equation 1, assuming a frictional bonding mechanism, and that the fiber fully debonds when the peak load P_{max} is achieved prior to slippage.

$$\tau = \frac{P_{max}}{\pi d L}$$ (eq. 1)

where: P_{max} - maximum pullout load
d - fiber diameter
L - fiber embedded length

The fiber embedded length, L, was made as large as possible in order to allow a complete pullout of the fiber, without rupture. To improve the accuracy the actual fiber diameter was measured under the microscope for each individual fiber.

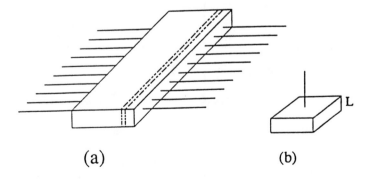

(a)   (b)

Fig. 2: Specimen after casting (a) and cutting (b)

TEST RESULTS

Pullout load - displacement curves

Typical pullout load-displacement curves of the various fibers are presented in Fig. 3. The dotted line in the figure represents the total embedded length of the fiber in the matrix.

The curves in Fig. 3 represent an average behavior of pullout of the fibers. Changes in the post peak behavior were observed for different matrices, as will be explained in the following:

i) **Fiber A** - high modulus polyethylene (Fig. 3a). Slip-hardening was observed for all the matrices, with the slip-hardening becoming greater for the denser matrices.

ii) **Fiber B** - steel (Fig. 3b). All fibers failed in tension prior to pullout. This was characteristic for all the matrices and even for short embedded length of 2-3 mm. The fracture load was also very low, much lower than the expected rupture load of such a fiber.

iii) **Fiber C** - carbon, 10 μm (Fig. 3c). The pullout load-displacement behavior of the fine carbon fiber was similar for the different matrices. For most cases, the decay of the load was linear, down to zero at complete pullout. For some cases of weak matrix, rapid decay was observed and the load was reduced down to zero slightly before complete pullout was achieved.

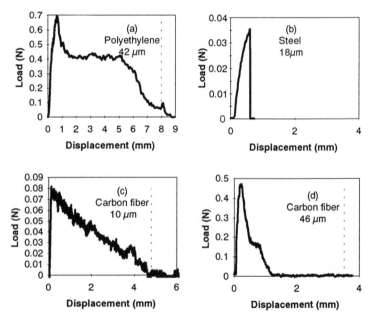

Fig. 3: Pullout load vs. displacement for: a. polyethylene fiber (Fiber A), b. steel fiber (Fiber B), c. thin carbon fiber (Fiber C), d. thick carbon fiber (Fiber D).

iv) **Fiber D** - carbon, 46 μm (Fig. 3d). The load decreased rapidly to zero after the maximum pullout load was achieved, much before complete pullout was reached. For the weak matrices, the decrease in load was moderated while more rapid decrease was observed for the denser matrices.

Bond Strength

Bond strength results from the pullout tests are presented in Table 2 and Fig. 4. The numbers in parentheses in Table 2 are the coefficients of variation. Test results are based usually on 6 specimens, but in some instances, where large coefficients of variation were observed, up to 20 specimens were tested. For Fiber B (steel fiber), all fibers failed in tension, regardless of matrix composition. Therefore, its bond properties could not be determined. In addition, the special surface structure of Fiber B (as will be discussed later in the paper) prevented the measurement of the true cross-section area of the fiber. Thus, an estimation of a minimum value for the bond strength, basing on the maximum load achieved, could not be obtained.

The bond strength values of fibers A and C are quite similar to those measured by Li et al. [7] for high modulus polyethylene fiber, and estimated by Katz and Bentur [8] for thin carbon fibers, confirming the reliability of the new testing procedure.

Table 2: Results of bond strength from pullout test

	Fiber A (Polyethylene)	Fiber C (Carbon 10 μm)	Fiber D (Carbon 46 μm)
0.35 LD	0.56 MPa (28%)	0.80 MPa (14%)	0.39 MPa (73%)
0.35 HD	0.61 MPa (39%)	1.29 MPa (14%)	>3.02 MPa* (20%)
0.50 LD	0.40 MPa (19%)	0.52 MPa (31%)	0.52 MPa (49%)
0.50 HD	0.63 MPa (19%)	0.66 MPa (16%)	>2.44 MPa* (23%)

* The results represent a minimum value only.

Effect of w/b ratio: Lowering the w/b ratio usually resulted in a better bond. The increase was greater for the small diameter carbon fibers. For this kind of fibers, lowering the w/b ratio from 0.5 to 0.35 resulted in an increase of 54% and 94% in the bond strength for matrices with no silica fume, and 10% silica fume, respectively. For the polyethylene fibers the changes were more moderate, and could not be identified clearly for the large diameter carbon fibers due to the high scattering of the results.

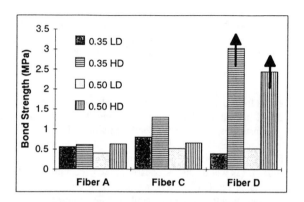

Fig. 4: Bond strength of fibers A, C and D for different matrices

Effect of silica fume: Introducing silica-fume into the matrix improved the bond strength of all the fibers. The strongest effect was on the bond strength of the thick carbon fibers (Fiber D) where an increase of 770% and 470% in the bond strength was observed for the low and high w/b ratio, respectively. When testing this type of fiber, embedded in matrices containing 10% of silica fume, many fibers broke in tension before pullout was achieved, and therefore the values presented in Table 2 are a minimum value only, calculated based on the results of the fibers which successfully pulled out and those which broke at the minimum embedded length.

Environmental Scanning Electron Microscopy (ESEM)
Typical ESEM photomicrographs of the fibers pulled from the cement matrix are presented in Figs. 5a through 5d for Fibers A to D, respectively. Fiber B looks more like a plexus of very fine filaments while the other fibers look like a solid single fiber. Special attention should be given to long grooves along the thick carbon fiber D. The grooves are relatively narrow (<4 μm) and partially filled with hydration products, marked by the arrow in Fig. 4d. Unfortunately, good quality observation on the surface of Fiber A, at the part which was embedded in the matrix, could not be achieved due to curling of the fiber. However, squashing marks caused by rigid particles could be seen on the surface of this type of fiber.

DISCUSSION AND CONCLUSIONS

The results of the pullout tests and ESEM observations indicate a bond mechanism based on friction and mechanical anchorage, with changes in the interface as the pullout process proceeds, depending on the matrix composition.

Fiber A and Fiber C (high modulus polyethylene and small diameter carbon) are very smooth fibers (Figs. 5a and 5c). However, the carbon fiber is much harder than the polyethylene fiber. The mechanism of bond, for the two fibers, is more of mechanical friction, with more damage to

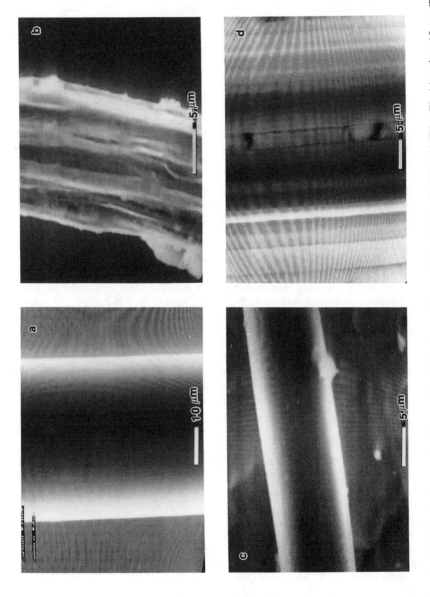

Fig. 5: ESEM photomicrograph of the surface of the fibers; (a) polyethylene (Fiber A), (b) steel (Fiber B), (c) carbon 10 μm (Fiber C) and (d) carbon 46 μm (Fiber D).

the polymeric fiber as the matrix becomes denser and stronger. Crumbs from the dense matrix trap at the fiber-matrix interface and squash into the fiber as it slips out, leading to a slip hardening behavior. The denser matrices (low w/b or matrices containing silica-fume) led to more damage to the fiber resulting in more pronounced slip-hardening behavior of this fiber.

Fiber C, which is denser than Fiber A, did not suffer from any damage to its surface. The linear decrease of the load after maximum pullout load is a characteristic of uniform friction mechanism of pullout.

Fibers B (steel) and D (thick carbon) show a characteristic behavior of a mechanical anchorage. It is possible that the structure of the steel fiber, which was made from a plexus of very fine filaments with an open structure (see Fig. 5b), allows the penetration of hydration products in-between the individual filaments resulting in good anchorage of the fiber in the matrix. This is probably the reason for the rupture of all the fibers of type B regardless of the cement matrix composition, and for the low loads needed for rupture.

For Fiber D, it seems that penetration of cement products into the long grooves observed on the fiber surface plays an important role in the mechanical connection between the fiber and matrix. Good penetration of the matrix into these grooves can provide a good mechanical anchoring between the fiber and the matrix. The drop of the pullout load after the peak, observed for this type of fiber, can be explained by breakage of the matrix penetrated into the grooves at the fiber surface, which leads to a severe reduction in the loads.

When using matrices of plain cement, with cement grain dimension comparable to or larger than groove width, the penetration of the matrix into the small grooves is poor, due to size effect and probably also due to the formation of a porous transition zone around the fiber, as shown for larger steel fibers (Bentur et al. [9], Wei et al. [10]) and aggregates (Goldman and Bentur [11]). This leads to poor and inhomogeneous mechanical bond, as can be seen in Table 2 by the low bond value and high coefficient of variation. In contrast, the use of ultra-fine particles such as silica fume (size <0.1 μm) is expected to cause densification of the transition zone as well as good penetration into the grooves. These effects may explain the remarkable increase in bond strength when silica fume is used with Fiber D, described earlier.

ACKNOWLEDGMENTS

This research was partially funded by a grant to the ACE-MRL from Conoco Inc. which also manufactured Fiber D especially for this research.

REFERENCES

1. Li, V.C. ASCE Journal of Materials in Civil Engineering, **4**, (1), 41-57 (1992).
2. Wang, Y., Li, V.C. and Backer, S. The International Journal of Cement Composites and Lightweight Concrete, **10**, (3),143-149 (1988).
3. Naaman, A.E. and Shah, S.P. Journal of the Structural Division, American Society of Civil Engineering, **102**, (ST8), 1537-1548 (1976).

4. Katz, A. and Li, V.C. Submitted for publication in the Journal of Materials Science Letters (1994).
5. Wang, Y., Backer, S. and Li, V.C. Journal of Materials Science Letters, 7, 842-844 (1988).
6. Larson, B.K., Drzal, L.T. and Sorousian, P. Composites, **21** (3), 205-215 (1990).
7. Li, V.C., Wu, H.C. and Chan, Y.W. "Advanced Technology on Design and Fabrication of Composite Materials and Structures", Ed. Carpinteri and Sih (1993).
8. Katz, A. and Bentur. Accepted for publication in the Advanced Cement Base Materials Journal (1994).
9. Bentur, A., Diamond, S. and Mindess, S., Journal of Materials Science, **20**, 3610-3620 (1985).
10. Wei, S., Mandel, J.A. and Said, S. ACI Materials Journal, **83**, 597-605 (1986).
11. Goldman, A. and Bentur, B. Cement and Concrete Research, **23** (4), 962-972 (1993).

BOND-SLIP MECHANISMS IN STEEL MICRO-FIBER REINFORCED CEMENT COMPOSITES

N. BANTHIA[1], N. YAN[1], C. CHAN[1], C. YAN[1] and A. BENTUR[2]
[1]Dept. of Civil Eng., University of British Columbia, Vancouver, B.C., Canada V6T 1Z4
[2]National Building Research Institute, Technion, Israel Institute of Technology, Technion City, Haifa, 32000 Israel

ABSTRACT

Bond-slip characteristics for steel micro-fibers bonded in cement-based matrices were investigated by conducting single fiber pull-out tests. The influence of the following factors was investigated: fiber inclination, fiber size, fiber embedded length and matrix refinement using silica fume. It was found that the bond-slip characteristics of fibers aligned with respect to the loading direction were necessarily superior than those inclined at an angle. Inclined fibers supported smaller peak pull-out loads and absorbed lesser pull-out energies than the aligned fibers. The use of silica fume in the matrix was found to improve both the *average interfacial bond strength* and the *maximum interfacial bond strength* between the fiber and the matrix.

Keywords: fiber reinforced cements, steel micro-fibers, bond-slip, pull-out tests, inclined fibers

INTRODUCTION

When a strong and ductile fiber such as steel is dispersed in a brittle cement-based matrix, significant changes in the microfracture processes can be expected. Cracks in the brittle matrix are bridged by the ductile fibers, which, in turn, undergo pull-out processes, and result in macroscopic fracture properties significantly superior to those of the parent matrix. As in any other composite, the influence of fiber-matrix bond-slip behavior on the properties of cement-based composites is fundamental.

The effectiveness of a given fiber as a medium of stress transfer is often assessed using a single fiber pull-out test where fiber slip is monitored as a function of the applied fiber load. And while some argue that single fiber pull-out data does not correlate well with the behavior of the real composite (1,2), fiber pull-out tests are routinely conducted as a means of optimizing fiber and matrix characteristics and to understand toughening mechanisms in these composites.

Pull-out characteristics of steel fibers embedded in cementitious matrices have been studied as a function of several variables including the rate of load application (3,4), temperature of the environment (5), processing variables and matrix quality (6,7,8) and other test variables such as fiber inclination, etc. (9). In addition, a number of matrix and fiber modifications have been examined as a way of improving the bond-slip characteristics of fibers in a cementitious matrix. These include matrix modifications such as silica fume and polymer additions (10, 11, 12) and fiber surface modifications such as coatings, surface indenting and notching (13, 14, 15).

For large fibers of steel (often called macro-fibers), the most effective means of improving the pull-out bond-slip characteristics is "mechanical deforming". Mechanical deformations in the form of a hook, cone or crimp placed at the end or along the fiber length to provide positive end-anchorage in concrete have proven to be very effective in improving the pull-out resistance (16, 17,

539

Mat. Res. Soc. Symp. Proc. Vol. 370 ® 1995 Materials Research Society

18). In steel micro-fiber reinforced composites, on the other hand, fibers are very fine with diameters in the micron range such that the critical lengths are very small and mechanical deformations are unnecessary and impractical. Limited data exist in the literature with regards to the bond-slip characteristics of steel micro-fibers.

EXPERIMENTAL PROGRAM

Pull-Out Specimens

Pull-out tests were performed using rectangular specimens shown in Figure 1. A very thin plastic separator with a hole of the size of the micro-fiber itself helped keep the fiber in place during casting and provided an artificial crack by disallowing the two parts of the specimen A and B to bond. Fibres were aligned at desired angles with respect to the loading direction under a magnifying glass using a specially prepared protractor. Although the micro-fibers had a tendency to curl, only the straight portions of the fibers were used.

Matrices and Fibres

Two cement-based matrices M1 (water:cement:silica-fume = 0.35:1:0.10) and M2 water:cement:silica-fume = 0.35:1.0:0.20) were investigated. Appropriate quantities of superplasticizer were added to both matrices. Stainless steel micro-fibers (composition: Si = 2.04%, Cr = 16.76%, Mn = 0.33%, Fe = 79.95% and Mo = 0.92%) with three rectangular sections: 40 μm x 100 μm; 60 μm x 120 μm and 80 μm x 220 μm were investigated. Fibers were cut to desired lengths from continuous spools produced by the process of mechanical shaving which led to non-uniform fiber surfaces with significant roughness as seen in Figure 2.

Pull-out Tests

Pull-out tests were carried out in a specially designed table mounted test apparatus shown in Figure 3. In a typical test, the pull-out specimen (Figure 1) was held between two grips G_1 and G_2 using two vertical screws S_1 and S_2. Grip G_1 is fixed while G_2 is mounted on rollers such that it could be moved in the horizontal direction by a small motor. An LVDT monitored the displacement of grip G_2 and a load cell (100 N capacity) measured the applied load. In a pull-out test, G_2 was pulled away from G_1 and the applied load vs. fiber slip curves were recorded. The values reported here were the average of at least seven pull-out curves with the coefficients of variations being in the range of 20 to 35%.

Three series of tests were conducted to investigate the following variables:
- Influence of fiber angle: Fibers, 60 μm x 120 μm in section, were embedded to a depth of 3 mm in matrix M1 at 0, 15, 30, 45, 60 and 75 degrees with respect to the loading direction and pull-out bond-slip curves were obtained.
- Influence of matrix refinement: Bond-slip curves were obtained for the fiber 60 μm x 120 μm in section when bonded in two matrices M1 (10% silica fume) and M2 (20% silica fume) to an embedment length of 2 mm.
- Influence of fiber embedment length: This was investigated for all three fiber sections. Fibers were embedded in Matrix M2 to lengths of 2 and 3 mm and bond-slip curves were obtained.

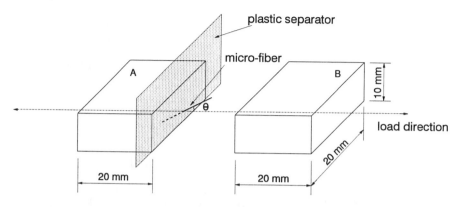

Figure 1.   Pull-out Test Specimen.

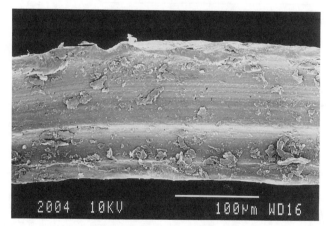

Figure 2.   Micro-Steel Fiber Investigated. Notice the Rough Surface of the Fiber.

RESULTS AND DISCUSSION

Influence of Fiber Inclination Angle

Representative bond-slip curves for the fiber with a section of 60 μm x 120 μm are given in Figure 4 for the various inclining angles with respect to the load direction. The data are analyzed in Figure 5 as a function of fiber angles. Notice that a fiber aligned with respect to the load direction ($\theta = 0°$) supported the highest pull-out load. Also, the energy absorbed to the peak pull-out load and the total pull-out energy (to a slip of 3 mm) were the highest for the aligned fiber. Interestingly, the measured slip at the peak load was also the lowest for the aligned fiber.

The observations here that the peak load and pull-out energy are maximum for the 0° inclination angle are not in accordance with the findings and theoretical analysis of ductile fibers in

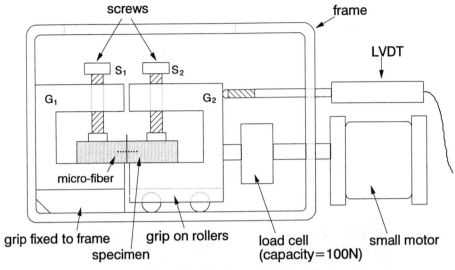

Figure 3. Test Set-up for Pull-out Tests.

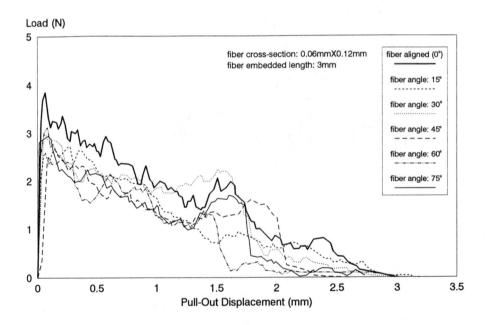

Figure 4. Bond-Slip Curves at Various Inclining Angles.

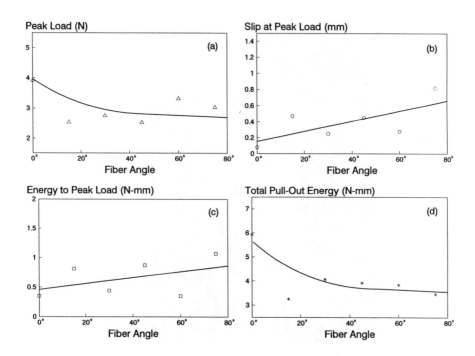

Figure 5.   Effect of Fiber Orientation on: (a) Peak Pull-out Load; (b) Slip at Peak Load; (c) Energy Up to Peak Load; (d) Total Pull-out Energy.

Table 1.   Fiber-Matrix Interfacial Bond Strength Values.

Matrix	Cross-section $(\mu m \times \mu m)$	Embedded Length ℓ_e (mm)	Average Bond Strength (MPa) (τ_{av})	Maximum Bond Strength (MPa) (τ_{max})
M1 (20% S.F.)	80×220	3	4.53	6.18
		5	3.23	
	40 × 100	2	4.60	7.10
		3	3.33	
	60 × 120	2	4.38	5.66
		3	3.73	
M2 (10% S.F.)	60 × 120	2	3.98	4.73
		3	3.60	

a brittle matrix (9,19,20,21). These studies have shown either a maximum in the value at an angle inbetween 0 to 45°, or at least a mild change only in the pull-out values at this range of orientation. The explanations given were based on the bending generated in the fiber emerging from the matrix, due to geometrical constraints induced as the crack over which the fiber is bridging is being opened in direction inclined to the fiber. In the case of a ductile fiber, the energy consumed in the bending

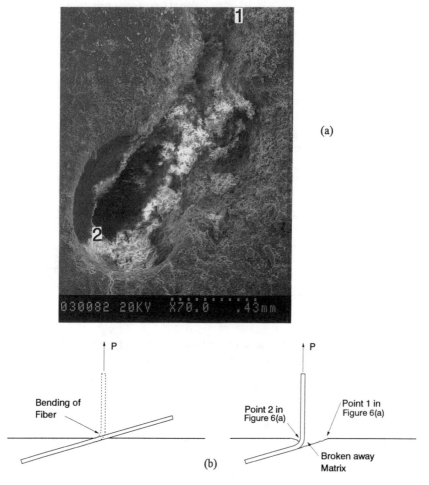

Figure 6. Observation of the Mode of Pull-Out Failure of an Inclined Fiber: (a) SEM Micrograph of the Surface; (b) Schematic Description of the Failure Mode Showing the Bending of the Fibre (i) Followed by Splitting and Breaking Away of Some of the Matrix (ii).

of the fiber and crushing of the matrix around it provide mechanisms which enhance the pull-out resistance. This is in contrast with a brittle fiber which cannot accommodate the large bending strains and breaks prematurely under such conditions (22). For the behavior of the steel micro-fiber reported here, which is ductile but showed a sharp decline in pull-out resistance in the oriented fibers, another mechanism may have to be considered: The bending stress, particularly when the fiber surface is rough, may lead to splitting in the matrix such that part of it breaks away from the fiber and loosens the fiber anchoring and impedes its capacity in the region of the bond. This is shown in the photograph in Fig. 6a and explained schematically in Fig. 6b. It can be seen that the exposed length of the fiber is about 0.8 mm which is a large portion of the 3 mm embedded length. In longer fibers (which is the case in large diameter fiber of the same aspect ratio of the micro-fiber) the break-away portion may be relatively small.

Influence of Matrix Refinement

Pull-out curves for aligned fibers (section = 60 μm x 120 μm) in the two matrices M1 (10% silica fume) and M2 (20% silica fume) are shown in Figure 7. Notice the improved pull-out resistance for the fiber in a matrix with a higher percentage of silica fume. This is well anticipated since a higher percentage of silica fume in the matrix is expected to lead to better packing and densification at the interface resulting in a higher bond.

Influence of Embedded Length

For a given fiber under a pull-out load, the average value of interfacial bond strength τ_{av} is given by:

$$\tau_{av} = \frac{P_u}{\pi \, d_f \, l_e} \tag{1}$$

where P_u is the peak pull-out load, d_f is the fiber diameter and l_e is the embedded length. For the three fiber sizes tested, values of t_{av} for the various values of l_e are tabulated in Table 1. Notice, as well expected, that there is a decrease in the average interfacial bond strength τ_{av} with an increase in the fiber embedded length.

Based on the shear lag model (23) with the assumption that the extensional stresses in the matrix are negligible relative to those in the fiber and shear stresses in the fiber are small compared to those in the matrix, it can be shown that the maximum shear stress occurs at the point where the fiber enters the matrix and its magnitude is given by:

$$\tau_{max} = \frac{P_u}{\pi \, d_f \, l_e} \{\alpha_1 l_e \coth \alpha_1 l_e\} \tag{2}$$

where,

$$\alpha_1 = \sqrt{\frac{4 \, G_i}{b_i \, d_f \, E_f}} \tag{3}$$

with G_i and b_i as the shear modulus and effective thickness of the interface, respectively. Using Eqn. 1, Eqn. 2 can be written as:

$$\tau_{max} = \tau_{av} \{\alpha_1 l_e \coth \alpha_1 l_e\} \tag{4}$$

As ℓ_e approaches zero, the function $\alpha_1 l_e \coth \alpha_1 l_e$ approaches unity and τ_{max} approaches τ_{av}. In other words, a given τ_{av} vs ℓ_e curve can be extrapolated to $\ell_e = 0$ and the resulting intercept on the τ_{av} axis must equal the maximum interfacial shear strength, τ_{max} (23).

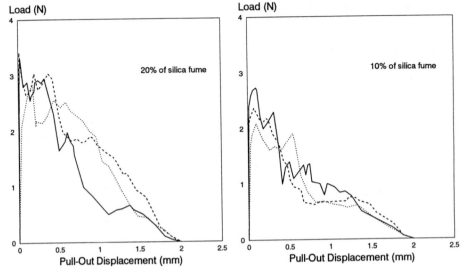

Figure 7. Influence of Silica Fume Content on the Bond-Slip Curves for an Embedment Length of 2 mm.

Such plots for the micro-fibers tested in this study are given in Figure 8. The resulting values of τ_{max} are given in Table 1. It can be seen that within permissible experimental variation, τ_{max} values for the various fiber sizes are reasonably similar, but increase with the silica fume content in the matrix.

CONCLUSIONS

1. For steel micro-fibers, the peak loads supported by fibers that are aligned in the direction of loading are higher than those supported by fibers inclined with respect to the loading direction. The peak loads for the aligned fibers also occur at smaller slips than the inclined fibers.

2. Even from the energy absorption point of view, a fiber aligned with respect to the loading direction absorbs greater amounts of energy at a certain slip than one that is inclined. A $0°$ inclination with respect to the loading direction is, therefore, the optimal inclination.

3. Both the average interfacial bond strength (τ_{av}) and the maximum interfacial bond strength (τ_{max}), increase with an increase in the content of silica fume in the matrix.

ACKNOWLEDGMENTS

The continued financial support of Natural Sciences and Engineering Research Council of Canada and Novocon International, Illinois, is greatly appreciated.

Interfacial Shear Strength (MPa)

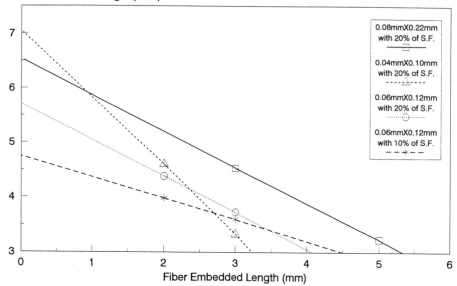

Figure 8.   Calculation of Maximum Interfacial Bond Strength (τ_{max}) Developed
Between Fiber and the Matrix.

REFERENCES

1.   Maage, M. Interaction between Steel Fibers and Cement-Based Matrices, Materials and Structures (RILEM), 10(59), 1977, pp. 297-301.
2.   Hughes, B.P. and Fattuhi, N.I., Fiber Bond Strengths in Cement and Concrete, Magazine of Concrete Research, 27 (92), 1975, pp. 161-166.
3.   Gokoz, Ulker, N., and Naaman, Antoine, A.E., Effect of Strain Rate on the Pull-out Behavior of Fibers in Mortar, International J. of Cement Composites and Lightweight Concrete, 3(3), 1981, pp. 187-202.
4.   Banthia, N. and Trottier, J.-F. Deformed Steel Fiber-Cementitious Matrix Bond under Impact, Cement and Concrete Research, 21(1), 1991, pp. 158-168.
5.   Banthia, N. and Trottier, J.-F., Micromechanics of Steel Fiber Pull-out, Rate Sensitivity at Very Low Temperatures, Cement and Concrete Composites, 14, 1992, pp. 119-130.
6.   Gray, R.J. and Johnston, C.D., The Effect of Matrix Composition of Fiber/Matrix Interfacial Bond Shear Strength in Fiber Reinforced Mortar, Cement and Concrete Research, 14, 1984, pp. 285-296.
7.   Banthia, N. and Trottier, J.-F., Effects of Curing Temperature and early Freezing on the Pull-out Resistance of Steel Fibers from a Cementitious Matrix, Cement and Concrete Research, 19(5), 1989, pp. 727-736.

8. Gopalaratnam, V.S. and Abu-Mathkour, H.J., Investigation of the Pull-out Characteristics of Steel Fibers from Mortar Matrices, Proc. International Symp. on Fiber Reinforced Concrete, Madras, India, 1987, pp. 2.201-2.211.

9. Naaman, Antoine E. and Shah, Surendra P., Pull-Out Mechanisms in Steel Fiber Reinforced Concrete, Jour. of the Structural Division, ASCE, Aug. 1976, pp. 1537-1548.

10. Banthia, N. "A Study of Some Factors Affecting the Fiber-Matrix Bond in Steel Fiber Reinforced Concrete", Can. Jour. of Civil Engineering, 17(4), 1990, pp. 610-620.

11. Wei, Sun, Mandel James A. and Said Samir, Study of the Interface Strength in Steel Fiber Reinforced Cement-Based Composites, ACI Jour., July-August 1986, pp. 597-605.

12. Igarashi, S. I. and Kawamura, M., Effects of the Addition of Silica Fume and Fine Aggregate on the Fracture Toughness for the Steel Fiber-Matrix Interfacial Zone, Proc. Fracture Processes in Concrete, Rock and Ceramics (Eds. J.G.M. van Mier, J.G. Rots and A. Bekker), E. & F. N. Spon, London, 1991, pp. 307-315.

13. Al Khalaf, M.N., Page, C.L. and Ritchie, A.G.B., Effects of Fiber Surface Composition on Mechanical Properties of Steel Fiber Reinforced Mortars, Cement and Concrete Research, 10(1), 1980, pp. 71-77.

14. Mayfield, Brian and Zelly, Brian M., Steel Fiber Treatment to Improve Bonds, Concrete, Concrete, 7(3), 1973, pp.35-37.

15. Tattersall, G.H., Urbanowicz, C.R., Bond Strength in Steel Fiber Reinforced Concrete, Magazine of Concrete Research, 26(87), 1974, pp. 105-113.

16. Naaman, Antoine E. and Najm, Husamuddin, N., Bond-slip Mechanisms of Steel Fibers in Concrete, ACI Materials Jour., March-April, 1992, pp. 135-145.

17. Rossi, P. and Chanvillard, G., A New Geometry of Steel Fiber for Fiber Reinforced Concrete, in High Performance Fiber Reinforced Cement Composites, (Eds. Reinhardt, H.W. and Naaman, A.E., RILEM, E. & F.N. SPON, London, pp. 129-139.

18. Chanvillard, G. and Aitcin, P.-C., Micromechanical Modeling of the Pull-out Behavior of Wiredrawn Steel Fibers from Cementitious Matrices, in Fiber Reinforced Cementitious Materials (Eds., S, Mindess and J. Skalny), Vol. 211, Materials Research Society, 1991, pp. 197-202.

19. Morton, J. and Groves, G.W., The Cracking of Composites Consisting of Discontinuous Ductile Fibers in a Brittle Matrix- Effect of Fiber Orientation, J. of Materials Science, 9, 1974, 1436-1445.

20. Li. V.C., Wang, Y. and Backer, S., Effect of Inclining Angle, Bundling and Surface Treatment on Synthetic Fiber Pull-out from a Cement Matrix, Composites, 21(2), 1990, pp. 132-140.

21. Brandt, A. M., On the Optimal Direction of Short Metal Fibers in Brittle matrix Composites, J. of Materials Science, 20, 1985, pp. 3831-3841.

22. Aveston, J., Mercer, R.A. and Sillwood, J.M., Fiber Reinforced Cements: Scientific Foundations for Specifications, Composites Standards Testing and Design, Proc. National Physical Laboratory Conference, U.K. 1974, pp. 95-103.

23. Gray, R.J., Analysis of the Effect of Embedded Fiber Length on Fiber Debonding and Pull-out from an Elastic Matrix, Part 1: Review of Theories, J. of Materials Science, 19, 1984, pp. 861-870.

BONDING AND INTERFACIAL MICROSTRUCTURE   IN CEMENTITIOUS
MATRICES REINFORCED BY WOVEN FABRIC

A.Peled, D.Yankelevsky and A.Bentur, National Building Research Institute,
Faculty of Civil Engineering, Technion, Israel Institute of Technology, Haifa, Israel

ABSTRACT

High performance cementitious composites can be produced by reinforcement with a high
volume of aligned fibers. One practical method of production of such composites would
be based on the use of woven fabrics, where the bonding may be different from that
predicted from evaluation of continuos aligned fibers. The present paper presents the
study of the pull-out behavior of a fabric from a cementitious matrix. The influence of
various fabric parameters were evaluated by comparing the pull-out of straight yarns
(fibers), crimped yarns untied from the fabric and the fabric itself. The influence of initial
tensioning of the yarns, which is an essential step in the production process, was also
evaluated. SEM observations were carried out to interpret the pull-out curves in terms of
microstructural characteristics. The crimped nature of the yarn was found to be an
important factor in enhancing the bond of the fabric, as it generated anchoring effects.
Tensioning of the yarn was detrimental to bond as it straightened the yarn and caused
Poisson effects which damaged the interface.

INTRODUCTION

There is considerable interest in recent years to develop high performance fiber reinforced
cementitious composites which exhibit strain hardening in the post cracking zone,
resulting in very high tensile strength and toughness. Several studies reported in this area
have been based on experimental evaluation of cementitious composites reinforced by
high volume (>5%) of aligned and continuos fibers to prove the potential of this
approach (1,2). However, for practical application of such concepts, to produce
composites in a technically viable process, there would be a need to explore procedures
other than hand lay up of individual longitudinal fibers. One alternative which is common
in the reinforced plastics industry is based on impregnation of woven textile fabrics, where
the fabric can be readily produced in a variety of geometries with different types of yarns
(fibers). Several studies reported the flexural properties of cementitious composites
reinforced with fabrics demonstrating that with this kind of reinforcement, strain
hardening and marked increases in flexural strength and toughness can be obtained, even if
low modulus polymeric fibers are being used (3,4,5). In a recent study indications were
obtained that the mechanics of such composites might be more complex than predicted on
the basis of simple composite materials approach assuming continuous and aligned fibers.
It was suggested that this may be due to a difference in the nature of bonding induced by
the special geometry of the yarns (fibers) in the textile. These indications were based on
evaluation of the flexural behavior of the composite. In order to resolve such influences a
systematic study of the nature of bonding between a woven fabric and a cementitious
matrix was initiated. The present paper reports the initial findings of this research pointing
out some of the mechanisms that may occur in such systems.

Mat. Res. Soc. Symp. Proc. Vol. 370 © 1995 Materials Research Society

Prior to presenting the testing program and the results it is essential to briefly discuss the structure of a woven fabric which is shown schematically in Figure 1. The basic unit in the fabric is a continuous yarn that can be made up of a single monofilament or several filaments bundled together. In the present paper the term yarn would be used rather than fiber, as the latter usually implies a short reinforcing unit. The density of the yarns in the warp (loading direction) and in the fill (weft) direction can be controlled and vary for the different directions. The warp and the fill yarns pass over and under each other, and thus the shape of each individual yarn can be described as a crimped geometry. Therefore they might be expected to be less efficient than straight, continuous yarns. However, in the case of cement reinforcement, where interfacial boding is limited, the crimped structure may provide enhanced bonding by an anchoring mechanism, which may compensate for the reduced load bearing capacity which is the result from the deviation from the straight geometry. Enhanced bonding is particularly important in such composites as it controls the multiple cracking process. The undulation and curvature of the woven yarn in the fabric depends on: (i) the spacing between yarns, (ii) the yarns' diameter, and (iii) the woven fabric structure, i.e. if the weave is such that the warp yarns passes over two or more fill yarns at a time.

EXPERIMENTAL

The fabrics used in this study were produced specially so that their geometry would be controlled to evaluate the influence of one parameter at a time. This was achieved with fabrics in which the density in the warp (loading) direction was kept the same, 22 yarns per cm, whereas in the fill direction the density varied between 0 (straight yarn, not in a fabric form) to 22 yarns per cm. In order to better resolve the interfacial interaction the yarns were all monofilament of polyethylene with 0.25mm diameter. The modulus of elasticity of the yarn was 892 MPa and its tensile strength 267 MPa.

The interfacial bond was evaluated by means of pull-out tests as shown schematically in Fig.2. The embedded length was 10 and 20mm. The specimens were prepared by a hand lay up technique using a matrix of 0.30 water/cement ratio paste with type I Portland cement. The curing was in water for a period of 7 days at which age the specimens were tested in saturated surface dry conditions. The pull-out test was carried out in an Instron testing machine at a crosshead rate of 0.25mm/minute and load-displacement curves were recorded. The of the curves for the two embedded length studied here were always found to be of the same shape, with the values in the curve of 20mm embedded length being almost twice as those of the 10mm embedded length. In the case of pull-out from the fabric embedded in matrix the bond strength was so high that yarn fracture occurred at 20mm embedded length. Thus, for this system only 10mm embedded length tests were recorded.

In view of the structure of the woven fabric discussed in the introduction, two influences should be considered: (i) the influence of the crimped nature of the yarn, (ii) the influence of the fills in providing anchoring to the pulled out yarn in the warp direction.

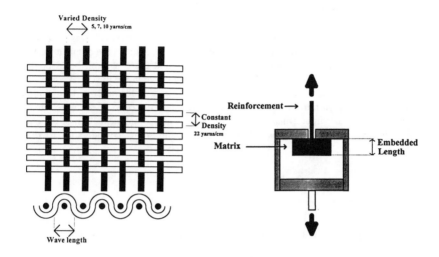

Figure1: Schematic description of the structure of the woven fabric.

Figure 2: Schematic description of of the pull-out test.

The first influence was resolved by pull-out testing of crimped yarns untied from the fabric, which were embedded in the in the matrix for the pull-out test and their performance was compared with that of a straight yarn. The second influence was evaluated by pull-out of a yarn from the fabric which was not embedded in paste matrix (to resolve the bonding in the fabric itself) and yarn from a fabric which was embedded in the matrix to resolve the combined effect of all the bonding mechanisms.

For the production of composite specimens of this kind there is a need to keep the yarns and fabrics in place by applying some tensile stress. As this stress may have an influence on the geometry of the crimped yarn, testing was carried out with specimens where different tensile loads were applied on the yarn while it was impregnated with the matrix: 10, 30 and 60 gm. The loads were maintained for 24 hours after casting, i.e. they were released after the matrix hardened. These loads are equivalent to stresses of about 2, 6 and 12MPa, i.e. 0.76, 2.29 and 4.58% of the strength of the yarn.

The present paper reports evaluation of fabrics and crimped yarns from a fabric with a density in the fill direction of 5 yarns per cm and 22 yarns per cm in the warp direction This fill density gave the composite with the highest flexural strength (5).

RESULTS AND DISCUSSION

Effect of Yarn Geometry

The load displacement curves of a straight yarn and a crimped yarn which was obtained by untying from the matrix are shown in Fig.3. The superior behavior of the crimped yarn can be clearly observed, both in enhancing the peak load and the post peak behavior. It is interesting to note the wavy nature of the crimped yarn curve in the post peak zone, with the "wave length" being in the range of 1 to 2 mm, which is of the order of magnitude of the density of the fill yarn from which the crimped fiber was untied (5 fill yarns per cm, i.e. one per mm). Another interesting feature is the almost horizontal inclination of the curve of the straight yarn in the post cracking zone, suggesting considerable frictional component.

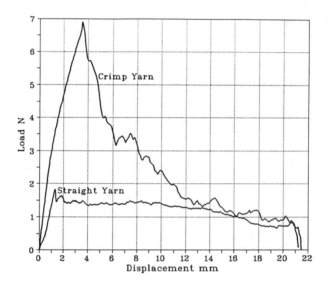

Figure 3: Pull-out load-displacement curves for crimped yarn (untied from the fabric) and straight yarn (20mm embedded length).

The improved bonding performance of the crimped fiber can be readily explained on the basis of a mechanical anchoring which is the result of the "wavy" nature in the fiber (Fig. 4a). The presence of such mechanism can be deduced also on the basis of observations of local damage in the matrix around the crimped zones (Fig.4b) created as the yarn is pulling out. This resistance also shows up in the yarn after pull out: roughening of the yarn surface with considerable local fibrillation which occurs in certain locations along the yarn (Fig.5).

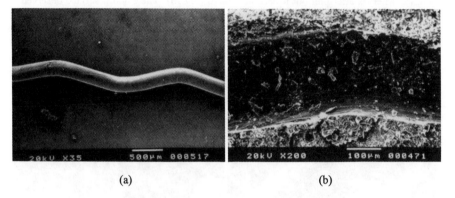

(a)   (b)

Figure 4:   The "wavy" nature of the crimped fiber (a) and its reflection in the shape of the
grove and local damage in the matrix around the crimp (b).

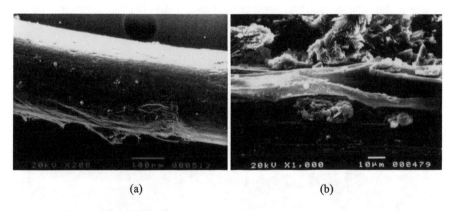

(a)   (b)

Figure 5:   Roughening and fibrillation along a crimped yarn as observed after pull-out.
(a) 200X (b) 1000X

The observation that in the straight yarn there is considerable frictional resistance in the post peak zone suggests that some anchoring mechanism is involved here, which is more than just interfacial friction. Observations prior to pull-out indicate that the yarn diameter is not uniform along its length (Fig. 6a) and this non uniformity is reflected in bulges in the matrix (Fig. 6b), which can provide an anchoring effect. Observations of the groove of the yarn in the matrix prior and after pull-out shows that initially the surface is smooth (Fig.7a) whereas after pull out it becomes rough (Fig.7b). The smooth surface observed initially is typical of the duplex film formed around an inclusion. After pull-out remnants of this film can be seen (Fig.7b) and its removal exposes underneath it the more porous matrix of the interfacial transition zone. The process of destruction of the duplex film can be ascribed to the bulges which occur along the straight yarn.

(a)   (b)

Figure 6: Non uniformity in the straight yarn diameter observed as bulges along the fiber (a) and in the matrix (b).

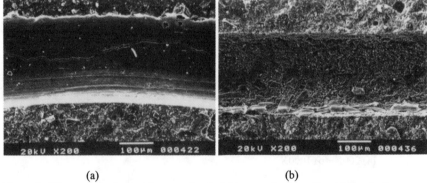

(a)   (b)

Figure 7: The grove in the matrix around a straight yarn prior to pull-out (a) and after pull-out (b) in composites produced by initial tensioning of 10gm.

Effect of the Initial Tensile Load in the Yarn

The effect of the initial load in the yarn, which is maintained for the first 24 hours during the production process, on the pull-out curve determined at 7 days is shown in Figs.8 and 9, for the crimped and straight yarns, respectively. In both systems the increase in the initial load resulted in a marked reduction in the pull-out performance. In the case of the crimped yarn, this can be explained on the basis of observations of the reduction in the waviness of the yarn as it is being stretched and straightened due to the increase in the initial load (photographs not shown here).

The above explanation can not apply for the straight yarn in which the bonding is not due to mechanical anchoring by crimps. SEM observations of the microstructure of the grooves prior to pull-out indicated considerable microstructural differences: in the 10gm initial tensioning, the groove surface is smooth as obtained by the duplex film (Fig.7a); in the 60 gm initial tensioning the duplex film is broken, exposing underneath the more porous material in the interfacial transition zone similar to the observation documented in Fig.7b for the 10gm system after loading. This difference in microstructure might be explained on the basis of the Poisson effect in the yarn (which can be quite significant in a low modulus polymeric material) : the release of the initial load at 24 hours occurs when the microstructural features are set, in particular the duplex film, but the supporting matrix around is still weak; the transverse expansion of the yarn due to the release of load, may cause damage to the surrounding microstructure. Apparently, such damage is evident in the system produced with 60gm initial loading, because of its greater Poisson effect and it may account for its poor bonding performance. This explanation can account for the marked reduction in the post peak performance of the systems at higher initial load (Figure 9); it would have been expected that it would also lead to reduction in the peak load. However, the peak load is maintained practically the same, suggesting that this portion of the bonding characteristic is controlled by another mechanism.

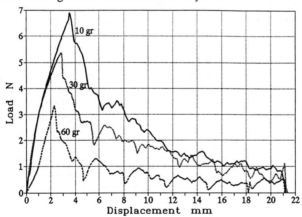

Figure 8: The influence of the initial load in the crimped yarn on the pull-out curve (20mm embedded length).

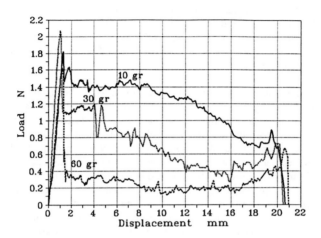

Figure 9: The influence of the initial load in the straight yarn on the pull-out curve (20mm embedded length).

<u>Pull-Out from Fabric</u>

The influence of the fabric on the pull-out behavior of the yarn is shown in Fig. 10. The pull-out of a yarn from the fabric itself (i.e. free fabric not embedded in a cementitious matrix) is met with some resistance which is the result of the interaction between the perpendicular weft and fill yarn at their zones of intersection. It can be seen that at the junctions some damage occurs in the pulled out weft yarn which is the result of its rubbing during movement against the fill yarn (Fig. 11). The resistance to pull out offered by this mechanism is considerably smaller than that obtained by the anchoring of a crimped fiber when it is pulled out from a cementitious matrix (Fig. 10). The yarn from the matrix and fabric curve shows values higher than the sum of the pull-out of the yarn from fabric and yarn from matrix. This implies a synergistic effect i.e. an additional mechanism on top of the mechanical anchoring of the crimped yarn and the resistance offered by the junction with the fill yarn. A possible additional mechanism is the fixation of the fill yarn when embedded in the matrix, in contrast with its flexibility in the free fabric. This hypothesis is the subject of additional research which is currently in progress.

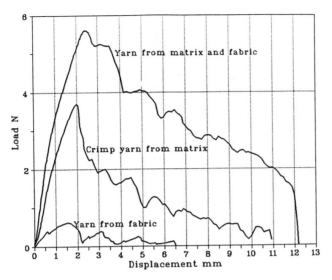

Figure 10: Pull-out curves of a yarn from a fabric not embedded in matrix, a crimped yarn from a cementitious matrix and a yarn from a fabric in a cementitious matrix (10mm embedded length).

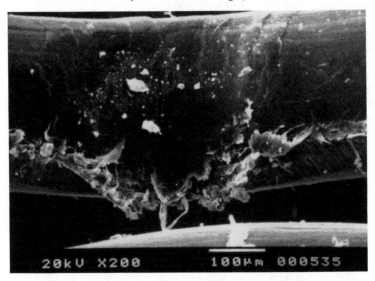

Figure 11: Damage to the surface of a weft yarn after the pull-out of the warp yarn over it, in a pull-out test of a warp yarn from a fabric not embedded in a matrix.

557

CONCLUSIONS

The results of the study presented here identified several mechanisms that can affect significantly the pull-out resistance in composites consisting of a fabric reinforcement and a cementitious matrix:

(a) The wavy nature of the crimped yarn in the woven fabric provides mechanical anchoring leading to a marked improvement in pull-out behavior compared to a straight yarn from the same material.

(b) The pull-out resistance of both, the straight and crimped yarn, decrease with increase in the tensile load applied initially during the production of the composite, which is required to hold the yarn in its position during the lay-up process. These influences are associated with straightening of the crimped fibers and with Poisson effect.

(c) The pull-out resistance of a yarn from a fabric in a matrix is superior to the sum of the resistance of the crimped yarn from the matrix and the yarn from a free fabric, suggesting a synergistic effect.

REFERENCES

1. S.P.Shah, Do Fibers Increase the Tensile Strength of Cement-Based Matrixes?, Amer. Concr. Inst. Materials J., 88, 595-602, 1991
2. B.Mobasher, A.Castro-Montero and S.P.Shah, Microcrack in Fiber Reinforced Concrete, Cem. Conc. Res., 20, 665-676, 1990.
3. R.N.Swamy and M.W.Hussin, pp.99-100 in Fiber Reinforced Cement and Concrete: Recent Developments, Eds. R.N.Swamy and B.Barr, Elsevier Applied Science, 1989.
4. R.N.Swamy and M.W.Hussin, pp.57-67 in Textile Composites in Building Construction, Proc. Int. Symp. Composite Materials with Textile Reinforcement for use in Building construction and Related Applications, Lyon, France 1990, Part 1, Eds. P.Hamelin and G.Verchey, Editions Pluralis: Paris, 1990.
5. A.Peled, A.Bentur and D.Yankelevsky, Woven Fabric Reinforcement of Cement Matrix, Advanced Cement Based Materials, 1, 216-223, 1994.

THE FIBER-MATRIX INTERFACE IN FIBER REINFORCED CONCRETE STUDIED BY CONTACT ELECTRICAL RESISTIVITY MEASUREMENT

XULI FU AND D.D.L. CHUNG
Composite Materials Research Laboratory, State University of New York at Buffalo, Buffalo, NY 14260-4400

ABSTRACT

The contact electrical resistivity of the fiber-matrix interface was found to correlate strongly with the shear debonding strength. In the case of the interface between stainless steel fiber and cement paste, the contact resistivity increased linearly with increasing debonding strength and interfacial phase(s) of volume resistivity higher than that of cement paste enhanced the bonding.

INTRODUCTION

The fiber-matrix interface is critical to the properties of fibrous composites. Interface evaluation has been conventionally conducted by single fiber pull-out testing [1], as this test provides a measure of the debonding shear strength. Detailed microstructural examination of the interface by transmission electron microscopy and surface analysis is also valuable, but it is tedious and does not provide a measure of the debonding strength. This paper provides a new method of interface evaluation. This method involves the measurement of the fiber-matrix contact electrical resistivity. The measurement is fast, sensitive and involves inexpensive equipment. It also provides an indirect measure of the debonding shear strength, due to the excellent correlation between the debonding shear strength and the contact resistivity for a given combination of fiber and matrix. The technique is demonstrated in this work by studying the interface between stainless steel fiber and cement paste (concrete without coarse or fine aggregate).

EXPERIMENTAL

The stainless steel fibers were of diameter 60 μm, volume electrical resistivity 6×10^{-5} Ω.cm, tensile strength 970 MPa and elongation at break 3.2%, as provided by International Steel Wool Corp., Springfield, Ohio. Portland cement (Type I) from Lafarge Corp., Southfield, MI, was used. The water/cement ratio was 0.35. The water reducing agent used in the amount of 0.5% by weight of cement was TAMOL SN (Rohm and Haas Co., Philadelphia, PA), which contained 93-96% sodium salt of a condensed naphthalenesulfonic acid. The volume electrical resistivity of the cement paste was 1.4×10^{5} Ω.cm at 1 day of curing, as measured by the four-probe method using silver paint for electrical contacts.

The contact electrical resistivity between the fiber and the cement paste was measured at 1 day of curing using the four-probe method and silver paint as electrical contacts, as

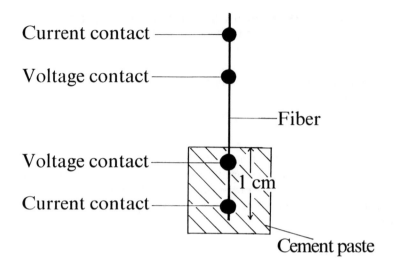

Fig. 1   Sample configuration for measuring the contact electrical resistivity of the interface between a fiber and cement paste.

illustrated in Fig. 1. One current contact and one voltage contact were on the fiber, while the other voltage and current contacts were on the cement paste embedding the fiber to a distance of 1 cm. The cement paste thickness was 1.5 mm on each side sandwiching the fiber. The fiber length was 5 cm. The current was 0.5-2 A; the voltage was 3-4 V. The resistance between the two voltage probes was measured; it corresponds to the sum of the fiber volume resistance, the interface contact resistance and the cement paste volume resistance. The measured resistance turned out to be dominated by the contact resistance, to the extent that the two volume resistance terms can be neglected. The contact resistivity (in $\Omega.cm^2$) is given by the product of the contact resistance (in Ω) and the contact (interface) area (in cm^2).

Single fiber pull-out testing was conducted on the same interface samples and at the same time as the contact resistivity was measured. For pull-out testing, one end of the fiber was embedded in cement paste, as in Fig. 1. Fig. 2 gives a typical plot of shear stress vs. displacement. The shear stress increased to a maximum which corresponds to the debonding shear strength.

Fig. 3 gives the corresponding values of the contact resistivity and the shear debonding strength for eight interface samples. Even though the eight interface samples were prepared identically, they varied in both the contact resistivity and debonding strength. (This may be related to the difference in interface cleanliness from sample to sample.) The greater the

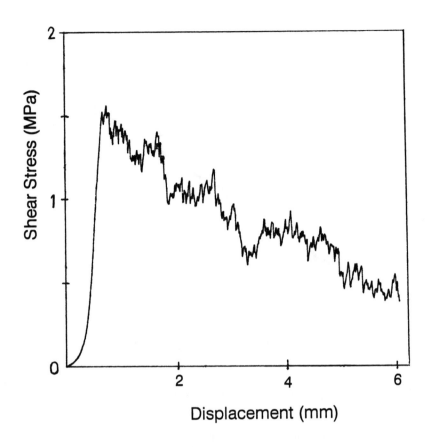

Fig. 2   Plot of shear stress vs. displacement during single fiber pull-out testing.

contact resistivity, the greater the debonding strength. The correlation is excellent and linear, indicating that the contact resistivity can be used to provide an indirect measure of the shear debonding strength for a given combination of fiber and matrix. As shown in Fig. 3, the fractional variation in contact resistivity is much larger than the fractional variation in debonding strength. Thus, the contact resistivity is a more sensitive indicator of the interface quality than the debonding strength. That the contact resistivity increases with increasing debonding strength indicates that interfacial phase(s) of volume resistivity higher than those of steel and cement paste enhanced the bonding between steel and cement paste. Such phases may be metal oxides or other compounds. If such interfacial phases did not play a role, the contact resistivity would have decreased with increasing debonding strength, as a physically tighter contact (with less void volume) is expected to have a lower contact resistivity. In the

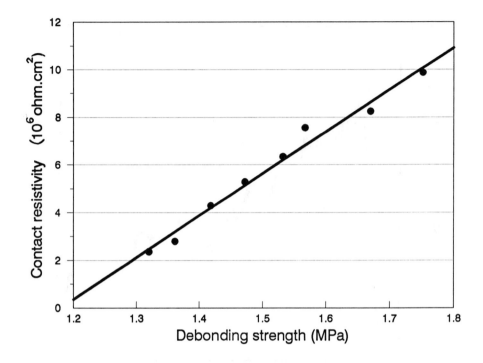

Fig. 3   Plot of contact electrical resistivity vs. shear debonding strength.

case of bonding between new and old mortars (no steel involved), the mortar-mortar contact resistivity was indeed observed in a companion work to increase with decreasing debonding strength [2].

CONCLUSION

Fiber-matrix contact electrical resistivity measurement was found to be a fast, inexpensive and sensitive method for characterizing the fiber-matrix interface. It gives an indirect measure of the debonding shear strength and provides information on the interfacial phase(s). By using stainless steel as the fiber and cement paste as the matrix, the technique was demonstrated. For this interface, the contact resistivity increased linearly with the debonding shear strength and interfacial phase(s) of volume resistivity higher than that of cement paste enhanced the bonding.

REFERENCES

1. Parviz Soroushian, Fadhel Aouadi and Mohamad Nagi, ACI Materials J. 88(1), 11 (1991).

2. Xuli Fu and D.D.L. Chung, to be published.

IMPROVING THE FLEXURAL STRENGTH OF FIBRE REINFORCED OIL WELL CEMENTS BY ADDITION OF A POLYMER LATEX

DEMOSTHENIS G. PAFITIS
Schlumberger Cambridge Research Limited, High Cross, Madingley Road, Cambridge,
CB3 0EL, United Kingdom

ABSTRACT

Fibre-reinforced cements are proving to be useful in various oilfield applications. Low cost and increased toughness render glass fibre reinforced cements of particular interest. In most cases, improvements in toughness are the result of extensive fibre pull-out and this can be clearly observed in electron micrographs of fracture surfaces. This observation implies that there is much scope for improving the interfacial shear strength between the hydrated cement and glass fibres. Experiments have shown that increases in flexural strength and in energy to fracture can be achieved by incorporating small amounts of a polymer latex. Improvements of a factor of four in energy to fracture have been measured when approximately 0.8% by volume of a styrene-butadiene copolymer latex is added to a glass fibre reinforced class G oilwell cement. Experimental results suggest that this effect is not due to improvements in the strength of the cement matrix but due to an enhancement of the interfacial shear strength between fibre and cement.

INTRODUCTION

In traditional oil and gas wells when drilling is complete, or has reached specified stages, a steel tube is placed into the wellbore and cement placed in the annular gap formed between the hole and steel tube. The cement forms a relatively impermeable seal which isolates the various zones which have been drilled and also prevents collapse of the wellbore under the action of earth stresses. The mechanical property criterion for the selection of oil well cements when applied to such well completions is currently based on the development of compressive strength only [1]. This has developed from the perception that when placed in an annulus as shown in figure 1, the cement must only withstand compressive radial and circumferential stresses. In many cases this is satisfactory. however, under certain conditions such as hole ovality, anisotropic earth stresses and high pressure within the steel liner, significant tensile and shear stresses may be encountered. Therefore, traditional *neat* cements may be unsuitable in these cases since generally they are much weaker in tension than in compression.

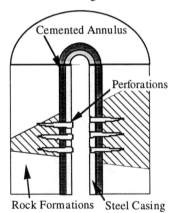

Figure 1. A Traditional Well Completion

In civil engineering, concrete and cement structures which must withstand some tensile or shear components of stress are commonly reinforced with long pretensioned steel rods or by the incorporation of various fibrous materials. In the wellbore, many of these techniques are clearly impossible, mostly because the cement must be pumpable over distances of up to kilometres. The incorporation of short chopped fibres into oilwell cements however, does appear to be a feasible technique for improving mechanical properties and this together with other cement toughening methods has been the subject of various studies. It has long been known that improvements in tensile strength and toughness of cements may be achieved by the incorporation of fine polymeric particles [2,3,4,5,6]. These are usually less than 1 μm in diameter and are introduced into the slurry during mixing as a water borne suspension known as a latex. Such latexes have been used in oilwell cements for more than a decade and may provide improved bonding to various rocks and steel casing in the wellbore [5,7,8,9]. They also improve the ability to prevent migration of reservoir fluids from one zone to another by decreasing permeability and preventing gas percolation through the setting slurry in the semi-solid state. Furthermore, the incorporation of polymer latexes where the polymer particle size is in the sub-micron range may lead to a significant reduction in the cement viscosity and therefore improve mixability and pumpability.

The combined effects of incorporating both fibrous materials and polymer latexes have been reported for various construction materials [10]. In this paper, the effects of combining a polymer latex together with short glass fibres on the mechanical properties of an oilwell cement are reported and discussed.

MATERIALS

The cement used in this study was a class G, high sulphate resistant oilwell cement supplied by Dykerhoff of Germany. This is a basic well cement intended for use at depths to 2440 metres. The cement is categorised by the American Petroleum Institute as class G and as such has a composition of up to 6% MgO, 3% SO_3, between 48% and 58% of tricalcium silicate and a maximum of 8% of tricalcium aluminate. The cement particle sizes range from less than 1 μm to approximately 100 μm with a mean of 30 μm.

The polymer latex was a styrene butadiene rubber latex which has a volumetric composition of 50% water and 50% polymer particles with a mean diameter of approximately 200 nm. The latex was stabilised with a low molecular weight non-ionic polglycol surfactant.

Glass fibres were 6 mm long Cem-FIL 70/30 water dispersible strands with a filament diameter of 20 μm.

METHODS

Mixing

Cement slurries were mixed using a 1 litre propeller type mixer capable of rotation speeds up to 12000 rpm. The required quantity of water was placed in the mixer and the rotation speed set to 4000 rpm. The entire volume of cement was then added in not more than 15 seconds and the mixer speed then set to 12000 rpm for a further 35 seconds.

In slurries containing glass fibres, they were dry blended with the cement prior to addition to the mix water.

In slurries containing polymer, the latex fluid was injected into the mix after all of the cement had been added and prior to mixing at the higher shear rate.

The temperature of the mix water and the additives prior to mixing was 22±3°C.

The range of slurry compositions used is presented in table 1. In order to maintain a constant water to cement ratio across all of the slurries, the mix water was reduced so as to account for the water introduced from the polymer latex.

Slurry Code.	Cement / g	Water / g	Glass / g	Latex / g	Vf glass	Vf SBR
A	792	349	0	0	0	0
B	792	349	8.9	0	0.005	0
C	792	344	0	10	0	0.008
D	792	344	8.9	10	0.005	0.008

Table 1. Slurry formulations

Specimen preparation

Immediately after mixing, the cement slurries were deaerated under vacuum for 15 minutes whilst being constantly agitated to aid degassing and then transferred into 5 sided molds made from PTFE. The filled molds were placed in an environment maintained at a relative humidity of 100% and at a temperature of 22±3°C. After 24 hours the samples were carefully removed from the molds and placed in the 100% relative humidity environment for a further 24 hours. The samples were then further pre-conditioned at room temperature and a relative humidity of 56±4% for a further 24 hours.

The sample surfaces were flattened by carefully grinding on a 600 grit abrasive paper and then all of the dimensions measured in various positions to an accuracy of 0.01 mm. A total of 10 specimens were produced for each slurry formulation.

Three point bend tests

A schematic of the three point bend apparatus is shown in figure 2. This apparatus was mounted on a Testometric Micro 350 universal testing machine with a maximum force measuring capacity of 5 kN. The knife edges were of hardened steel and had a radius of curvature of approximately 2 mm. The distance between the two lower knife edge tips was 100 mm and the upper knife edge positioned equidistant between the lower two. The as cast cement samples were approximately 10 mm thick, 30 mm wide and 140 mm long. Each sample was carefully positioned on the apparatus using a jig designed to facilitate correct alignment with respect to the knife edges. Displacement was measured using an LVDT on the lower face of the sample directly below the loading point. Load was measured to an accuracy of 0.1N and the displacement rate set at a constant value 1.0±0.1 mm/min.

Displacement and load were simultaneously recorded during each test using computerised data

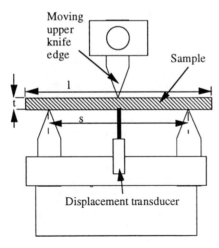

Figure 2. Schematic of 3 point bend apparatus.

logging equipment. This data was transformed and corrections made for machine stiffness to give stress, strain and absorbed energy data.

RESULTS AND DISCUSSION

Strength and energy data for the four slurries are presented in table 2 and are discussed in the following paragraphs. The data shows the mean and standard deviations for a total of ten samples of each of the slurry formulations.

Slurry code	Mean Peak Stress / MPa	Standard deviation	Mean Energy at peak / g.cm	Standard deviation
A	47	14	499	215
B	62	8	1280	230
C	43	16	597	180
D	73	5	2100	220

Table 2. Strength and peak energy data

Stress versus strain curves for representative samples of each of the cement formulations tested in 3 point bending are shown in figure 3. The letters shown on the graph correspond to the slurry codes provided in tables 1 and 2. The neat cement as shown with line A exhibited linear behaviour to fracture, at which point the sample failed instantaneously across the entire thickness of the beam.

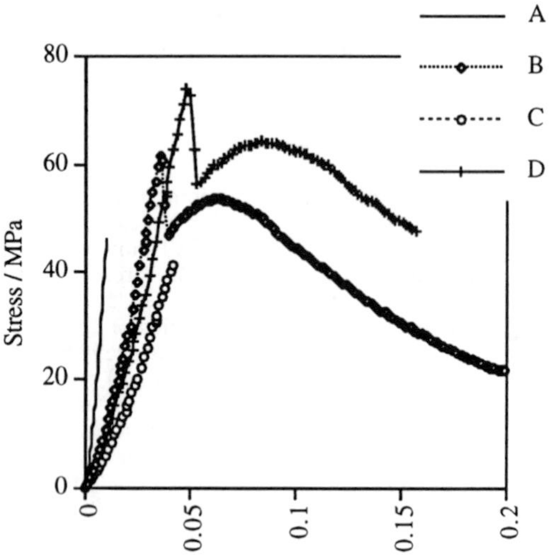

Figure 3. Typical stress-strain curves.

The mean peak stress after the 72 hour precondition period was 47 MPa and the deviation of values about this mean was reasonably broad as indicated by the standard deviation figure presented in table 2.

The slurry containing glass fibre but no latex is shown by line B and differs significantly from the neat cement stress-strain curves. Firstly, the bending modulus was found to be lower than that of the neat cement and the maximum stress was higher with a mean value of 62 MPa. The reduced modulus value for the composite was in some ways surprising since the glass fibres have a modulus of approximately 70 GPa and might therefore have been expected to lead to an increase in modulus when incorporated into a matrix which has a stiffness of only about 8 GPa. The reduced modulus could be indicative of poor bonding between the cement matrix and the glass fibres although this hypothesis is not supported by any strong evidence. The variability about the mean peak stress was reduced for the fibre reinforced cement when compared to the neat cement as indicated by the calculated standard deviation presented in table 2. Indeed, the variability was very low for a brittle matrix composite with a low fibre volume fraction and was largely a result of careful sample preparation and little compositional variation between cement batches. Less careful preparation during the deaeration stage was found to significantly affect the range of strength data since crack initiation from voids became very common. Unlike the neat cement samples represented by line A, the fibre reinforced cement did not fail by the immediate propagation of a fracture across the entire width of the sample. This is indicated by the post-peak stress region of the stress-strain curve in which significant stresses are still maintained even when a matrix crack has propagated from the lower side of the beam. After the maximum stress was reached, the load rapidly dropped but then was found to increase once again as fibres which bridged the cracked cement matrix where pulled from the fracture surfaces or where fractured themselves. Following the second maximum load, which was in all cases lower than the first, the load was found to successively decrease until the sample was broken into two separate pieces. In practice, the samples were never forced to complete separation in order to avoid damaging the displacement transducer positioned below the loading point. The energy to peak stress for the glass reinforced samples was an average of 156% greater than the neat samples, a result of both the increased stress and failure strain.

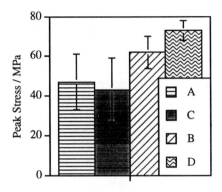

Figure 4. Peak stress for 4 cements

Consider now the samples in which the polymer latex was incorporated as shown by the stress-strain curves C and D. The unreinforced latex modified cement is represented by line C and clearly is similar to the neat sample represented by line A. However, the failure strain was on average a factor of 1.7 times greater than that of the neat cement samples. The peak stress for the latex modified cement was slightly lower than that of the neat cement and this might be explained in terms of the reduced number of strong cementitious bonds which could be produced.

There is clearly a trade off between improved toughness as indicated by the higher energy to peak stress value, and the absolute strength. The addition of the polymer particles allows the cement to withstand greater strains prior to fracture and therefore render it a more compliant and tough material. Similar observations have been made by Dingley [6] and have been explained in terms of the ability of the polymer phase to prevent or delay micro-crack coalescence during loading. The choice of using a material with either higher strength and modulus or lower modulus and strength but higher strain to failure is one which has lead to confusion amongst many oilfield engineers. Certainly, for applications where failure is controlled by the imposition of a

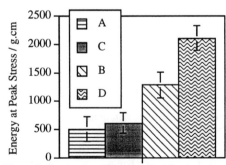

Figure 5. Energy at peak stress data for 4 cements.

critical strain, as is the case for most oilwell applications, the latex modified cement would be preferable.

Finally, line D is representative of a cement in which both glass fibres and the polymer latex were incorporated and there appears to be synergy between these two additives. Both the peak stress and energy to peak stress have been increased with respect to all of the previously mentioned materials. The mean peak stress was in some cases twice as high as the neat and unreinforced latex modified cements and the failure strain almost identical to the mean of the unreinforced latex cement. The combination of these two factors resulted in a material which exhibited an energy at peak stress which on average was four times greater than the neat cement. Furthermore, the variability in strength from sample to sample was reduced to the lowest of all of the materials. Charts showing the mean strengths and energies together with indications of the variability in these two parameters for all of the materials tested are shown in figures 4 and 5.

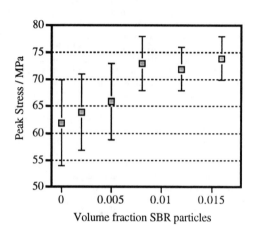

Figure 6. Effect of latex concentration on peak stress.

As a result of the observed synergy when fibres and latex were added, further experiments were performed in which the latex concentrations were varied up to a maximum volume fraction of 1.5%. Tests were not carried out in which the glass fibre content was increased since further addition of fibres was found to significantly effect the rheological characteristics of the slurry to such an extent that pumping would become impossible. The data from these tests, again in which ten samples were tested at each concentration, are shown in figure 6. The mean strength of the latex modified composites was found to increase from approximately 65 MPa to 75 MPa when the latex concentration was increased from 0 to 1.5% by total volume. Further additions were not made in order to establish an optimum concentration since some separation of the latex from the slurry was observed at concentration greater than 1.8% by volume.

CONCLUSIONS

The following conclusions have been drawn from the experimental work described:

* Significant improvements in the flexural strength of a glass fibre reinforced oilwell cement may be achieved by the incorporation of a stable styrene butadiene rubber latex.

* The addition of SBR latex at volume fractions up to 1.8% to a neat oilwell cement with water:cement ratio of 0.44 was found to slightly decrease strength but increase the failure strain in bending. The net effect was to increase the energy to fracture by approximately 20% compared to a neat cement.

* Addition of 0.5% volume fraction of water dispersible glass fibres was found to increase the strength of a neat class G oilwell cement by 32% and the energy to first fracture by 156%

* The combined addition of glass fibres and an SBR polymer latex at low concentrations has a significant effect on the mechanical properties of class G cement. In particular, both strength and energy to first fracture are increased to values greater than may be found by the incorporation of either fibres or latex alone.

REFERENCES

[1] Specification for Materials and Testing of Well Cements. API Specification 10, Fifth edition, July (1990).
[2] R. Dennis, .in Construction Materials Reference Book, edited by. D.K. Doran, (Butterworth Heinemann, 1992.), p. 39/1.
[3] Isenburg J.E., Vanderhoff J.W., Hypothesis for reinforcement of portland cement by polymer latexes. Journal of the American Ceramics Society . 57, no. 5, pp. 242 - 245.
[4] Lavelle J.A., Acrylic Latex-modified portland cement. ACI Materials journal, Jan- Feb (1988).
[5] Nakayama M., Beaudoin J.J., Bond strength development between latex-modified cement paste and steel. Cement and concrete research, 17, pp 562-572, (1987).
[6] Dingley R.G. The structure and properties of hydraulic cement pastes modified with polymer latex. PhD thesis, University of Southampton, UK. 1974.
[7] Su, Z., Bijen J.M., Larbi J.A. The influence of polymer modification on the adhesion of cement pastes to aggregates. Cement and Concrete research. 21, pp. 727-736, (1991).
[8] Carter L.G., Evans G.W. A study of Cement-Pipe bonding. SPE 764. Feb 1964.
[9] Carpenter R.B., Brady J.L., Blount C.G. Effect of temperature and cement admixes on bond strength. Journal of petroleum technology August 1992.
[10] Majumdar A.J., Laws V. Glass Fibre Reinforced Cement. (published by BSP Professional books,1991).

Author Index

Subject Index

4Cao.3Al$_2$O$_3$.SO$_3$, 135

added alkali, 67
admixtures, 319
advanced coal technology by-products, 179
AFt, 143
aggregate, 319
air voids, 217
alinite, 169
alkali silica reaction, 57, 89
alkali-activated cements, 199
ancient concrete, 159
andesite, 295, 367
asr gel, 57

backscattered electron imaging, 3, 13, 217, 417, 419
bending strength, 497
bi-layer, 329
bond cracks, 337
bonding, 309, 347, 419
bredigite, 135
brittleness number, 377
bulk modulus, 413

calcium
    aluminate cement, 279
    hydroxide morphology, 99
    sulfoaluminate
        cement, 135
        hydrates, 143
    sulfosilicate, 135
        cement, 135
capillary
    pores, 217, 227, 237
    porosity, 227
carbonate, 199
carbonation, 471
cement hydration, 143, 279
charge-dispersal model, 329
chemical deterioration, 191
chevron-notched beam, 295
chloride penetration, 465
class F fly ash, 191
clay, 209
composite, 309, 429
compressive yield stress, 279
computer model, 33, 429, 437
confocal microscopy, 107
corrosion rate, 471
crack
    propagation, 387
    stability, 377
cracks, 83, 463
crystallization pressure, 83
curing time, 99
C$_x$AHy, 143

damage, 465

debonding
    energy, 507, 519
    strength, 559
degradation, 457
delayed ettringite formation (DEF), 57, 67
deterioration, 49
diabase, 377
dielectric amplification, 255
diffusion, 449
digital image analysis, 347
dispersion, 89
dolomite, 295, 367
drying shrinkage, 33
DSP matrix, 507
durability, 89, 237, 465
dynamic
    elastic modulus, 407
    shear modulus, 407

elastic modulus, 367, 397, 413
electrical
    conductivity, 429
    properties, 255
    resistivity, 559
electrically-induced shape changes, 265
electrochemical monitoring, 471
electromechanical behaviour, 265
embedment length, 217
ettringite, 143, 179
ettringite-thaumasite solid solution, 179
evolution, 159
expansion, 57, 67
    mechanism, 67
expansive cements, 169

feature analysis, 23
ferrite-perovskite solid solution, 135
fibres, 159, 479, 497, 507
    asbestos, 507
    carbon, 295
    coir, 479
    glass, 565
    inclined, 539
    interactions, 497
    malva, 479
    nylon, 519
    polyethylene, 295, 549
    polypropylene, 479, 519
    pullout, 507, 519, 529, 539, 549, 565
    sisal, 479
    stainless steel, 539, 559
    steel, 217, 227, 529
    vegetable, 479
    wood, 487
finite elements, 419
flocculated suspensions, 279
fluid permeability, 429
fluidized bed combustion, 179